PHILIP'S
CERTIFICATE
ATLAS

Great Britain from the Air

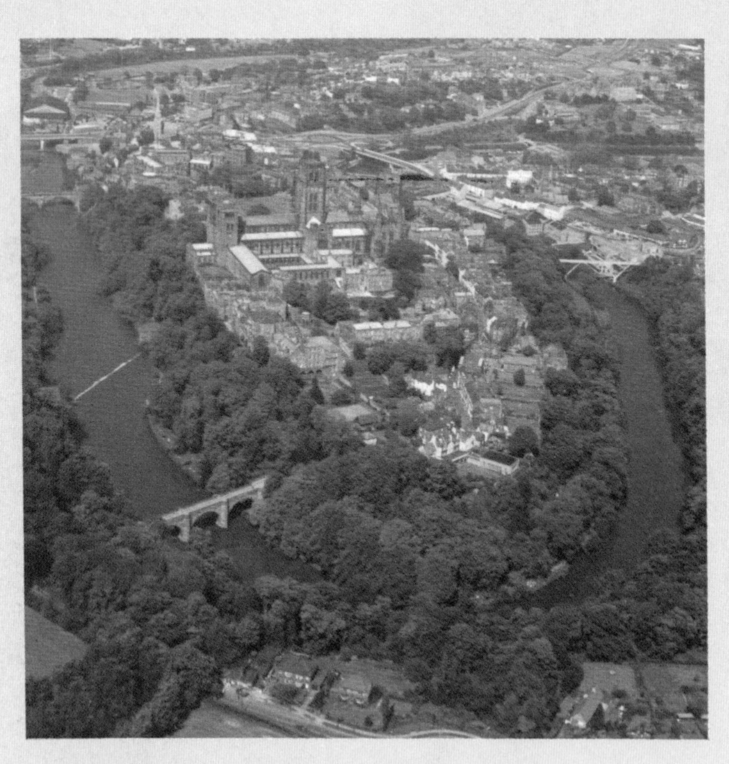

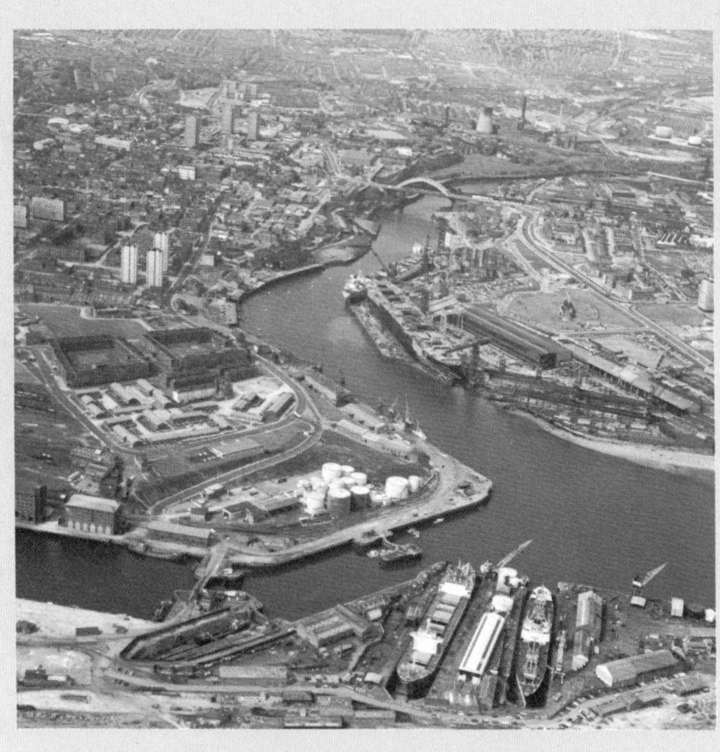

1. DURHAM
The old city centre of Durham is on a protected but constricted site within an incised meander of the River Wear. Later developments have been forced to spread on to flatter land away from the river and have a more sprawling, less dense, character.

2. SUNDERLAND (TYNE & WEAR)
The growth of Sunderland has not been restricted by physical features. The availability of flat land next to the river has promoted heavy industrial developments close to the city centre. Large scale residential developments are evident on the higher slopes away from the river.

3. MILTON KEYNES (BUCKINGHAMSHIRE)
The recent growth of Milton Keynes has been planned and controlled resulting in a formal open pattern of development. Care has been taken to provide necessary services (such as shops and schools), plenty of car access and open spaces (both private and communal).

4. POLPERRO (CORNWALL)
The old, picturesque village of Polperro lies in a very restricted site which has severely limited growth. The harbour, whitewashed houses and inns make this village a great tourist attraction.

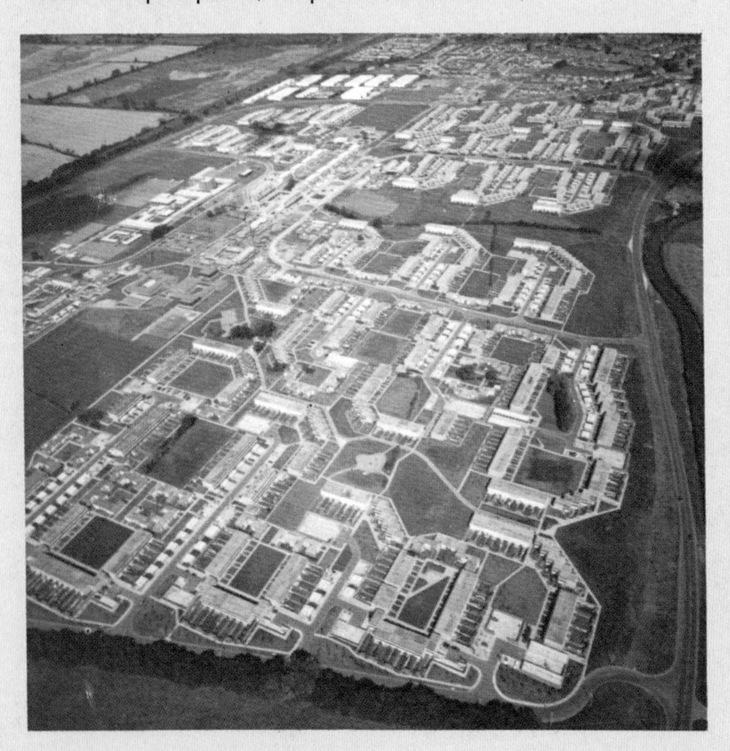

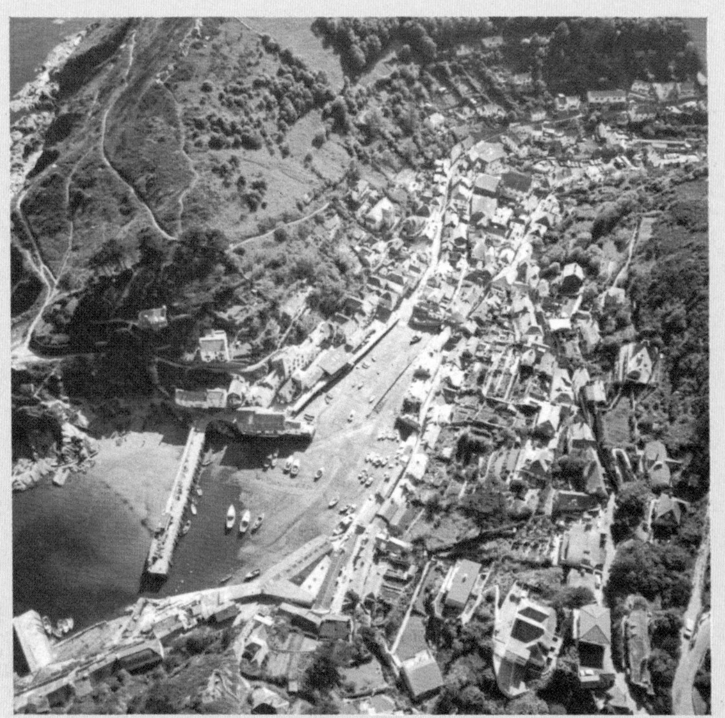

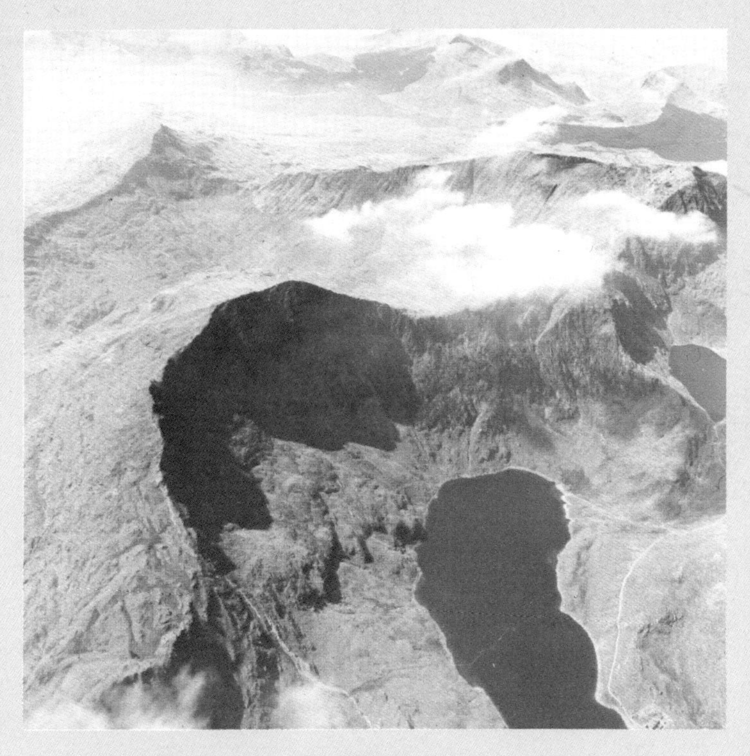

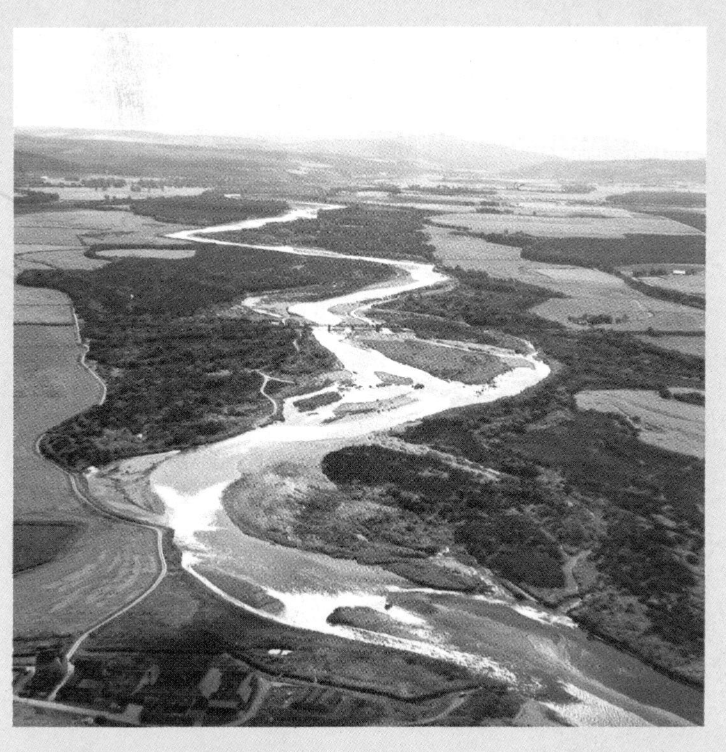

5. LLYN LLYDAW AND SNOWDON

The Snowdonia region, like many highland areas in Scotland and Wales, is characterized by deep glacial erosion resulting in a dramatic landscape and scenery which has attracted tourism.

6. RIVER SPEY (GRAMPIAN)

In this picture the River Spey flows over an area of glacial deposition, which is characteristically flat. The river meanders within a belt of natural vegetation suggesting that its course is too unstable to risk cultivating the nearby floodplain.

7. EGLWYSEG MOUNTAINS (CLWYD)

This south-west facing limestone scarp is formed by differential weathering of the rocks. The poor soils and lack of surface water on the limestone plateau make this a poor farming area so that cultivation is restricted to the lowland areas.

8. RUMPS POINT (CORNWALL)

The processes of erosion and deposition on the hard rock of these promontories have resulted in the formation of near vertical cliffs and sandfilled bays, including a treacherous bank of sand across the estuary of the River Camel which hindered the development of the port at Padstow.

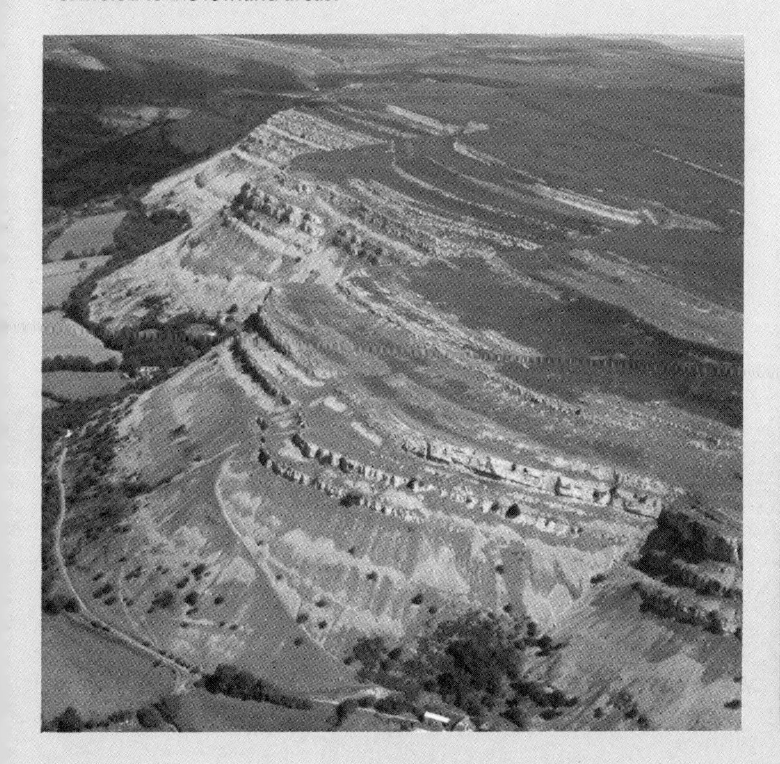

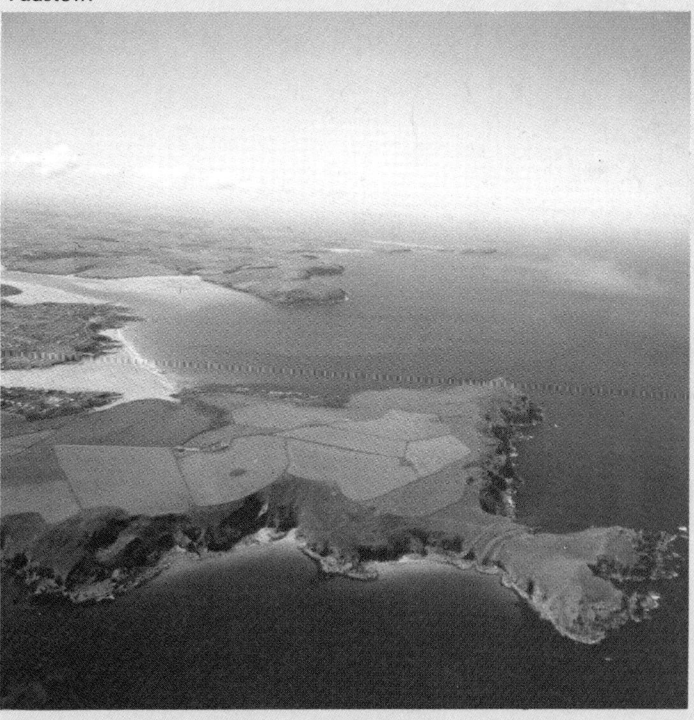

PHILIP'S

CERTIFICATE ATLAS

HEINEMANN-PHILIP
ATLASES

Edited by

B. M. Willett, B.A.   *Cartographic Editor*

H. Fullard, M.Sc.   *Consultant Cartographer*

D. Gaylard   *Assistant Cartographic Editor*
George Philip Ltd.

Maps and index produced by

G. Atkinson, B.A., Prudence Davson, B.Sc.,
Margaret Emslie, B.Sc., D. Fairbairn, B.A.,
N. Harris, B.A., P. Pearce, B.Sc., Elizabeth Prince-
Smith, B.Sc., Mary Spence, M.A., R. Smith, B.Sc.,
S. Smith, B.A., Helen Stenhouse, B.Sc., and the
Cartographic Staffs of George Philip Printers Ltd.,
London, Cartographic Services (Cirencester) Ltd.,
Fairey Surveys Ltd., and David L. Fryer &
Company

and prepared under the direction of

A. G. Poynter, M.A.   *Director of Cartography*

First Edition 1979	Reprinted with revision 1985
Reprinted 1981	Fourth Edition 1987
Second Edition 1983	Reprinted 1992
Third Edition 1984	

ISBN 0 540 05667 7 (paperback)
     0 540 05666 9 (hardback)

© 1987 George Philip Ltd., London

Printed in Hong Kong

Preface

The prime function of a map is to portray an objective reality – topographical details and the patterns of human occupance that have developed upon it. In the tight limits of a school atlas, this requires a fine balance to show what is essential and yet maintain precision, legibility and a pleasing whole.

A school atlas must not be solely an arid enumeration of local detail and names. This century has seen an unparalleled expansion in the demand for specialized map information of all types and maps are expected to be not only more accurate and more detailed but also more interesting and more easily understood. The atlas should break new ground in the presentation and portrayal of essential facts of geography. It must give information on the matters which affect our environment and our daily lives. It must enable students to visualize landscape and understand the complex natural and human interrelationships which are the basis of regional, national and world problems. It must provide data for effective comparison of our own country and society with that of others in the world and between rich, poor and developing nations.

Philips' Certificate Atlas attempts to fulfil these aims by including topographic maps at various scales and on various projections, thematic maps, tables of statistics, diagrams, cross-sections, graphs, satellite, air and land photographs and an index. Every attempt has also been made to include up-to-date information, otherwise difficult to obtain by hard-pressed teachers and students, and to keep pace with the present-day greater rate of change of economies and societies.

The content is arranged with the world scene as the opening chapter followed by continental and regional studies.

Because many of the decisions which vitally affect our daily lives are taken at state government level, the world chapter starts with summary tables of the principal countries and cities of the world giving their areas and populations. Thematic maps, diagrams and graphs provide an overall view of the composition of the earth, and its surrounding atmosphere, its topography and climate, leading to a series of maps and graphics of spatial patterns of water and food resources, minerals and power resources, population and political organizations.

As befits an atlas intended for U.K. schools, there is emphasis on Great Britain and Ireland with thematic maps of the topography, geology, climate, land use, industry, power resources and population distribution. This is not, however, allowed to detract from a proper emphasis on our neighbouring countries in Europe and, in particular, on the E.E.C.

Throughout the regional and continental studies, topographic maps are complemented by thematic maps, diagrams and illustrations, which attempt to epitomise the character and problems of each country or region. For example, in West Germany, the importance of Ruhr industry and the significance of Rhine traffic, regional contrasts in Italy, the phenomenal industrial growth of Japan, the primary products of Australia and New Zealand, the economic production of North America.

Following the map section, there are tables of the production and trade of the principal countries and climatic statistics for representative stations throughout the world. A list of geographical terms precedes the index which gives the latitude and longitude co-ordinates for approximately 11 000 names.

H. FULLARD

Contents

Topographical Maps
Layer-coloured topographical maps showing principal settlements, relief features, communications and international boundaries.

World

Europe

Asia

Contents

Principal Countries of the World

Country	Area in thousands of square km.	Population in thousands	Density of population per sq. km.	Capital Population in thousands
Afghanistan	647	17 222	27	Kābul (1 127)
Albania	29	2 841	98	Tiranë (202)
Algeria	2 382	20 500	9	Algiers (1 740)
Angola	1 247	8 339	7	Luanda (700)
Argentina	2 767	29 627	11	Buenos Aires (9 927)
Australia	7 687	15 369	2	Canberra (256)
Austria	84	7 549	90	Vienna (1 531)
Bangladesh	144	94 651	657	Dhaka (3 459)
Belgium	31	9 856	318	Brussels (989)
Belize	23	156	5	Belmopan (3)
Benin	113	3 720	33	Porto-Novo (208)
Bhutan	47	1 360	29	Thimphu (60)
Bolivia	1 099	6 082	5	Sucre (64) La Paz (881)
Botswana	600	1 007	2	Gaborone (79)
Brazil	8 512	129 662	15	Brasilia (1 177)
Brunei	6	209	35	Bandar Seri Begawan (58)
Bulgaria	111	8 946	81	Sofia (1 094)
Burkina Faso	274	6 607	24	Ouagadougou (286)
Burma	677	36 750	54	Rangoon (2 459)
Burundi	28	4 540	162	Bujumbura (141)
Cameroon	475	9 165	19	Yaoundé (485)
Canada	9 976	24 907	2	Ottawa (738)
Central African Rep.	623	2 450	4	Bangui (382)
Chad	1 284	4 789	4	Ndjamena (303)
Chile	757	11 682	15	Santiago (4 132)
China	9 597	1 039 677	108	Peking (9 330)
Colombia	1 139	27 190	24	Bogotá (4 486)
Congo	342	1 651	5	Brazzaville (422)
Costa Rica	51	2 379	47	San José (245)
Cuba	115	9 884	86	Havana (1 951)
Cyprus	9	655	73	Nicosia (161)
Czechoslovakia	128	15 415	120	Prague (1 190)
Denmark	43	5 118	119	Copenhagen (1 366)
Djibouti	22	332	15	Djibouti (150)
Dominican Republic	49	5 962	121	Santo Domingo (1 313)
Ecuador	284	9 251	32	Quito (1 110)
Egypt	1 001	45 915	46	Cairo (6 818)
El Salvador	21	5 232	249	San Salvador (884)
Equatorial Guinea	28	381	14	Rey Malabo (37)
Ethiopia	1 222	33 680	28	Addis Ababa (1 478)
Fiji	18	670	37	Suva (71)
Finland	337	4 863	14	Helsinki (932)
France	547	54 652	99	Paris (8 510)
French Guiana	91	78	1	Cayenne (38)
Gabon	268	1 127	4	Libreville (350)
Gambia	11	696	63	Banjul (103)
Germany, East	108	16 864	156	East Berlin (1 173)
Germany, West	249	61 638	248	Bonn (293)
Ghana	239	12 700	53	Accra (965)
Greece	132	9 848	75	Athens (3 027)
Greenland	2 176	52	0.02	Godthaab (10)
Guatemala	109	7 699	71	Guatemala (1 300)
Guinea	246	5 704	23	Conakry (763)
Guinea-Bissau	36	836	23	Bissau (109)
Guyana	215	922	4	Georgetown (188)
Haiti	28	5 201	186	Port-au-Prince (888)
Honduras	112	4 092	37	Tegucigalpa (534)
Hong Kong	1	5 313	5 313	Hong Kong (1 184)
Hungary	93	10 702	115	Budapest (2 064)
Iceland	103	236	2	Reykjavík (124)
India	3 288	732 256	223	Delhi (5 729)
Indonesia	2 027	156 442	77	Jakarta (6 503)
Iran	1 648	42 070	26	Tehrān (4 589)
Iraq	435	14 654	34	Baghdād (2 969)
Irish Republic	70	3 508	50	Dublin (915)
Israel	21	4 097	195	Jerusalem (429)
Italy	301	56 836	189	Rome (2 831)
Ivory Coast	322	9 300	29	Abidjan (1 850)
Jamaica	11	2 260	205	Kingston (671)
Japan	372	119 259	320	Tōkyō (11 676)
Jordan	98	3 489	36	'Ammān (681)
Kampuchea	181	6 981	39	Phnom Penh (500)
Kenya	583	18 784	32	Nairobi (1 200)
Korea, North	121	19 185	158	Pyŏngyang (1 500)
Korea, South	98	39 951	408	Seoul (8 367)
Kuwait	18	1 672	93	Kuwait (775)
Laos	237	4 209	18	Vientiane (120)
Lebanon	10	2 739	274	Beirut (702)
Lesotho	30	1 444	48	Maseru (45)
Liberia	111	2 113	19	Monrovia (425)
Libya	1 760	3 356	2	Tripoli (980)
Luxembourg	3	365	121	Luxembourg (79)
Madagascar	587	9 400	16	Antananarivo (400)
Malawi	118	6 429	54	Lilongwe (99)
Malaysia	330	14 860	45	Kuala Lumpur (938)
Mali	1 240	7 528	6	Bamako (419)
Malta	0.3	377	1 256	Valletta (14)
Mauritania	1 031	1 779	2	Nouakchott (135)
Mauritius	2	993	496	Port Louis (150)
Mexico	1 973	75 103	38	Mexico (14 750)
Mongolia	1 565	1 803	1	Ulan Bator (419)
Morocco	447	22 110	49	Rabat (842)
Mozambique	783	13 311	17	Maputo (384)
Namibia	824	1 040	1	Windhoek (61)
Nepal	141	15 738	112	Katmandu (235)
Netherlands	41	14 362	350	Amsterdam (994)
New Zealand	269	3 203	12	Wellington (343)
Nicaragua	130	3 058	23	Managua (615)
Niger	1 267	6 040	5	Niamey (225)
Nigeria	924	89 022	96	Lagos (1 477)
Norway	324	4 129	13	Oslo (643)
Oman	212	1 131	5	Muscat (25)
Pakistan	804	89 729	112	Islamabad (201)
Panama	76	2 089	27	Panamá (655)
Papua New Guinea	462	3 190	7	Port Moresby (124)
Paraguay	407	3 472	8	Asunción (708)
Peru	1 285	18 790	15	Lima (5 258)
Philippines	300	52 055	173	Manila (1 630)
Poland	313	36 571	117	Warsaw (1 649)
Portugal	92	10 056	109	Lisbon (1612)
Puerto Rico	9	3 350	372	San Juan (1 086)
Romania	238	22 638	95	Bucharest (1 979)
Rwanda	26	5 700	219	Kigali (157)
Saudi Arabia	2 150	10 421	5	Riyadh (667)
Senegal	196	6 316	32	Dakar (799)
Sierra Leone	72	3 672	51	Freetown (316)
Singapore	0.6	2 529	4 170	Singapore (2 517)
Somali Republic	638	5 269	8	Mogadishu (600)
South Africa	1 221	31 008	25	Pretoria (739) Cape Town (1 491)
Spain	505	38 228	76	Madrid (3 188)
Sri Lanka	66	15 416	234	Colombo (1 412)
Sudan	2 506	20 362	8	Khartoum (476)
Surinam	163	407	2	Paramaribo (151)
Swaziland	17	605	36	Mbabane (23)
Sweden	450	8 331	19	Stockholm (1 420)
Switzerland	41	6 482	158	Bern (301)
Syria	185	9 660	52	Damascus (1 112)
Taiwan	36	18 700	519	Taipei (2 271)
Tanzania	945	20 378	22	Dodoma (46)
Thailand	514	49 459	96	Bangkok (5 468)
Togo	56	2 756	49	Lomé (283)
Trinidad and Tobago	5	1 202	240	Port of Spain (60)
Tunisia	164	6 886	42	Tunis (774)
Turkey	781	47 279	61	Ankara (2 276)
Uganda	236	14 625	62	Kampala (332)
United Arab Emirates	84	1 206	14	Abu Dhabi (243)
U.S.S.R.	22 402	272 500	12	Moscow (8 537)
United Kingdom	245	56 377	230	London (6 767)
United States	9 363	234 496	25	Washington (3 429)
Uruguay	178	2 968	17	Montevideo (1 362)
Venezuela	912	16 394	18	Caracas (2 944)
Vietnam	330	57 181	173	Hanoi (2 571)
Western Samoa	3	159	53	Apia (32)
Yemen, North	195	6 232	32	Sana' (278)
Yemen, South	288	2 158	7	Aden (264)
Yugoslavia	256	22 800	89	Belgrade (1 407)
Zaïre	2 345	31 151	13	Kinshasa (2 444)
Zambia	753	6 242	8	Lusaka (538)
Zimbabwe	391	7 740	20	Harare (681)

Principal Cities of the World

The population figures used are from censuses or more recent estimates and are given in thousands for towns and cities over 500,000 (over 750,000 in China, India, the U.S.S.R. and the U.S.A.) Where possible the population of the metropolitan area is given e.g. Greater London, Greater New York, etc.

AFRICA

ALGERIA (1977)
Algiers	1 740
Oran	543

ANGOLA (1982)
Luanda	700

CAMEROON (1983)
Douala	708

EGYPT (1976)
Cairo	6 818
Alexandria	2 318
El Giza	1 230

ETHIOPIA (1983)
Addis Ababa	1 478

GHANA (1984)
Accra	965

GUINEA (1980)
Conakry	763

IVORY COAST (1982)
Abidjan	1 850
Bouaké	640

KENYA (1985)
Nairobi	1 200

LIBYA (1982)
Tripoli	980
Benghazi	650

MOROCCO (1981)
Casablanca	2 409
Rabat-Salé	842
Fès	562
Marrakesh	549

NIGERIA (1975)
Lagos	1 477
Ibadan	847

SENEGAL (1979)
Dakar	799

SOMALI REP. (1982)
Mogadishu	600

SOUTH AFRICA (1980)
Johannesburg	1 726
Cape Town	1 491
Durban	961
Pretoria	739
Port Elizabeth	585

SUDAN (1983)
Omdurman	526

TANZANIA (1978)
Dar-es-Salaam	757

TUNISIA (1984)
Tunis	774

ZAÏRE (1976)
Kinshasa	2 444
Kananga	704

ZAMBIA (1980)
Lusaka	538

ZIMBABWE (1983)
Harare	681

ASIA

AFGHANISTAN (1982)
Kābul	1 127

BANGLADESH (1982)
Dhaka	3 459
Chittagong	1 388
Khulna	623

BURMA (1983)
Rangoon	2 459

KAMPUCHEA (1983)
Phnom Penh	500

CHINA (1982)
Shanghai	11 940
Peking	9 330
Tientsin	7 850
Shenyang	4 080
Wuhan	3 280
Kwangchow	3 160
Chungking	2 690
Harbin	2 560
Chengtu	2 510
Tzepo	2 264
Sian	2 220
Nanking	2 170
Taiyuan	1 790
Changchun	1 770
Dairen	1 520
Chengchow	1 517
Lanchow	1 430
Tsinan	1 360
Tangshan	1 351
Kweiyang	1 330
Kunming	1 320
Anshan	1 240
Tsitsihar	1 232
Tsingtao	1 210
Hangchow	1 201
Fushun	1 200
Foochow	1 142
Changsha	1 100
Kirin	1 099
Shihkiachwang	1 098
Nanchang	1 061
Paotow	1 051
Hwainan	1 017
Wulumuchi	944
Suchow	793

HONG KONG (1981)
Kowloon	2 450
Hong Kong	1 184
Tsuen Wan	599

INDIA (1981)
Calcutta	9 194
Bombay	8 243
Delhi	5 729
Madras	4 289
Bangalore	2 922
Ahmadabad	2 548
Hyderabad	2 546
Pune	1 686
Kanpur	1 639
Nagpur	1 302
Jaipur	1 015
Lucknow	1 008
Coimbatore	920
Patna	919
Surat	914
Madurai	908
Indore	829
Varanasi	797
Jabalpur	757

INDONESIA (1980)
Jakarta	6 503
Surabaya	2 028
Bandung	1 462
Medan	1 379
Semarang	1 026
Palembang	787
Ujung Pandang	709
Malang	512

IRAN (1976)
Tehrān	4 589
Esfahān	842
Mashhad	743
Tabrīz	715

IRAQ (1970)
Baghdād	2 969

JAPAN (1982)
Tōkyō	11 676
Yokohama	2 848
Ōsaka	2 623
Nagoya	2 093
Kyōto	1 480
Sapporo	1 465
Kobe	1 383
Fukuoka	1 121
Kitakyūshū	1 065
Kawasaki	1 055
Hiroshima	898
Sakai	809
Chiba	756
Sendai	662
Okayama	551
Kumamoto	522
Kagoshima	514
Amagasaki	510
Higashiōsaka	501

JORDAN (1981)
'Ammān	681

KOREA, NORTH (1972)
Pyŏngyang	1 500

KOREA, SOUTH (1980)
Seoul	8 367
Pusan	3 160
Taegu	1 607
Inchŏn	1 085
Kwangju	728
Taejon	652

KUWAIT (1980)
Kuwait	775

LEBANON (1980)
Beirut	702

MALAYSIA (1980)
Kuala Lumpur	938

PAKISTAN (1981)
Karachi	5 103
Lahore	2 922
Faisalabad	1 092
Rawalpindi	806
Hyderabad	795
Multan	730
Gujranwala	597
Peshawar	555

PHILIPPINES (1980)
Manila	1 630
Quezon City	1 166
Davao	610

SAUDI ARABIA (1974)
Riyadh	667
Jedda	561

SINGAPORE (1983)
Singapore	2 517

SRI LANKA (1982)
Colombo	1 412

SYRIA (1982)
Damascus	1 112
Aleppo	985

TAIWAN (1981)
Taipei	2 271
Kaohsiung	1 227
Taichung	607
Tainan	595

THAILAND (1982)
Bangkok	5 468

TURKEY (1982)
İstanbul	2 949
Ankara	2 276
İzmir	1 083
Adana	864
Konya	691
Bursa	658
Gaziantep	526

VIETNAM (1979)
Ho Chi Minh City	3 420
Hanoi	2 571
Haiphong	1 279

AUSTRALIA AND NEW ZEALAND

AUSTRALIA (1983)
Sydney	3 335
Melbourne	2 865
Brisbane	1 138
Adelaide	969
Perth	969

NEW ZEALAND (1983)
Auckland	864

EUROPE

AUSTRIA (1984)
Vienna	1 531

BELGIUM (1983)
Brussels	989

BULGARIA (1984)
Sofia	1 094

CZECHOSLOVAKIA (1984)
Prague	1 190

DENMARK (1984)
Copenhagen	1 366

FINLAND (1983)
Helsinki	932

FRANCE (1982)
Paris	8 510
Lyons	1 170
Marseilles	1 080
Lille	935
Bordeaux	628
Toulouse	523

GERMANY, EAST (1982)
East Berlin	1 173
Leipzig	557
Dresden	521

GERMANY, WEST (1983)
West Berlin	1 860
Hamburg	1 618
Munich	1 284
Cologne	953
Essen	635
Frankfurt	615
Dortmund	595
Düsseldorf	580
Stuttgart	571
Bremen	545
Duisburg	542
Hanover	524

GREECE (1981)
Athens	3 027
Thessaloníki	871

HUNGARY (1984)
Budapest	2 064

IRISH REPUBLIC (1981)
Dublin	915

ITALY (1983)
Rome	2 831
Milan	1 561
Naples	1 209
Turin	1 069
Genoa	747
Palermo	712

NETHERLANDS (1984)
Rotterdam	1 025
Amsterdam	994
The Hague	672
Utrecht	501

NORWAY (1984)
Oslo	643

POLAND (1984)
Warsaw	1 649
Łodz	849
Kraków	740
Wrocław	636
Poznań	574

PORTUGAL (1981)
Lisbon	1 612
Oporto	1 315

ROMANIA (1982)
Bucharest	1 979

SPAIN (1981)
Madrid	3 188
Barcelona	1 755
Valencia	752
Seville	654
Zaragoza	591
Málaga	503

SWEDEN (1983)
Stockholm	1 420

SWITZERLAND (1983)
Zürich	840

U.S.S.R. (1983-84)
Moscow	8 537
Leningrad	4 827
Kiev	2 409
Tashkent	1 986
Baku	1 661
Kharkov	1 536
Minsk	1 442
Gorki	1 392
Novosibirsk	1 384
Sverdlovsk	1 286
Kuybyshev	1 250
Dnepropetrovsk	1 140
Tbilisi	1 140
Yerevan	1 114
Odessa	1 113
Omsk	1 094
Chelyabinsk	1 086
Donetsk	1 064
Perm	1 048
Ufa	1 048
Alma-Ata	1 046
Kazan	1 039
Rostov	983
Volgograd	969
Saratov	893
Riga	875
Krasnoyarsk	857
Zaporozhye	844
Voronezh	840

UNITED KINGDOM (1981)
London (1985)	6 767
Birmingham (1985)	1 007
Glasgow	762
Liverpool	510

YUGOSLAVIA (1981)
Belgrade	1 407
Zagreb	1 175
Skopje	507

NORTH AMERICA

CANADA (1983)
Toronto	3 067
Montréal	2 862
Vancouver	1 311
Ottawa	738
Edmonton	699
Calgary	634
Winnipeg	601
Québec	580
Hamilton	548

CUBA (1982)
Havana	1 951

DOMINICAN REP. (1981)
Santo Domingo	1 313

EL SALVADOR (1983)
San Salvador	884

GUATEMALA (1983)
Guatemala	1 300

HAITI (1982)
Port-au-Prince	888

HONDURAS (1982)
Tegucigalpa	534

JAMAICA (1980)
Kingston	671

MEXICO (1979)
Mexico	14 750
Guadalajara	2 468
Netzahualcóyotl	2 331
Monterrey	2 019
Puebla	711
Ciudad Juárez	625
León	625
Tijuana	566

NICARAGUA (1981)
Managua	615

PANAMA (1981)
Panama	655

PUERTO RICO (1980)
San Juan	1 086

UNITED STATES (1984)
New York	17 807
Los Angeles	12 373
Chicago	8 035
Philadelphia	5 755
San Francisco	5 685
Detroit	4 577
Boston	4 027
Houston	3 566
Washington	3 429
Dallas	3 348
Miami	2 799
Cleveland	2 788
St. Louis	2 398
Atlanta	2 380
Pittsburgh	2 372
Baltimore	2 245
Minneapolis-St. Paul	2 231
Seattle	2 208
San Diego	2 064
Tampa	1 811
Denver	1 791
Phoenix	1 715
Cincinnati	1 673
Milwaukee	1 568
Kansas City	1 477
Portland	1 341
New Orleans	1 319
Columbus	1 279
Sacramento	1 220
Buffalo	1 205
Indianapolis	1 195
San Antonio	1 188
Providence	1 095
Norfolk	1 026
Salt Lake City	1 025
Rochester	989
Louisville	963
Oklahoma	963
Memphis	935
Dayton	930
Birmingham	895
Nashville-Davidson	890
Greensboro	886
Albany	843
Orlando	824
Honolulu	805
Richmond	796
Jacksonville	795

SOUTH AMERICA

ARGENTINA (1980)
Buenos Aires	9 927
Córdoba	982
Rosario	955
Mendoza	597
La Plata	560

BOLIVIA (1982)
La Paz	881

BRAZIL (1980)
São Paulo	8 493
Rio de Janeiro	5 091
Belo Horizonte	1 781
Salvador	1 502
Fortaleza	1 308
Recife	1 204
Brasilia	1 177
Pôrto Alegre	1 125
Nova Iguaçu	1 095
Curitiba	1 025
Belém	933
Goiánia	717
Campinas	665
Manaus	633
São Gonçalo	615
Duque de Caxias	576
Santo André	553
Guarulhos	533

CHILE (1983)
Santiago	4 132

COLOMBIA (1980)
Bogotá	4 486
Medellin	1 812
Cali	1 232
Barranquilla	900

ECUADOR (1982)
Guayaquil	1 301
Quito	1 110

PARAGUAY (1983)
Asunción	708

PERU (1983)
Lima	5 258

URUGUAY (1981)
Montevideo	1 362

VENEZUELA (1980)
Caracas	2 944
Maracaibo	901
Valencia	506

Map Projections

As the Earth is spherical in shape, it cannot be represented on a plane surface without some distortion. The map projection is a system for attempting to represent the sphere on a two-dimensional plane. A projection has certain properties: the representation of correct area, true shape or true bearings. The preservation of one property can only be secured at the expense of the other qualities.

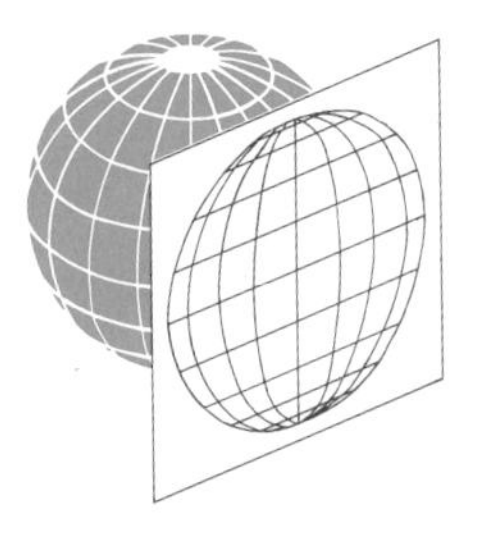

Azimuthal

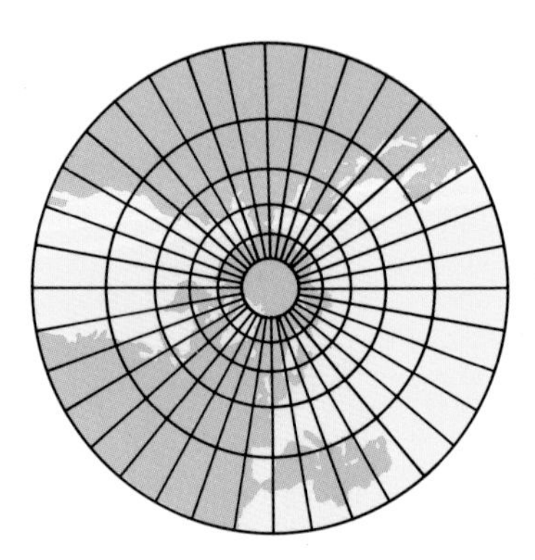

Cylindrical

Conical

An Azimuthal projection is constructed by the projection of part of the globe onto a plane tangential to a pole, the equator, or any other single point on the globe. The zenithal gnomonic projection shown below (A), has a plane touching the pole. It is ideal for showing polar air-routes because the shortest distance between any two points is a straight line. Air-route distances from one point (e.g. Capetown) are best shown by the Oblique Zenithal Equidistant (B).

Cylindrical projections are constructed by projecting a portion of the globe onto a cylinder tangential to the globe. The cylinder may be tangential to the equator. The tangential line is the only one true to scale, with distortion of size and shape increasing towards the top and bottom of the cylinder. The Mercator projection below (A) is a modification of a cylindrical projection. It avoids distortion of shape by making an increase in scale along the parallels. There is still size distortion, but its great use is for navigation since bearings can be plotted as straight lines. The Mollweide projection (B) is a 'conventional' cylindrical projection on which the meridians are no longer parallel. This is an equal area projection, useful for mapping distributions. In this case, it has been 'interrupted'.

Conical projections use the projection of the globe onto a cone which is tangential to a parallel (or any other small circle) on the globe. The scale is correct along this tangential line and can be made to be correct along the parallel or the meridians. In the simple conic (A) below, the scale is correct along the parallel indicated and the meridians. Bonnes projection (B) is equal area, the parallels being correctly divided, although there is shape distortion, especially at the edges.

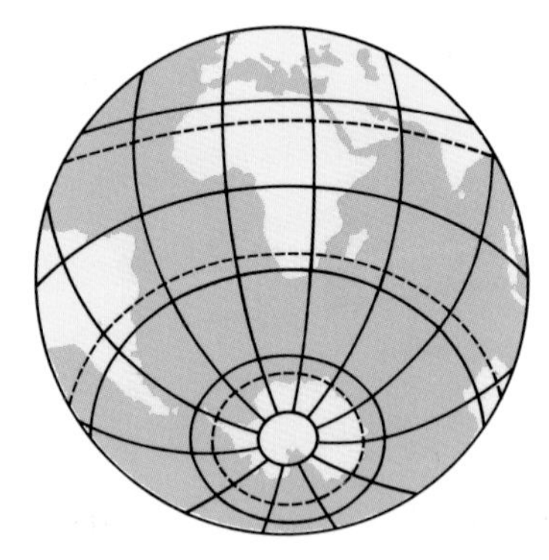

(A) **Zenithal Gnomonic**

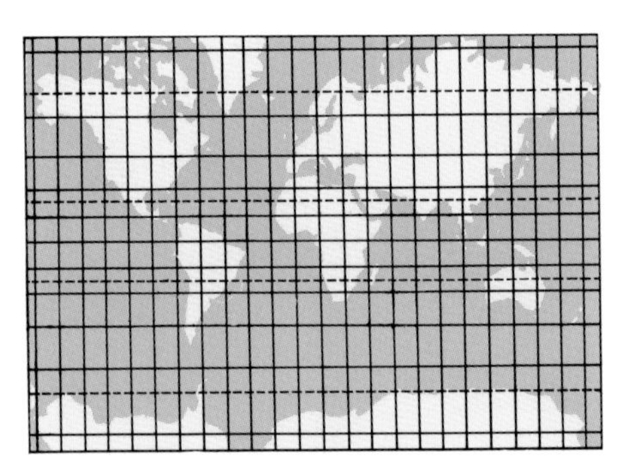

(A) **Mercator**

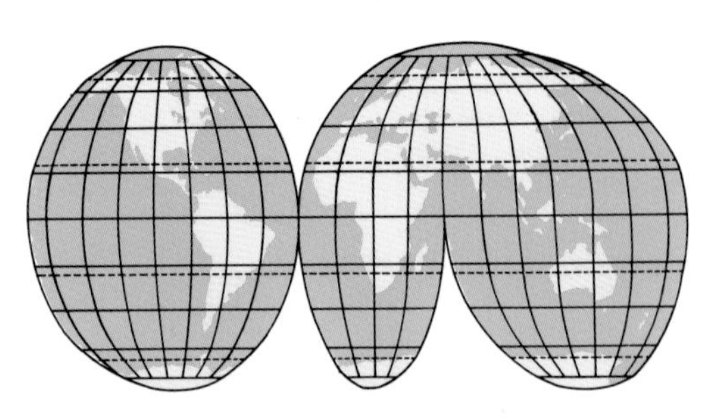

(A) **Simple Conic**

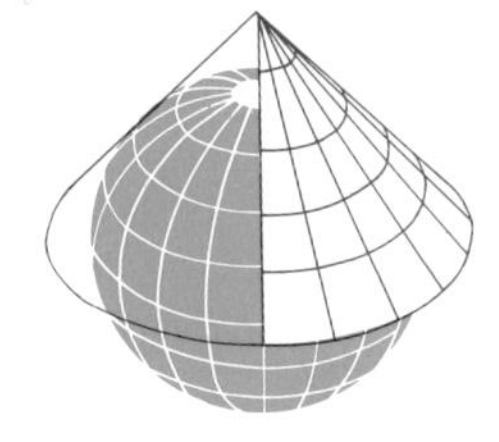

(B) **Oblique Zenithal Equidistant**

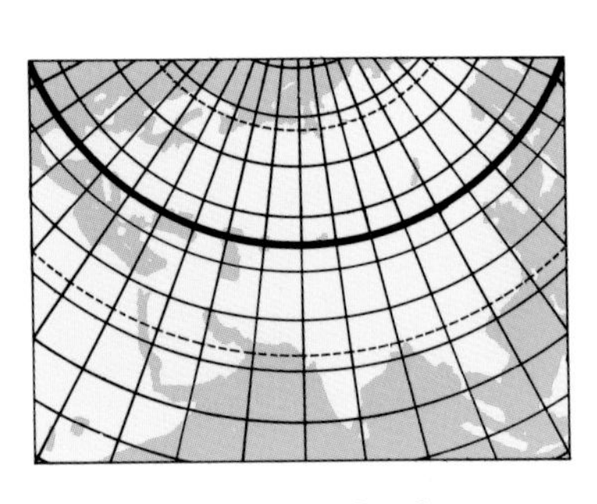

(B) **Interrupted Mollweide**

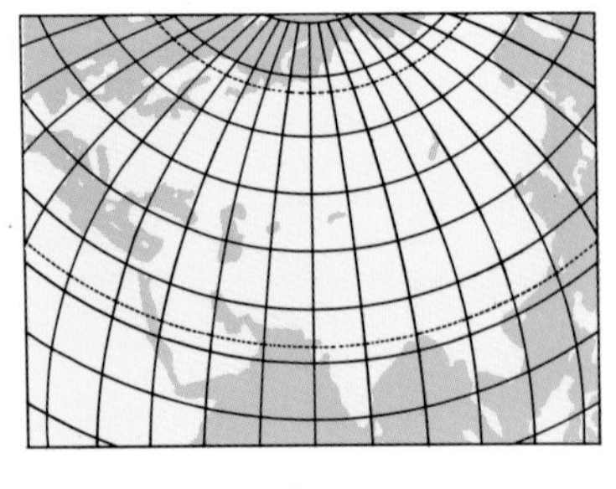

(B) **Bonne**

Many projections can be obtained if the surface onto which the graticule is developed cuts the sphere, rather than being tangential to it. These projections, called secant, can reduce distortion.
The choice of map projection depends largely on the property required — correct area, scale, shape or bearings.
The area and location to be covered is also important. Conical projections cannot cover the entire globe and are most suited for temperate areas of large longitudinal extent. Azimuthal projections are best for larger scale maps of small areas so that distortion around the edges is not too great. Cylindrical projections are best to map the whole world.

KEY TO SYMBOLS

These are the symbols used in the general maps in the atlas. The thematic symbols are explained in a key alongside each map and an attempt has been made to keep them consistent throughout the atlas.

Settlement

⬠ **BERLIN**   ⬡ DETROIT   ■ Fukuoka   ● Rosario   ◉ Gaza   ◎ Kananga   ⊙ *Berhampur*   ○ *Laurel*   ○ *Riverton*

⬡ **BIRMINGHAM**   ▣ **Derby**   ◉ **Swansea**   ○ Greenock   ○ Herne Bay   ○ Minehead

The settlement symbols and their related type are decided upon by the population of the town or city for each area mapped. These limits vary from one map to another so the population size for each group is not given in this table.

Administration

———————— International Boundaries

- - - - - - International Boundaries
(Undemarcated or undefined)

—·—·—·—·— Internal Boundaries

∴ Sites of Archaeological or Historical Importance

[ ] National Parks

International boundaries are drawn to show the 'de facto' situation where there are rival claims to territory.

Communications

——M4—— Motorways

- - - - - - Motorways under construction

———————— Principal roads

- - - - - - - Tracks, seasonal and other roads

⊣- - - -⊢ Road tunnels

———————— Principal railways

———————— Other railways

· · · · · · · · · Railways under construction

⊣- - - -⊢ Railway tunnels

⌣ Passes

⊢—·—·—·—⊣ Principal canals

⊢—┼—┼—⊣ Principal oil pipelines

✿ ✦ Principal airports

Hydrology

⌇ Perennial streams

- - - - - - Seasonal streams

⊣—┼—⊢ Dams

Perennial lakes

Seasonal lakes

Swamps and marshes

Permanent ice

⌣ Wells in desert

Relief

m   6000   4000   2000   1000   200   0

0   200   400   1000   1500   2000   3000   4000   6000   m

The coloured height reference is repeated on the edge of the page.

▲ 4507   Height above sea level in metres   ▼ 1710   Depth below sea level in metres   *1134*   Height of lakes

Abbreviations of measures used;   mm {Millimetres / Millimeters} ;   m {Metres / Meters} ;   km {Kilometres / Kilometers} ;   mb Millibars ;   tons Metric Tonnes ;   M.S.L. Mean Sea Level

The origin of the earth is still open to much conjecture although the most widely accepted theory is that it was formed from a solar cloud consisting mainly of hydrogen. Under gravitation the cloud condensed and shrank to form our planets orbiting around the sun. Gravitation forced the lighter elements to the surface of the earth where they cooled to form a crust while the inner material remained hot and molten. Earth's first rocks formed over 3 500 million years ago but since then the surface has been constantly altered.

Until comparatively recently the view that the primary units of the earth had remained essentially fixed throughout geological time was regarded as common sense, although the concept of moving continents has been traced back to references in the Bible of a break up of the land after Noah's floods. The continental drift theory was first developed by Antonio Snider in 1858 but probably the most important single advocate was Alfred Wegener who, in 1915, published evidence from geology, climatology and biology. His conclusions are very similar to those reached by current research although he was wrong about the speed of break-up.

The measurement of fossil magnetism found in rocks has probably proved the most influential evidence. While originally these drift theories were openly mocked, now they are considered standard doctrine.

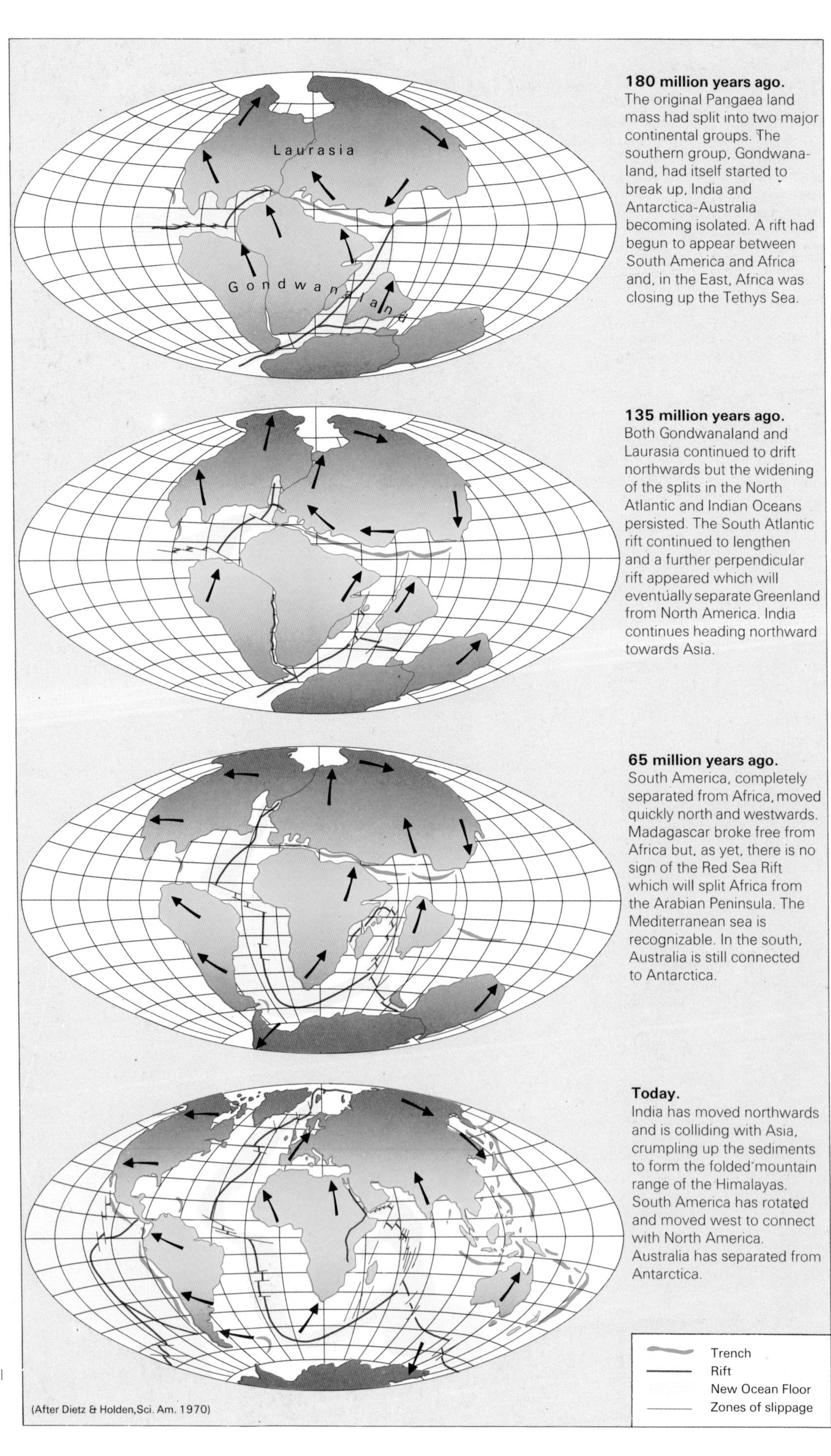

The jigsaw
As knowledge of the shape and structure of the earth's surface grew, several of the early geographers noticed the great similarity in shape of the coasts bordering the Atlantic. It was this remarkable similarity which led to the first detailed geological and structural comparisons. Even more accurate fits can be made by placing the edges of the continental shelves in juxtaposition.

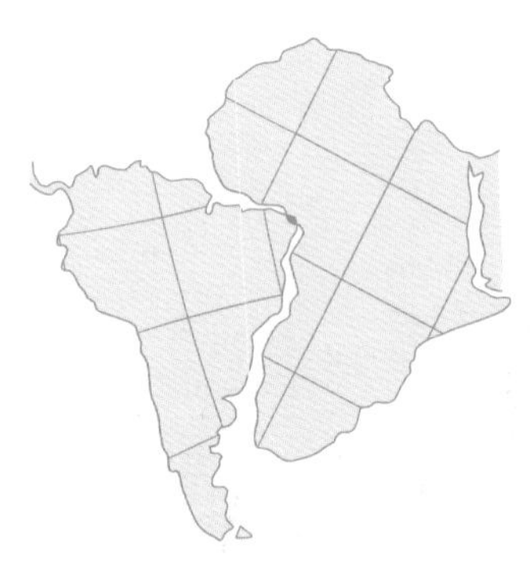

180 million years ago.
The original Pangaea land mass had split into two major continental groups. The southern group, Gondwanaland, had itself started to break up, India and Antarctica-Australia becoming isolated. A rift had begun to appear between South America and Africa and, in the East, Africa was closing up the Tethys Sea.

135 million years ago.
Both Gondwanaland and Laurasia continued to drift northwards but the widening of the splits in the North Atlantic and Indian Oceans persisted. The South Atlantic rift continued to lengthen and a further perpendicular rift appeared which will eventually separate Greenland from North America. India continues heading northward towards Asia.

65 million years ago.
South America, completely separated from Africa, moved quickly north and westwards. Madagascar broke free from Africa but, as yet, there is no sign of the Red Sea Rift which will split Africa from the Arabian Peninsula. The Mediterranean sea is recognizable. In the south, Australia is still connected to Antarctica.

Today.
India has moved northwards and is colliding with Asia, crumpling up the sediments to form the folded mountain range of the Himalayas. South America has rotated and moved west to connect with North America. Australia has separated from Antarctica.

(After Dietz & Holden, Sci. Am. 1970)

	Trench
	Rift
	New Ocean Floor
	Zones of slippage

Plate tectonics

The original debate about continental drift was only a prelude to a more radical idea; plate tectonics. The basic theory is that the earth's crust is made up of a series of rigid plates which float on a soft layer of the mantle and are moved about by convection currents in the earth's interior. These plates converge and diverge along margins marked by earthquakes, volcanoes and other seismic activity. Plates diverge from mid-ocean ridges where molten lava pushes upwards and forces the plates apart at a rate of up to 40 mm a year. Converging plates form either a trench, where the oceanic plate sinks below the lighter continental rock, or mountain ranges where two continents collide. This explains the paradox that while there have always been oceans none of the present oceans contain sediments more than 150 million years old.

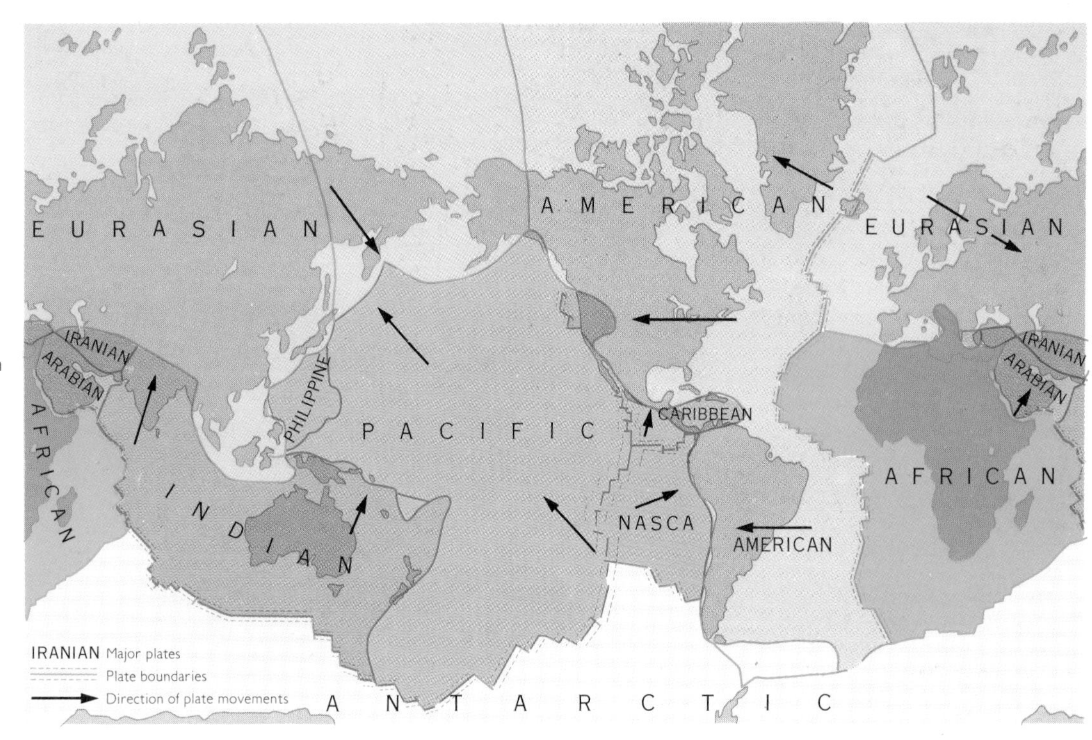

IRANIAN Major plates
Plate boundaries
Direction of plate movements

Trench boundary

The present explanation for the comparative youth of the ocean floors is that where an ocean and a continent meet the ocean plate dips under the less dense continental plate at an angle of approximately 45°. All previous crust is then ingested by downward convection currents. In the Japanese trench this occurs at a rate of about 120 mm a year.

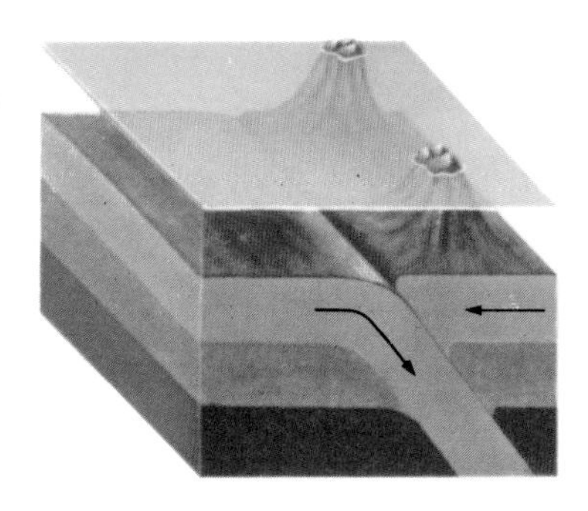

Transform fault

The recent identification of the transform, or transverse, fault proved to be one of the crucial preliminaries to the investigation of plate tectonics. They occur when two plates slip alongside each other without parting or approaching to any great extent. They complete the outline of the plates delineated by the ridges and trenches and demonstrate large scale movements of parts of the earth's surface

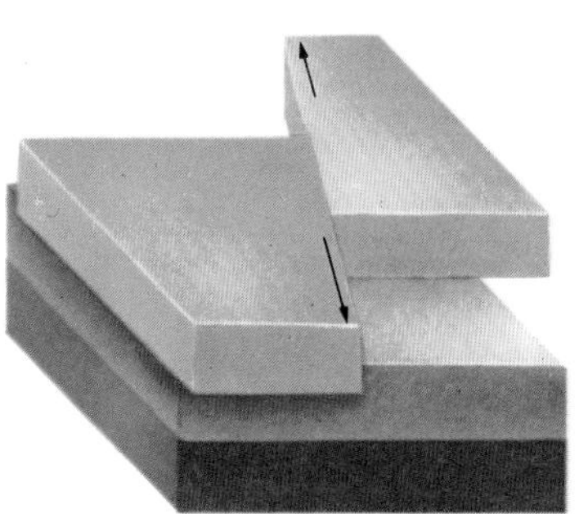

Ridge boundary

Ocean rises or crests are basically made up from basaltic lavas for although no gap can exist between plates, one plate can ease itself away from another. In that case hot, molten rock instantly rises from below to fill in the incipient rift and forms a ridge. These ridges trace a line almost exactly through the centre of the major oceans.

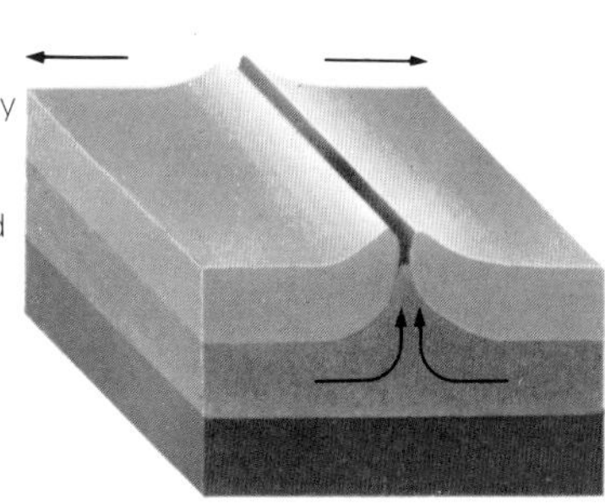

Destruction of ocean plates.

As the ocean plate sinks below the continental plate some of the sediment on its surface is scraped off and piled up on the landward side. This sediment is later incorporated in a folded mountain range which usually appears on the edge of the continent, such as the Andes. Similarly if two continents collide the sediments are squeezed up into new mountains.

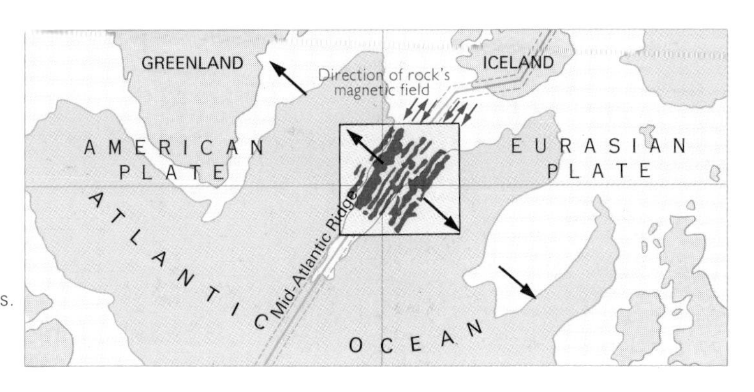

Sea floor spreading

Reversals in direction of the earth's magnetic field have occurred throughout history. As molten rock emerges at the ocean ridges it cools and is magnetised in the direction of the earth's magnetic field. By mapping the ocean floor across the ridge a striped pattern of solidified rock magnetised in alternate directions is produced (see inset area). As the dates of the last few reversals are known the rate of spreading can be calculated (40mm per year in the North Atlantic).

The earth's surface is slowly but continually being rearranged. Some changes such as erosion and deposition are extremely slow but they upset the balance which causes other more abrupt changes often originating deep within the earth's interior. The constant movements vary in intensity, often with stresses building up to a climax such as a particularly violent volcanic eruption or earthquake.

The crust (below and right)
The outer layer or crust of the earth consists of a comparatively low density, brittle material varying from 5 km to 50 km deep beneath the continents. This consists predominately of silica and aluminium; hence it is called 'sial'. Extending under the ocean floors and below the sial is a basaltic layer known as 'sima', consisting mainly of silica and magnesium.

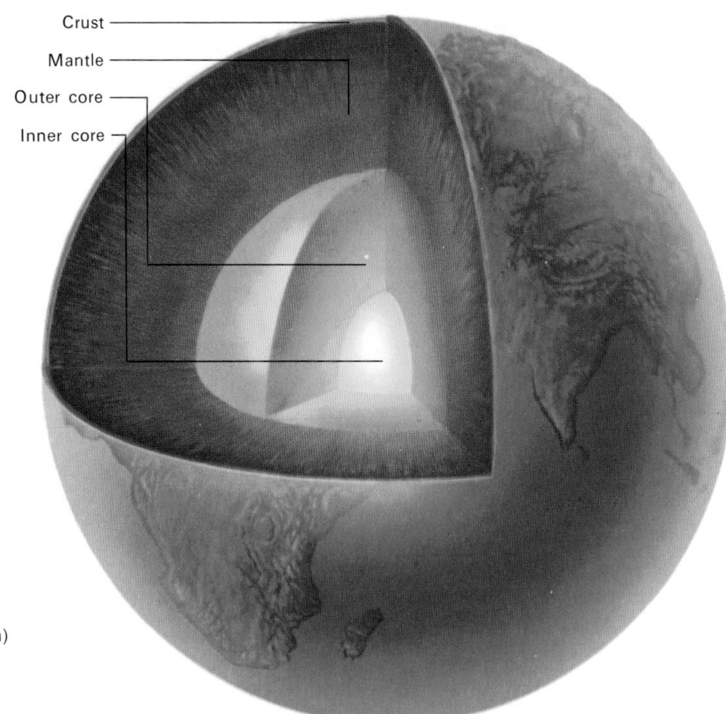

Crust
Mantle
Outer core
Inner core

Continental crust   Ocean crust

Sediment
Granite rock (sial)
Basaltic layer (sima)
Mantle

Volcanoes (right, below and far right)
Volcanoes occur when hot liquefied rock beneath the crust reaches the surface as lava. An accumulation of ash and cinders around a vent forms a cone. Successive layers of thin lava flows form an acid lava volcano while thick lava flows form a basic lava volcano. A caldera forms when a particularly violent eruption blows off the top of an already existing cone.

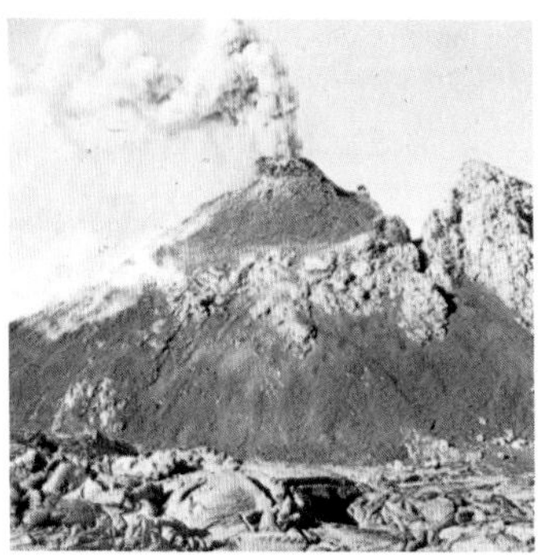

The mantle (above)
Immediately below the crust, at the mohorovicic discontinuity line, there is a distinct change in density and chemical properties. This is the mantle - made up of iron and magnesium silicates - with temperatures reaching 1 600 °C. The rigid upper mantle extends down to a depth of about 1 000 km below which is the more viscous lower mantle which is about 1 900 km thick.

The core (above)
The outer core, approximately 2 100 km thick, consists of molten iron and nickel at 2 000 °C to 5 000 °C possibly separated from the less dense mantle by an oxidised shell. About 5 000 km below the surface is the liquid transition zone, below which is the solid inner core, a sphere of 2 740 km diameter where rock is three times as dense as in the crust.

Shield volcano   **Cinder cone**   **Hornit cone**   **Caldera**

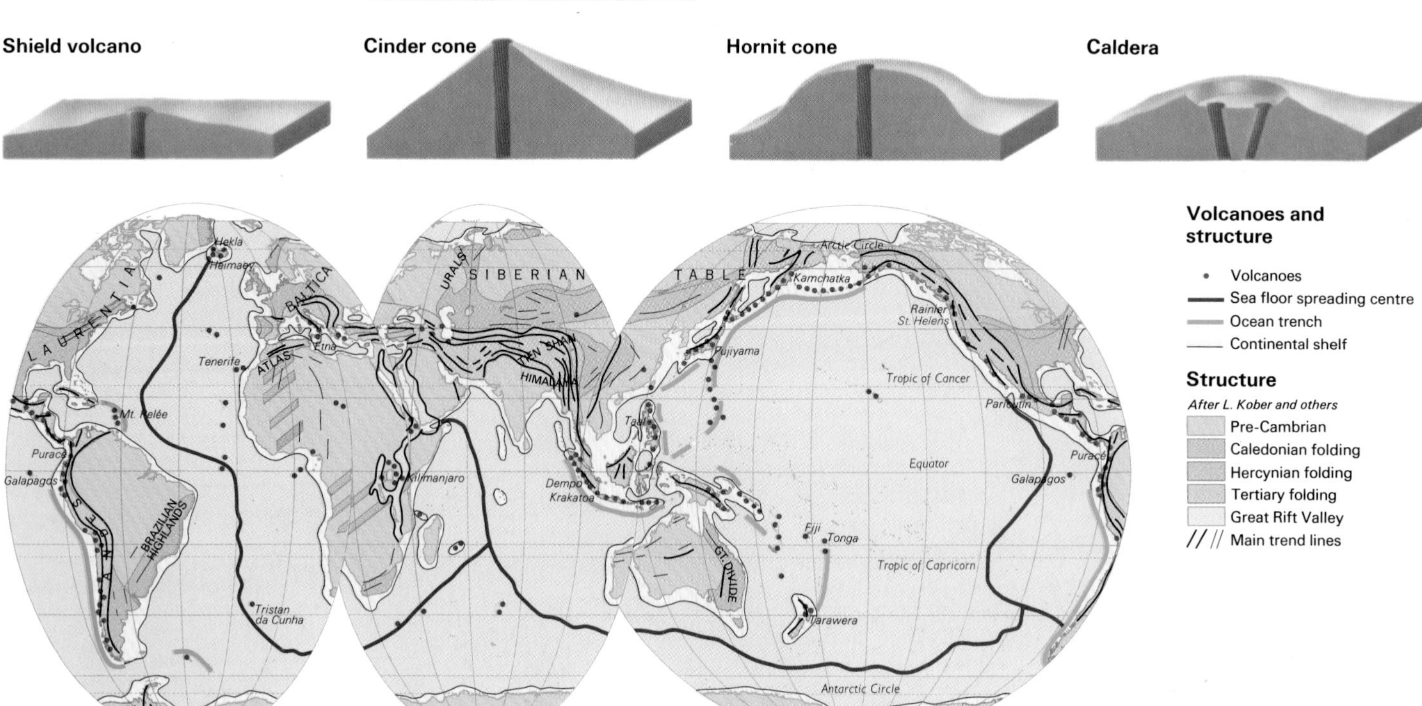

Volcanoes and structure

- Volcanoes
— Sea floor spreading centre
— Ocean trench
— Continental shelf

Structure
After L. Kober and others

Pre-Cambrian
Caledonian folding
Hercynian folding
Tertiary folding
Great Rift Valley
/// // Main trend lines

Projection: *Interrupted Mollweide's Homolographic*

5

Major earthquakes in the last 100 years and numbers killed

1896	Japan (tsunami)	22 000	1923	Japan, Tokyo	143 000
1906	San Francisco	destroyed	1930	Italy, Naples	2 100
1906	Chile, Valparaiso	22 000	1931	New Zealand, Napier	destroyed
1908	Italy, Messina	77 000	1931	Nicaragua, Managua	destroyed
1920	China, Kansu	180 000	1932	China, Kansu	70 000
			1935	India, Quetta	60 000
			1939	Chile, Chillan	20 000

1939/40	Turkey, Erzincan	30 000
1948	Japan, Fukui	5 100
1956	N. Afghanistan	2 000
1957	W. Iran	10 000
1960	Morocco, Agadir	12 000
1962	N.W. Iran	10 000
1963	Yugoslavia, Skopje	1 000

1966	U.S.S.R., Tashkent	destroyed
1970	N. Peru	66 800
1974	N. Pakistan	10 000
1976	China, Tangshan	650 000
1978	Iran, Tabas	11 000
1980	Algeria, El Asnam	20 000
1985	Mexico City, Mexico	20 000

World distribution of earthquakes

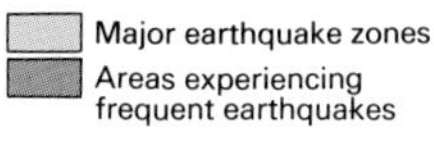

Major earthquake zones
Areas experiencing frequent earthquakes

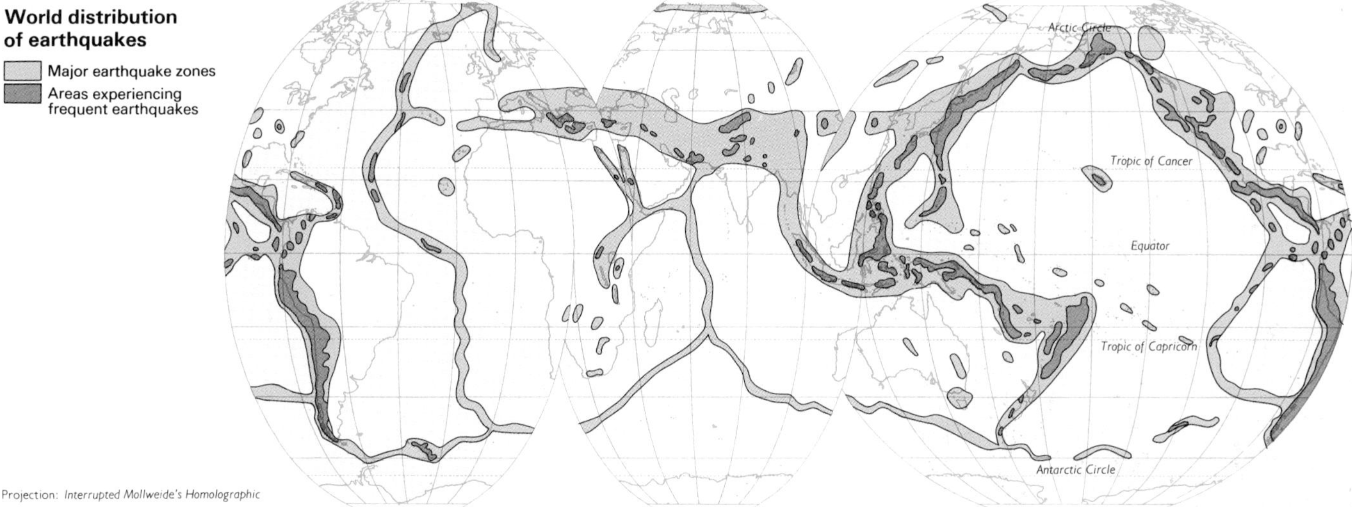

Projection: Interrupted Mollweide's Homolographic

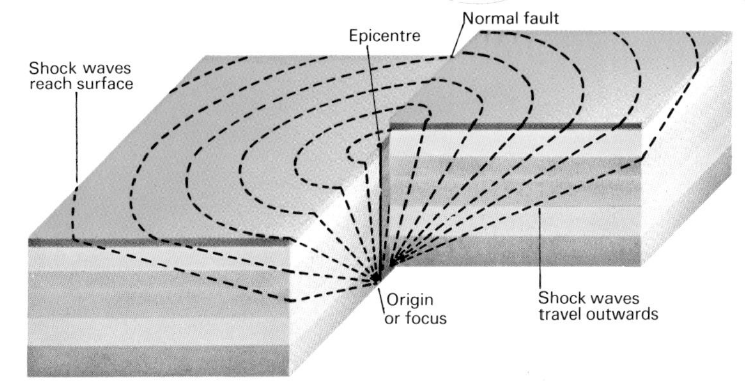

Earthquakes *(right and above)*
Earthquakes are a series of rapid vibrations originating from the slipping or faulting of parts of the earth's crust when stresses within build up to breaking point. They usually happen at depths varying from 8 km to 30 km. Severe earthquakes cause extensive damage when they take place in populated areas destroying structures and severing communications. Most loss of life occurs due to secondary causes i.o. falling masonry, fires or tsunami waves.

Alaskan earthquake, 1964

Seismic Waves *(right)*
The shock waves sent out from the focus of an earthquake are of three main kinds each with distinct properties. Primary (P) waves are compressional waves which can be transmitted through both solids and liquids and therefore pass through the earth's liquid core. Secondary (S) waves are shear waves and can only pass through solids. They cannot pass through the core and are reflected at the core-mantle boundary taking a concave course back to the surface. The core also refracts the P waves causing them to alter course, and the net effect of this reflection and refraction is the production of a shadow zone at a certain distance from the epicentre, free from P and S waves. Due to their different properties P waves travel about 1,7 times faster than S waves. The third main kind of wave is a long (L) wave, a slow wave which travels along the earth's surface, its motion being either horizontal or vertical.

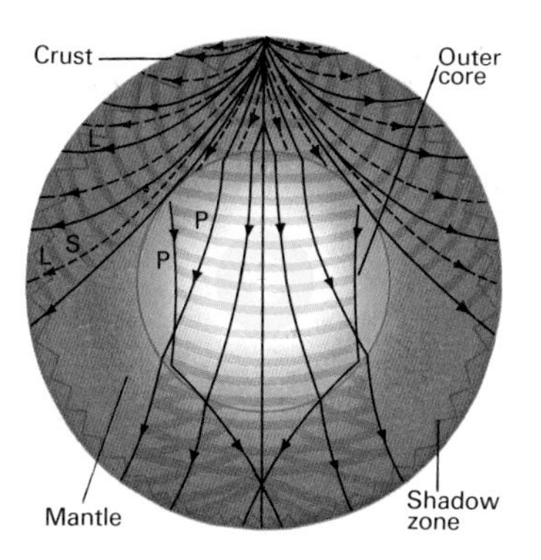

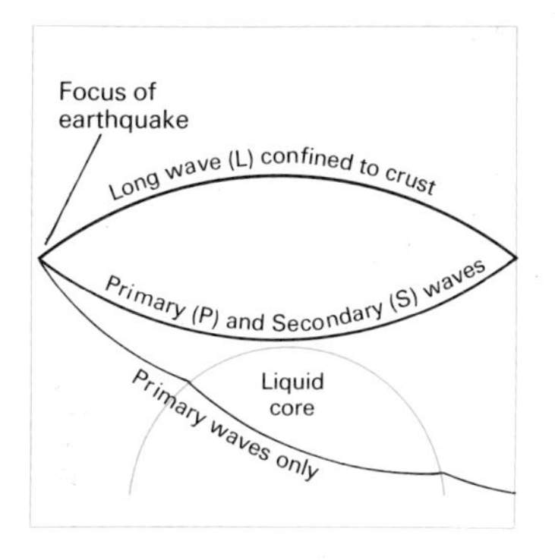

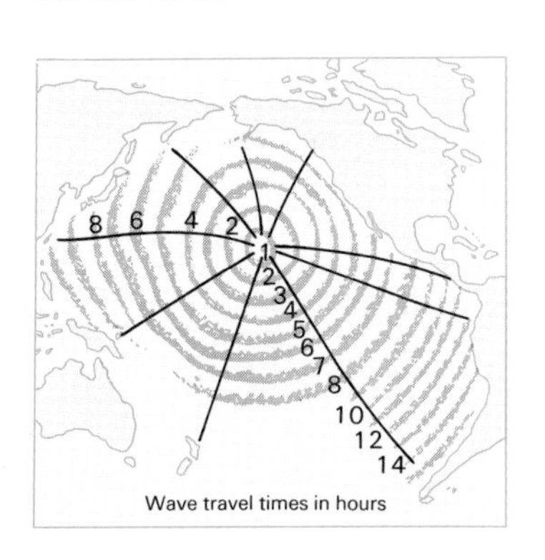

Wave travel times in hours

Tsunami waves *(left)*
A sudden slump in the ocean bed during an earthquake forms a trough in the water surface subsequently followed by a crest and smaller waves. A more marked change of level in the sea bed can form a crest, the start of a Tsunami which travels up to 600 km/h with waves up to 60 m high. Seismographic detectors continuously record earthquake shocks and warn of the Tsunami which may follow it.

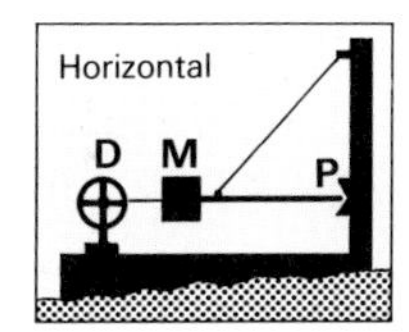
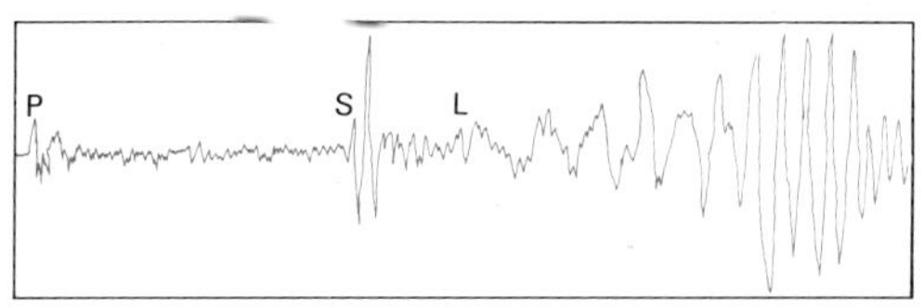
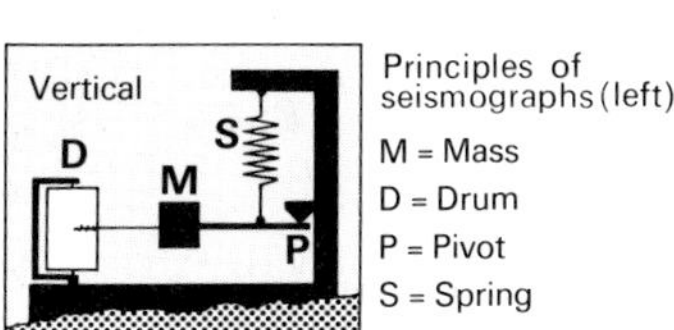

Principles of seismographs (left)
M = Mass
D = Drum
P = Pivot
S = Spring

Seismographs
are delicate instruments capable of detecting and recording vibrations due to earthquakes thousands of kilometres away. P waves cause the first tremors. S the second, and L the main shock.

Its surface

Highest point on the earth's surface: Mt. Everest, Tibet - Nepal boundary   8 848 m
Lowest point on the earth's surface: The Dead Sea, Jordan   below sea level 395 m
Greatest ocean depth,: Challenger Deep, Mariana Trench   11 022 m
Average height of land   840 m
Average depth of seas and oceans   3 808 m

Dimensions

Superficial area	510 000 000 km²
Land surface	149 000 000 km²
Land surface as % of total area	29,2 %
Water surface	361 000 000 km²
Water surface as % of total area	70,8 %
Equatorial circumference	40 077 km
Meridional circumference	40 009 km
Equatorial diameter	12 756,8 km
Polar diameter	12 713,8 km
Equatorial radius	6 378,4 km
Polar radius	6 356,9 km
Volume of the Earth	1 083 230 x 10⁶ km³
Mass of the Earth	5,9 x 10²¹ tonnes

The Figure of Earth

An imaginary sea-level surface is considered and called a geoid. By measuring at different places the angles from plumb lines to a fixed star there have been many determinations of the shape of parts of the geoid which is found to be an oblate spheriod with its axis along the axis of rotation of the earth. Observations from satellites have now given a new method of more accurate determinations of the figure of the earth and its local irregularities.

Land and Sea Hemispheres.

About 85% of the total land area is contained in the hemisphere centred on a point between Paris and Brussels.

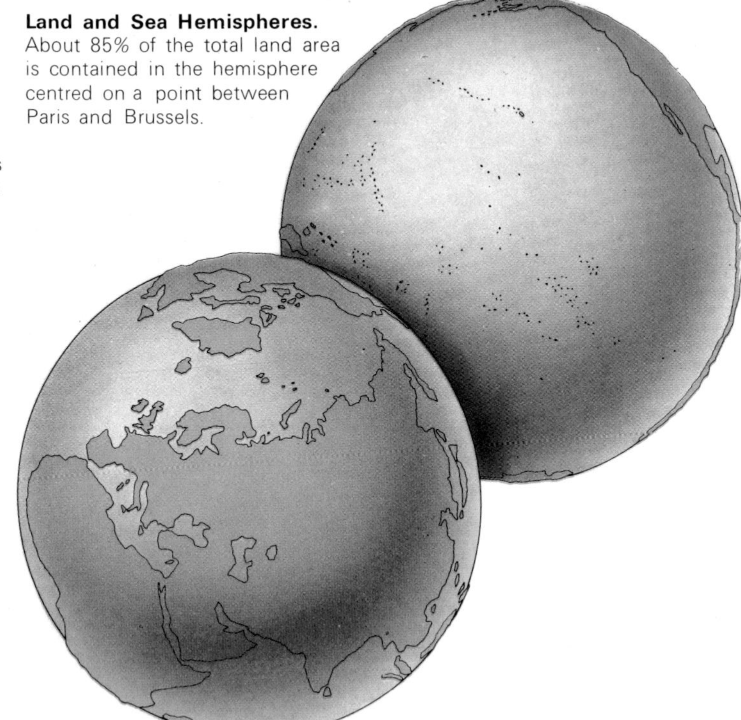

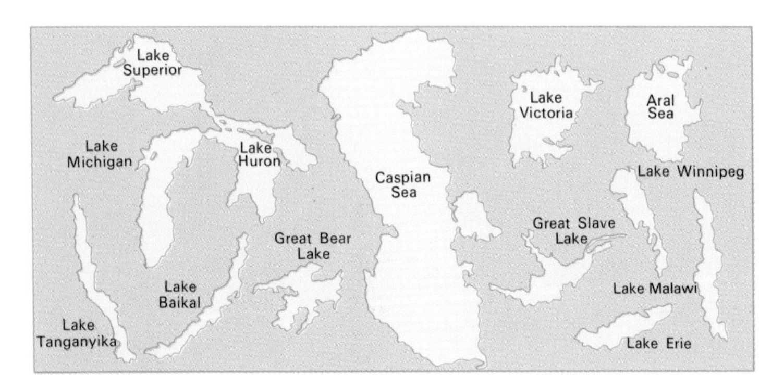

Oceans and Seas
Area in 1000 km²

Pacific Ocean	165 721	North Sea	575
Atlantic Ocean	81 660	Black Sea	448
Indian Ocean	73 442	Red Sea	440
Arctic Ocean	14 351	Baltic Sea	422
Mediterranean Sea	2 966	The Gulf	238
Bering Sea	2 274	St. Lawrence, Gulf of	236
Caribbean Sea	1 942	English Channel & Irish Sea	179
Mexico, Gulf of	1 813	California, Gulf of	161
Okhotsk, Sea of	1 528		
East China Sea	1 248		
Hudson Bay	1 230		
Japan, Sea of	1 049		

Lakes and Inland Seas
Areas in 1000 km²

Caspian Sea, Asia	424,2	Lake Ontario, N.America	19,5
Lake Superior, N.America	82,4	Lake Ladoga, Europe	18,4
Lake Victoria, Africa	69,5	Lake Balkhash, Asia	17,3
Aral Sea (Salt), Asia	63,8	Lake Maracaibo, S.America	16,3
Lake Huron, N.America	59,6	Lake Onega, Europe	9,8
Lake Michigan, N.America	58,0	Lake Eyre (Salt), Australia	9,6
Lake Tanganyika, Africa	32,9	Lake Turkana (Salt), Africa	9,1
Lake Baikal, Asia	31,5	Lake Titicaca, S.America	8,3
Great Bear Lake, N.America	31,1	Lake Nicaragua, C.America	8,0
Great Slave Lake, N.America	28,9	Lake Athabasca, N.America	7,9
Lake Malawi, Africa	28,5	Reindeer Lake, N.America	6,3
Lake Erie, N.America	25,7	Issyk-Kul, Asia	6,2
Lake Winnipeg, N.America	24,3	Lake Torrens (Salt), Australia	6,1
Lake Chad, Africa	20,7	Koko Nor (Salt), Asia	6,0
		Lake Urmia, Asia	6,0
		Vänern, Europe	5,6

Longest rivers

	km.
Nile, Africa	6 690
Amazon, S.America	6 280
Mississippi-Missouri, N.America	6 270
Yangtze, Asia	4 990
Zaïre, Africa	4 670
Amur, Asia	4 410
Hwang Ho (Yellow), Asia	4 350
Lena, Asia	4 260
Mekong, Asia	4 180
Niger, Africa	4 180
Mackenzie, N.America	4 040
Ob, Asia	4 000
Yenisei, Asia	3 800

0   1000   2000   3000   4000   5000   6000   700

The Highest Mountains and The Greatest Depths

Mount Everest defied the world's greatest mountaineers for many years and in 1953 the first successful attempt was made. The world's highest mountains have now been climbed but there are many as yet unexplored in the Himalaya.

The greatest depths are the ocean trenches representing less than 2% of the total area of the sea-bed. They are of great interest as lines of structural weakness in the earth's crust where earthquakes are frequent.

Mountain heights in metres

1 Kosciusko 2 230
2 Mt. Cook (N.Z.) 3 764
3 Kinabalu 4 101
4 Jaya (Irian) 5 029
5 Thabana Ntlenyana 3 482
6 Ruwenzori 5 109
7 Cameroon Peak 4 070
8 Dj. Toubkal 4 165
9 Ras Dashen 4 620
10 Kilimanjaro 5 895
11 Roraima 2 810
12 Chimborazo 6 267
13 Illimani 6 462
14 Huascaran 6 768
15 Ojos del Salado 6 863
16 Aconcagua 6 960
17 Galdhøppigen 2 469
18 Mont Blanc 4 807
19 Mulhacen 3 478
20 Elbrus 5 633
21 Fujiyama 3 776
22 Communism Peak 7 495
23 8 598
24 K2 8 611
25 Muztagh 7 723
26 Everest 8 848
27 Mt. Elbert 4 399
28 Mt. Logan 6 050
29 Mt. Whitney 4 418
30 Mt. McKinley 6 194

Kanchenjunga

Oceania — Africa — South America — Europe and Asia — North America

Ocean depths in metres

Sea level

31 Mauritius basin 6 400
32 W. Australian basin 6 459
33 Java trench 7 450
34 Mindanao trench 10 497
35 Mariana trench 11 022
36 Japan trench 10 554
37 Bougainville deep 9 140
38 Kuril trench 10 542
39 Aleutian trench 7 822
40 Kermadec trench 10 047
41 Tonga trench 10 822
42 Cayman trough 7 680
43 Puerto Rico trough 9 200
44 S. Sandwich trench 8 428
45 Romanche deep 7 758

Indian Ocean — Pacific Ocean — Atlantic Ocean

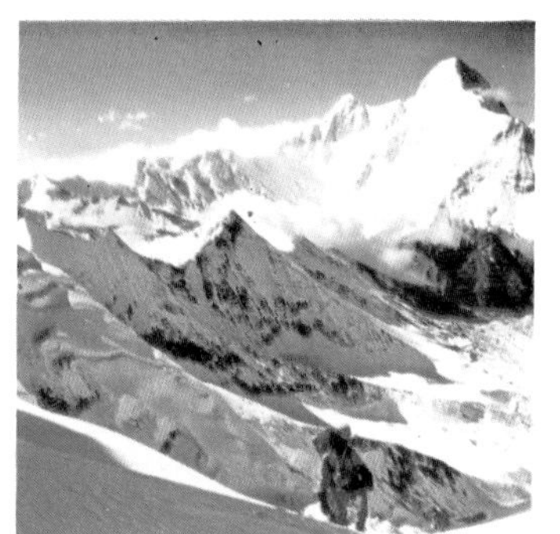

High mountains in the Himalaya

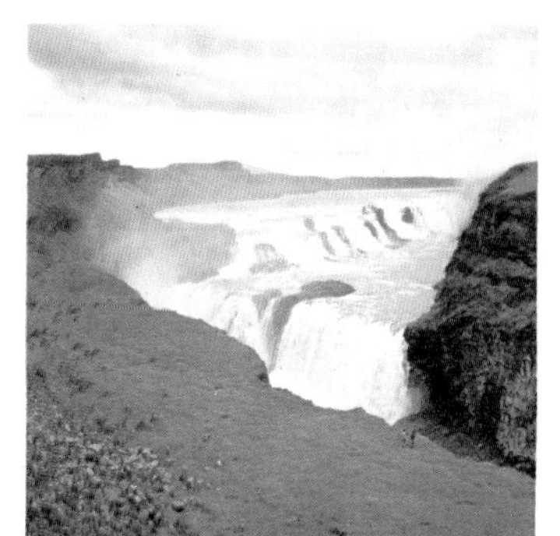

Waterfall in Iceland

Highest Waterfalls	height in metres
Angel, Venezuela	980
Tugela, South Africa	853
Yosemite, California	738
Mardalsfossen, Norway	655
Sutherland, New Zealand	579
Reichenbach, Switzerland	548
Wollomombi, Australia	518
Ribbon, California	491
Gavarnie, France	422
Tyssefallene, Norway	414
Krimml, Austria	370
King George VI, Guyana	366
Silver Strand, California	356

Highest Dams	height in metres
Africa	
Cabora Bassa, Zambezi R.	168
Akosombo Main Dam, Volta R.	141
Asia	
Nurek, Vakhsh R., U.S.S.R.	317
Bhakra, Sutlej R., India	226
Europe	
Grand Dixence, Switzerland	284
Vajont, Vajont R., Italy	261
America	
Oroville, Feather R.	235
Hoover, Colorado R.	221
Australasia	
Warragamba, N.S.W., Australia	137

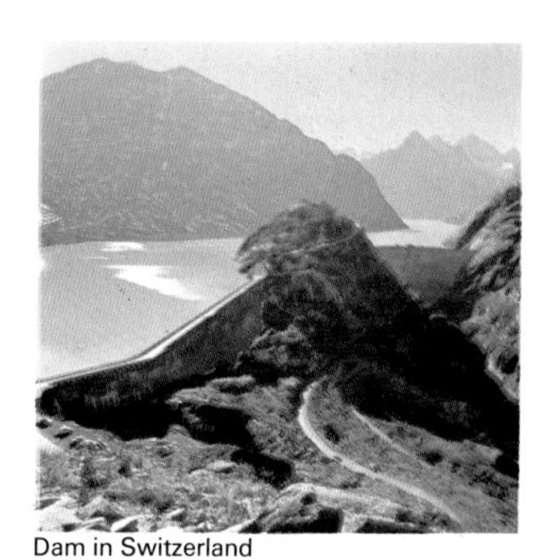

Dam in Switzerland

The World: Physical

The numbers on the map refer to the mountains and ocean depths named in the above diagrams

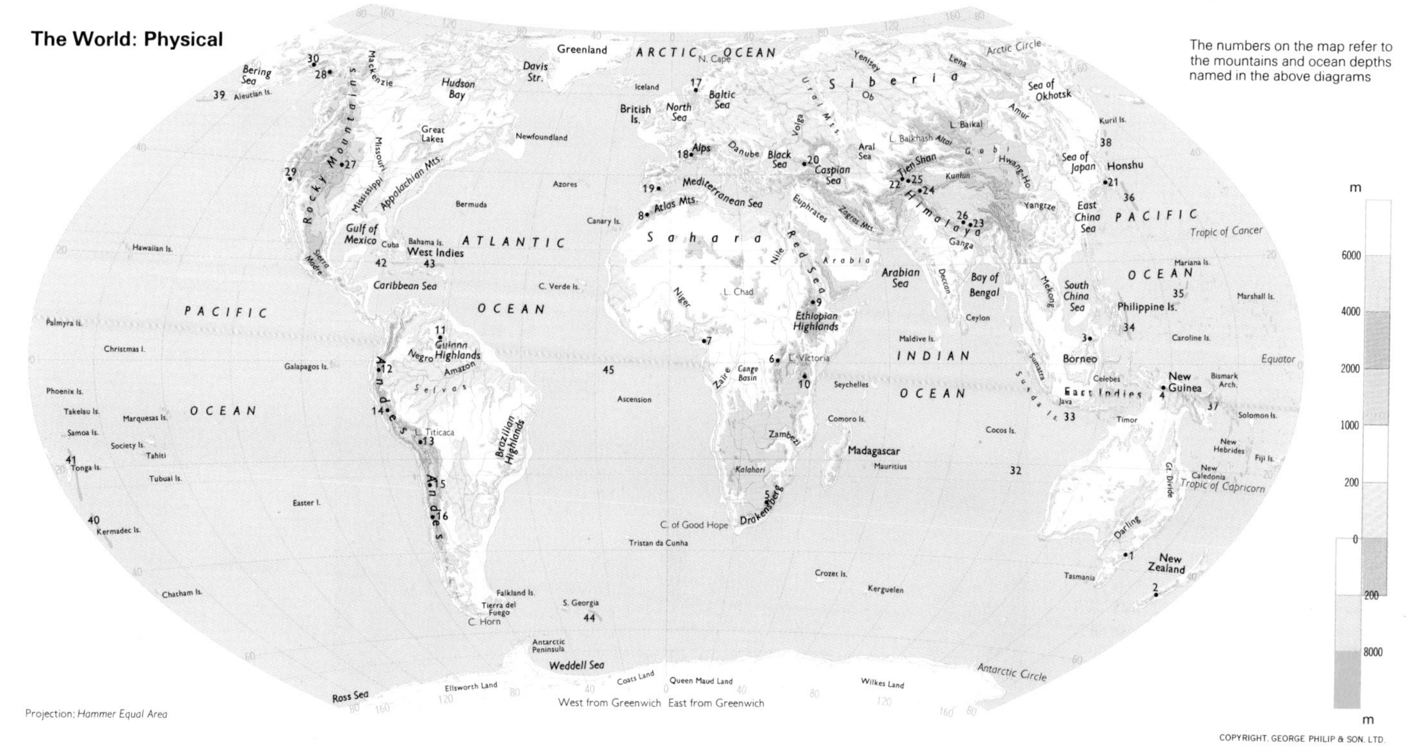

Projection: Hammer Equal Area

West from Greenwich   East from Greenwich

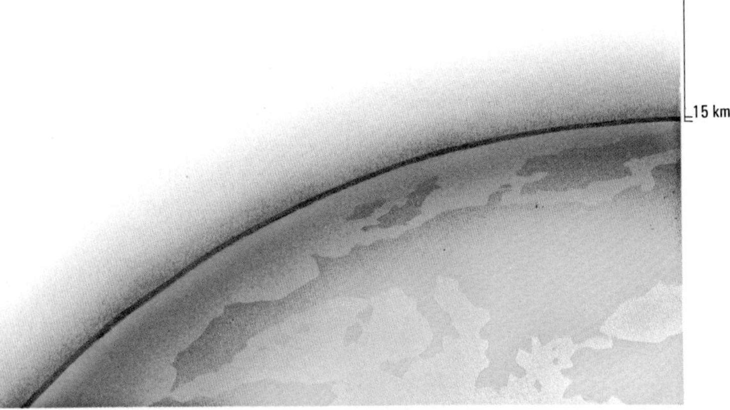

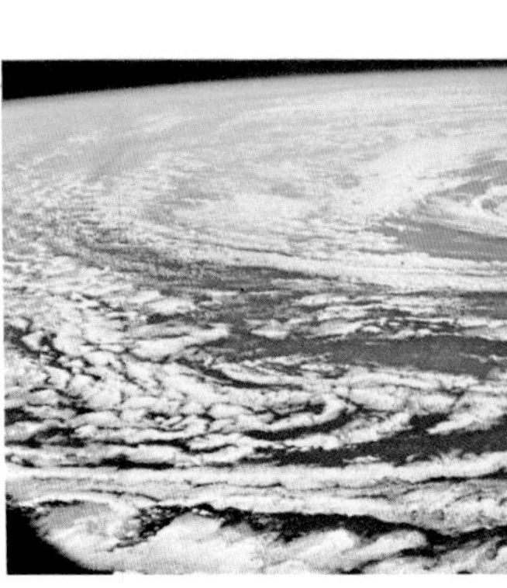

Earth's thin coating (right)
The atmosphere is a blanket of protective gases around the earth providing insulation against otherwise extreme alternations in temperature. The gravitational pull increases the density nearer the earth's surface so that 80% of the atmospheric mass is in the first 15 km. It is a very thin layer in comparison with the earth's diameter of 12 680 km, like the cellulose coating on a globe.

Exosphere(1)
The exosphere merges with the interplanetary medium and although there is no definite boundary with the ionosphere it starts at a height of about 600 km. The rarified air mainly consists of a small amount of atomic oxygen up to 600 km and equal proportions of hydrogen and helium with hydrogen predominating above 2 400 km.

Ionosphere(2)
Air particles of the ionosphere are electrically charged by the sun's radiation and congregate in four main layers, D, E, F1 and F2, which can reflect radio waves. Aurorae, caused by charged particles deflected by the earth's magnetic field towards the poles, occur between 65 km and 965 km above the earth. It is mainly in the lower ionosphere that meteors from outer space burn up as they meet increased air resistance.

Stratosphere(3)
A thin layer of ozone contained within the stratosphere absorbs ultra-violet light and in the process gives off heat. The temperature ranges from about -55°C at the tropopause to about -60°C in the upper part, known as the mesosphere, with a rise to about 2°C just above the ozone layer. This portion of the atmosphere is separated from the lower layer by the tropopause.

Troposphere(4)
The earth's weather conditions are limited to this layer which is relatively thin, extending upwards to about 8 km at the poles and 15 km at the equator. It contains about 85% of the total atmospheric mass and almost all the water vapour. Air temperature falls steadily with increased height at about 1°C for every 100 metres above sea level.

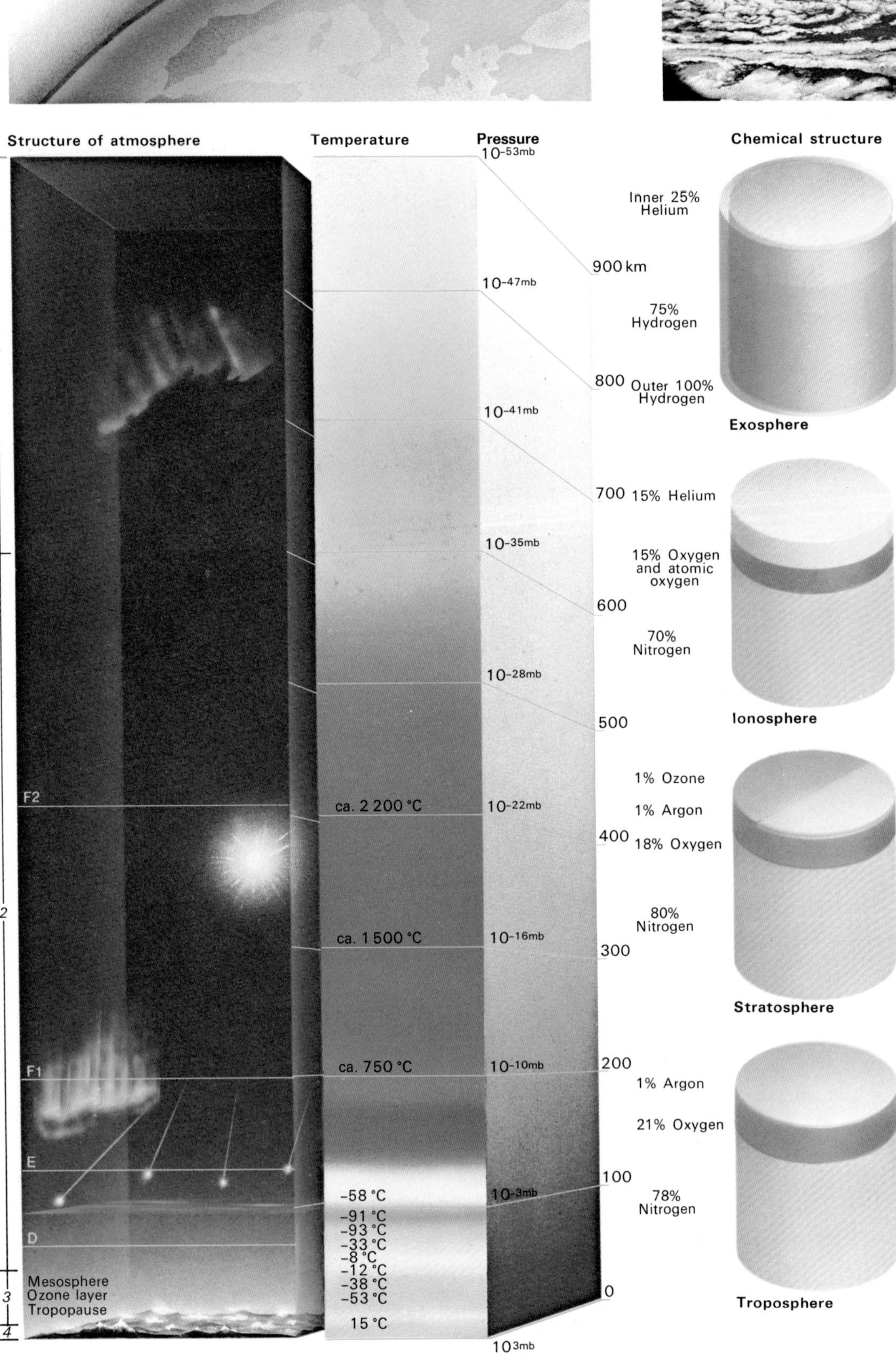

Structure of atmosphere — Temperature — Pressure — Chemical structure

Exosphere: Inner 25% Helium; 75% Hydrogen; Outer 100% Hydrogen

Ionosphere: 15% Helium; 15% Oxygen and atomic oxygen; 70% Nitrogen

Stratosphere: 1% Ozone; 1% Argon; 18% Oxygen; 80% Nitrogen

Troposphere: 1% Argon; 21% Oxygen; 78% Nitrogen

Pacific Ocean
Cloud patterns over the Pacific show the paths of prevailing winds.

Circulation of the air

30°N

Equator

30°S

Circulation of the air
Owing to high temperatures in equatorial regions the air near the ground is heated, expands and rises producing a low pressure belt. It cools, causing rain, spreads out then sinks again about latitudes 30° north and south forming high pressure belts.

High and low pressure belts are areas of comparative calm but between them, blowing from high to low pressure, are the prevailing winds. These are deflected to the right in the northern hemisphere and to the left in the southern hemisphere (Coriolis effect). The circulations appear in three distinct belts with a seasonal movement north and south following the overhead sun.

Cloud types

Clouds form when damp air is cooled, usually by rising. This may happen in three ways: when a wind rises to cross hills or mountains; when a mass of air rises over, or is pushed up by another mass of denser air; when local heating of the ground causes convection currents.

Cirrus *(1)* are detached clouds composed of microscopic ice crystals which gleam white in the sun resembling hair or feathers. They are found at heights of 6 000 to 12 000 metres.

Cirrostratus *(2)* are a whitish veil of cloud made up of ice crystals through which the sun can be seen often producing a halo of bright light.

Cirrocumulus *(3)* is another high altitude cloud formed by turbulence between layers moving in different directions.

Altostratus *(4)* is a grey or bluish striated, fibrous or uniform sheet of cloud producing light drizzle.

Altocumulus *(5)* is a thicker and fluffier version of cirro cumulus, it is a white and grey patchy sheet of cloud.

Nimbostratus *(6)* is a dark grey layer of cloud obscuring the sun and causing almost continuous rain or snow.

Cumulus *(7)* are detached heaped up, dense low clouds. The sunlit parts are brilliant white while the base is relatively dark and flat.

Stratus *(8)* forms dull overcast skies associated with depressions and occurs at low altitudes up to 1500 metres.

Cumulonimbus *(9)* are heavy and dense clouds associated with storms and rain. They have flat bases and a fluffy outline extending up to great altitudes.

High clouds / Middle clouds / Low clouds

Thousands of metres

1 Cirrus

2 Cirrostratus

3 Cirrocumulus

4 Altostratus

5 Altocumulus

6 Nimbostratus

7 Cumulus

8 Stratus

9 Cumulonimbus

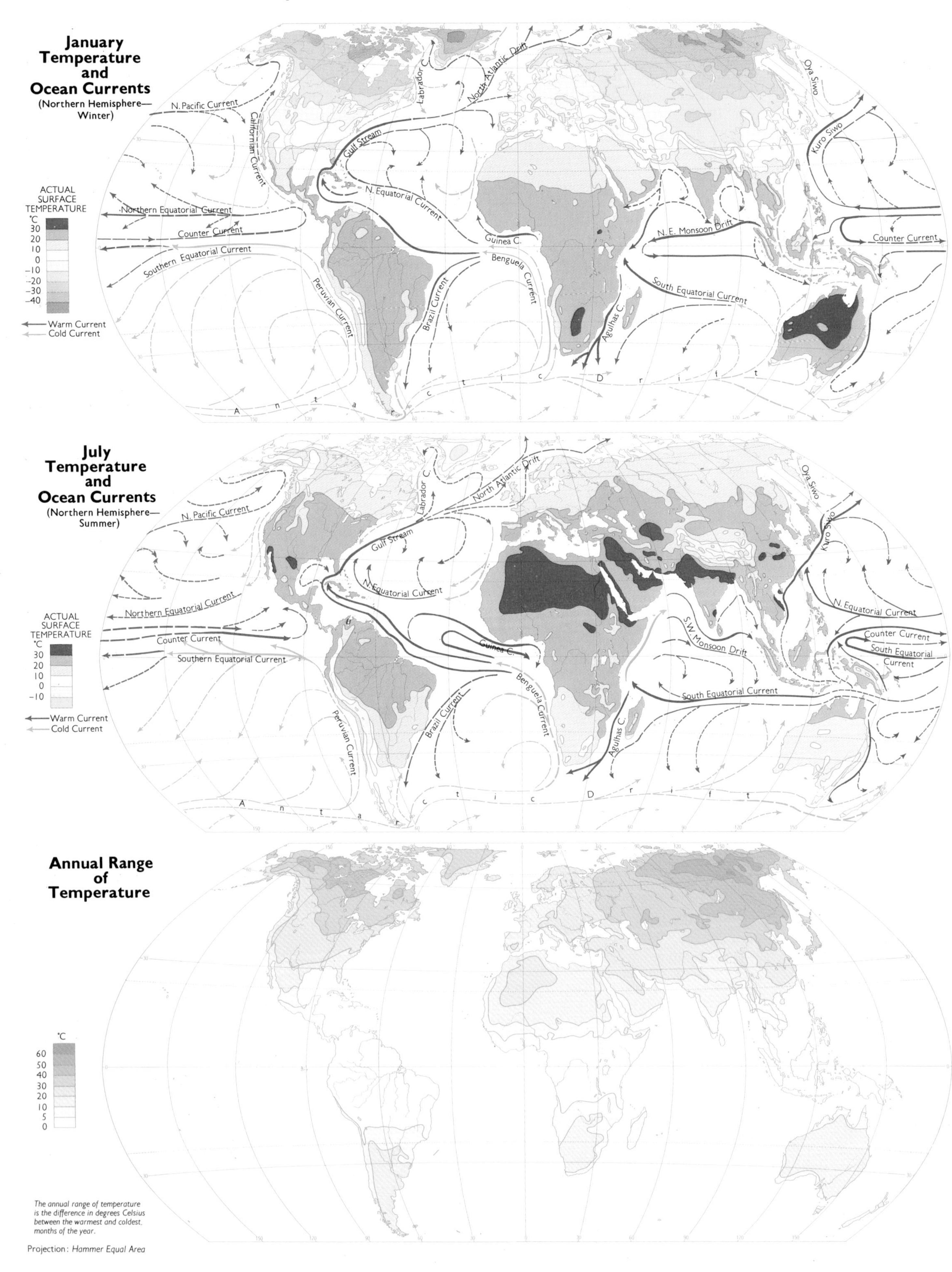

January Temperature and Ocean Currents
(Northern Hemisphere— Winter)

ACTUAL SURFACE TEMPERATURE
°C
30
20
10
0
-10
-20
-30
-40

⟵ Warm Current
⟵ Cold Current

July Temperature and Ocean Currents
(Northern Hemisphere— Summer)

ACTUAL SURFACE TEMPERATURE
°C
30
20
10
0
-10

⟵ Warm Current
⟵ Cold Current

Annual Range of Temperature

°C
60
50
40
30
20
10
5
0

The annual range of temperature is the difference in degrees Celsius between the warmest and coldest months of the year.

Projection: Hammer Equal Area

1:190 000 000

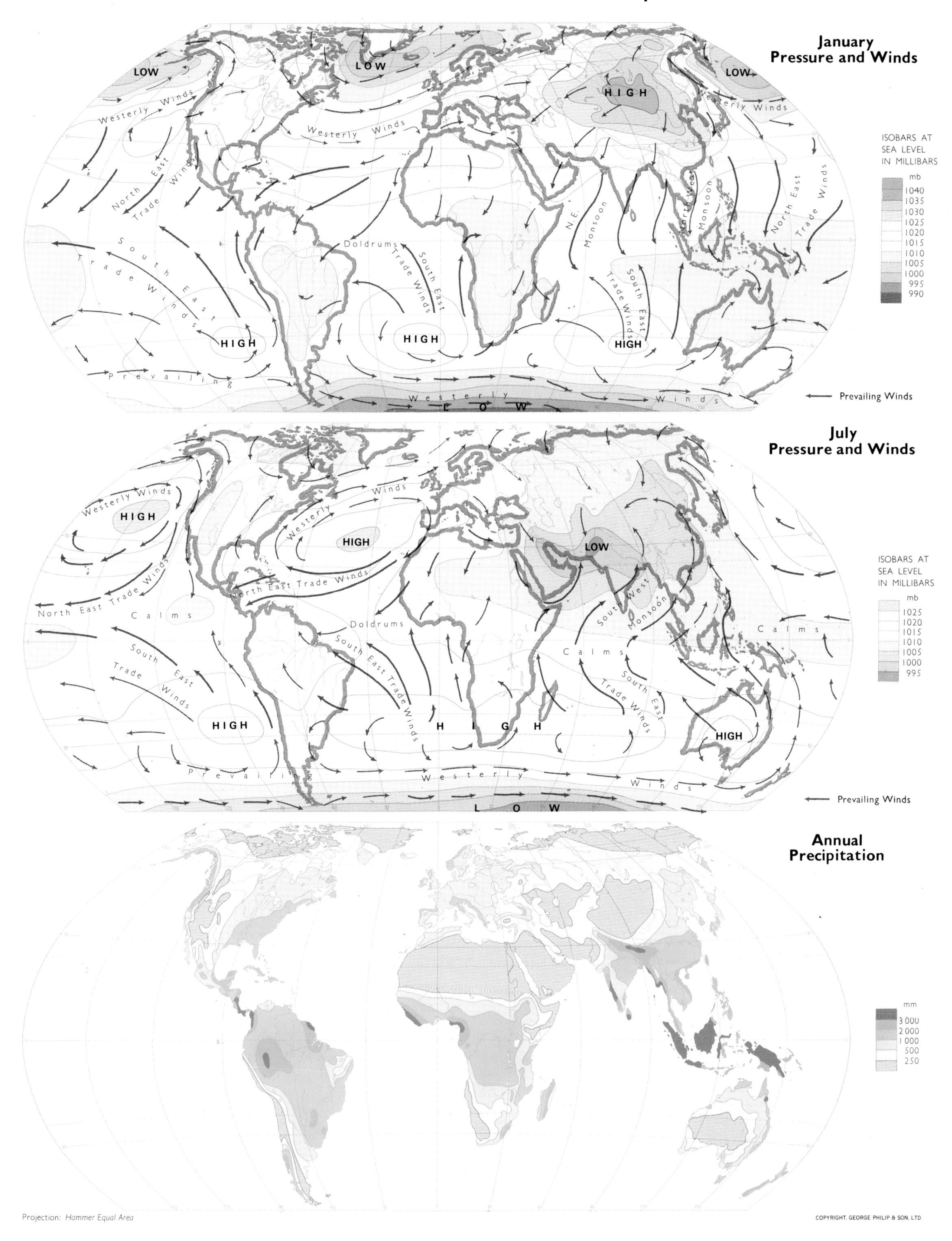

January Pressure and Winds

ISOBARS AT SEA LEVEL IN MILLIBARS

mb
1040
1035
1030
1025
1020
1015
1010
1005
1000
995
990

⟵ Prevailing Winds

July Pressure and Winds

ISOBARS AT SEA LEVEL IN MILLIBARS

mb
1025
1020
1015
1010
1005
1000
995

⟵ Prevailing Winds

Annual Precipitation

mm
3 000
2 000
1 000
500
250

Projection: *Hammer Equal Area*

All weather occurs over the earth's surface in the lowest level of the atmosphere, the troposphere. Weather has been defined as the condition of the atmosphere at any place at a specific time with respect to the various elements: temperature, sunshine, pressure, winds, clouds, fog, precipitation. Climate, on the other hand, is the average of weather elements over previous months and years.

Climate graphs *right*
Each graph typifies the climatic conditions experienced in the related coloured region on the map. The red columns show the mean monthly temperature range; the upper limit is the mean monthly maximum and the lower limit the mean monthly minimum. From these the mean monthly median temperature can be determined as the mid-point of the column. The blue columns show the average monthly rainfall.

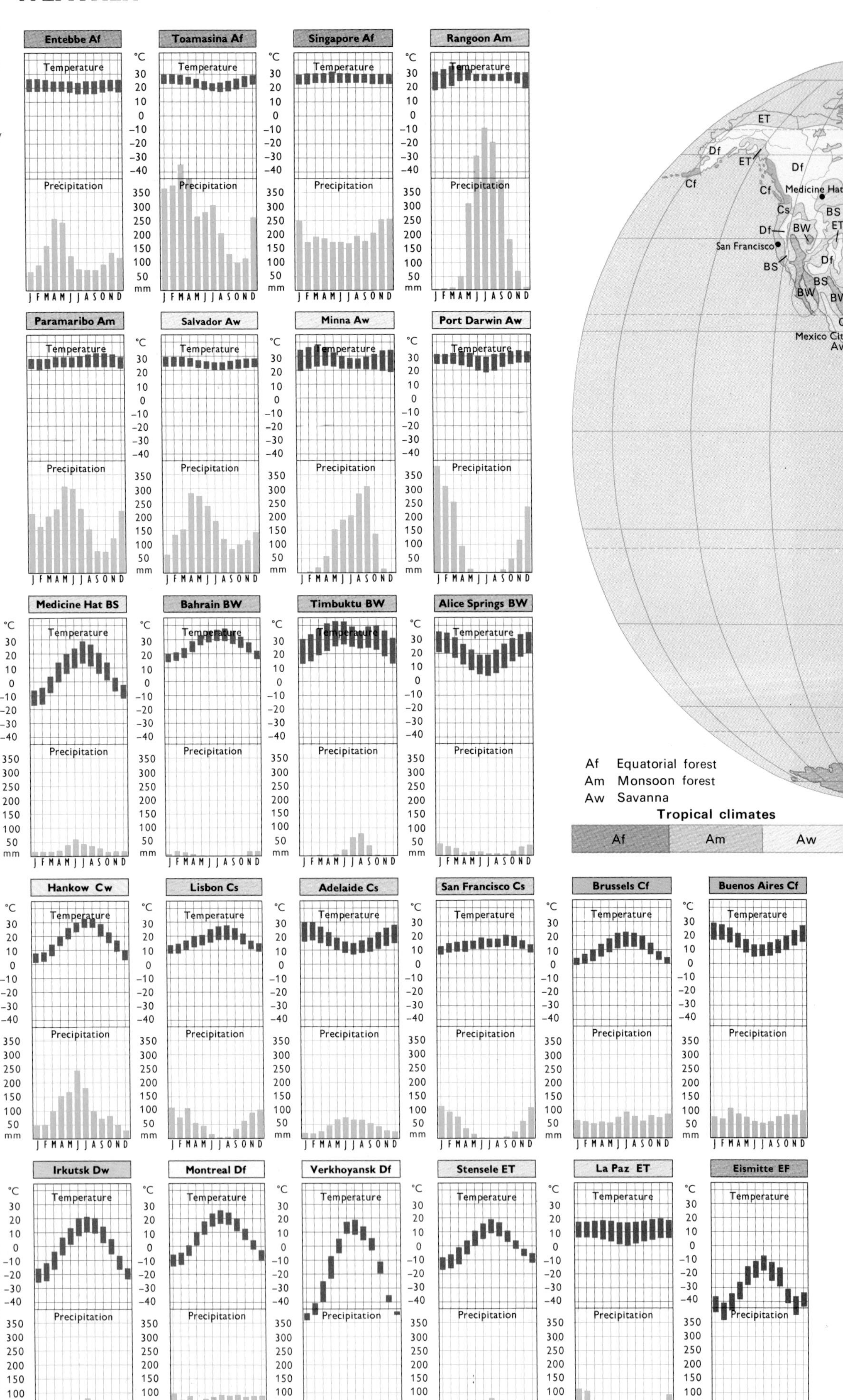

Af Equatorial forest
Am Monsoon forest
Aw Savanna

Tropical climates

| Af | Am | Aw |

BS Steppe	
BW Desert	Dw Dry winters
Cw Dry winters	Df Rain at all seasons
Cs Dry summers	ET Tundra
Cf Rain at all seasons	EF Polar

*These stations are marginal and show characteristics of two or more climatic groups

Dry climates	Warm temperate climates	Cool temperate climates	Cold climates
BS	BW	Cw	Cs	Cf	Dw	Df	ET	EF

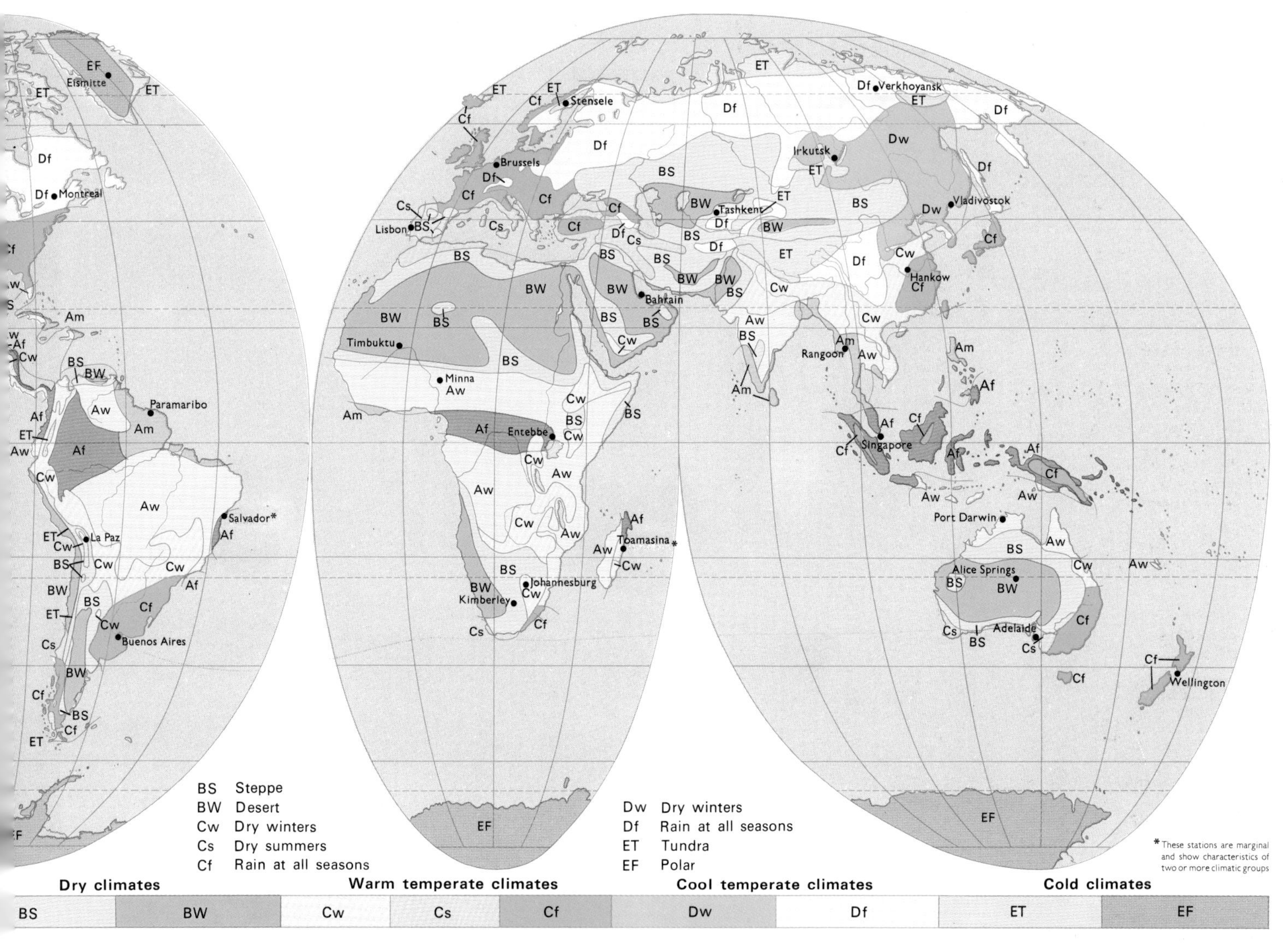

Tropical storm tracks *below*

A tropical cyclone, or storm, is designated as having winds of gale force (16 m/s) but less than hurricane force (33 m/s). It is a homogenous air mass with upward spiralling air currents around a windless centre, or eye. An average of 65 tropical storms occur each year, over 50% of which reach hurricane force. They originate mainly during the summer over tropical oceans.

Extremes of climate & weather *right*

Tropical high temperatures and polar low temperatures combined with wind systems, altitude and unequal rainfall distribution result in the extremes of tropical rain forests, inland deserts and frozen polar wastes. Fluctuations in the limits of these extreme zones and extremes of weather result in occasional catastrophic heat-waves and drought, floods and storms, frost and snow.

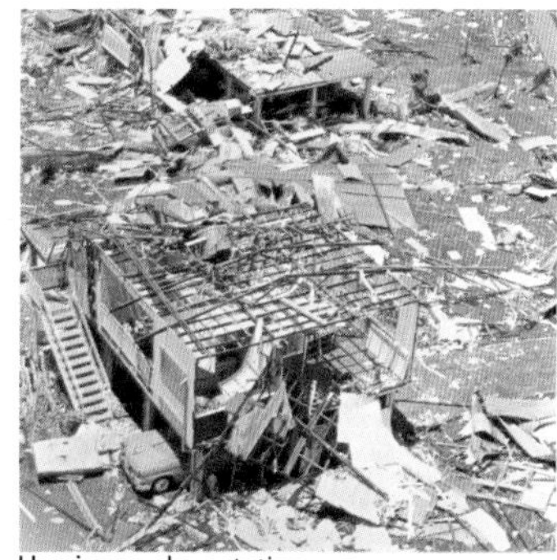

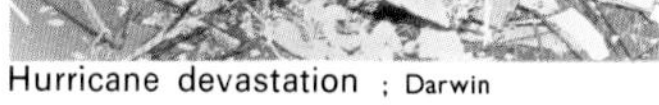

Hurricane devastation ; Darwin

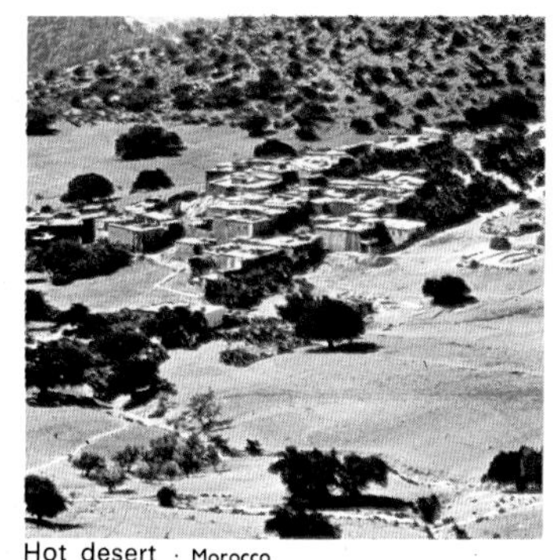

Hot desert ; Morocco

Tropical cyclone tracks
(Intense cyclones are called typhoons in the N.W. Pacific and hurricanes in the W. Atlantic)

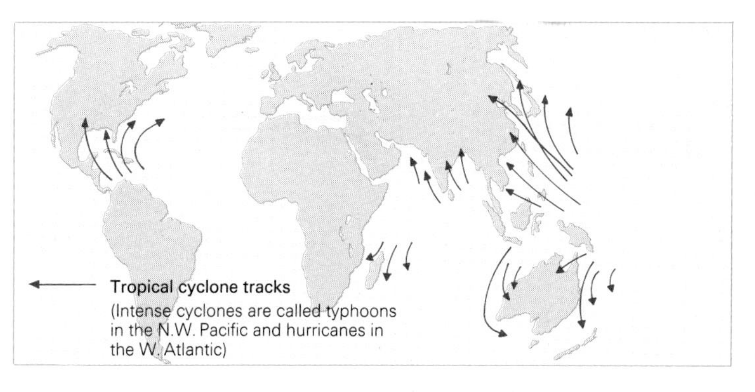

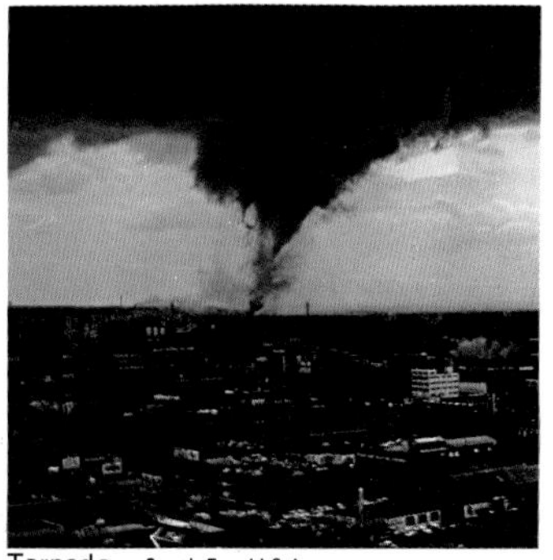

Tornado ; South East U.S.A.

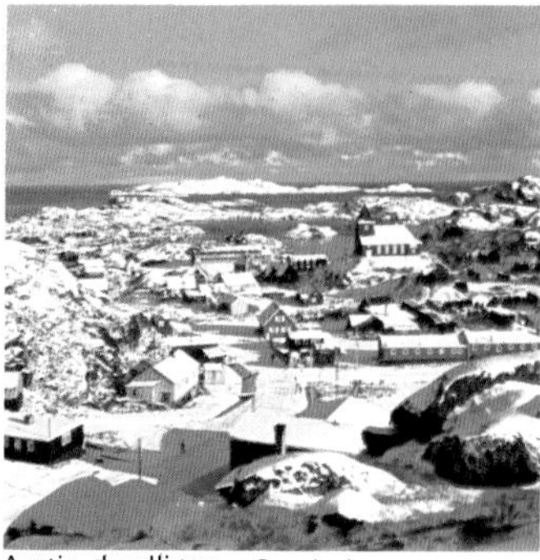

Arctic dwellings ; Greenland

Water resources and vegetation

Fresh water is essential for life on earth and in some parts of the world it is a most precious commodity. On the other hand it is very easy for industrialised temperate states to take its existence for granted, and man's increasing demand may only be met finally by the desalination of earth's 1250 million cubic kilometres of salt water. 70% of the earth's fresh water exists as ice.

The hydrological cycle

Water is continually being absorbed into the atmosphere as vapour from oceans, lakes, rivers and vegetation transpiration. On cooling the vapour either condenses or freezes and falls as rain, hail or snow. Most precipitation falls over the sea but one quarter falls over the land of which half evaporates again soon after falling while the rest flows back into the oceans.

Distribution of water

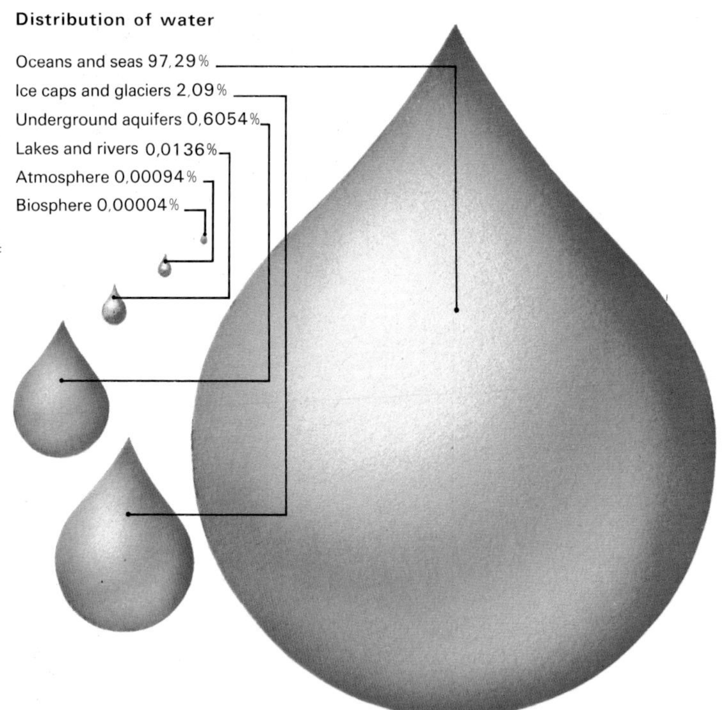

Oceans and seas 97,29 %
Ice caps and glaciers 2,09 %
Underground aquifers 0,6054 %
Lakes and rivers 0,0136 %
Atmosphere 0,00094 %
Biosphere 0,00004 %

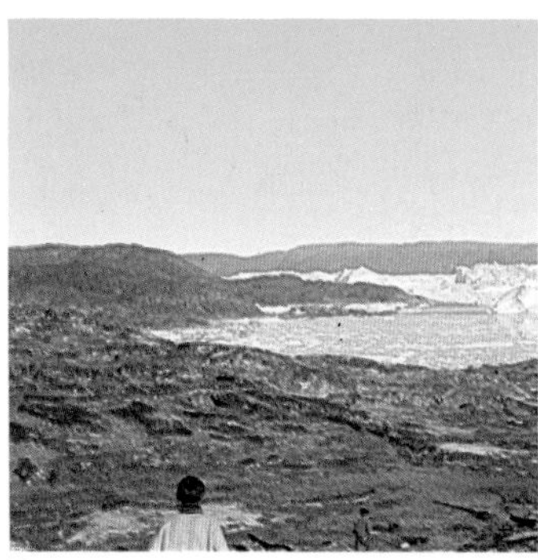

Tundra

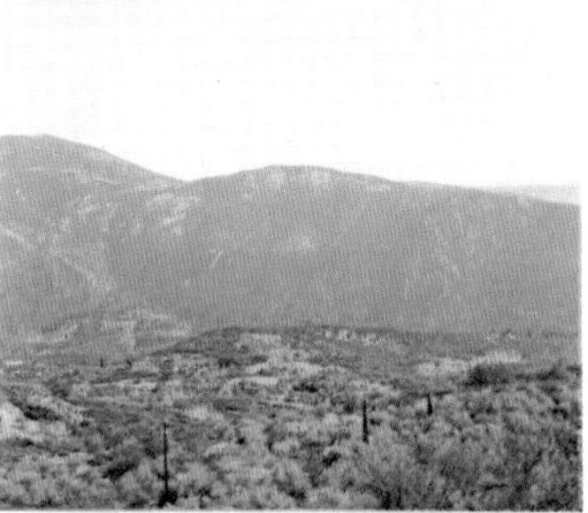

Mediterranean scrub

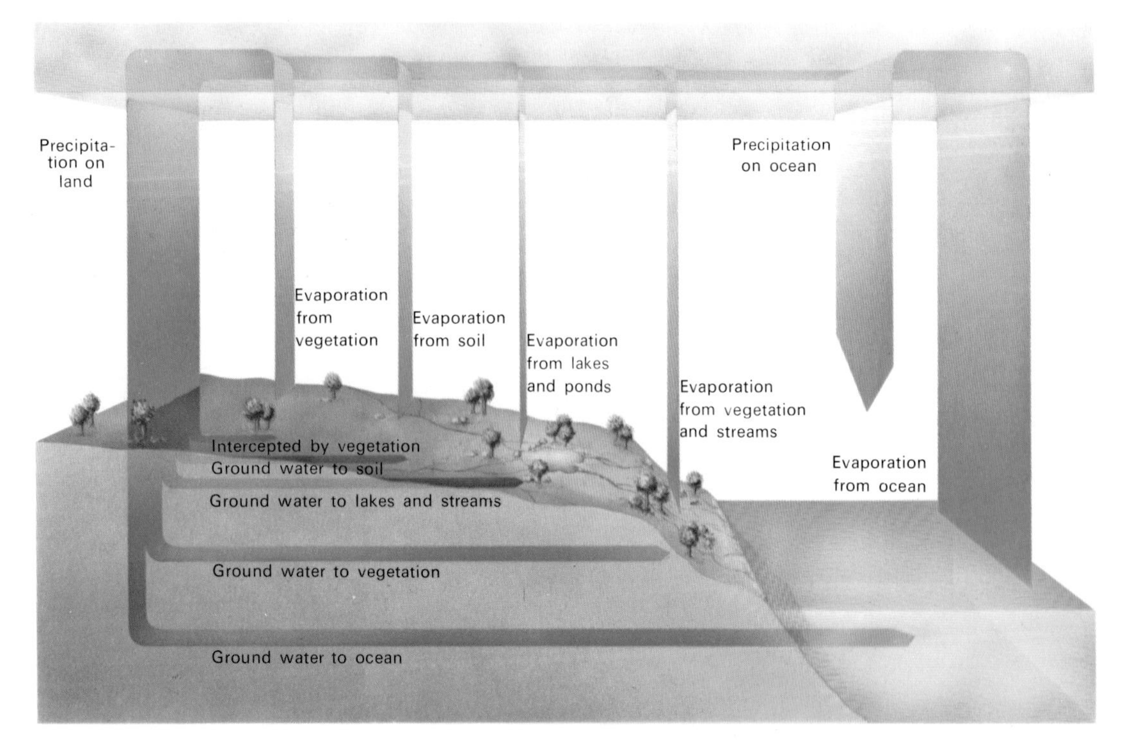

Precipitation on land

Precipitation on ocean

Evaporation from vegetation

Evaporation from soil

Evaporation from lakes and ponds

Evaporation from vegetation and streams

Evaporation from ocean

Intercepted by vegetation
Ground water to soil
Ground water to lakes and streams
Ground water to vegetation
Ground water to ocean

Domestic consumption of water

An area's level of industrialisation, climate and standard of living are all major influences in the consumption of water. On average Europe consumes 636 litres per head each day of which 180 litres is used domestically. In the U.S.A. domestic consumption is slightly higher at 270 litres per day. The graph (right) represents domestic consumption in the U.K. in 1970.

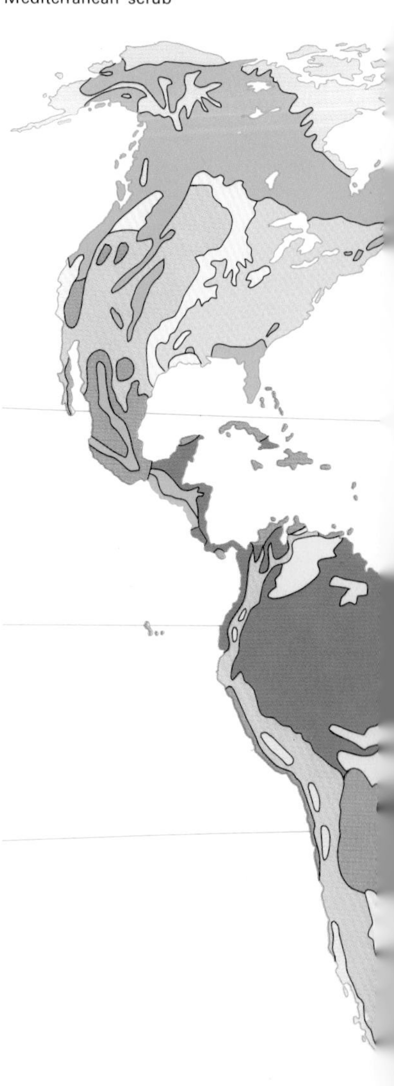

Drinking and eating 3%
Garden 3%
Laundry 9%
Dishwashing 10%
Waste in distribution 13%
Personal hygiene 30%
Toilet 32%

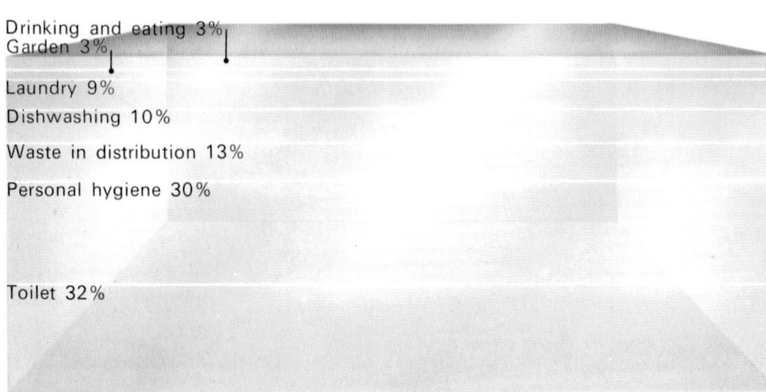

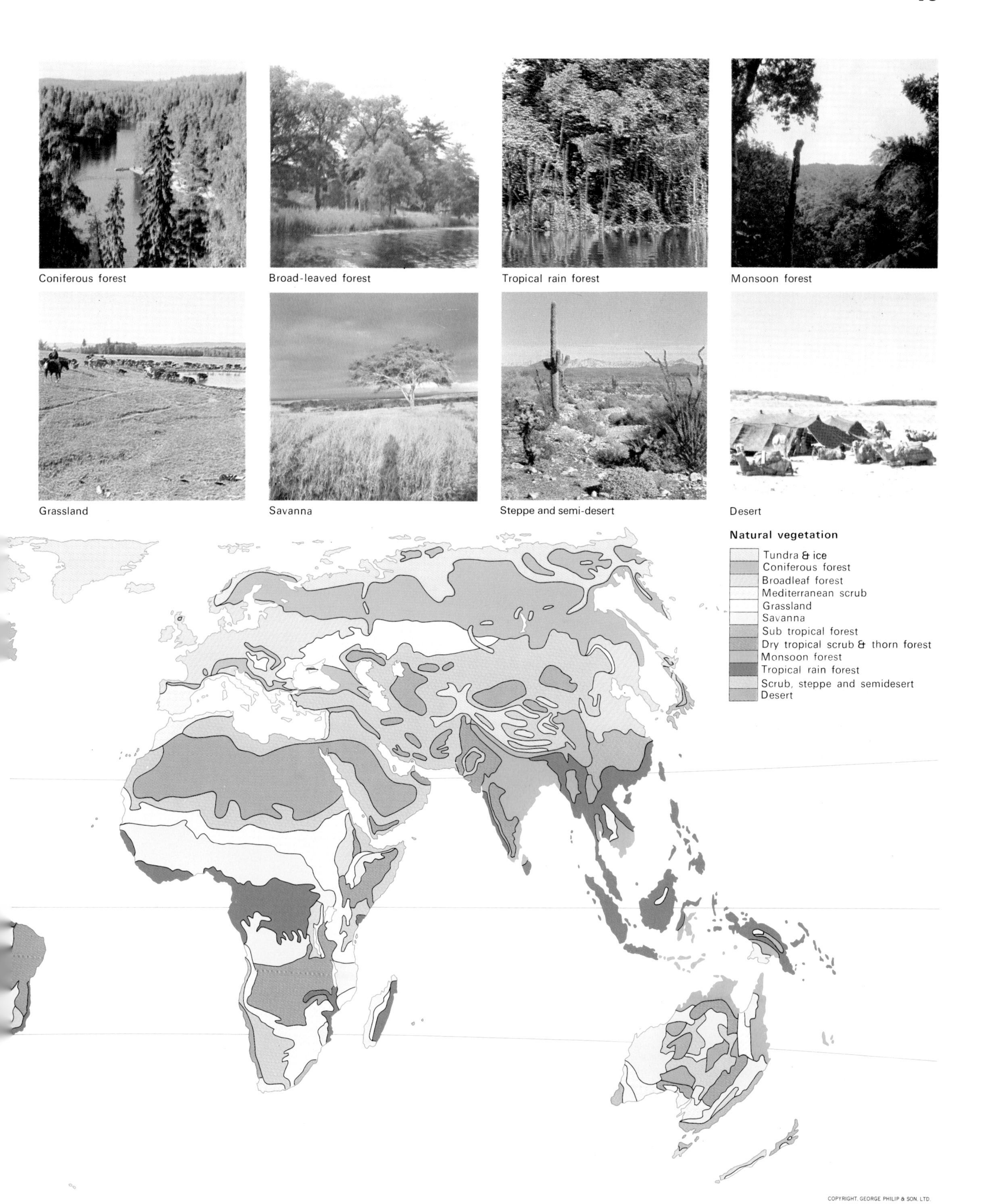

Coniferous forest

Broad-leaved forest

Tropical rain forest

Monsoon forest

Grassland

Savanna

Steppe and semi-desert

Desert

Natural vegetation

Tundra & ice
Coniferous forest
Broadleaf forest
Mediterranean scrub
Grassland
Savanna
Sub tropical forest
Dry tropical scrub & thorn forest
Monsoon forest
Tropical rain forest
Scrub, steppe and semidesert
Desert

Wheat

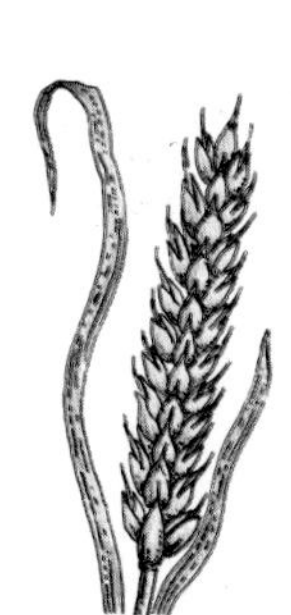

The most important grain crop in the temperate regions though it is also grown in a variety of climates e.g. in Monsoon lands as a winter crop.

World production 1984
522.9 million tonnes

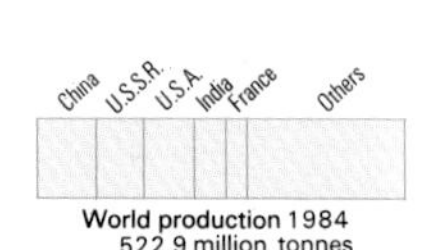

Oats

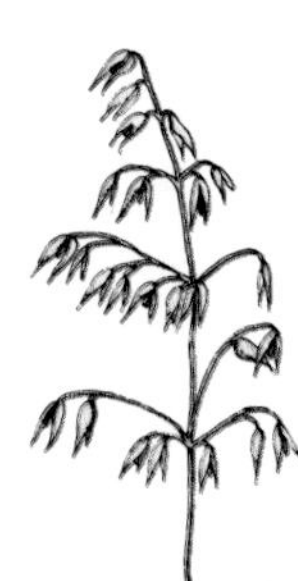

Widely grown in temperate regions with the limit fixed by early autumn frosts. Mainly fed to cattle. The best quality oats are used for oatmeal, porridge and breakfast foods.

World production 1984
43.3 million tonnes

- Wheat
- Oats

1 dot represents
2 million tonnes

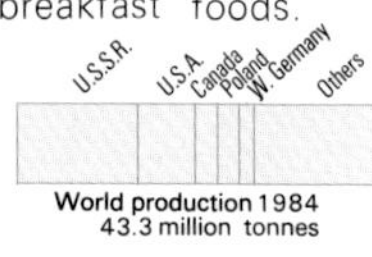

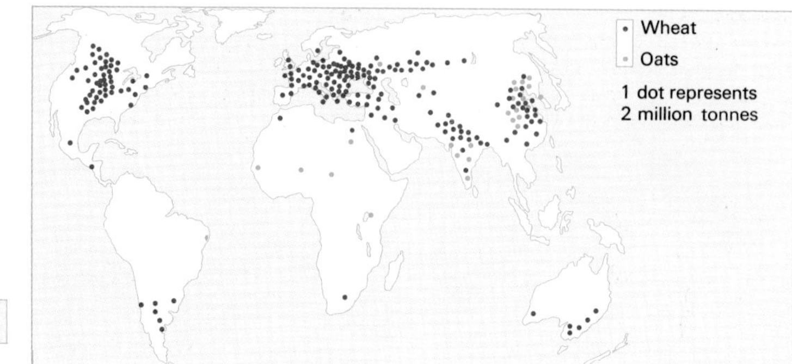

Rye

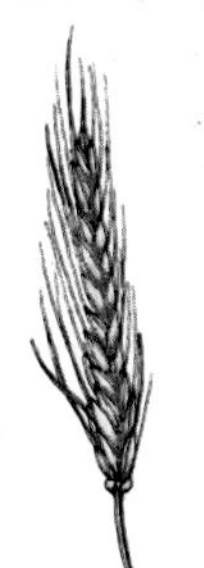

The hardiest of cereals and more resistant to cold, pests and disease than wheat. An important foodstuff in Central and E. Europe and the U.S.S.R.

World production 1984
31.3 million tonnes

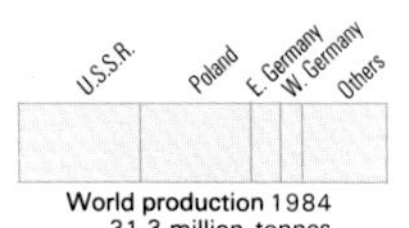

Maize (or Corn)

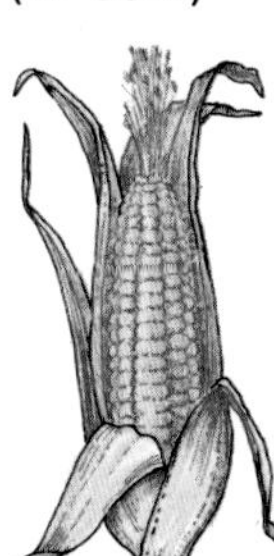

Needs plenty of sunshine, summer rain or irrigation and frost free for 6 months. Important as animal feed and for human food in Africa, Latin America and as a vegetable and breakfast cereal.

World production 1984
450.7 million tonnes

- Rye
- Maize

1 dot represents
2 million tonnes

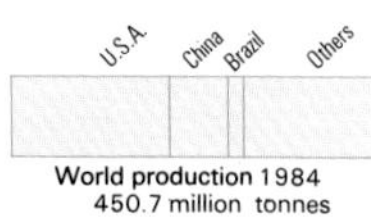

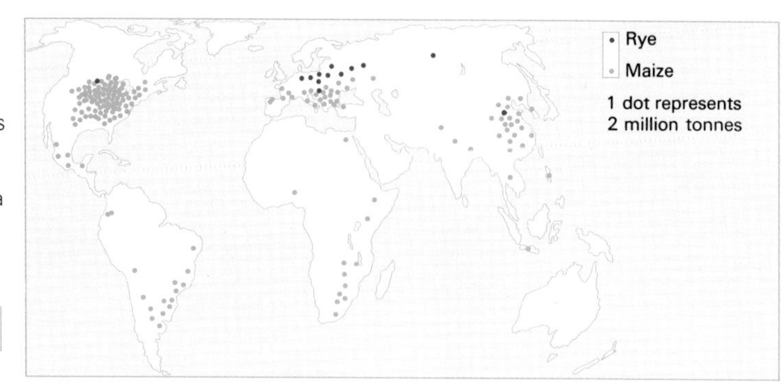

Barley

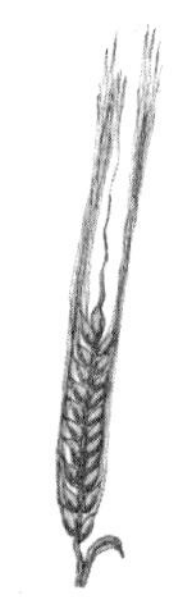

Has the widest range of cultivation requiring only 8 weeks between seed time and harvest. Used mainly as animal-feed and by the malting industry.

World production 1984
171.3 million tonnes

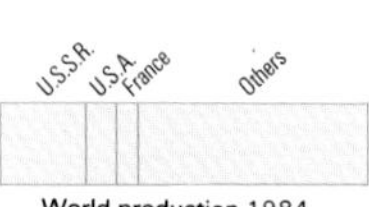

Rice

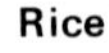

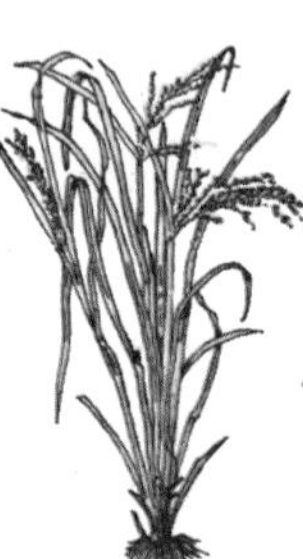

The staple food of half the human race. The main producing areas are the flood plains and hill terraces of S. and E. Asia where water is abundant in the growing season.

World production 1984
471.2 million tonnes

- Barley
- Rice

1 dot represents
2 million tonnes

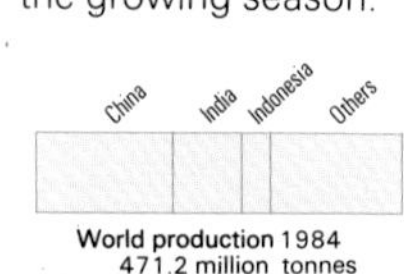

Millets

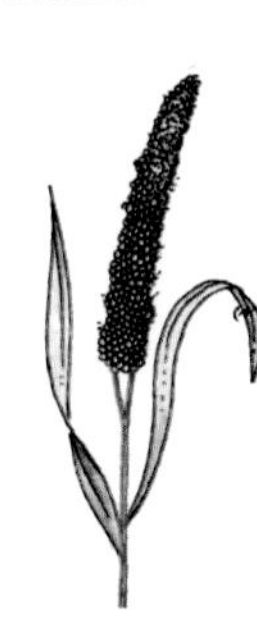

The name given to a number of related members of the grass family, of which sorghum is one of the most important. They provide nutritious grain.

World production 1984
103.2 million tonnes

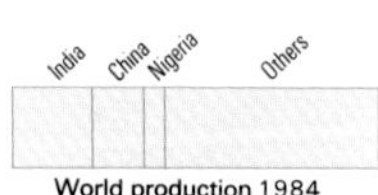

Potatoes

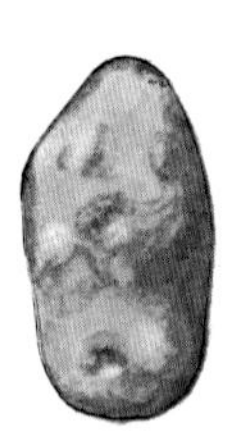

An important food crop though less nutritious weight for weight than grain crops. Requires a temperate climate with a regular and plentiful supply of rain.

World production 1984
312.2 million tonnes

- Millets
- Potatoes

1 dot represents
2 million tonnes

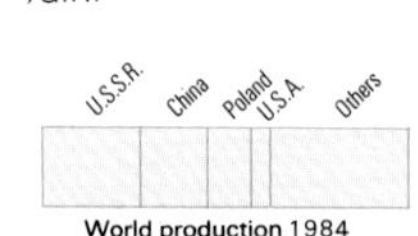

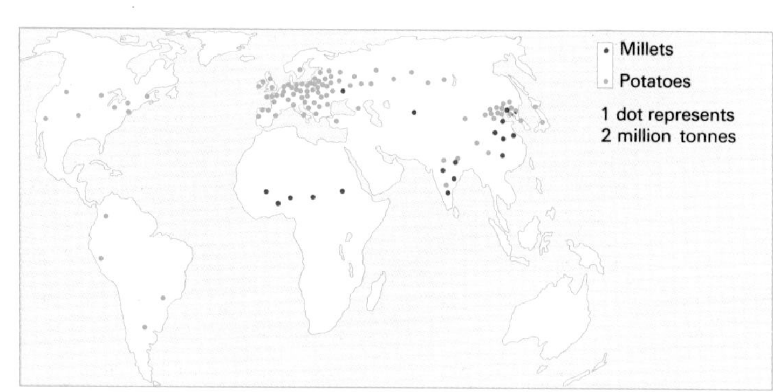

Vegetable oilseeds and oils

Despite the increasing use of synthetic chemical products and animal and marine fats, vegetable oils extracted from these crops grow in quantity, value and importance. Food is the major use- in margarine and cooking fats.

Groundnuts are also a valuable subsistence crop and the meal is used as animal feed. Soya-bean meal is a growing source of protein for humans and animals. The Mediterranean lands are the prime source of olive oil.

Groundnut

Soya bean

Sunflower

- Groundnuts
- Soya beans
- Sunflower seed

1 dot represents
1 million tonnes

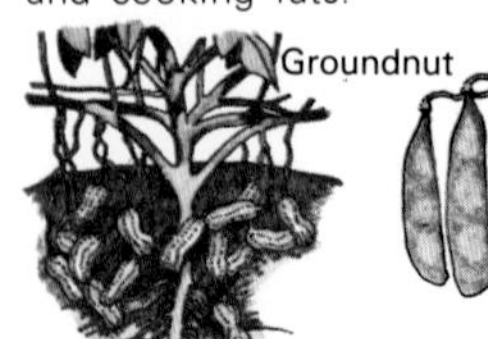

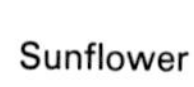

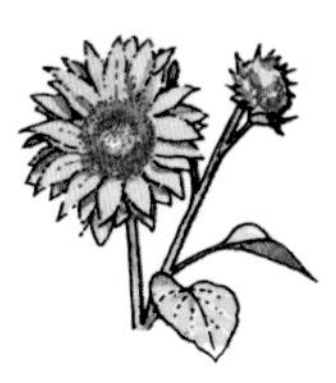

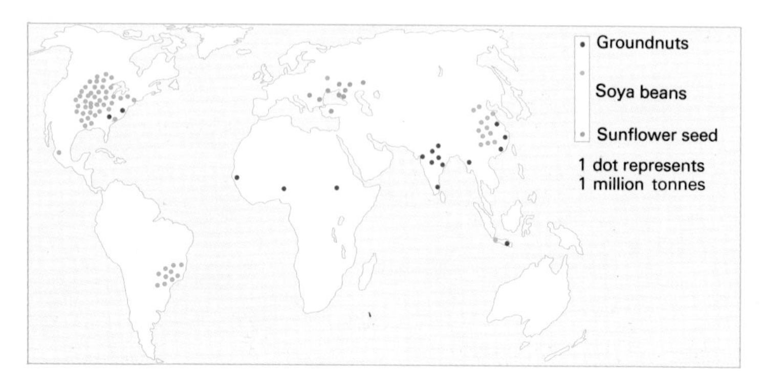

Tea and cacao

Tea requires plentiful rainfall and well-drained, sloping ground, whereas cacao prefers a moist heavy soil. Both are grown mainly for export.

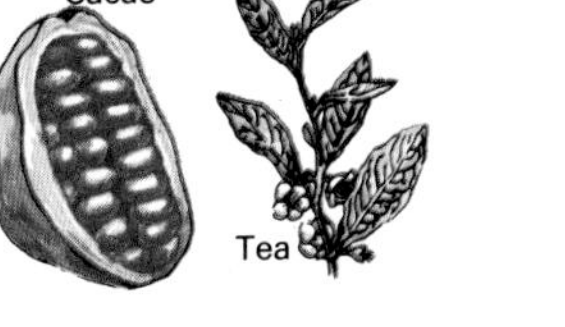

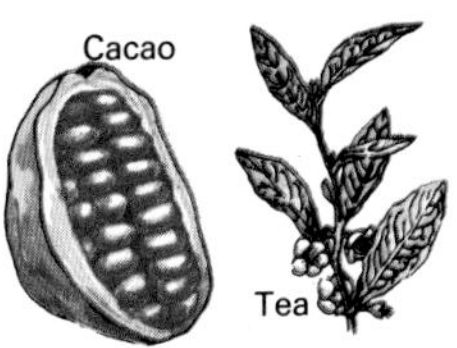

Coffee

Prefers a hot climate, wet and dry seasons and an elevated location. It is very susceptible to frost, drought and market fluctuations.

Brazil Colombia Indonesia Mexico Others

World production 1984
5.2 million tonnes

- • Tea
- • Cacao
- • Coffee

1 dot represents 100 000 tonnes

Sugar beet

Requires a deep, rich soil and a temperate climate. Europe produces over 90 % of the world's beets mainly for domestic consumption.

U.S.S.R. France U.S.A. W. Germany Poland Others

World production 1984
293.5 million tonnes

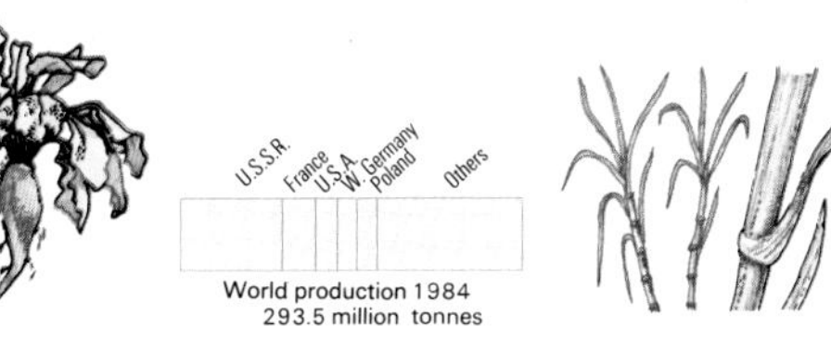

Sugar cane

Also requires deep and rich soil but a tropical climate. It produces a much higher yield per hectare than beet and is grown primarily for export.

Brazil India Cuba China Others

World production 1984
935.8 million tonnes

- • Sugar beet
- • Sugar cane

1 dot represents 10 million tonnes

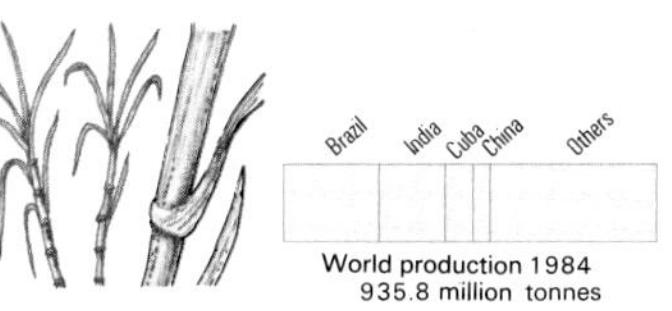

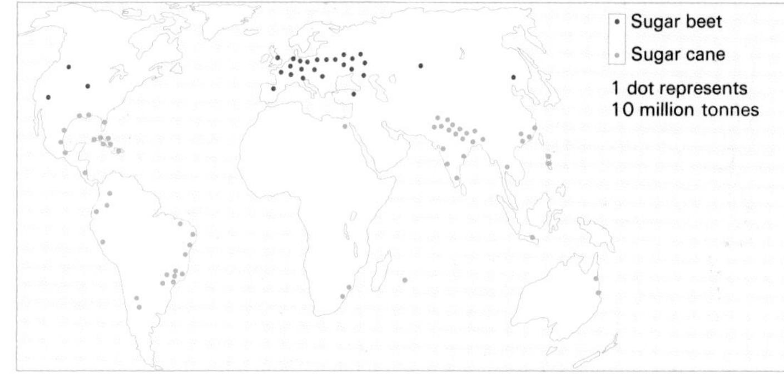

Fruit

With the improvements in canning, drying and freezing, and in transport and marketing, the international trade and consumption of deciduous and soft fruits, citrus fruits and tropical fruits has greatly increased. Recent developments in the use of the peel will give added value to some of the fruit crops.

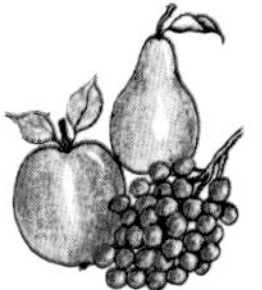

Fish

Commercial fishing requires large shoals of fish of one species within reach of markets. Freshwater fishing is also important. A rich source of protein, fish will become an increasingly valuable food source.

Japan U.S.S.R. China U.S.A. Others

World catch 1983
76.5 million tonnes

- ▭ Temperate fruit
- ▭ Citrus fruit
- ▭ Principal fishing grounds

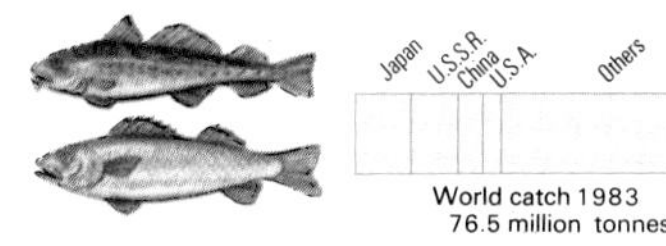

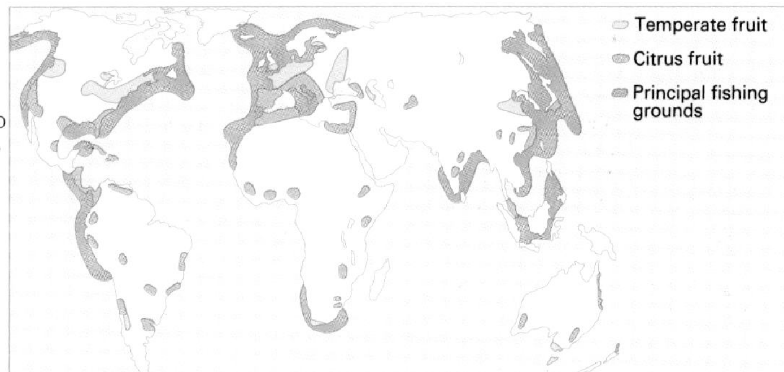

Beef cattle

Australia, New Zealand and Argentina provide the major part of international beef exports. Western U.S.A. and Europe have considerable production of beef for their local high demand.

U.S.A. U.S.S.R. Argentina Brazil France Others

World production 1984
46 million tonnes

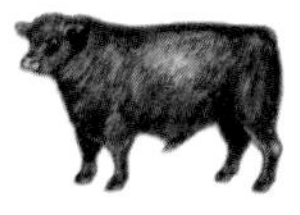

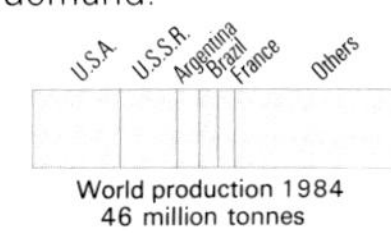

Dairy cattle

The need of herds for a rich diet and for nearby markets result in dairying being characteristic of densely-populated areas of the temperate zones - U.S.A., N.W. Europe, and S.E. Australia.

U.S.S.R. U.S.A. W. Germany France U.K. Others

World production 1984
373.5 million tonnes

- • Cattle

1 dot represents 10 million head

- ▭ Dairy produce

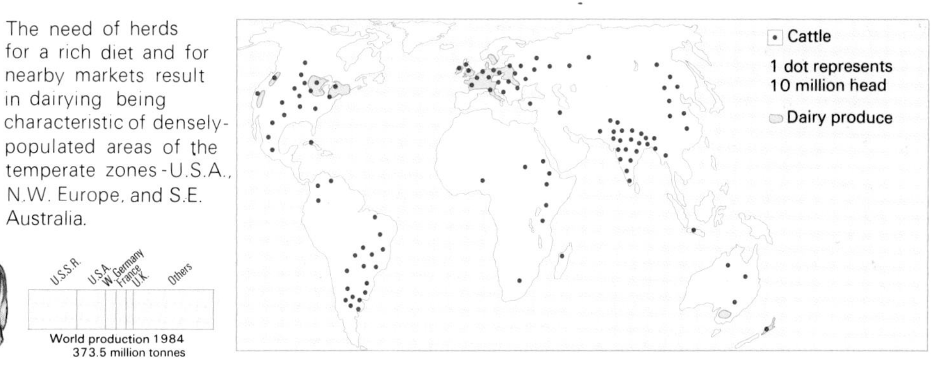

Sheep

Raised mostly for wool and meat, their skins and the cheese from their milk are important products in some countries. The merino yields a fine wool and crossbreeds are best for meat.

U.S.S.R. N.Z. China India Australia Pakistan Others

World production 1984
8 million tonnes

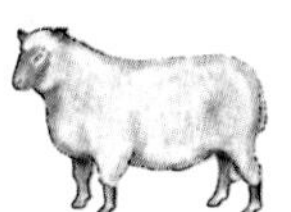

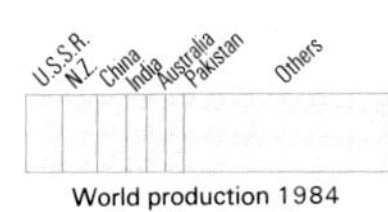

Pigs

Can be reared in most climates from monsoon to cool temperate. They are abundant in China, the Corn Belt of the U.S.A. N.W. and C. Europe, Brazil and U.S.S.R.

China U.S.A. U.S.S.R. W. Germany Others

World production 1984
55 million tonnes

- • Sheep
- • Pigs

1 dot represents 10 million head

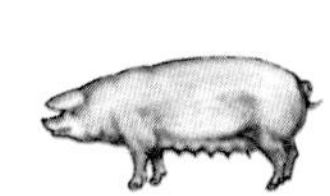

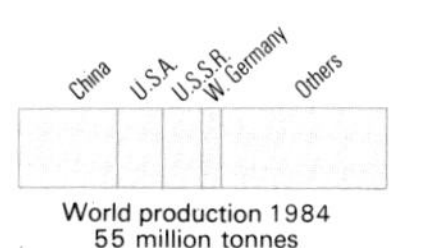

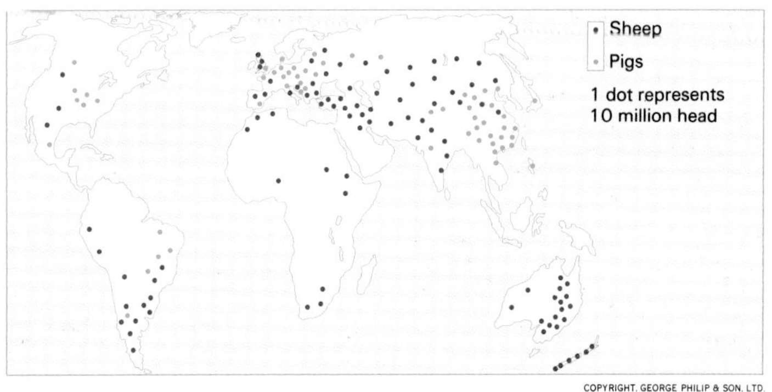

Production of ferro-alloy metals

Steel is refined iron with the addition of other minerals and ferro-alloys. The ferro-alloys give the steel their own special properties; for example resistance to corrosion (chromite and nickel), hardness (tungsten and vanadium), elasticity (molybdenum), magnetic properties (cobalt), high tensile strength (manganese) and high ductility (molybdenum).

Chromite
U.S.S.R. | S.Africa | Albania | Zimbabwe | Turkey | Others
World production 1983 7,7 million tonnes

Nickel
U.S.S.R. | Canada | Australia | New Caledonia | Cuba | Indonesia | Others
World production 1983 651 600 tonnes

Manganese
U.S.S.R. | S. Africa | Brazil | Gabon | China | Australia | Others
World production 1983 22 million tonnes

Tungsten
China | U.S.S.R. | Bolivia | Brazil | Austria | Others
World production 1983 38 900 tonnes

Molybdenum
Chile | U.S.A. | U.S.S.R. | Canada | Others
World production 1983 60 400 tonnes

Vanadium
U.S.S.R. | S. Africa | China | Finland | U.S.A. | Others
World production 1983 28 800 tonnes

World production of pig iron and ferro-alloys

World production 1984 481,0 million tonnes

Others 10%
Romania 2%
S.Korea 2%
Belgium 2%
India 2%
Czech. 2%
Poland 2%
U.K.2%
Canada 2%
Italy 2.5%
France 3%
Brazil 4%
W.Germany 6%
China 8.5%
U.S.A.10%
Japan 17%
U.S.S.R. 23%

World production of iron ore (Fe content)

World production 1984 495 million tonnes

U.S.S.R. 30% | China 12% | Australia 10% | Brazil 9% | U.S.A. 7% | India 5% | Canada 5% | S. Africa 3% | Sweden 2% | Liberia 2% | Others 15%

Development of world production of pig iron and ferro alloys

'000 million tonnes
0 100 200 300 400 500 550

1920
1925
1930
1935
1940
1945
1950
1955
1960
1965
1970
1975
1976
1977
1978
1979
1980
1981
1982
1983
1984

Production of non-ferrous metals and diamonds

Tin
Malaysia | Indonesia | Thailand | Brazil | Bolivia | China | Others
World production 1984 205 900 tonnes

Gold
S. Africa | U.S.S.R. | Canada | U.S.A. | Chile | Brazil | Others
World production 1983 1385, 1 tonnes

Copper
Chile | U.S.A. | U.S.S.R. | Canada | Zambia | Zaire | Others
World production 1984 8,3 million tonnes

Zinc
Canada | U.S.S.R. | Australia | Peru | Others
World production 1984 6,7 million tonnes

Silver
Mexico | Peru | U.S.S.R. | U.S.A. | Canada | Australia | Poland | Others
World production 1984 12 789 tonnes

Lead
U.S.S.R. | Australia | U.S.A. | Canada | Peru | Mexico | China | Others
World production 1984 3,1 million tonnes

Bauxite
Australia | Guinea | Jamaica | Brazil | U.S.S.R. | Others
World production 1984 92,5 million tonnes

Diamonds
Zaire | Botswana | U.S.S.R. | S. Africa | Australia | Others
World production 1984 63,9 million carats

World consumption of non-ferrous metals

Copper
1949/51
1963/65
1983/84 36% from scrap

Lead
1949/51
1963/65
1983/84 47% from scrap

Zinc
1949/51
1963/65
1983/84 22% from scrap

0 1 2 3 4 5 6 7 8 9 10 million tonnes

Nickel
1949/51
1963/65
1983/84

Tin
1949/51
1963/65
1983/84 22% from scrap

'000 tonnes
0 100 200 300 400 500 600 700 800

Aluminium (from Bauxite)
1949/51
1963/65
1983/84 20% from scrap

0 5 10 15 20 25 million tonnes

Tropic of Cancer

Equator

Tropic of Capricorn

Principal sources of iron ore and ferro-alloys

- ■ Iron ore
- ◨ Chromite
- ◠ Cobalt
- ◻ Manganese
- ◺ Molybdenum
- ◉ Nickel
- + Tungsten
- ◞ Vanadium
- Iron ore trade flow

Principal sources of non-ferrous metals and other minerals

Base metals
- ▲ Antimony
- ◠ Copper
- ◠ Lead
- ◒ Mercury
- ● Tin
- ◆ Zinc

Light metals
- ● Bauxite
- ▼ Beryllium
- ■ Lithium
- ▼ Titanium

Rare metals
- ✦ Uranium

Precious metals
- ▲ Gold
- ● Platinum
- ■ Silver

Precious stones
- △ Diamonds

Mineral fertilizers
- ◉ Nitrates
- ◻ Phosphates
- ◆ Potash
- ◆ Sulphur
- ◆ Pyrites

Other industrial minerals
- ■ Asbestos
- △ Mica

Structural regions

- Pre-Cambrian shields
- Sedimentary cover on Pre-Cambrian shields
- Primary (Caledonian and Hercynian) folding
- Sedimentary cover on Primary folding
- Secondary folding
- Sedimentary cover on Secondary folding
- Tertiary (Alpine) folding
- Sedimentary cover on Tertiary folding

Coal
World production 1984
4 123 million tonnes
28% of World energy production

- Others 15%
- S. Africa 3%
- Czech. 3%
- India 4%
- Australia 4%
- W. Germ. 5%
- Poland 6%
- E. Germany 7%
- U.S.S.R. 15%
- U.S.A. 20%
- China 18%

- Coal
- Lignite

1 dot represents 20 million tonnes

Crude petroleum
World production 1984
2 698 million tonnes
42% of World energy production

- Iraq 2%
- Libya 2%
- Kuwait 2%
- Nigeria 2%
- U.A.E. 2%
- Indon. 2%
- Canada 3%
- Venezuela 4%
- Iran 4%
- U.K. 4%
- China 4%
- Mexico 5%
- Saudi Arabia 9%
- U.S.A. 16%
- U.S.S.R. 22%
- Others 18%

1 dot represents 20 million tonnes

Natural gas
World production 1984
59 084 petajoules
19% of World total energy production

- Others 14%
- Mexico 2%
- Norway 2%
- Indonesia 3%
- U.K. 3%
- Romania 3%
- Neth. 4%
- Canada 5%
- U.S.A. 29%
- U.S.S.R. 35%

1 dot represents 10 milliard m³

Hydro-electric power
World production 1983
1,9 billion kWh
7% of World energy production

- Others 23.5%
- Switz. 2%
- Italy 2%
- Sweden 3%
- India 3%
- France 4%
- China 4.5%
- Japan 4.5%
- Norway 5.5%
- Brazil 8%
- U.S.S.R. 9%
- Canada 14%
- U.S.A. 17%

1 dot represents 10 000 million kWh

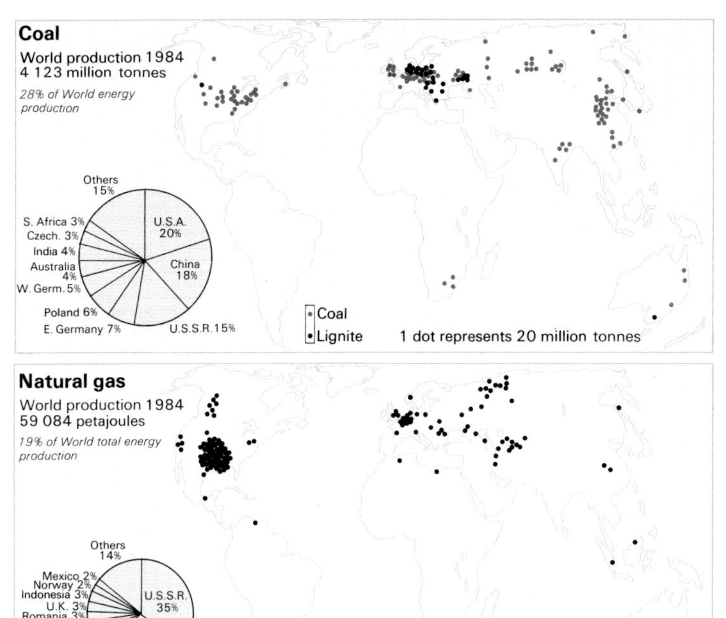

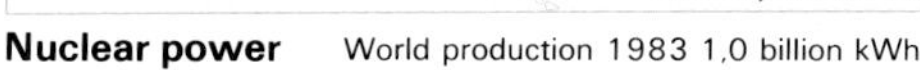

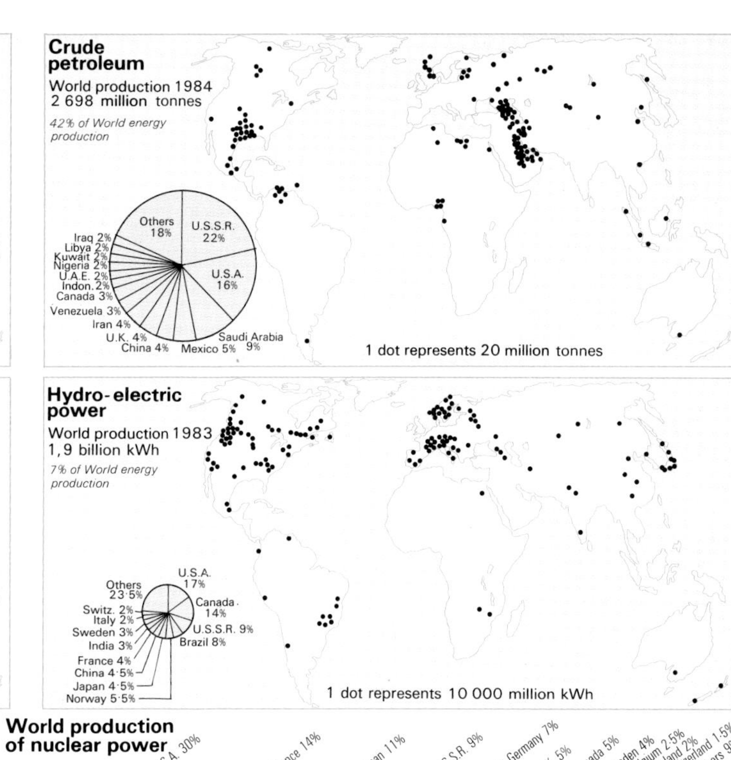

Nuclear power
World production 1983 1,0 billion kWh

Only 4% of world energy production is from nuclear power but it is expected to rise to 15% by the year 2000.

World production of nuclear power

| U.S.A. 30% | France 14% | Japan 11% | U.S.S.R. 9% | W. Germany 7% | U.K. 5% | Canada 5% | Sweden 4% | Belgium 2.5% | Finland 2% | Switzerland 1.5% | Others 9% |

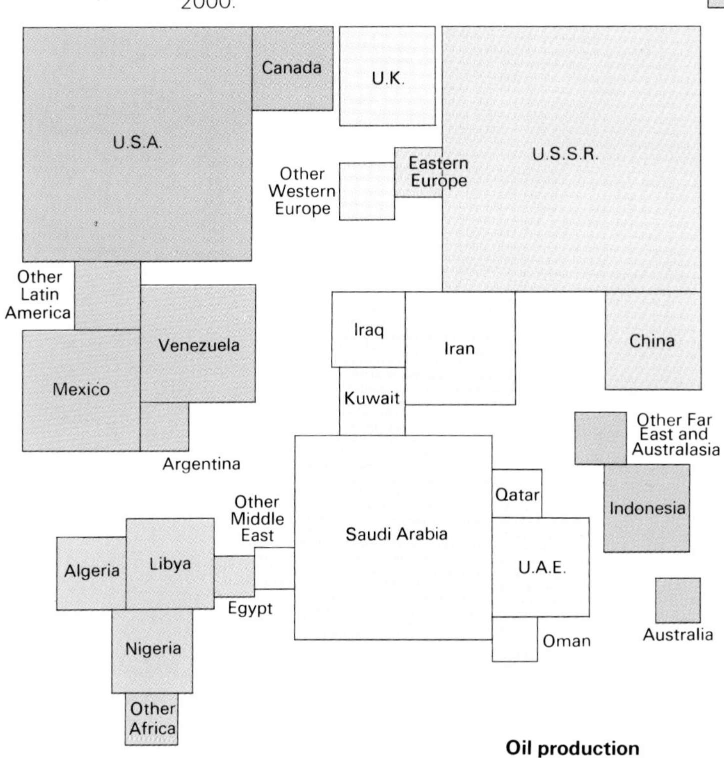

Oil production

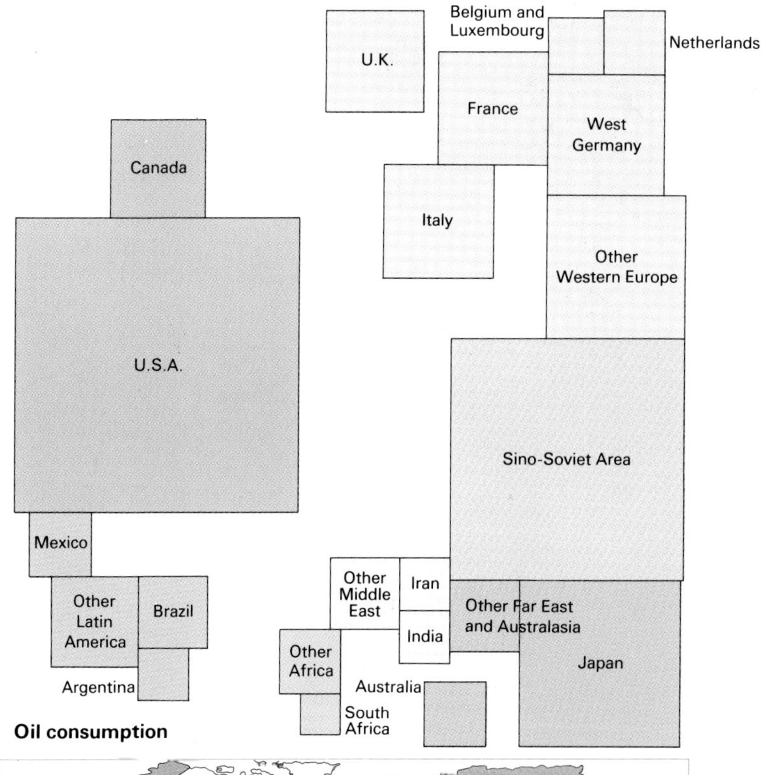

- U.S.A.
- Canada
- U.K.
- Eastern Europe
- U.S.S.R.
- Other Western Europe
- Other Latin America
- Venezuela
- Mexico
- Iraq
- Iran
- China
- Kuwait
- Argentina
- Other Far East and Australasia
- Indonesia
- Other Middle East
- Qatar
- Saudi Arabia
- Algeria
- Libya
- Egypt
- U.A.E.
- Nigeria
- Oman
- Australia
- Other Africa

Oil consumption

- U.K.
- Belgium and Luxembourg
- Netherlands
- France
- West Germany
- Canada
- Italy
- Other Western Europe
- U.S.A.
- Sino-Soviet Area
- Mexico
- Other Latin America
- Brazil
- Other Middle East
- Iran
- India
- Other Far East and Australasia
- Japan
- Other Africa
- Australia
- Argentina
- South Africa

Oil's new super-powers *above*

When countries are scaled according to their production and consumption of oil they take on new dimensions. At present, large supplies of oil are concentrated in a few countries of the Caribbean, the Middle East and North Africa, except for the vast indigenous supplies of the U.S.A. and U.S.S.R. The Middle East, with 55% of the world's reserves, produces 20% of the world's supply and yet consumes less than 3%. The U.S.A.,

despite its great production, has a deficiency of nearly 225 million tons a year, consuming 28% of the world's total. Estimates show that Western Europe, at present consumes 591 million tons or 21% of the total each year.

Energy balance

millions of tons of coal equivalent

- −500 to −200
- −200 to −50
- −50 to 0
- 0 to +50
- +50 to +200
- +200 to +500

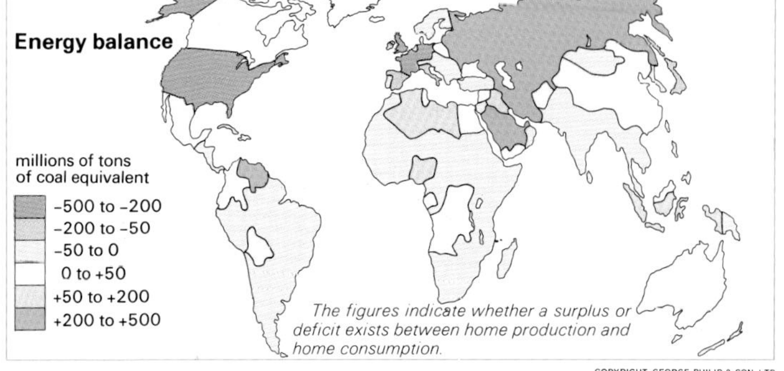

The figures indicate whether a surplus or deficit exists between home production and home consumption.

Population distribution
(right and lower right)
People have always been unevenly distributed in the world. Europe has for centuries contained nearly 20% of the world's population but after the 16-19th century explorations and consequent migrations this proportion has rapidly decreased. In 1750 the Americas had 2% of the world's total: in 2000 AD they are expected to contain 14%

The most densely populated regions are in India, China and Europe where the average density is between 100 and 200 per km² although there are pockets of extremely high density elsewhere. In contrast French Guiana has less than 1,0 persons per km². The countries in the lower map have been redrawn to make their areas proportional to their populations.

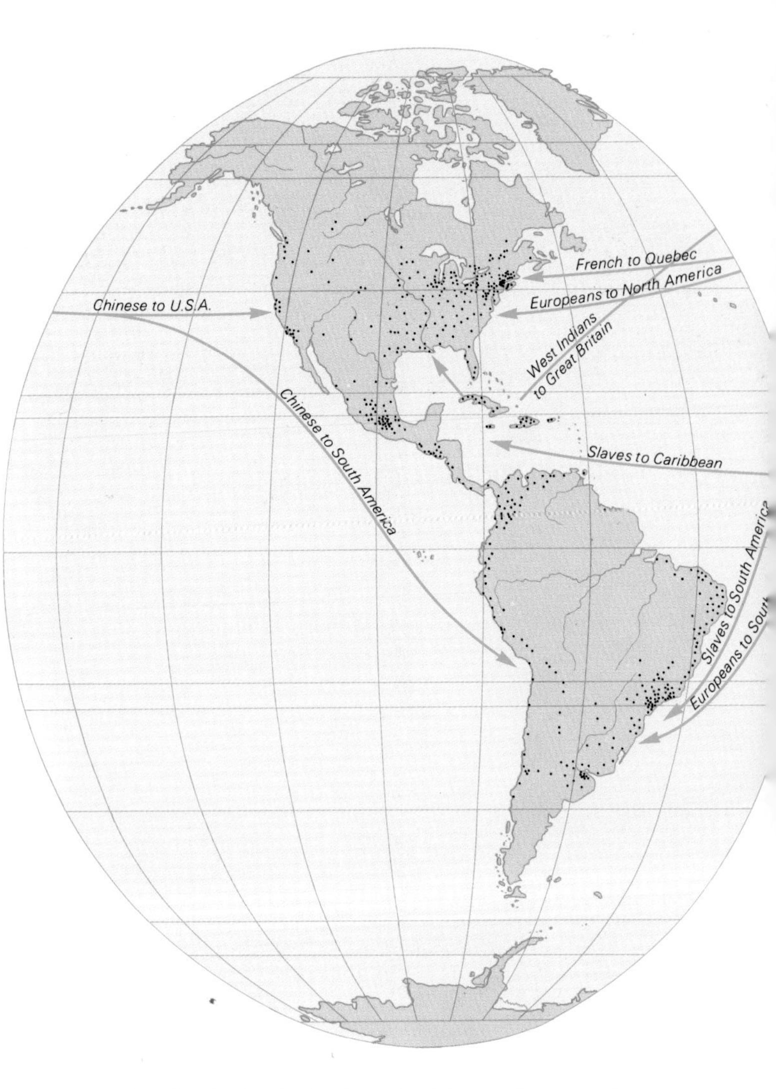

U.S.A. 1985
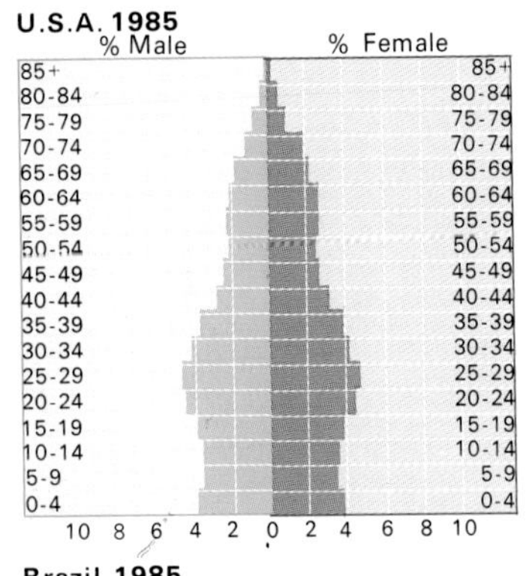

U.K. 1985
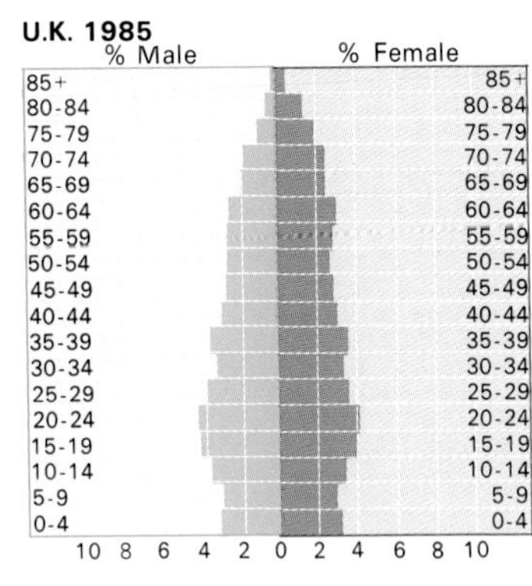

Brazil 1985
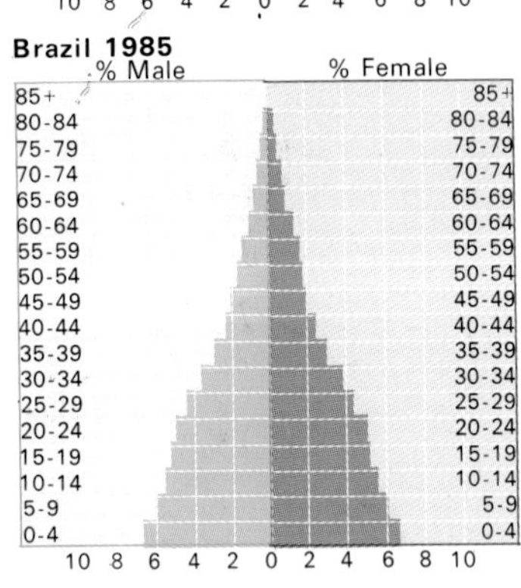

U.S.S.R. 1985
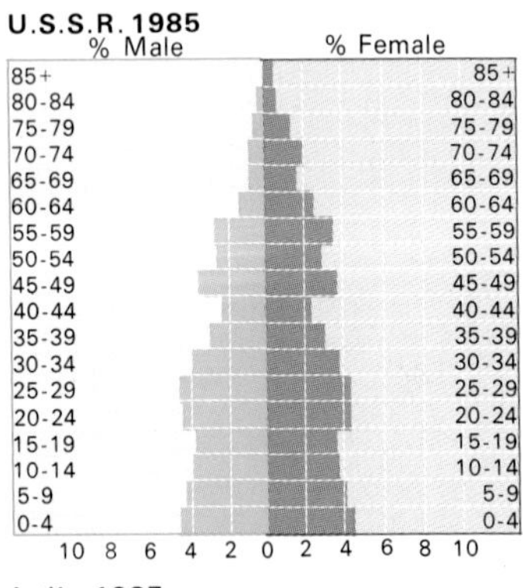

South Africa 1985
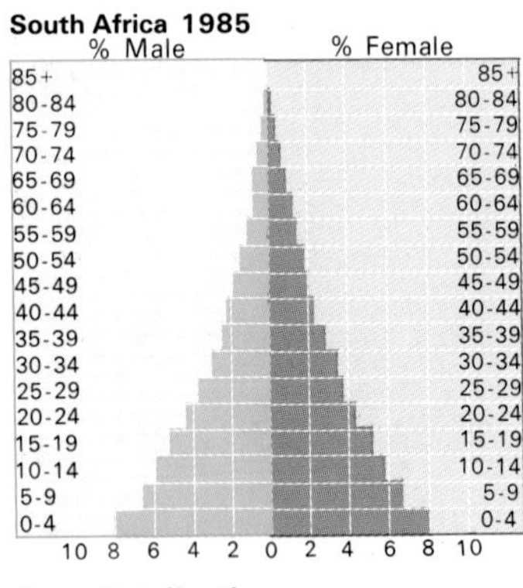

India 1985
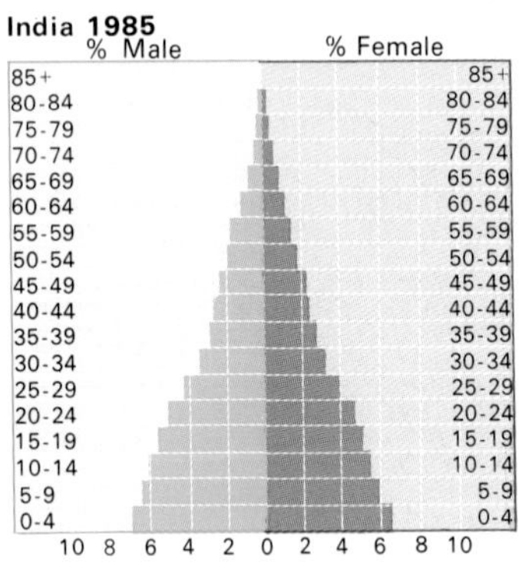

Increase in urbanisation in developed and developing countries

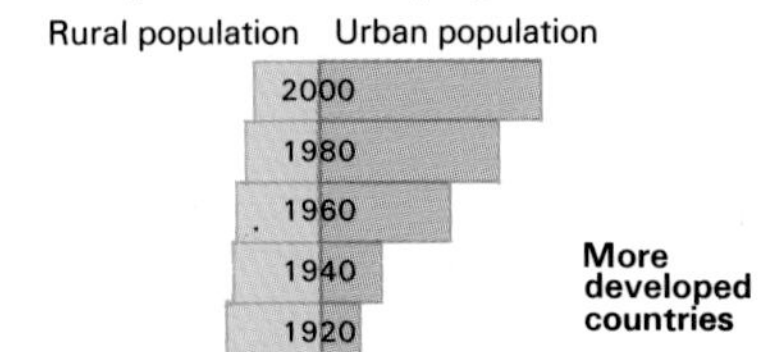

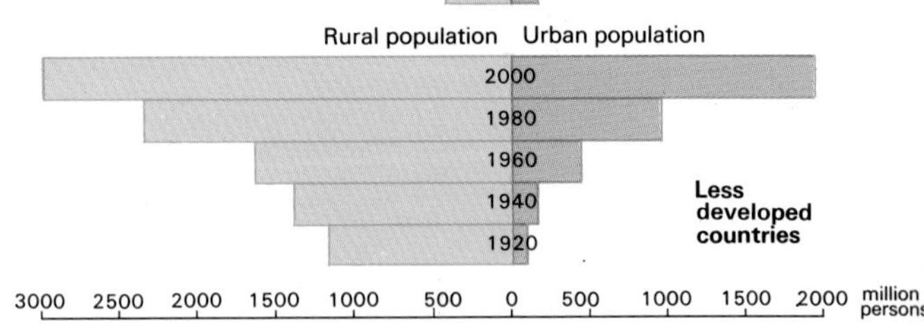

Age distribution
The U.K. shows many demographic features characteristic of European countries. Birth and death rates have declined with a moderate population growth - there are nearly as many old as young. In contrast, India and several other countries have few old and many young because of the high death rates and even higher birth rates. It is this excess that is responsible for the world's population explosion.

1650   1700   1750   1800

World population distribution

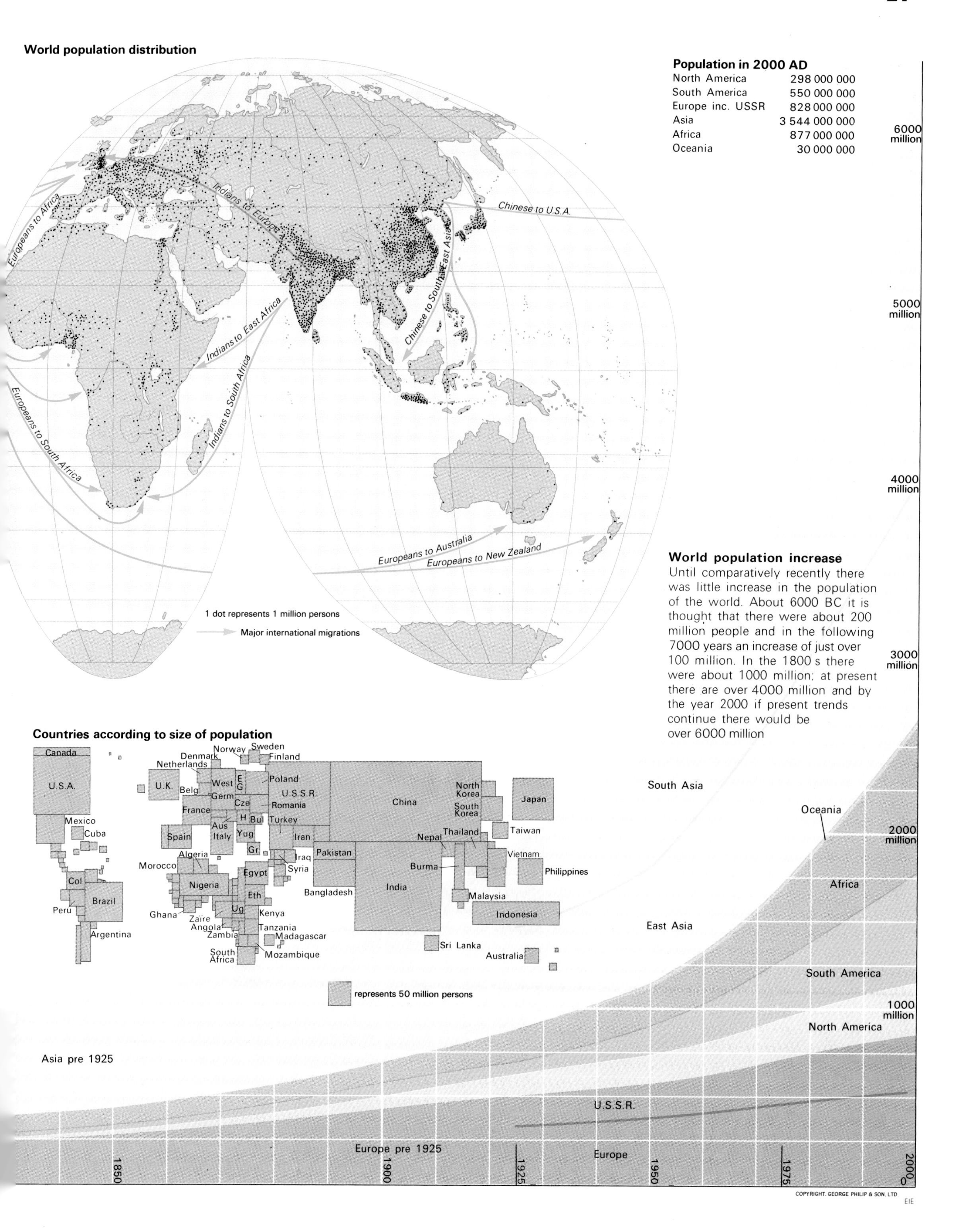

Europeans to Africa

Indians to Europe

Chinese to U.S.A.

Indians to East Africa

Chinese to South East Asia

Europeans to South Africa

Indians to South Africa

Europeans to Australia

Europeans to New Zealand

1 dot represents 1 million persons

→ Major international migrations

Population in 2000 AD

North America	298 000 000
South America	550 000 000
Europe inc. USSR	828 000 000
Asia	3 544 000 000
Africa	877 000 000
Oceania	30 000 000

World population increase

Until comparatively recently there was little increase in the population of the world. About 6000 BC it is thought that there were about 200 million people and in the following 7000 years an increase of just over 100 million. In the 1800 s there were about 1000 million; at present there are over 4000 million and by the year 2000 if present trends continue there would be over 6000 million

Countries according to size of population

Canada

U.S.A.

Mexico

Cuba

Col

Peru

Brazil

Argentina

Denmark
Netherlands
Norway
Sweden
Finland

U.K.

Belg

West Germ

E G

Cze

H Bul

Poland

U.S.S.R.

Romania

Turkey

China

North Korea

South Korea

Japan

France

Spain

Aus Italy

Yug

Gr

Iran

Nepal

Thailand

Taiwan

Algeria

Iraq

Pakistan

Vietnam

Morocco

Egypt

Syria

Burma

India

Philippines

Nigeria

Eth

Bangladesh

Malaysia

Ghana

Ug

Kenya

Zaïre

Tanzania

Indonesia

Angola

Zambia

Madagascar

South Africa

Mozambique

Sri Lanka

Australia

□ represents 50 million persons

6000 million

5000 million

4000 million

3000 million

South Asia

Oceania

2000 million

Africa

East Asia

South America

1000 million

North America

Asia pre 1925

Europe pre 1925

U.S.S.R.

Europe

1850 1900 1925 1950 1975 2000

0

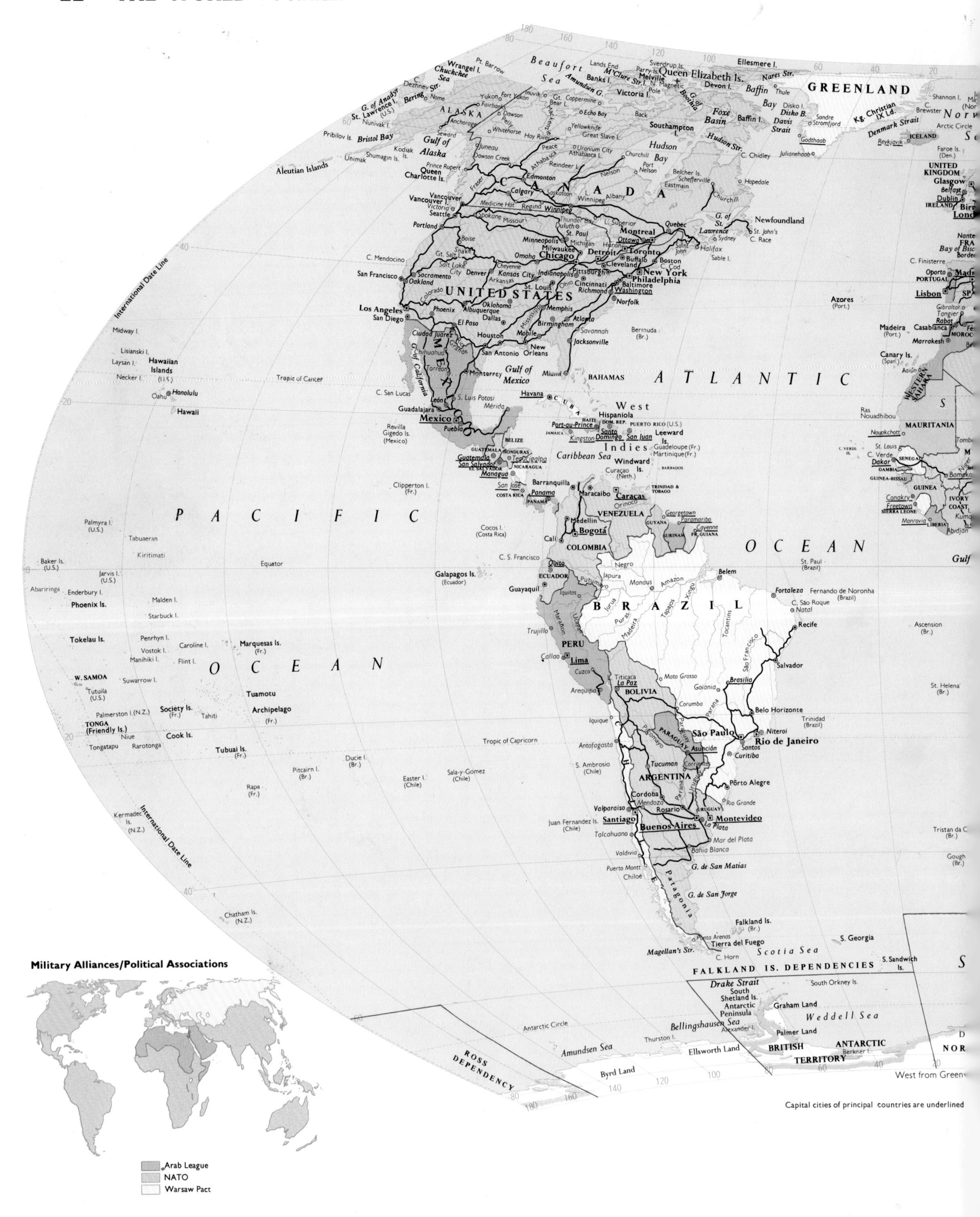

Military Alliances/Political Associations

- Arab League
- NATO
- Warsaw Pact

Capital cities of principal countries are underlined

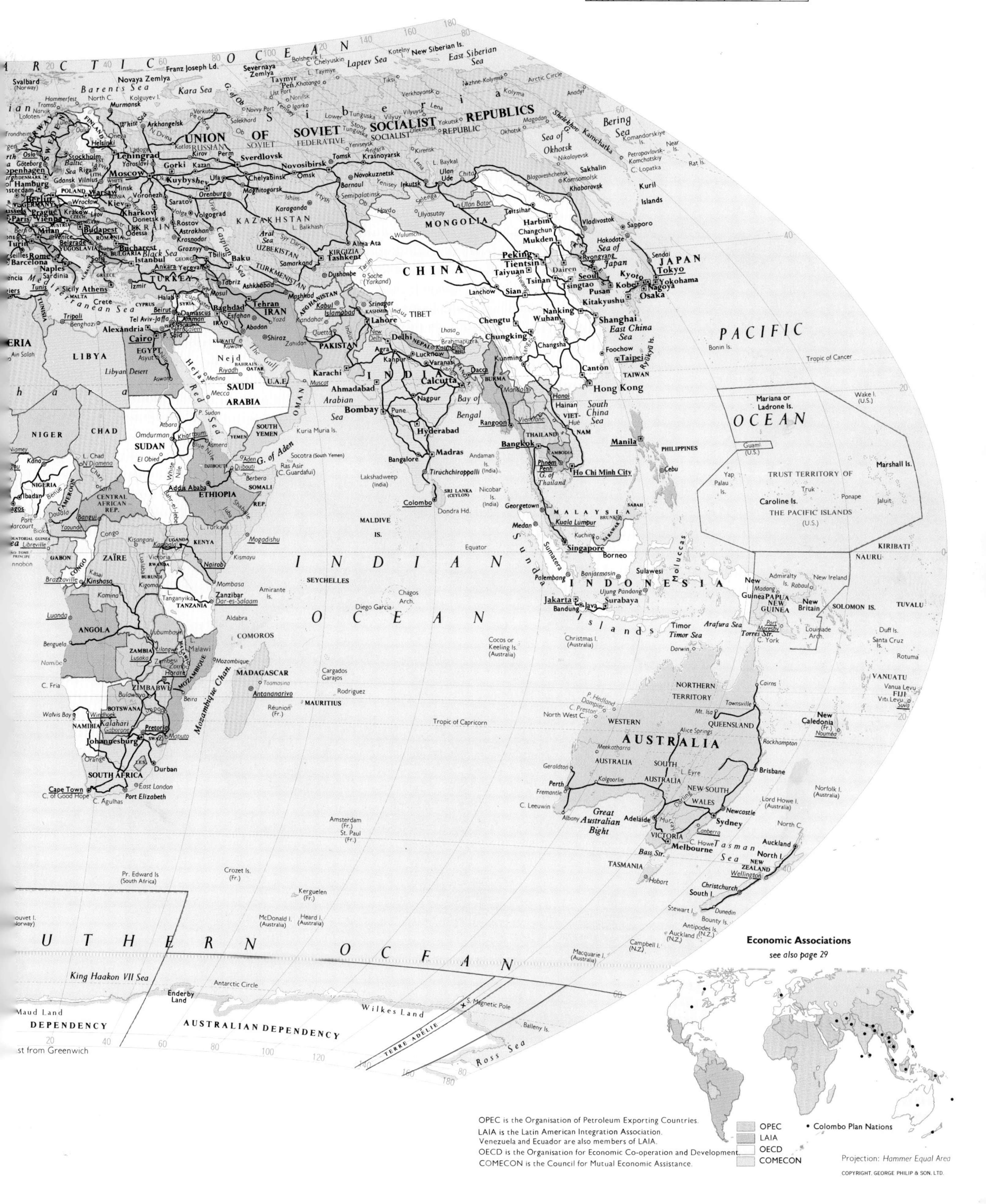

1:80 000 000

500   0   500   1000 1500 2000 2500 3000 3500 km

ARCTIC OCEAN

Svalbard (Norway)   Franz Joseph Ld.   Severnaya Zemlya   Kotelny New Siberian Is.
Novaya Zemlya   Taymyr   Bolshevik   C. Chelyuskin   Laptev Sea   East Siberian Sea
Barents Sea   Kara Sea   G. of Ob   Khatanga   Tiksi   Nizhne-Kolymsk   Kolyma   Anadyr   Arctic Circle

Hammerfest   North C   Kolguyev I.   Vorkuta   Norilsk   Verkhoyansk   Lena   Nizhne Port
Narvik   White Sea   Arkhangelsk   S. Dvina   Yenisey   S i b e r i a   Yakutsk   Nizhne-Kamchatsk
Tromsø   Murmansk   Onega   Tunguska   Vilyuy   Olekminsk   Okhotsk   Kamchatka

UNION OF SOVIET SOCIALIST REPUBLICS

FINLAND   Ladoga   RUSSIAN   SOVIET   FEDERATIVE   SOCIALIST   REPUBLIC   Sea of   Komandorskie Is.
Helsinki   Leningrad   Gorki   Kazan   Sverdlovsk   Novosibirsk   Tomsk   Krasnoyarsk   Irkutsk   Okhotsk   Petropavlovsk-Kamchatskiy

Stockholm   Moscow   Kuybyshev   Ufa   Chelyabinsk   Omsk   Barnaul   Tenisey   Angara   L. Baykal   Ulan Ude   Blagoveshchensk   Komsomolsk   Rat I.
Baltic   Riga   Minsk   Perm   Magnitogorsk   Ishim   Irtysh   Semipalatinsk   Ulan Bator   Chita   Khabarovsk   Sakhalin

POLAND   Kiev   Voronezh   Saratov   Orenburg   Karaganda   Balkhash   MONGOLIA   Amur   Vladivostok   Sapporo
Warsaw   Kharkov   Volga   Volgograd   Astrakhan   Aral Sea   Syr Darya   L. Balkhash   Ulyasutay   Harbin   Changchun   Mukden   Hokkaido
Wrocław   U K R A I N E   Rostov   Krasnodar   KAZAKHSTAN   Alma Ata   Wulumchi   Peking   Tientsin   Dairen   Ryongyang   Sea of   Sendai
Kraków   Donetsk   Grozny   Tbilisi   UZBEKISTAN   Tashkent   C H I N A   Taiyuan   Tsingtao   Seoul   Japan   JAPAN
Budapest   Odessa   Black Sea   Baku   Samarkand   KIRGIZIA   Soche (Yarkand)   Lanchow   Sian   Tsinan   Pusan   Kobe   Tokyo
ROMANIA   BULGARIA   Istanbul   Ankara   Yerevan   TURKMENISTAN   Dushanbe   Tarim   TIBET   Chengtu   Chungking   Nanking   Shanghai   Kyoto   Yokohama

Bucharest   Sofia   GREECE   TURKEY   Izmir   Mosul   Ashkhabad   AFGHANISTAN   Kabul   KASHMIR   Lhasa   Wuhan   East China   Nagoya   Osaka
YUGOSLAVIA   Athens   CYPRUS   Beirut   SYRIA   Baghdad   Tehran   Islamabad   Srinagar   Indus   Brahmaputra   Changsha   Foochow   Kitakyushu
Crete   Damascus   IRAQ   IRAN   Esfahan   Lahore   Delhi   NEPAL   Katmandu   Kunming   Canton   Taipei
MALTA   Amman   Jerusalem   Abadan   Yazd   Kandahar   New Delhi   Agra   Lucknow   BURMA   Hanoi   Hong Kong   TAIWAN

Tripoli   Benghazi   Alexandria   Port Said   KUWAIT   Shiraz   Quetta   PAKISTAN   Kanpur   Varanasi   Dacca   Mandalay   VIET-   Hué   Hainan   PACIFIC   OCEAN
Cairo   Suez   BAHRAIN   QATAR   Zahidan   Karachi   Agra   I N D I A   Calcutta   NAM   South   Bonin Is.   Tropic of Cancer
EGYPT   Nejd   QATAR   Muscat   Ahmadabad   Nagpur   Bay of   Rangoon   THAILAND   China   Wake I. (U.S.)
LIBYA   Asyut   Riyadh   U.A.E.   OMAN   Bombay   Pune   Bengal   Bangkok   CAMBODIA   Sea   Manila   Mariana or Ladrone Is.   20

Libyan Desert   Medina   SAUDI   Hyderabad   Andaman Is. (India)   Phnom   Ho Chi Minh City   Cebu   Guam (U.S.)   Marshall Is.
P. Sudan   Mecca   ARABIA   Bangalore   Madras   G. of   PHILIPPINES   Yap   Palau
Aswan   YEMEN   Arabian   Tiruchchirappalli   Thailand   TRUST TERRITORY OF
SUDAN   SOUTH   Sea   SRI LANKA   Nicobar Is.   Medan   MALAYSIA   SABAH   Caroline Is.   Truk   Ponape   THE PACIFIC ISLANDS
Omdurman   YEMEN   (CEYLON)   (India)   Georgetown   BRUNEI   (U.S.)   Jaluit
Khartoum   El Obied   Aden   G. of Aden   Colombo   Dondra Hd.   Kuala Lumpur   KIRIBATI
Atbara   Asmera   DJIBOUTI   Socotra (South Yemen)   MALDIVE   Singapore   Borneo   NAURU
Asmera   SOMALI   IS.   Sulawesi   New   Admiralty   New Ireland

White Nile   ETHIOPIA   Berbera   C. Guardafui   Palembang   Banjarmasin   Guinea   Is.   Rabaul   Duff Is.
NIGER   CHAD   Addis Ababa   REP.   Lakshadweep (India)   Sumatera   INDONESIA   Moluccas   Madang   PAPUA   New   TUVALU
Kano   SUDAN   SOMALI   Equator   Jakarta   Java   Ujung Pandang   NEW   Britain   SOLOMON IS.
Niamey   CENTRAL   Ras Asir   Bandung   Surabaya   GUINEA   Louisiade   Santa Cruz
NIGERIA   AFRICAN   KENYA   Mombasa   SEYCHELLES   Chagos Arch.   Port Moresby   Arch.   Is.   Rotuma
Ibadan   CAMEROON   Nairobi   Timor   Arafura Sea   VANUATU
Douala   ZAÏRE   Kisangani   Victoria   Diego Garcia   Timor Sea   Darwin   Vanua Levu   FIJI

EQUATORIAL GUINEA   UGANDA   Kampala   I N D I A N   Aldabra   Christmas I. (Australia)   C. York   Viti Levu   Suva
Libreville   Brazzaville   BURUNDI   Tanganyika   O C E A N   COMOROS   Cocos or   Cairns   NORTHERN   Rotuma
GABON   Kinshasa   RWANDA   Kigoma   Cargados   Keeling Is. (Australia)   TERRITORY   Townsville

Congo   Kasai   TANZANIA   Zanzibar   Garajos   P. Hedland   Mt. Isa   QUEENSLAND   New Caledonia (Fr.)
Kananga   Dar-es-Salaam   Amirante   Rodriguez   Dampier   C. Preston   Alice Springs   Noumea
ANGOLA   Lubumbashi   Malawi   MADAGASCAR   Réunion (Fr.)   MAURITIUS   Meekatharra   WESTERN   SOUTH   Rockhampton

Luanda   ZAMBIA   Antananarivo   North West C   AUSTRALIA   L. Eyre   Brisbane
Benguela   Lusaka   Harare   Toamasina   Geraldton   AUSTRALIA   AUSTRALIA   NEW SOUTH
C. Fria   ZIMBABWE   Tropic of Capricorn   Perth   Kalgoorlie   WALES   Newcastle   Norfolk I. (Australia)
Namibe   Bulawayo   Beira   Fremantle   C. Leeuwin   Great   Adelaide   Murray   Sydney   Lord Howe I. (Australia)
Walvis Bay   BOTSWANA   MOZAMBIQUE   Australian   Albany   Bight   VICTORIA   Canberra   Auckland   North I.
Windhoek   Kalahari   Maputo   Mozambique Chan   Bass Str.   Melbourne   Tasman   NEW
NAMIBIA   Gaborone   Pretoria   TASMANIA   Sea   ZEALAND   Wellington

Orange   Johannesburg   SWAZ   Durban   Amsterdam (Fr.)   Hobart   Christchurch   South I.
SOUTH AFRICA   LES   St. Paul (Fr.)   Stewart I.   Dunedin
Cape Town   East London   Kerguelen (Fr.)   Antipodes Is. (N.Z.)
C. of Good Hope   Port Elizabeth   Crozet Is. (Fr.)   Bounty Is. (N.Z.)
C. Agulhas   Pr. Edward Is. (South Africa)   McDonald I.   Heard I.   Campbell I. (N.Z.)   Macquarie I. (Australia)

S O U T H E R N   O C E A N

King Haakon VII Sea   Antarctic Circle   S. Magnetic Pole
Maud Land   Enderby Land   Wilkes Land   Balleny Is.
DEPENDENCY   AUSTRALIAN DEPENDENCY   TERRE ADÉLIE   Ross Sea
East from Greenwich   20   40   60   80   100   120   160   80

Economic Associations
see also page 29

OPEC is the Organisation of Petroleum Exporting Countries.
LAIA is the Latin American Integration Association.
Venezuela and Ecuador are also members of LAIA.
OECD is the Organisation for Economic Co-operation and Development.
COMECON is the Council for Mutual Economic Assistance.

OPEC
LAIA
OECD
COMECON

• Colombo Plan Nations

Projection: *Hammer Equal Area*

COPYRIGHT, GEORGE PHILIP & SON, LTD.

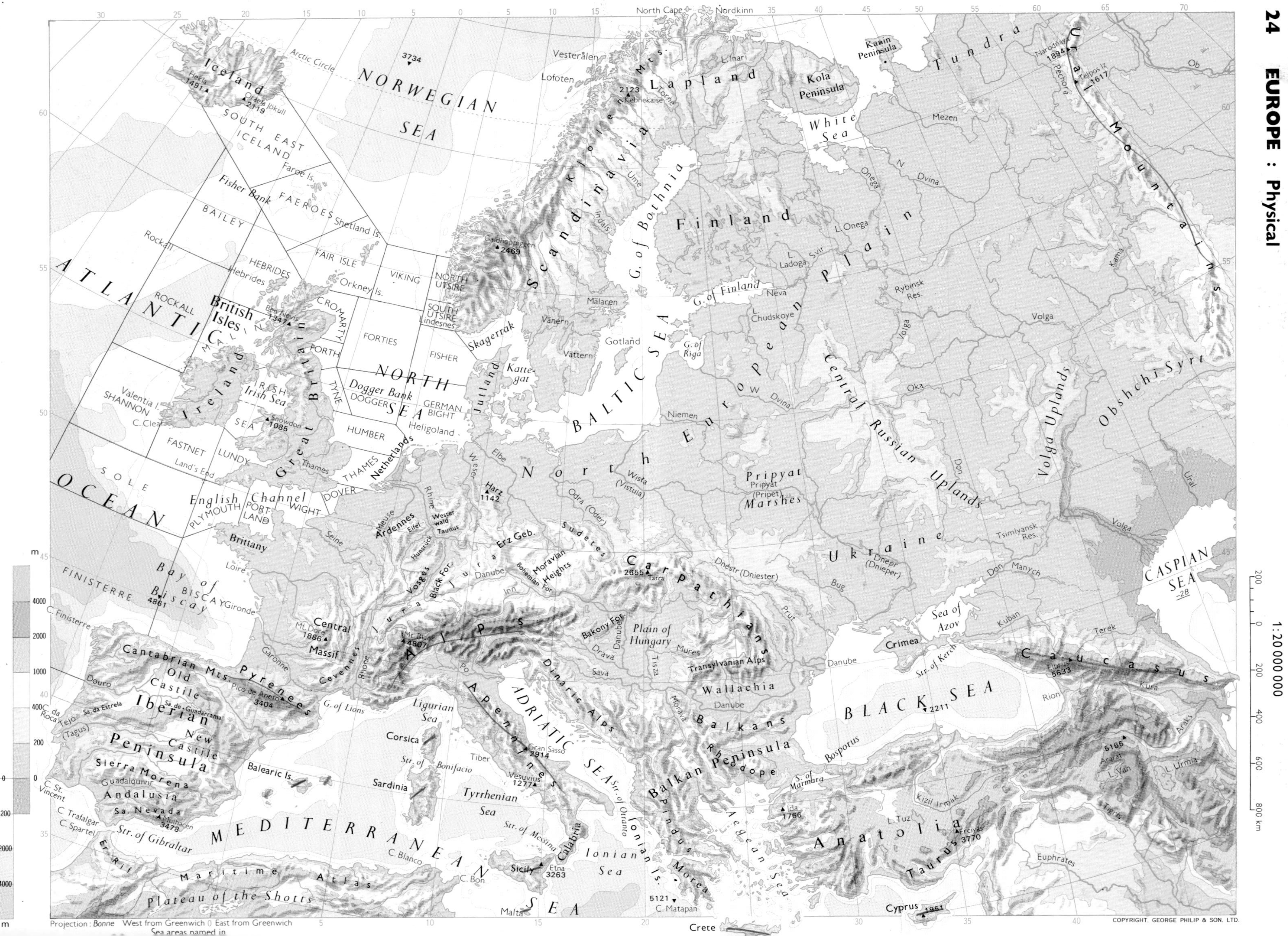

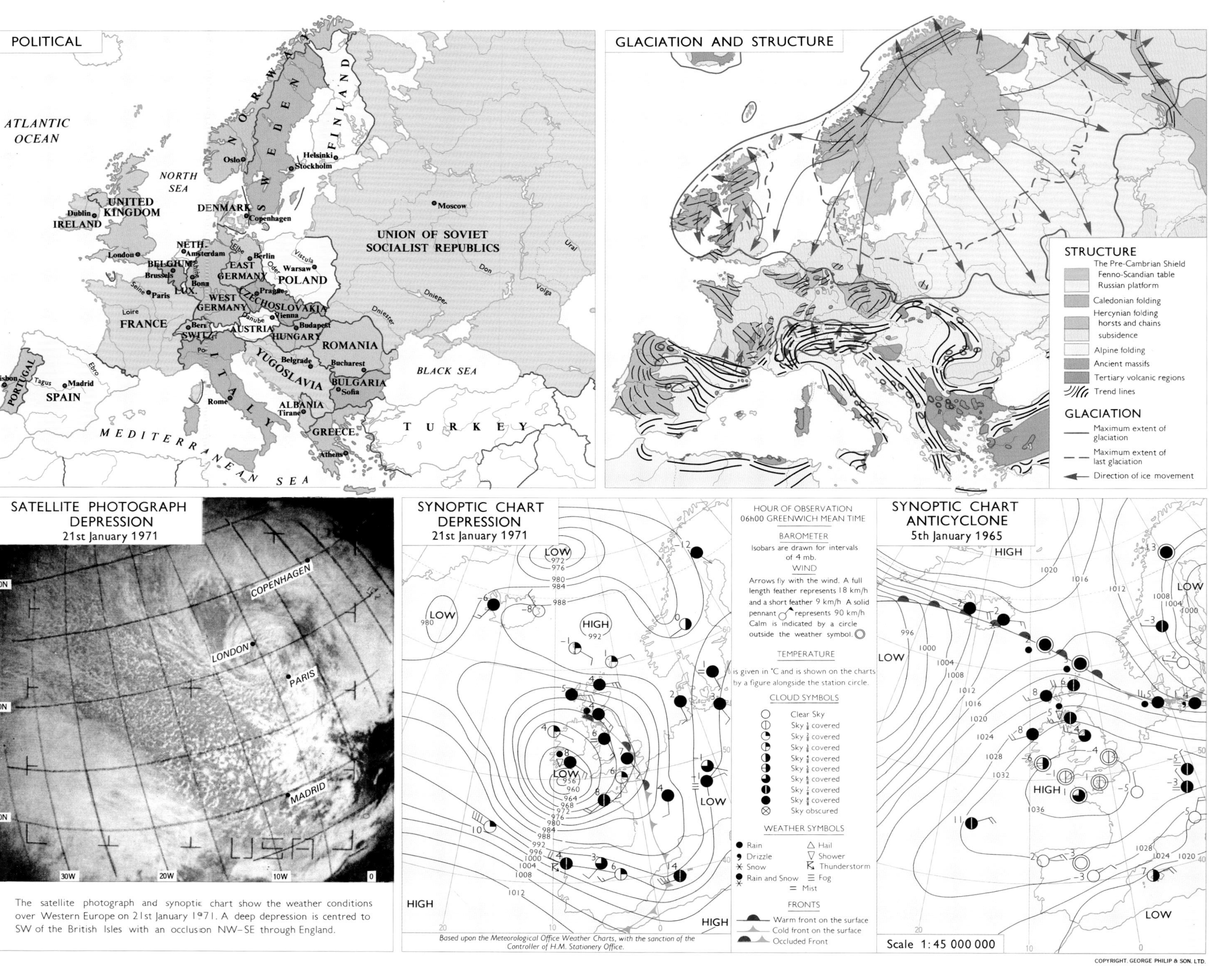

1:40 000 000

400 0 400 800 1200 1600 km

GLACIATION AND STRUCTURE

STRUCTURE
The Pre-Cambrian Shield
Fenno-Scandian table
Russian platform
Caledonian folding
Hercynian folding
horsts and chains
subsidence
Alpine folding
Ancient massifs
Tertiary volcanic regions
Trend lines

GLACIATION
Maximum extent of glaciation
Maximum extent of last glaciation
Direction of ice movement

POLITICAL

ATLANTIC OCEAN

IRELAND
Dublin
UNITED KINGDOM
London

NORWAY
Oslo
SWEDEN
Stockholm
FINLAND
Helsinki

UNION OF SOVIET SOCIALIST REPUBLICS
Moscow

NORTH SEA
DENMARK
Copenhagen

NETH.
Amsterdam
BELGIUM
Brussels
LUX.
Paris
FRANCE
Seine
Loire

EAST GERMANY
WEST GERMANY
Berlin
Bonn
Elbe
Oder
SWITZ.
Bern
POLAND
Warsaw
Vistula

CZECHOSLOVAKIA
Prague
AUSTRIA
Vienna
Danube

HUNGARY
Budapest
ROMANIA
Bucharest
YUGOSLAVIA
Belgrade
BULGARIA
Sofia
Dnieper
Dniester

ITALY
Rome

ALBANIA
Tirane
GREECE
Athens

SPAIN
Madrid
PORTUGAL
Lisbon
Tagus
Ebro

MEDITERRANEAN SEA

BLACK SEA
TURKEY

Ural
Volga
Don

SYNOPTIC CHART ANTICYCLONE 5th January 1965

Scale 1: 45 000 000

COPYRIGHT GEORGE PHILIP & SON LTD.

HOUR OF OBSERVATION
06h00 GREENWICH MEAN TIME

BAROMETER
Isobars are drawn for intervals of 4 mb.

WIND
Arrows fly with the wind. A full length feather represents 18 km/h and a short feather 9 km/h A solid pennant represents 90 km/h. Calm is indicated by a circle outside the weather symbol.

TEMPERATURE
is given in °C and is shown on the charts by a figure alongside the station circle.

CLOUD SYMBOLS
Clear Sky
Sky ¼ covered
Sky ½ covered
Sky ¾ covered
Sky ⅞ covered
Sky ¼ covered
Sky ½ covered
Sky ¾ covered
Sky covered
Sky obscured

WEATHER SYMBOLS
Rain
Drizzle
Snow
Rain and Snow
Hail
Shower
Thunderstorm
Fog
Mist

FRONTS
Warm front on the surface
Cold front on the surface
Occluded Front

SYNOPTIC CHART DEPRESSION 21st January 1971

Based upon the Meteorological Office Weather Charts, with the sanction of the Controller of H.M. Stationery Office.

SATELLITE PHOTOGRAPH DEPRESSION 21st January 1971

COPENHAGEN
PARIS
MADRID
LONDON

The satellite photograph and synoptic chart show the weather conditions over Western Europe on 21st January 1971. A deep depression is centred to SW of the British Isles with an occlusion running NW-SE through England.

1 : 40 000 000

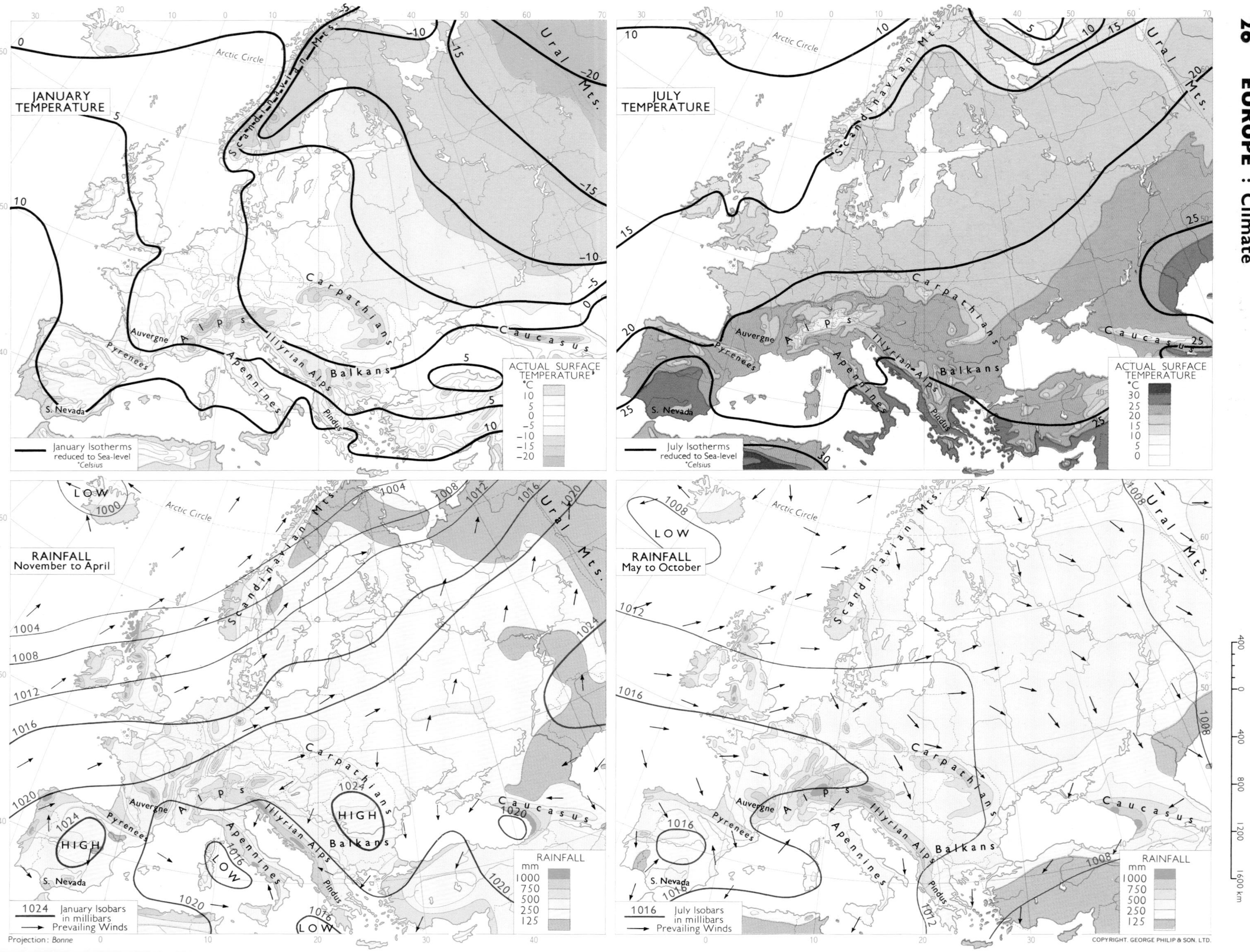

COPYRIGHT. GEORGE PHILIP & SON, LTD.

ACTUAL SURFACE
TEMPERATURE
°C
30
25
20
15
5
0

JULY
TEMPERATURE

July Isotherms
reduced to Sea-level
Celsius

RAINFALL
mm
1000
750
500
250
125

LOW

RAINFALL
May to October

1024   July Isobars
in millibars
Prevailing Winds

ACTUAL SURFACE
TEMPERATURE
°C
10
5
-5
-10
-15
-20

JANUARY
TEMPERATURE

January Isotherms
reduced to Sea-level
Celsius

RAINFALL
mm
1000
750
500
250
125

RAINFALL
November to April

1024   January Isobars
in millibars
Prevailing Winds

Projection: Bonne

1 : 40 000 000

400   0   400   800   1200   1600 km

Map B — DENSITY OF POPULATION

Towns with over 500 000 inhabitants

Inhabitants per km²
- under 1
- 1 – 6
- 6 – 12
- 12 – 25
- 25 – 50
- 50 – 100
- 100 – 200
- over 200

Labels: Saratov, Moscow, Leningrad, Helsinki, Riga, Kiev, Lvov, Odessa, Bucharest, Sofia, Istanbul, Athens, Oslo, Stockholm, Copenhagen, Warsaw, Wroclaw, Budapest, Belgrade, Hamburg, Berlin, Leipzig, Prague, Vienna, Amsterdam, Cologne, Milan, Rome, Naples, Palermo, London, Liverpool, Dublin, Paris, Marseilles, Barcelona, Valencia, Algiers, Madrid, Lisbon

Arctic Circle

Urals, Scandinavian Mts, Carpathians, Balkans, Pindus, Dinaric Alps, Alps, Apennines, Auvergne, Pyrenees, S. Nevada, Caucasus, Atlas Mts.

Map D — RELIGIONS

Christianity
- Protestant
- Roman Catholic
- Greek and Russian Orthodox

Mohammedanism
Others

✧ Judaism : major centres

Labels: Urals, Caucasus, Carpathians, Balkans, Scandinavian, Dinaric Alps, Apennines, Pindus, Auvergne, Pyrenees, S. Nevada, Atlas Mts.

Arctic Circle

East from Greenwich 10 · 20 · 30

Map A — ANNUAL RAINFALL

mm
- 1500
- 1000
- 750
- 500
- 250

Labels: Urals, Scandinavian Mts, Caucasus, Carpathians, Balkans, Pindus, Dinaric Alps, Alps, Apennines, Auvergne, Pyrenees, S. Nevada, Atlas Mts.

Arctic Circle

Map C — LINGUISTIC DIVISIONS

- Teutons
- Greco-Latins
- Slavs
- Celts
- Lithuanians and Letts
- Basques
- Caucasians
- Ural-Altai (Finns, Mogyars etc.)
- Turki
- Arabs and Berbers

Labels: Vogul, Siryan, Permyak, Votyak, Cheremis, Kazan, Chuvash, Mordvin, Kirgiz, Karelians, FINNS, Karels, Ests, Leningrad, Moscow, RUSSIANS, White, Russians, Ukrainians, Kiev, Odessa, Nogai, OSMANLI TURKS, Lithus, Lapps, SCANDINAVIANS, Swedes, Norwegians, Danes, Frisians, Dutch, GERMANS, Berlin, Poles, Czechs, Ruthenians, Vienna, Magyars, Croats, Roumanians, Bulgars, Serbs, Albans, GREEKS, Istanbul, ITALIANS, Rome, FRENCH, Paris, Lyons, ENGLISH, London, Scots, Irish, Manx, Welsh, Bretons, Icelanders, SPANIARDS, Madrid, Catalans, BASQUES, Portuguese, Berbs, Algiers, Tunis, ITALIANS

Arctic Circle

East from Greenwich 10 · 20 · 30 · 40

Projection: Bonne

1 : 40 000 000

400　　0　　400　　800　　1200　　1600 km

ELECTRICITY
Sources of power, by percentage

	'000 millions of kWh generated 1984	Thermal	Hydro-electric	Nuclear
West Germany	394.2	77	5	18
U.K.	306.4	80	2	18
France	302.4	27	25	48
Italy	180.4	71	24	5
Poland	135.2	97	3	—
Sweden	119.6	4	59	37
Spain	117.0	68	24	8
East Germany	110.1	87	1	12
Norway	106.6	0.3	99.7	—
Czechoslovakia	78.2	87	5	8
Yugoslavia	73.8	54	30	16
Romania (1983)	68.0	86	14	—
Netherlands	62.8	94	6	—
Belgium	54.6	52	2	46
Switzerland	47.8	2	69	29

The burning of coal is still important for electricity generation but West European countries must import oil to meet their energy requirements. These imports will be reduced by the exploitation of North Sea oil and gas. Most of the potential sites for hydro-electric power have been harnessed and nuclear power is still in the development stages.

POWER STATION CAPACITY

Oil and gas
◨ > 500 MW
◩ < 500 MW

Coal and lignite
▣ > 500 MW
▨ < 500 MW

Nuclear
◆ > 500 MW
◇ < 500 MW

Hydro-electric
✿ > 500 MW
❀ < 500 MW

< less than > more than

ELECTRICITY

MINERAL PRODUCTION, 1984
The tables show the leading producers of the industrially important minerals in Europe.

OIL
'000 tons
U.K.	121 236
Norway	37 776
Romania	12 504
West Germany	4 056
Yugoslavia	4 044
Hungary	2 004

NATURAL GAS
Petajoules
Netherlands	2 444
U.K.	1 590
Romania (1983)	1 680
Norway	1 166
West Germany	587
Italy	533

LIGNITE
'000 tons
East Germany	296 340
W. Germany	127 296
Czechoslovakia	104 748
Yugoslavia	54 540
Poland	50 376
Romania	36 000

COAL
'000 tons
Poland	191 592
West Germany	84 036
U.K.	51 252
Czechoslovakia	26 424
France	16 596
Spain	15 132

IRON ORE
'000 tons
Sweden	11 325
France	4 504
Spain	3 726
Norway	2 418
Yugoslavia	1 861
Austria	1 116

▲ Salt ● Potash
● Bauxite ● Lead

Oil · Natural gas · Coal · Lignite · Iron ore

MINERALS

Upper Silesia
North Sea
Ruhr
Lorraine
South Yorkshire

LIGHT INDUSTRIAL PRODUCTS, 1984

TEXTILES
Cotton yarn
'000 tons
West Germany	194.4
France	193.2
Romania (1983)	178.8
Italy (1983)	150.0
Czech. (1983)	140.4

Wool yarn
'000 tons
U.K.	126.1
France	107.0
Belgium	91.9
West Germany	79.2
East Germany	48.0
	36.6

MOTOR VEHICLES
Cars
'000
West Germany	3 788
France	2 910
Italy	1 174
U.K.	910
Sweden	314

Commercial vehicles
'000
France	424
West Germany	264
U.K.	224
Italy	160
Czech. (1983)	85
Sweden	46

ELECTRONICS
'000 televisions
West Germany	3 840
U.K.	2 849
France	2 009
Italy	1 650
Spain	1 117
Belgium	799

Textiles
⊕ Electrical engineering
Motor vehicles

LIGHT INDUSTRY

HEAVY INDUSTRIAL PRODUCTS, 1984

CRUDE STEEL
'000 tons
West Germany	39 384
Italy	23 076
France	19 020
Poland	16 536
U.K.	15 120
Czechoslovakia	14 832

MERCHANT VESSELS
'000 gross registered tons launched
West Germany	489
Denmark	376
Poland	317
Finland	315
Italy	233
France	221

SULPHURIC ACID
'000 tons
France	4 520
West Germany	4 399
Spain	3 407
Poland	2 770
U.K.	2 654
U.K. (1983)	2 339

Sulphuric acid is a fundamental basis for most branches of the chemical industry.

Heavy engineering
◁ Iron and steel plants
Shipbuilding
Chemicals

HEAVY INDUSTRY

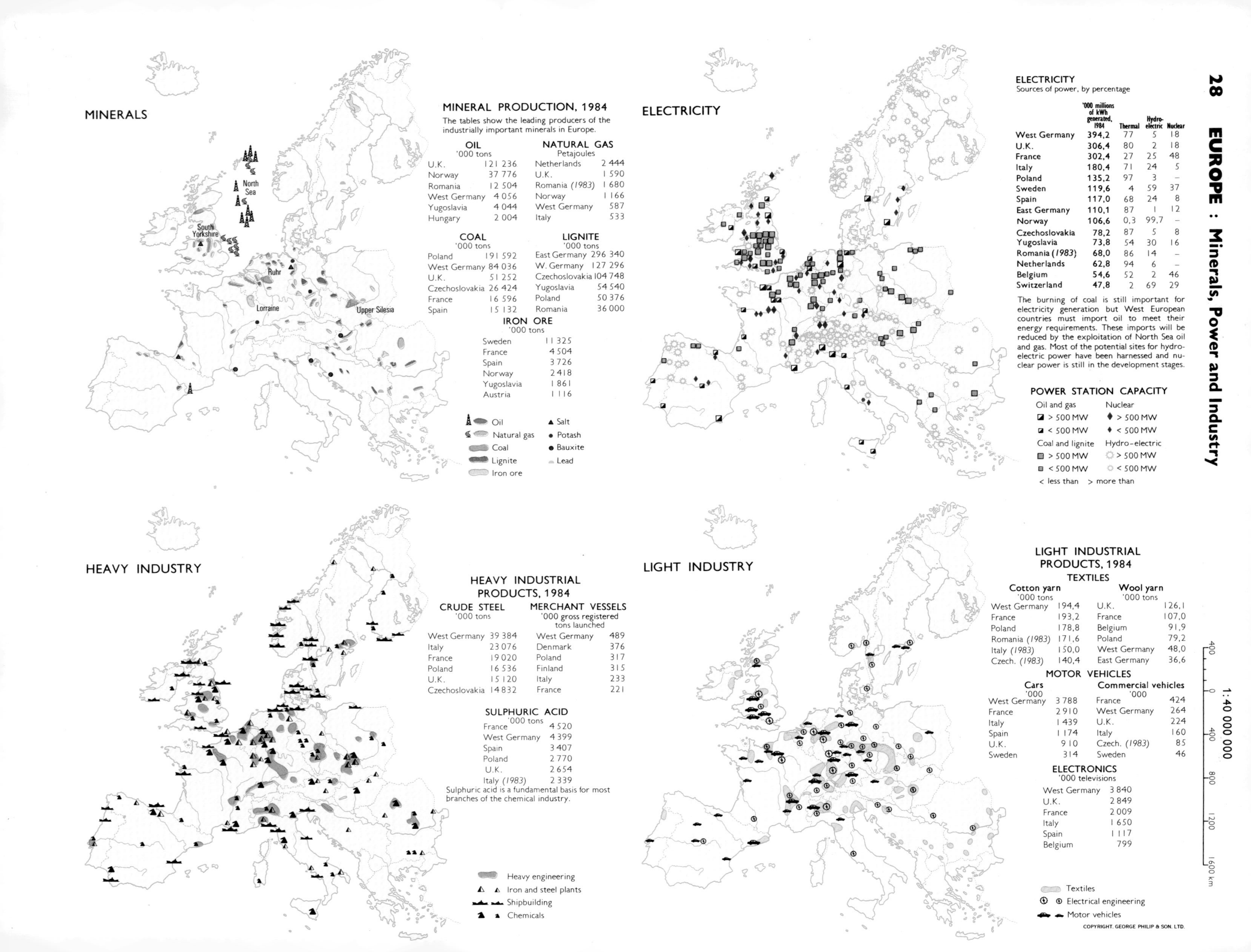

1:40 000 000

400   0   400   800   1200   1600 km

STANDARDS OF LIVING

Gross Domestic Product (GDP) is a measure of a country's total production of goods and services.

Gross Domestic Product per person in 1981

£7-8000
£6-7000
£5-6000
£4-5000
£3-4000
£2-3000
£1-2000

FINLAND £34.0 bn. (2.7%)

NORWAY £37.9 bn. (3.7%)

SWEDEN £63.3 bn. (1.3%)

DENMARK £38.8 bn. (1.8%)

NETH. £94.1 bn. (1.5%)

WEST GERMANY £449.9 bn. (2.1%)

AUSTRIA £45.9 bn. (2.8%)

SWITZERLAND £66.9 bn. (0.7%)

YUGOSLAVIA £32.3 bn. (5.3%)

GREECE £21.2 bn. (3.0%)

ITALY £243.1 bn. (2.2%)

SPAIN £94.1 bn. (1.8%)

U.K. £313.6 bn. (1.1%)

IRISH REP. £12.4 bn. (3.2%)

BELGIUM £55.2 bn. (1.8%)

FRANCE £357.7 bn. (2.5%)

PORTUGAL £14.0 bn. (4.0%)

Gross Domestic Product in 1983 in £ billions
(% annual average growth 1973-83 is given in brackets)

The number of people employed in manufacturing in Europe has steadily declined since 1970. Service industries such as tourism are becoming increasingly important.

RECEIPTS FROM TOURISM
as percentage of G.D.P. 1982
(Total receipts in $ U.S. million are given in brackets)

Austria	8.5%	(5 649)
Greece	4.5%	(1 527)
Portugal	4.1%	(878)
Spain	3.9%	(7 126)
Switzerland	3.1%	(3 015)
Ireland	2.8%	(477)
Italy	2.4%	(8 234)
Denmark	2.3%	(1 305)
Belgium	1.9%	(1 578)
Norway	1.5%	(815)
France	1.3%	(6 991)
Yugoslavia	1.2%	(844)
Finland	1.2%	(579)
U.K.	1.1%	(5 144)
U.S.A.	0.4%	(11 293)
Hong Kong	6.0%	(1 457)

EMPLOYMENT BY ECONOMIC ACTIVITY

% of the labour force, 1984

Services
Industry
Agriculture

For example:
U.K. 2.6% Agriculture
33.0% Industry
64.4% Services
GREECE 29.5% Agriculture
27.8% Industry
42.7% Services

NORWAY 43%
SWEDEN 47%
FINLAND 48%
U.K. 42%
WEST GERMANY 39%
FRANCE 41%
YUGOSLAVIA 37%
ITALY 33%
GREECE 33%
IRISH REPUBLIC 31%
SPAIN 29%

U.K. 42% % of labour force who are female (for selected countries)

TRADE ORGANISATIONS

E.F.T.A. was founded in 1960 to establish free trade of industrial products between its members and to promote peace in Western Europe.

COMECON was formed in 1949 to develop the trade and resources of the Soviet bloc countries.

The E.E.C. or Common Market was founded in 1957 by the Treaty of Rome. Members share common agricultural and industrial policies and there are tariffs on external trade. The E.E.C. has over 320 million inhabitants and is one of the wealthiest markets in the world.

E.F.T.A. (European Free Trade Association) H.Q. Geneva

COMECON (Council for Mutual Economic Assistance) H.Q. Moscow

E.E.C. (The European Economic Community) H.Q. Brussels

The Six in 1957 (Pop. 170 000 000)
The Nine in 1973 (Pop. 256 000 000)
The Ten in 1981 (Pop. 270 000 000)
The Twelve in 1986 (Pop. 321 000 000)

ICELAND
NORWAY
SWEDEN
FINLAND
DENMARK
NETH.
U.K.
IRISH REP.
BELG.
Brussels
EAST GER.
WEST GERMANY
POLAND
CZECH.
Geneva
SWITZ.
AUSTRIA
HUNGARY
ROMANIA
BULGARIA
YUGOSLAVIA
ALB.
GREECE
ITALY
FRANCE
SPAIN
PORT.

IMPORTANCE OF TRADE TO THE E.E.C.

E.E.C.'s share of each country's total imports 1983

over 60%
50-60%
40-50%
30-40%
20-30%
10-20%
0-10%

DENMARK I 3 322 E 3 196
GREECE I 960 E 488
NETH. I 4 309 E 4 549
WEST GER. I 5 395 E 5 049
BELG. I 2 489 E 2 758
FRANCE I 1 888 E 1 698
ITALY I 1 414 E 1 192
U.K. I 1 775 E 1 628
IRISH REP. I 2 730 E 2 720
PORTUGAL I 767 E 510
SPAIN I 744 E 608

Trade in E.E.C. in U.S. $ per capita 1984
I = Imports   E = Exports

WORLD IMPORTS BY REGION 1985

W. EUROPE 38.5% (of which E.E.C. 32.3%)
ASIA 22.2% (of which Japan 6.4%)
OCEANIA 1.5%
AFRICA 3.8%
S. AMERICA 4.2%
N. AMERICA 21.4% (of which U.S.A. 17.7%)
E. EUROPE (incl. U.S.S.R.) 8.4%

1 : 40 000 000

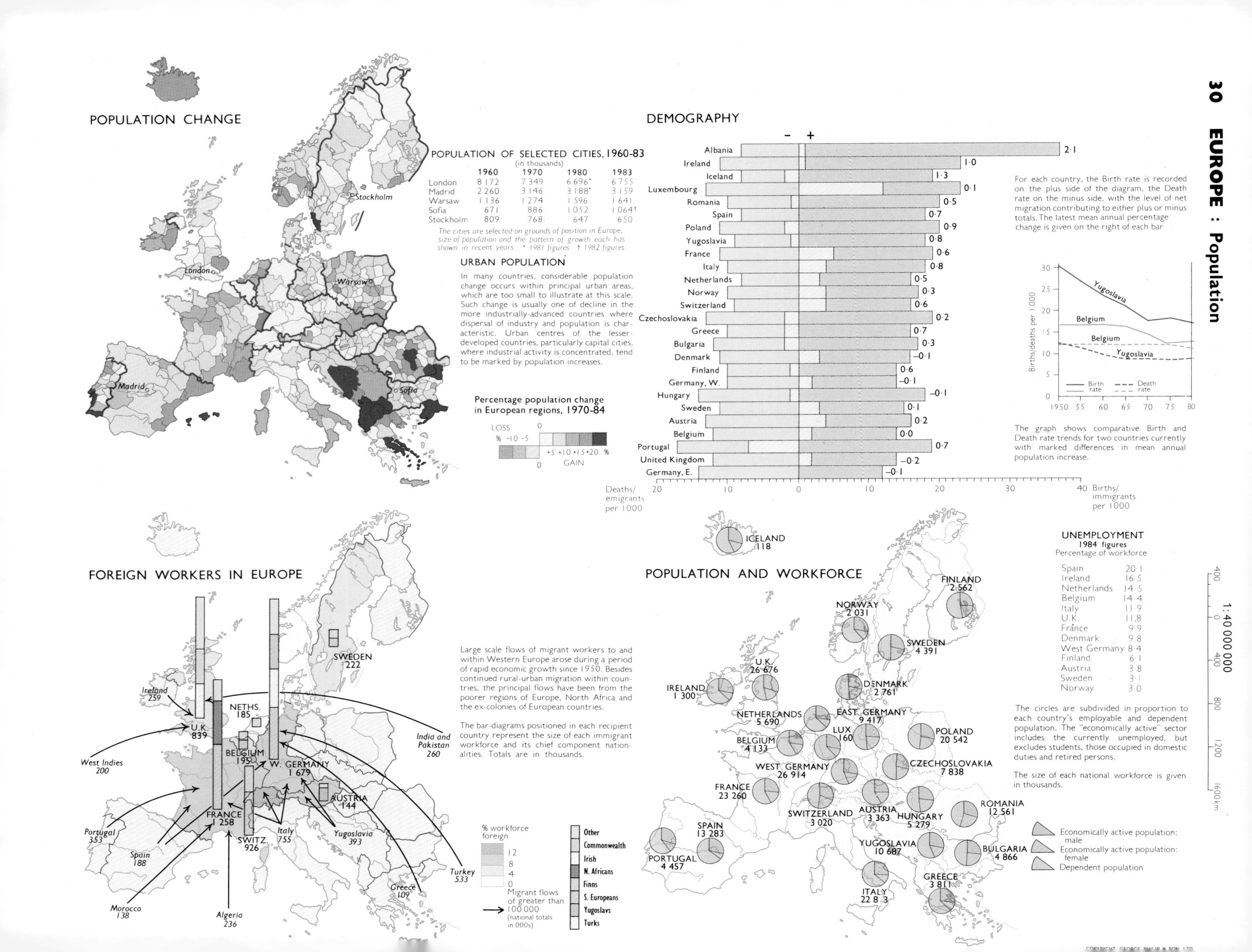

DEMOGRAPHY

For each country, the Birth rate is recorded on the plus side of the diagram, the Death rate on the minus side, with the level of net migration contributing to either plus or minus totals. The latest mean annual percentage change is given on the right of each bar

The graph shows comparative Birth and Death rate trends for two countries currently with marked differences in mean annual population increase.

Birth rate
Death rate

POPULATION OF SELECTED CITIES, 1960-83
(in thousands)

	1960	1970	1980	1983
London	8 172	7 349	6 696*	6 755
Madrid	2 260	3 146	3 188*	3 159
Warsaw	1 136	1 274	1 596	1 641
Sofia	809	886	1 052	1 064†
Stockholm	809	768	647	650

*1981 figures † 1982 figures

URBAN POPULATION

In many countries, considerable population change occurs within principal urban areas, which are too small to illustrate at this scale. Such change is usually one of decline in the more industrially-advanced countries where dispersal of industry and population is characteristic. Urban centres of the lesser-developed countries, particularly capital cities, where industrial activity is concentrated, tend to be marked by population increases.

The cities are selected on grounds of position in Europe, size of population and the pattern of growth each has shown in recent years.

Percentage population change in European regions, 1970-84

LOSS
% -10 -5   0   +5 +10 +5 +20 %
GAIN

UNEMPLOYMENT
1984 figures
Percentage of workforce

Spain	20.1
Ireland	16.5
Netherlands	14.5
Belgium	14.4
Italy	11.9
U.K.	11.8
France	9.9
Denmark	9.8
West Germany	8.4
Finland	6.1
Austria	3.8
Sweden	3.1
Norway	3.0

The circles are subdivided in proportion to each country's employable and dependent population. The 'economically active' sector includes the currently unemployed, but excludes students, those occupied in domestic duties and retired persons.

The size of each national workforce is given in thousands.

Economically active population: male
Economically active population: female
Dependent population

POPULATION AND WORKFORCE

FINLAND 2 562
NORWAY 2 031
SWEDEN 4 391
DENMARK 2 761
POLAND 20 542
EAST GERMANY 9 417
CZECHOSLOVAKIA 7 838
ROMANIA 12 561
BULGARIA 4 866
HUNGARY 5 279
AUSTRIA 3 363
YUGOSLAVIA 10 687
GREECE 3 811
ITALY 22 813
LUX. 160
SWITZERLAND 3 020
WEST GERMANY 26 914
FRANCE 23 260
BELGIUM 4 133
NETHERLANDS 5 690
U.K. 26 676
IRELAND 1 300
SPAIN 13 283
PORTUGAL 4 457
ICELAND 118

POPULATION CHANGE

Stockholm
Warsaw
Sofia
London
Madrid

FOREIGN WORKERS IN EUROPE

Large scale flows of migrant workers to and within Western Europe arose during a period of rapid economic growth since 1950. Besides continued rural-urban migration within countries, the principal flows have been from the poorer regions of Europe, North Africa and the ex-colonies of European countries.

The bar-diagrams positioned in each recipient country represent the size of each immigrant workforce and its chief component nationalities. Totals are in thousands.

% workforce foreign
12
8
4
0

Migrant flows of greater than 100 000 (national totals in 000s)

Other
Commonwealth
Irish
N. Africans
Finns
S. Europeans
Yugoslavs
Turks

India and Pakistan 260
Turkey 533
Greece 109
Yugoslavia 393
Austria 144
Italy 755
SWEDEN 222
W. GERMANY 1 679
SWITZ. 926
NETHS. 185
BELGIUM 195
FRANCE 258
U.K. 839
Ireland 259
West Indies 200
Portugal 353
Spain 188
Morocco 138
Algeria 236

1 : 6 000 000

50   0   50   100   150   200   250 km

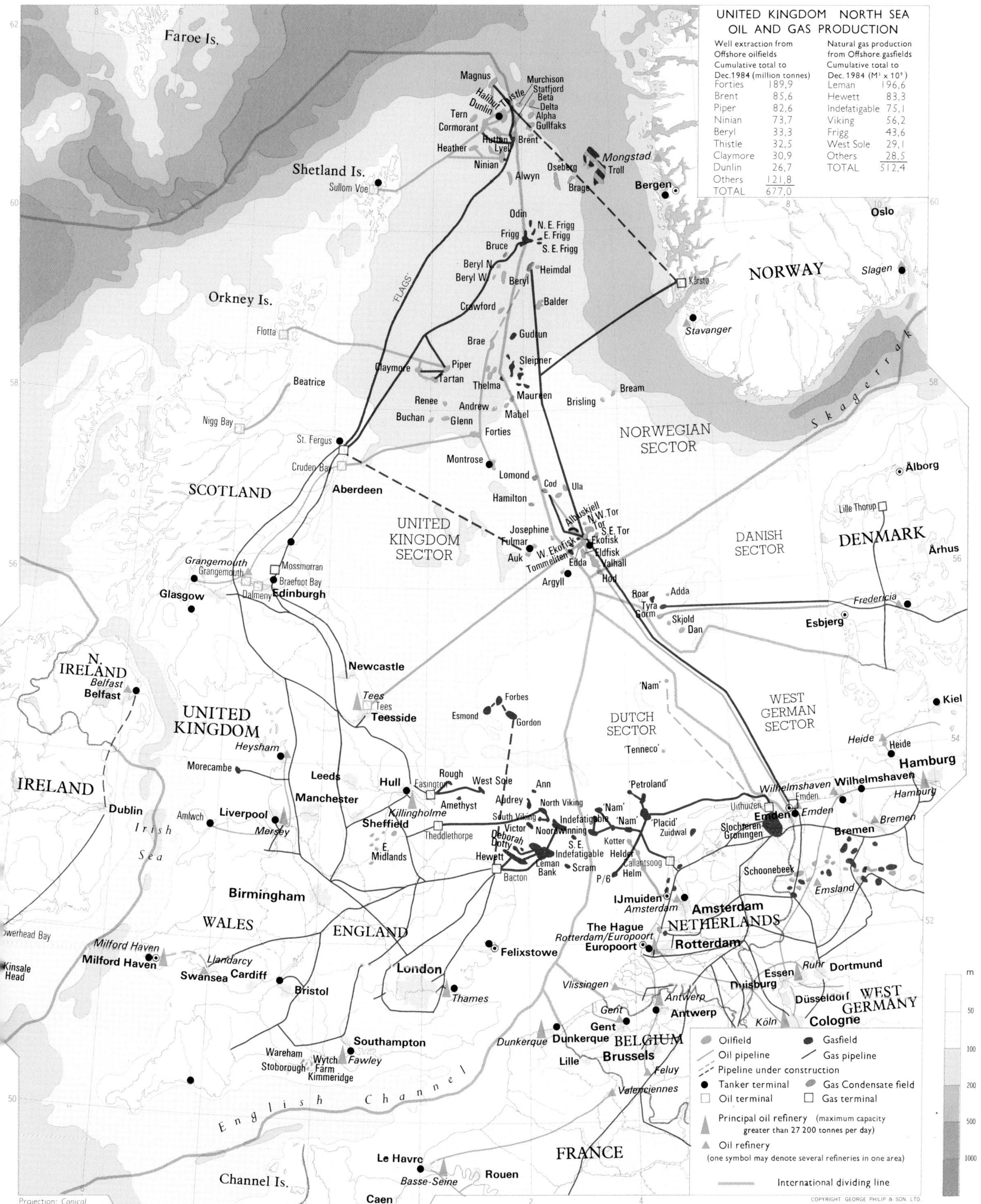

Projection: Conical
with two standard parallels

COPYRIGHT GEORGE PHILIP & SON LTD

UNITED KINGDOM NORTH SEA OIL AND GAS PRODUCTION

Well extraction from Offshore oilfields Cumulative total to Dec.1984 (million tonnes)		Natural gas production from Offshore gasfields Cumulative total to Dec. 1984 (M³ x 10⁹)	
Forties	189,9	Leman	196,6
Brent	85,6	Hewett	83,3
Piper	82,6	Indefatigable	75,1
Ninian	73,7	Viking	56,2
Beryl	33,3	Frigg	43,6
Thistle	32,5	West Sole	29,1
Claymore	30,9	Others	28,5
Dunlin	26,7	TOTAL	512,4
Others	121,8		
TOTAL	677,0		

Legend:

- ⬬ Oilfield
- ⬮ Gasfield
- ── Oil pipeline
- ── Gas pipeline
- ─ ─ ─ Pipeline under construction
- ● Tanker terminal
- ⬮ Gas Condensate field
- ☐ Oil terminal
- ☐ Gas terminal
- ▲ Principal oil refinery (maximum capacity greater than 27 200 tonnes per day)
- ▲ Oil refinery (one symbol may denote several refineries in one area)
- ──── International dividing line

Elevation scale (m): 50, 100, 200, 500, 1000

1:5 000 000

50   0   50   100   150   200   250 km

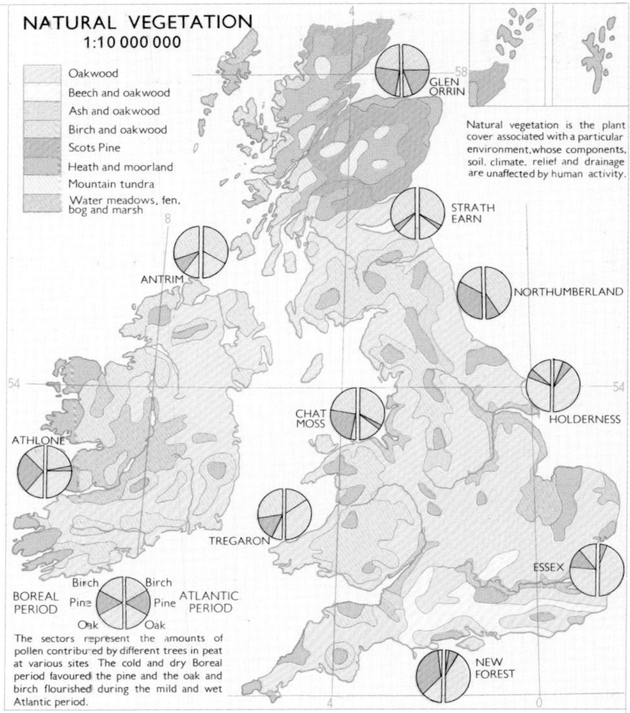

NATURAL VEGETATION 1:10 000 000

Natural vegetation is the plant cover associated with a particular environment, whose components, soil, climate, relief and drainage are unaffected by human activity.

NORTHUMBERLAND
HOLDERNESS
ESSEX
STRATH EARN
GLEN ORRIN
NEW FOREST
CHAT MOSS
TREGARON
ANTRIM
ATHLONE

Oakwood
Beech and oakwood
Ash and oakwood
Birch and oakwood
Scots Pine
Heath and moorland
Mountain tundra
Water meadows, fen, bog and marsh

BOREAL PERIOD — Birch, Pine, Oak
ATLANTIC PERIOD — Birch, Pine, Oak

The sectors represent the amounts of pollen contributed by different trees in peat at various sites. The cold and dry Boreal period favoured the pine and the oak and birch flourished during the mild and wet Atlantic period.

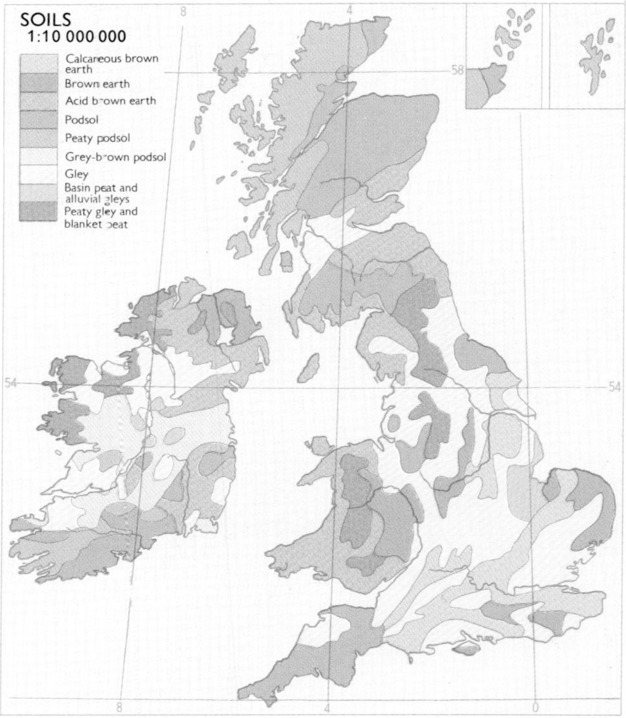

SOILS 1:10 000 000

Calcareous brown earth
Brown earth
Acid brown earth
Podsol
Peaty podsol
Grey-brown podsol
Gley
Basin peat and alluvial gleys
Peaty gley and blanket peat

Main map labels

NORTH SEA
ATLANTIC OCEAN
IRISH SEA
CELTIC SEA
North Channel
St. George's Channel
Bristol Channel
English Channel
FRANCE

Dogger Bank
Shetland Is.   Foula   Sumburgh Hd.
Orkney Is.   Pentland Firth
Fair I.
Butt of Lewis   Lewis 789   North Minch   Outer Hebrides   Inner Hebrides
C. Wrath   Duncansby Hd.   Dunnet Hd.
Buchan Ness
Ben More Assynt 998   L. Shin   Moray Firth   Dee   Don
L. Maree   Ben Wyvis 1045   L. Ness   Glen Mor   Grampians   F. of Tay   Fife Ness   Bass Rock
Skye 1009   Eigg   Coll   Cairn Gorm 1182   Ben Macdhui 1311   Sidlaw Hills   Ochil Hills
Ardnamurchan Pt.   Mull   Ben Nevis 1343   974   L. Lomond   Firth of Forth   Lammermuir Hills
Firth of Lorn   North West Highlands   Clyde   Pentland Hills
Islay   Arran   Firth of Clyde   Southern Uplands   Broad Law 840   The Cheviot 816
Mull of Kintyre   Ayr   Nith   Eden   Cheviot Hills
Fair Hd.   Mull of Galloway   Solway Firth   Cross Fell 893   Pennines   737
Giant's Causeway   I. of Man   Snaefell 620   Bee's Hd.   Sca Fell 978   Cumbrian Mts.
Antrim Mts.   Merrick 843   Morecambe Bay   Formby Pt.   Liverpool Bay
Bann   Neagh   Gt. Ormes Hd.   Cheshire Plain   Wrekin 407
Malin Hd.   Mourne Mts. 852   Anglesey   Holy I.   Menai St.   Snowdon 1085   Cambrian Mts. 892
Tory I.   Errigal 752   Derryveagh Mts.   L. Erne   Bann   Plynlimmon 752   Cotswolds
Donegal Bay   L. Erne   Cardigan Bay   Tywi   Brecon Beacons   Mendip Hills
Errris Hd.   Achill I.   Mweelrea 819   Connemara   Bog of Allen   Wicklow Mts. 926   Barrow
L. Mask   L. Corrib   Central Plain   Slieve Bloom Mts.   Boyne   St. David's Hd.
Galway Bay   Shannon   L. Derg   Golden Vale   Galty Mts. 920   Carnsore Pt.
Dunmore Hd.   Macgillycuddy's Reeks 1041   Blackwater   Lee   Cork Harbour   C. Clear
Milford Haven   Hartland Point   Lundy   Exmoor 520   Yes Tor 618   Dartmoor   Bodmin Moor
Land's End   Isles of Scilly   Start Pt.   Tamar   Exe   N. Dorset Downs
Flamborough Hd.   Spurn Hd.   Holderness   Humber   Yorkshire Wolds   Lincolnshire Wolds
N. York Moors 454   Vale of York   Lincoln Heath   Trent   The Peak 636   The Derwent
Tees   Swale   Ouse   Wharfe   Aire   Don
The Naze   The Wash   Breckland   The Fens   Gt. Ouse   Nene   Welland   Witham
North Foreland   Dungeness   Beachy Hd.   Str. of Dover   I. of Wight   Needles   Portland Bill
North Downs   South Downs   The Weald   Chiltern Hills   Thames   Medway   Lee   Stour
Marlboro Downs   Berks Downs   Hampshire Downs   Salisbury Plain   Avon   Kennet   Cherwell
Severn   Wye   Avon   Parrett   238   30 m

West from Greenwich   East from Greenwich

m   1000   400   200   100   0

m   0   100   200   400   600

Projection: Conical with two standard parallels

1:5 000 000

50 0 50 100 150 200 250 km

Cross-section A — CENTRAL SCOTLAND

A₁ SOUTHERN UPLANDS — R. TWEED — CENTRAL VALLEY — LENNOX HILLS — L. LOMOND — GRAMPIANS
A — Sea Level — 20 km

Cross-section B — NORTH PENNINES

B₁ NEWCASTLE — DURHAM COALFIELD — WEARDALE — CROSS FELL — EDEN VALLEY — PENRITH — LAKE DISTRICT
B — Sea Level — 20 km

Cross-section C — SOUTH SHROPSHIRE

C₁ CLEE HILLS — WENLOCK EDGE — CAER CARADOC — CHURCH STRETTON — LONG MYND
C — Sea Level — 5 km / 20 km

Cross-section D — SOUTH WALES COALFIELD

D₁ BRECON — BRECON BEACONS — HIRWAUN — RHONDDA — MAESTEG — VALE OF GLAMORGAN — BRISTOL CHANNEL
D — Sea Level — 10 km

Cross-section E — SOUTH EAST ENGLAND

E₁ SOUTH DOWNS — WEALD — NORTH DOWNS — LONDON — WATFORD — CHILTERNS
E — Sea Level — 20 km

Cross-section F — SOUTH EAST IRELAND

F₁ WEXFORD — BLACKSTAIRS MOUNTAIN — R. BARROW — LEINSTER COALFIELD — SLIEVE BLOOM
F — Sea Level — 20 km

COPYRIGHT GEORGE PHILIP & SON LTD.

Legend

RECENT
- Alluvium

TERTIARY (Cainozoic)
- Sands and Clays

SECONDARY (Mesozoic)
- Cretaceous–Chalk
- Jurassic and Cretaceous –Clays and Sands
- Liassic (Jur.), Triassic and Permian –Sandstones and Clays

PRIMARY (Paleozoic)
- Carboniferous–Coal Measures, Limestone and Millstone Grit
- Old Red Sandstone and Devonian
- Ordovician, Silurian and Cambrian

ANCIENT (Pre-Cambrian)
- Torridonian, Longmyndian etc.
- Metamorphic

IGNEOUS (Various ages)
- Volcanic (e.g. Basalt)
- Intrusive (e.g. Granite)

Major faults
A — A₁ Cross sections

Map labels

Shetland Is.
Orkney Is.
Outer Hebrides
Skye
Mull
North West Highlands
Grampians
Glen Mor Fault
Moine Thrust
Highland Boundary Fault
Southern Upland Fault
Southern Uplands
Cheviot Hills
Pennines
Cumbrian Mts.
N. York Moors
Lincolnshire Wolds
I. of Man
Anglesey
Cambrian Mts.
Cotswolds
Chilterns
North Downs
South Downs
I. of Wight
Exmoor
Dartmoor
Antrim Mts.
Mourne Mts.
Wicklow Mts.
Derryveagh Mts.
Connemara
Central Plain
Slieve Bloom Mts.
Galty Mts.
Mts. of Kerry

SOUTHWARD LIMIT OF GLACIATION

West from Greenwich 0 East from Greenwich

Projection: Conical with two standard parallels

1:10 000 000

100   0   100   200   300   400 km

WIND

Lerwick 4.5
Dyce 15
Wick 3.1
Dungeness 2.1
Mildenhall 6.4
Manby 6.5
Heathrow 6.0
Elmdon 9.2
Tynemouth 4.7
Ringway 9.2
Turnhouse 15.2
Exeter 13.5
Aberporth 5.6
Stornoway 6.5
Tiree 6.8
Aldergrove 5.0
Dublin 5.7
Rosslare 1.1
Belmullet 2.1
Shannon 2.3
Valentia 1.1

(34) N E S W

% calms in a year
Direction the wind blows from
% frequency of wind from a direction
Force of wind (Beaufort Scale)

BEAUFORT FORCE	SPEED (K.P.H.)	CATEGORY
1-3	1-20	Light breeze
4	21-29	Moderate breeze
5-6	30-50	Fresh to strong wind
7	51-61	Moderate gale
8-12	over 62	Gale, storm or hurricane

SNOW

Average number of mornings with snow cover per year

more than 50
20-50
15-20
10-15
5-10
less than 5

(after Manley, 1970)

RAIN DAYS

Average number of days per year in which 1mm or more of precipitation was recorded

over 250 days
225-250 days
200-225 days
175-200 days
less than 175 days

RAINFALL EXTREMES AND IRRIGATION

Average number of years in ten when irrigation is theoretically necessary for crops (in England and Wales only)

over 9
8-9
7-8
6-7
5-6
under 5

(after Pearl and others, 1954)

Regions of reliably high rainfall (more than 250mm in at least 70% of the years)

Regions of occasionally low rainfall (less than 750mm in at least 30% of the years)

ANNUAL RAINFALL AND ISOBARS

1008 mb.
1012 mb.
1013 mb.
1016 mb.
1016 mb.

mm
2500
2000
1500
1000
750
625

January Isobars
July Isobars

VARIABILITY OF RAIN

The percentage frequency with which rainfall varies from the normal rainfall regime in an area: the higher the percentage figure, the more variable the rainfall

over 20%
18-20%
16-18%
14-16%
12-14%
10-12%
under 10%

(after Gregory, 1955)

Rainfall is least variable in the wetter northern and western areas and most variable in the drier eastern and southern areas

1:10 000 000

100   0   100   200   300   400 km

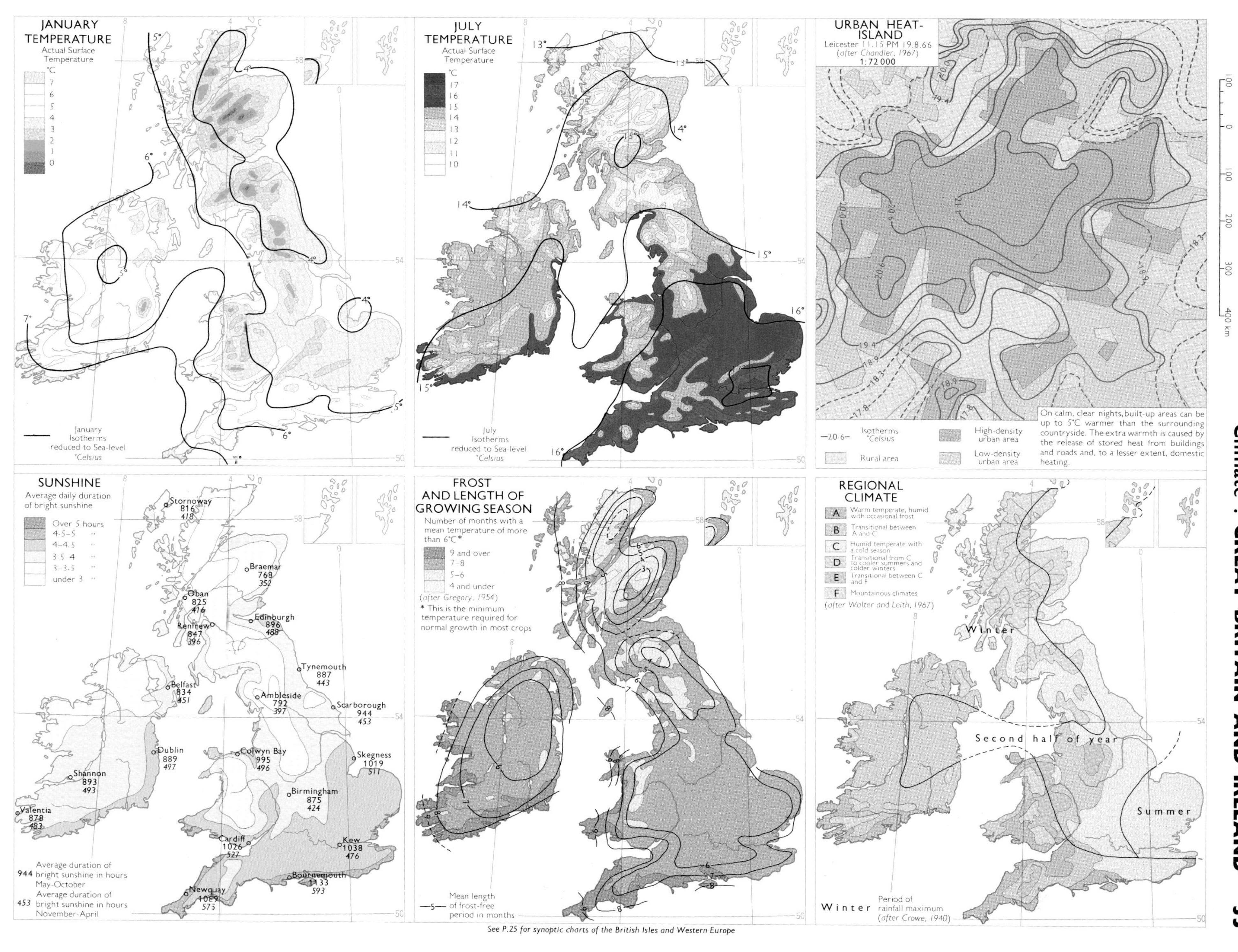

URBAN HEAT-ISLAND
Leicester 1.1.15 PM 19.8.66
(after Chandler, 1967)
1:72 000

On calm, clear nights, built-up areas can be up to 5°C warmer than the surrounding countryside. The extra warmth is caused by the release of stored heat from buildings and roads and, to a lesser extent, domestic heating.

—20.6— Isotherms Celsius

High-density urban area

Low-density urban area

Rural area

REGIONAL CLIMATE
(after Walter and Leith, 1967)

A Warm temperate, humid with occasional frost
B Transitional between A and C
C Humid temperate with a cold season
D Transitional from C to cooler summers and colder winters
E Transitional between C and F
F Mountainous climates

Period of Winter rainfall maximum
(after Crowe, 1940)

JULY TEMPERATURE
Actual Surface Temperature
°C
17
16
15
14
13
12
11
10

—— July Isotherms reduced to Sea-level °Celsius

FROST AND LENGTH OF GROWING SEASON
Number of months with a mean temperature of more than 6°C*

9 and over
7–8
5–6
4 and under

(after Gregory, 1954)
*This is the minimum temperature required for normal growth in most crops

—5— Mean length of frost-free period in months

See P. 25 for synoptic charts of the British Isles and Western Europe

JANUARY TEMPERATURE
Actual Surface Temperature
°C
7
6
5
4
3
2
1
0

—— January Isotherms reduced to Sea-level °Celsius

SUNSHINE
Average daily duration of bright sunshine

Over 5 hours
4.5–5 „
4–4.5 „
3.5–4 „
3–3.5 „
under 3 „

944 Average duration of bright sunshine in hours May–October
453 Average duration of bright sunshine in hours November–April

Stornoway 816 / 418
Braemar 768 / 352
Edinburgh 896 / 488
Tynemouth 887 / 443
Scarborough 944 / 453
Skegness 1019 / 511
Oban 825 / 416
Renfrew 844 / 396
Ambleside 792 / 397
Colwyn Bay 995 / 495
Birmingham 875 / 424
Kew 1038 / 476
Belfast 834 / 451
Bournemouth 1133 / 593
Dublin 889 / 497
Cardiff 1026 / 527
Shannon 893 / 493
Newquay 1052 / 575
Valentia 878 / 483

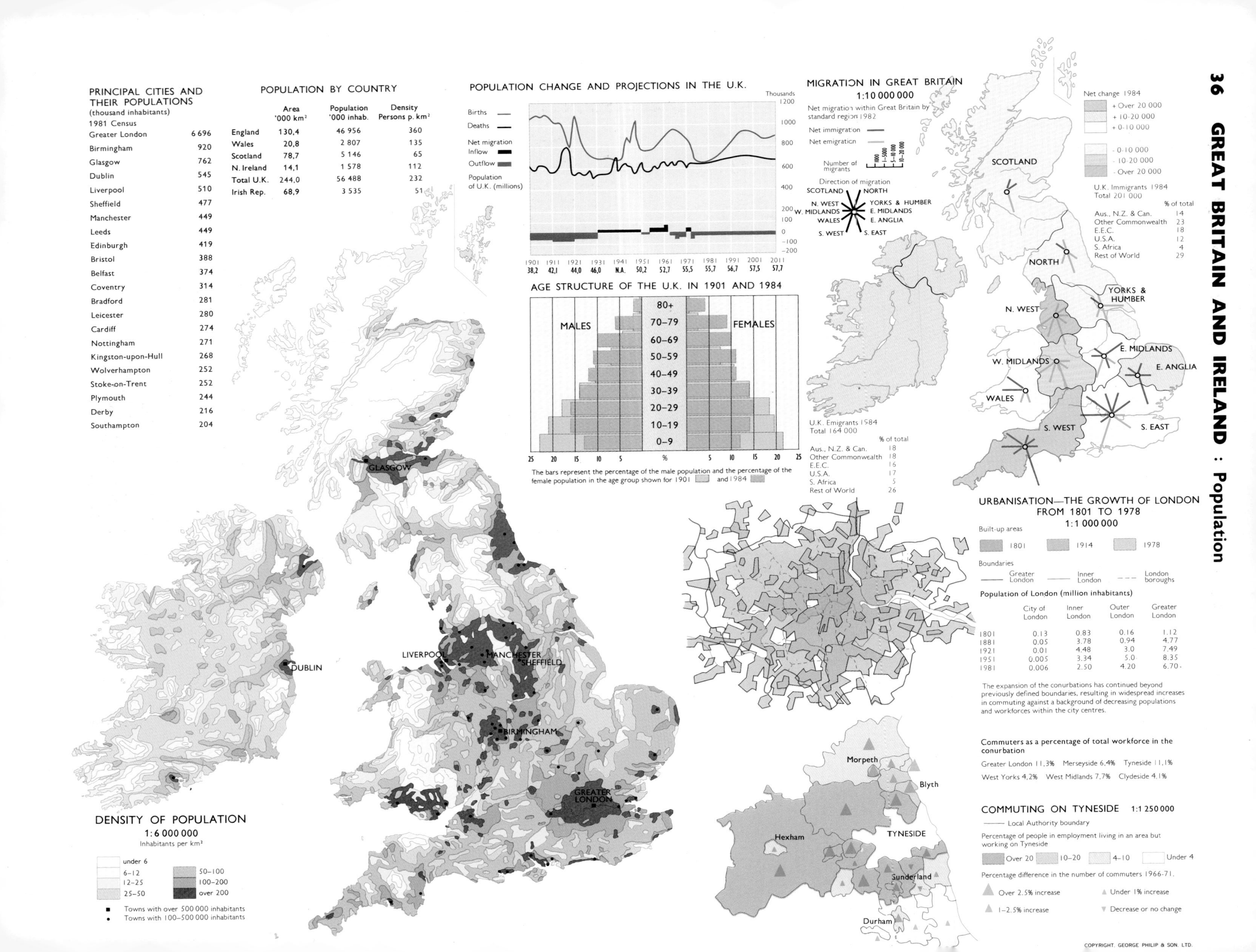

Net change 1984

+ Over 20 000	
+ 10-20 000	
+ 0-10 000	
- 0-10 000	
- 10-20 000	
- Over 20 000	

U.K. Immigrants 1984
Total 201 000

	% of total
Aus., N.Z. & Can.	14
Other Commonwealth	23
E.E.C.	18
U.S.A.	12
Rest of World	29

URBANISATION—THE GROWTH OF LONDON FROM 1801 TO 1978
1:1 000 000

Built-up areas: 1801 | 1914 | 1978

Boundaries:
Greater London
Inner London
London boroughs

Population of London (million inhabitants)

	City of London	Inner London	Outer London	Greater London
1801	0.13	0.83	0.16	1.12
1881	0.05	3.78	0.94	4.77
1921	0.01	4.48	3.0	7.49
1951	0.005	3.34	5.0	8.35
1981	0.006	2.50	4.20	6.70

The expansion of the conurbations has continued beyond previously defined boundaries, resulting in widespread increases in commuting against a background of decreasing populations and workforces within the city centres.

Commuters as a percentage of total workforce in the conurbation

Greater London 11.3%	Merseyside 6.4%	Tyneside 1.1%
West Yorks 4.2%	West Midlands 7.7%	Clydeside 4.1%

COMMUTING ON TYNESIDE
1:1 250 000

— Local Authority boundary

Percentage of people in employment living in an area but working on Tyneside

Under 4	4-10	10-20	Over 20

Percentage difference in the number of commuters 1966-71.

▲ Over 20	▼ Under 1% increase
▲ Over 2.5% increase	▼ Decrease or no change
▲ 1-2.5% increase	

MIGRATION IN GREAT BRITAIN
1:10 000 000

Net migration within Great Britain by standard region 1982
— Net immigration
— Net emigration

Number of migrants:
1000
1-5000
5-10 000
10-20 000

Direction of migration

SCOTLAND | NORTH
N. WEST | YORKS & HUMBER
W. MIDLANDS | E. MIDLANDS
WALES | E. ANGLIA
S. WEST | S. EAST

U.K. Emigrants 1984
Total 164 000

	% of total
Aus., N.Z. & Can.	18
Other Commonwealth	16
E.E.C.	15
U.S.A.	5
S. Africa	
Rest of World	26

Regions (map): SCOTLAND, NORTH, N. WEST, YORKS & HUMBER, W. MIDLANDS, E. MIDLANDS, E. ANGLIA, WALES, S. WEST, S. EAST

Tyneside map labels: Morpeth, Blyth, TYNESIDE, Sunderland, Durham, Hexham

POPULATION CHANGE AND PROJECTIONS IN THE U.K.

Thousands: 1200, 1000, 800, 600, 400, 200, 100, 0, -100, -200

Years: 1901 1911 1921 1931 1941 1951 1961 1971 1981 1991 2001 2011

Births	—
Deaths	—
Net migration	
Inflow	
Outflow	
Population of U.K. (millions)	

AGE STRUCTURE OF THE U.K. IN 1901 AND 1984

FEMALES / MALES

Age groups: 80+, 70-79, 60-69, 50-59, 40-49, 30-39, 20-29, 10-19, 0-9

Year	Population
1901	38.2
1911	42.1
1921	44.0
1931	46.0
1951	50.2
1961	52.7
1971	55.5
1981	56.7
1991	57.5
2001	57.5
2011	57.7

% scale: 25 20 15 10 5 0 5 10 15 20 25

The bars represent the percentage of the male population and the percentage of the female population in the age group shown for 1901 ▢ and 1984 ▢.

POPULATION BY COUNTRY

		Area '000 km²	Population '000 inhab.	Density Persons p. km²
England		130.4	46 956	360
Wales		20.8	2 807	135
Scotland		78.7	5 146	65
N. Ireland		14.1	1 578	112
Total U.K.		244.0	56 488	232
Irish Rep.		68.9	3 535	51

PRINCIPAL CITIES AND THEIR POPULATIONS
(thousand inhabitants)
1981 Census

Greater London	6 696
Birmingham	920
Glasgow	762
Dublin	545
Liverpool	510
Sheffield	477
Manchester	449
Leeds	449
Edinburgh	419
Bristol	388
Belfast	374
Coventry	314
Bradford	281
Leicester	280
Cardiff	274
Nottingham	271
Kingston-upon-Hull	268
Wolverhampton	252
Stoke-on-Trent	252
Plymouth	244
Derby	216
Southampton	204

DENSITY OF POPULATION
1:6 000 000

Inhabitants per km²

under 6	6-12	12-25	25-50	50-100	100-200	over 200

● Towns with over 500 000 inhabitants
● Towns with 100–500 000 inhabitants

Density map labels: GLASGOW, LIVERPOOL, MANCHESTER, SHEFFIELD, BIRMINGHAM, GREATER LONDON, DUBLIN

Population density map city labels: MANCHESTER, LIVERPOOL, SHEFFIELD, BIRMINGHAM, GREATER LONDON

ECONOMIC PLANNING AND DEVELOPMENT REGIONS 1:10 000 000

NORTH EAST
YORKSHIRE AND HUMBERSIDE
NORTH WEST
EAST MIDLANDS
WEST MIDLANDS
WALES
SOUTH EAST
SOUTH WEST
SCOTLAND
NORTHERN IRELAND

WALES Economic Planning Region and Boundary
Development Areas
Intermediate Areas
Highlands and Islands Development Board Area
Special Development Areas

Nationalised industries and public corporations, such as the Electricity, Gas, Water and Coal Boards, British Rail and the G.P.O., are subdivided into regions for the purpose of administration and planning. These regions vary from industry to industry but the Electricity Board (.......) is a typical example.

THE ADMINISTRATIVE STRUCTURE OF THE U.K.

This diagram shows the principal administrative and planning areas in the U.K. and their functions. The boxes have borders which match those on the accompanying maps. The county of Cheshire is given as an example of local government areas.

CENTRAL GOVERNMENT
Parliament and the Civil Service determine and administer national policies.

PARLIAMENTARY CONSTITUENCIES

ECONOMIC PLANNING REGIONS

DEVELOPMENT AREAS
Encourage growth in problem areas

PUBLIC SERVICES & NATIONALISED INDUSTRIES SUB-REGIONS.

SCOTLAND | WALES | ENGLAND | N. IRELAND

REGIONS & ISLAND AREAS
Fire, police, public health, education, libraries, civil defence, refuse disposal,

COUNTIES | COUNTIES
DISTRICTS | DISTRICTS

DISTRICTS
Housing, sewage disposal, cemeteries, parks, museums, recreation, local planning

PARISHES & WARDS

COMMUNITY COUNCILS

PARISHES
Street lighting, paving, footpaths and tracks, playgrounds, allotments

WARDS

DISTRICTS

PARLIAMENTARY AREAS

Parliamentary Constituency Boundary
Borough Constituency
County Constituency
(B C)
(C C)

LOCAL GOVERNMENT ADMINISTRATIVE AREAS

County Boundary
District Boundary
Parish Boundary
Urban areas with wards, not parishes

CHESHIRE 1:550 000

CHEADLE (BC)
MACCLESFIELD
CONGLETON
CREWE (CC)
KNUTSFORD
VALE ROYAL
CREWE AND NANTWICH
NANTWICH (CC)
NORTHWICH (CC)
NEWTON
WARRINGTON (BC)
RUNCORN
HALTON (CC)
WIDNES (CC)
ELLESMERE PORT
CITY OF CHESTER (CC)
CHESTER
WIRRAL (CC)
GREATER MANCHESTER
MERSEYSIDE
CHESHIRE

ADMINISTRATIVE AREAS 1:6 000 000

International Boundary
National Boundary
County, Region and N. Ireland District Boundary
oTruro Administrative headquarters

On 1st April 1986 the administrative functions of the six metropolitan counties, such as planning, education, transportation, libraries and social services, were transferred to the city and town boroughs and various non-elected residuary bodies.

The districts of Northern Ireland have been numbered and can be identified by reference to this table

1. Londonderry
2. Limavady
3. Coleraine
4. Ballymoney
5. Moyle
6. Larne
7. Ballymena
8. Magherafelt
9. Cookstown
10. Strabane
11. Omagh
12. Fermanagh
13. Dungannon
14. Craigavon
15. Armagh
16. Newry & Mourne
17. Banbridge
18. Down
19. Lisburn
20. Antrim
21. Newtownabbey
22. Carrickfergus
23. North Down
24. Ards
25. Castlereagh
26. Belfast

SHETLAND
Lerwick
ORKNEY
Kirkwall
WESTERN ISLES
Stornoway
Lewis
Harris
North Uist
Benbecula
South Uist
Barra
HIGHLAND
Inverness
GRAMPIAN
Aberdeen
Rhum
Eigg
Coll
Tiree
Mull
Colonsay
Jura
Islay
Arran
TAYSIDE
Dundee
FIFE
Glenrothes
CENTRAL
Stirling
STRATHCLYDE
Glasgow
LOTHIAN
Edinburgh
BORDERS
Newtown St. Boswells
DUMFRIES & GALLOWAY
Dumfries
SCOTLAND
Douglas
ISLE OF MAN

NORTHUMBERLAND
Newcastle*
TYNE & WEAR
Durham
DURHAM
Middlesbrough
CLEVELAND
NORTH YORKSHIRE
Northallerton
Carlisle
CUMBRIA
LANCASHIRE
Preston
Hull
HUMBERSIDE
WEST YORKS
Wakefield
SOUTH YORKS
Barnsley
Lincoln
LINCOLN
MERSEYSIDE
Liverpool
GREATER MANCHESTER
Manchester
CHESHIRE
Chester
DERBY
Matlock
NOTTS
Nottingham
Mold
CLWYD
GWYNEDD
Caernarfon
Anglesey
STAFFORD
Stafford
SALOP
Shrewsbury
LEICESTER
Leicester
NORFOLK
Norwich
SUFFOLK
Ipswich
POWYS
Llandrindod Wells
HEREFORD & WORCESTER
Worcester
WARWICK
Birmingham
WEST MIDLANDS
NORTHAMPTON
Northampton
CAMBRIDGE
Cambridge
DYFED
Carmarthen
GLOUCESTER
Gloucester
OXFORD
Oxford
BEDFORD
Bedford
HERTS
Hertford
ESSEX
Chelmsford
GWENT
Cwmbran
AVON
Bristol
WILTS
Trowbridge
BERKS
Reading
GREATER LONDON
Kingston*
SURREY
Aylesbury
BUCKS
WEST GLAM
Swansea
SOUTH GLAM
Cardiff
MID GLAM
SOMERSET
Taunton
DORSET
Dorchester
HAMPSHIRE
Winchester
WEST SUSSEX
Chichester
EAST SUSSEX
Lewes
KENT
Maidstone
ISLE OF WIGHT
Newport
DEVON
Exeter
CORNWALL
Truro
Isles of Scilly
Lundy I.
ENGLAND
WALES
SCOTLAND

* Newcastle is also Admin. H.Q. for Northumberland
Cardiff is also Admin. H.Q. for Mid Glamorgan
Kingston is Admin. H.Q. for Surrey

DONEGAL
Lifford
LONDONDERRY
NORTHERN IRELAND
Dundalk
Monaghan
MONAGHAN
Omagh
CAVAN
Cavan
LEITRIM
Carrick on Shannon
SLIGO
Sligo
MAYO
Castlebar
ROSCOMMON
Roscommon
LONGFORD
Longford
WESTMEATH
Mullingar
MEATH
An Uaimh
LOUTH
Drogheda
DUBLIN
Dublin
KILDARE
Naas
OFFALY
Tullamore
GALWAY
Galway
CLARE
Ennis
LAOIS
Port Laoise
CARLOW
Carlow
WICKLOW
Wicklow
WEXFORD
Wexford
KILKENNY
Kilkenny
TIPPERARY
Clonmel
LIMERICK
Limerick
WATERFORD
Waterford
KERRY
Tralee
CORK
Cork
IRISH REPUBLIC

The United Kingdom of Great Britain and Northern Ireland does not include the Channel Islands or the Isle of Man which are direct dependencies of the Crown with their own parliaments and taxes.

Projection: Conical with two standard parallels

1:10 000 000

100   0   100   200   300   400 km

LEADING AGRICULTURAL ENTERPRISES

Crops
Dairy
Beef
Sheep
Pigs
Horticulture
Crofting

The leading enterprises shown on this map are those enterprises which use the most man-days in each district.

POWER

Coal production
Oilfields
Gas terminals
Oil terminals
Oil refineries
Power stations
Hydro-electric
Thermal
Nuclear

Sullom Voe
Flotta
Dounreay
Nigg
St. Fergus
Cruden Bay
FIFE
Grangemouth
CENTRAL
Dalmeny
MIDLOTHIAN
AYR
Chapel Cross
Hunterston
NORTHUMBERLAND AND DURHAM
Barrow-in-Furness
Sellafield
Wylfa
Trawsfynydd
Milford Haven
SOUTH WALES
Berkeley
Oldbury
Hinkley Point
Winfrith
Easington
Theddlethorpe
NOTTS AND DERBY
YORKS
LANCS
N. STAFFS
S. STAFFS
WARWICK
LEICESTER
Seaton Carew
Teeside
Bacton
Sizewell
Bradwell
Dungeness
Dinorwig

For further information on oil and natural gas see P.31

MINERALS

Iron ore
Potash
Limestone
Gravel
Fluorspar
Salt
Tin
China clay
Lead
Zinc

Bantry Bay

LAND AREA UNDER DIFFERING AGRICULTURAL USES IN A SELECTION OF REGIONS

Percentage of total land area

Arable
Pasture
Rough grazing
Others

WESTERN ISLES 290 000 ha
NORTHUMBERLAND 504 000 ha
DERBYSHIRE 263 000 ha
MERSEYSIDE 24 000 ha
LINCOLNSHIRE 589 000 ha
POWYS 508 000 ha
DEVON 672 000 ha
NORTHERN IRELAND 1 413 000 ha
IRISH REPUBLIC 6 895 000 ha

AGRICULTURE, FORESTRY AND FISHING EMPLOYEES AS A PERCENTAGE OF TOTAL WORKING POPULATION IN THE U.K.

Total numbers of employees are given in thousands

'84	340
'80	370
'76	393
'72	427
'68	423
'64	540
'60	610
'56	661
'52	724
1948	842

4% 3% 2% 1% 0

NUMBER AND SIZE OF AGRICULTURAL HOLDINGS IN THE U.K.

under 2 ha
2–5 ha
5–20 ha
20–40 ha
40–50 ha
50–100 ha
over 100 ha

'82
'80
'78
'76
'74
'72
'70
1968

300 000 holdings
200 000 holdings
100 000 holdings

LAND USE

Arable farming
Permanent pasture
Upland pasture
Hill farming and crofting
Forestry
Market gardening
Urban areas

PRINCIPAL CROPS

Wheat
Oats
Barley
Potatoes
Fruit
Hops

AGRICULTURAL PROCESSING

Canning and freezing
Principal areas of milk manufacture
Principal livestock markets
Sugar factories
Flour milling

FISHING

Principal fishing ports
Other important fishing ports

1:16 000 000

100 0 100 200 300 400 km

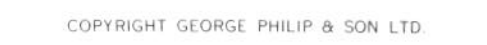

COPYRIGHT GEORGE PHILIP & SON LTD

IRON AND STEEL

Consett, Teesside, Scunthorpe, Sheffield, Rotherham, Workington, Port Talbot, Ebbw Vale, Belfast

TEXTILES
Cotton
Wool
Synthetic fibres
Silk
Linen and Jute

Kirkcaldy, Glasgow, Preston, Manchester, Stoke-on-Trent, Nottingham, Birmingham, Axminster, Belfast, Dublin

SHIPBUILDING AND MARINE ENGINEERING
○ Shipbuilding
□ Marine Engineering

Newcastle, Hull, Birkenhead, Sheffield, London, Southampton, Greenock and Port Glasgow, Belfast, Dublin

MOTOR VEHICLES
● Car Manufacture
■ Lorry Manufacture
□ Component Manufacture
△ Steel Strip Mills

Bathgate, Liverpool, Ellesmere Port, Birmingham, Luton, Oxford, Dagenham, Belfast, Dublin

TOTAL NUMBER OF EMPLOYEES IN EACH INDUSTRY 1984

Paper, Printing and Publishing
Timber and Furniture
Bricks, Pottery, Glass, Cement
Clothing and Footwear
Leather goods
Textiles
Motor Vehicles
Shipbuilding, Marine Engineering
Electrical Engineering
Instrument Engineering
Mechanical Engineering
Other metals e.g. Copper
Aluminium
Iron and Steel
Chemicals
Food, Drink and Tobacco

Number of employees 1978

1 000 000   800 000   600 000   400 000   200 000   0

ELECTRICAL ENGINEERING

Newcastle, Nottingham, Chelmsford, London, Rugby, Liverpool, Manchester, Stafford, Birmingham, Glasgow, Belfast, Dublin

ALUMINIUM
○ Aluminium Processing
□ Aluminium Smelting
△ Aluminium Production

Fort William, Kinlochleven, Lynemouth, Invergordon, St. Albans, London, Holyhead, Anglesey

Boundaries on all maps denote Standard U.K. Planning Regions. These are coloured to show the number of people employed in a region as a percentage of national employment in that industry. Figures for the Irish Republic are not available.

0–5%
5–10%
10–20%
over 20%

Primary (Agriculture, Mining and Fishing)
Secondary (Manufacturing)
Tertiary (Services and Government)

EMPLOYMENT BY REGION

SCOTLAND, NORTH, YORKSHIRE AND HUMBERSIDE, EAST MIDLANDS, EAST ANGLIA, SOUTH EAST, WEST MIDLANDS, WALES, SOUTH WEST, NORTH WEST, NORTHERN IRELAND, IRISH REPUBLIC

7 000 000 people
5 000 000
2 500 000
1 000 000
500 000

MECHANICAL ENGINEERING

Dundee, Newcastle, Teesside, Manchester, Coventry, Birmingham, Dublin, Glasgow, Londonderry, Belfast

CHEMICALS

Teesside, Hull, Grangemouth, Ardrossan, Manchester, Nottingham, Birmingham, Liverpool, London, Belfast

PAPER, PRINTING AND PUBLISHING
■ Paper manufacture
○ Printing and Publishing

Newcastle, Edinburgh, Leeds, Manchester, Nottingham, Norwich, London, Stoke-on-Trent, Birmingham, Oxford, Bristol, Cardiff, Glasgow, Belfast

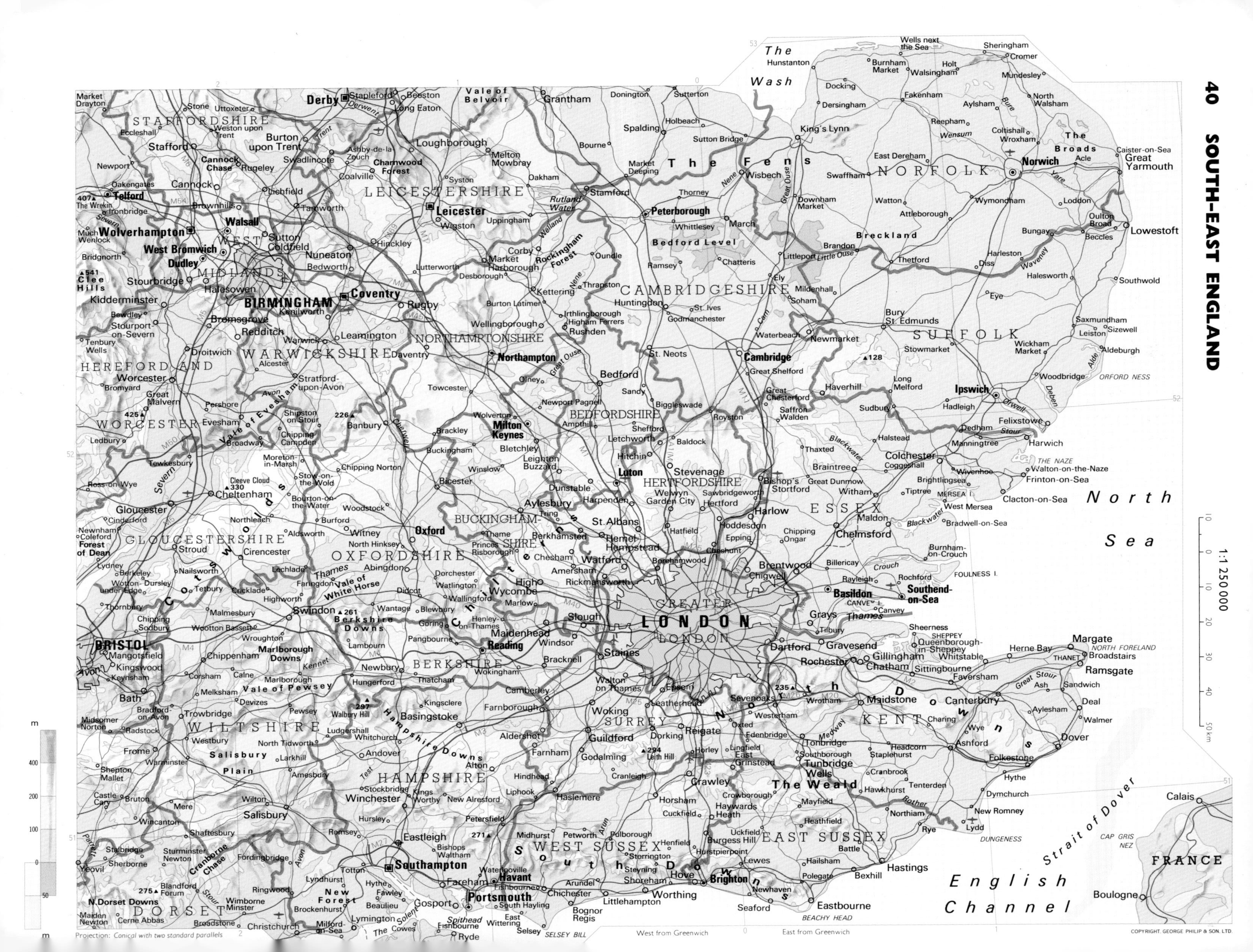
1:1 250 000
10 0 10 20 30 40 50 km
North Sea
The Wash
English Channel
Strait of Dover
FRANCE
Calais
Boulogne
CAP GRIS NEZ
NORTH FORELAND
DUNGENESS
BEACHY HEAD
SELSEY BILL
ORFORD NESS
THE NAZE
FOULNESS I.
MERSEA I.
SHEPPEY
THANET
CANVEY I.
NORFOLK
SUFFOLK
ESSEX
KENT
SURREY
EAST SUSSEX
WEST SUSSEX
HAMPSHIRE
BERKSHIRE
WILTSHIRE
DORSET
OXFORDSHIRE
GLOUCESTERSHIRE
BUCKINGHAMSHIRE
BEDFORDSHIRE
HERTFORDSHIRE
CAMBRIDGESHIRE
NORTHAMPTONSHIRE
LEICESTERSHIRE
WARWICKSHIRE
STAFFORDSHIRE
WEST MIDLANDS
HEREFORD AND WORCESTER
GREATER LONDON
LONDON
BIRMINGHAM
BRISTOL
The Broads
The Fens
Breckland
Bedford Level
North Downs
South Downs
The Weald
Chiltern Hills
Cotswolds
Salisbury Plain
Vale of Pewsey
Vale of White Horse
New Forest
Cranborne Chase
Marlborough Downs
Berkshire Downs
Hampshire Downs
Vale of Evesham
Vale of Belvoir
Rockingham Forest
Charnwood Forest
Cannock Chase
Forest of Dean
Clee Hills
N. Dorset Downs
Norwich
Ipswich
Cambridge
Peterborough
Northampton
Leicester
Coventry
Oxford
Reading
Southampton
Portsmouth
Brighton
Luton
Milton Keynes
Southend-on-Sea
Basildon
Wolverhampton
Walsall
Dudley
West Bromwich
Derby
Hastings
Eastbourne
Dover
Folkestone
Canterbury
Margate
Ramsgate
Colchester
Chelmsford
Great Yarmouth
Lowestoft
King's Lynn
Bedford
Swindon
Gloucester
Cheltenham
Salisbury
Winchester
Guildford
Maidstone
Bath
East from Greenwich
West from Greenwich
Projection: Conical with two standard parallels
COPYRIGHT, GEORGE PHILIP & SON, LTD.
m
400
200
100
0
50
m

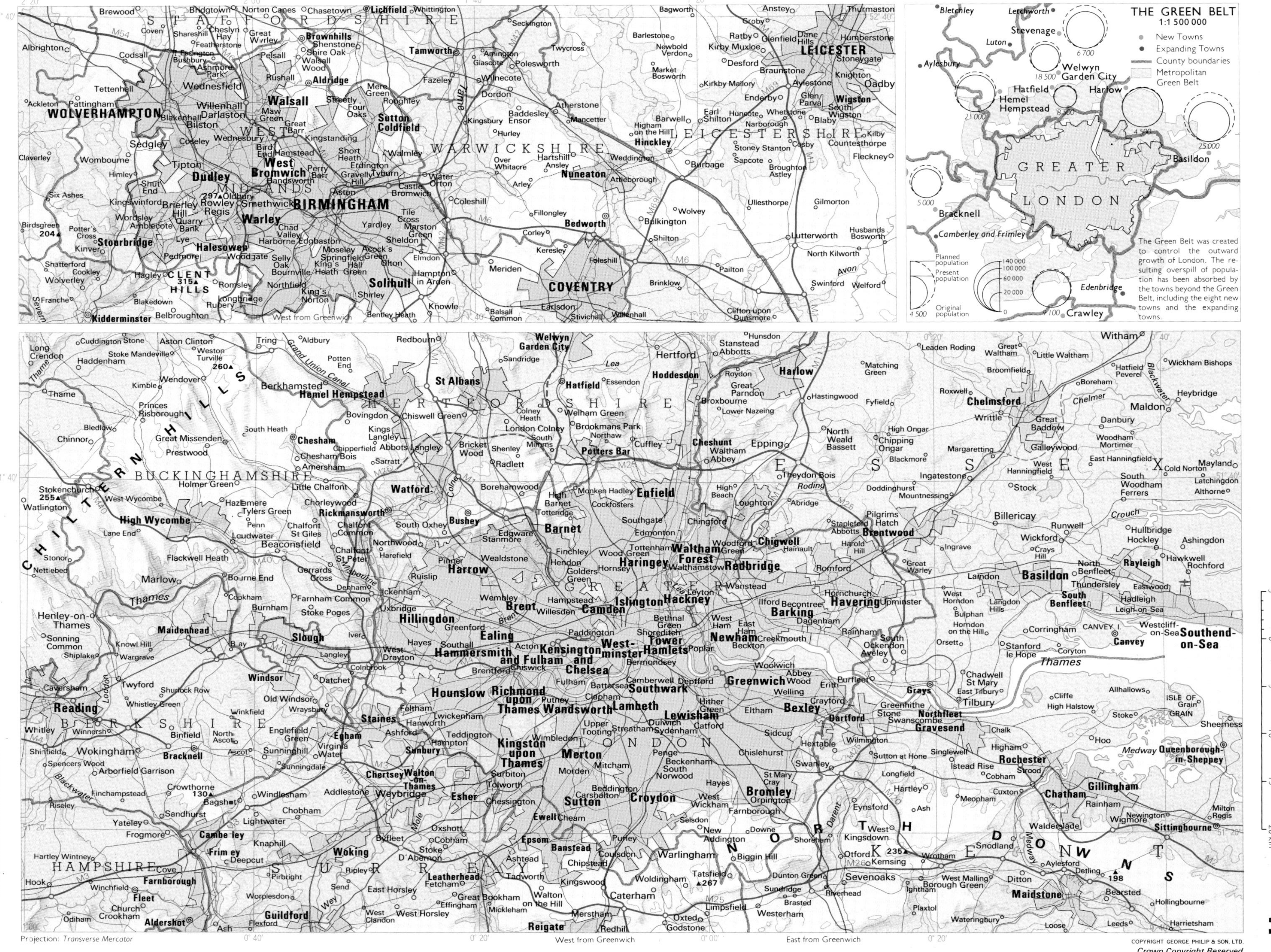

THE GREEN BELT
1:1 500 000

New Towns
Expanding Towns
County boundaries
Metropolitan
Green Belt

The Green Belt was created to control the outward growth of London. The resulting overspill of population has been absorbed by the towns beyond the Green Belt, including the eight new towns and the expanding towns.

GREATER LONDON

Leechworth
Stevenage
Welwyn Garden City
Hatfield
Hemel Hempstead
Luton
Harlow
Basildon
Bletchley
Aylesbury
Bracknell
Camberley and Frimley
Edenbridge
Crawley

Planned population
Present population
Original population

Projection: Transverse Mercator

West from Greenwich   East from Greenwich

1:1 250 000

10 0 10 20 30 40 50 km

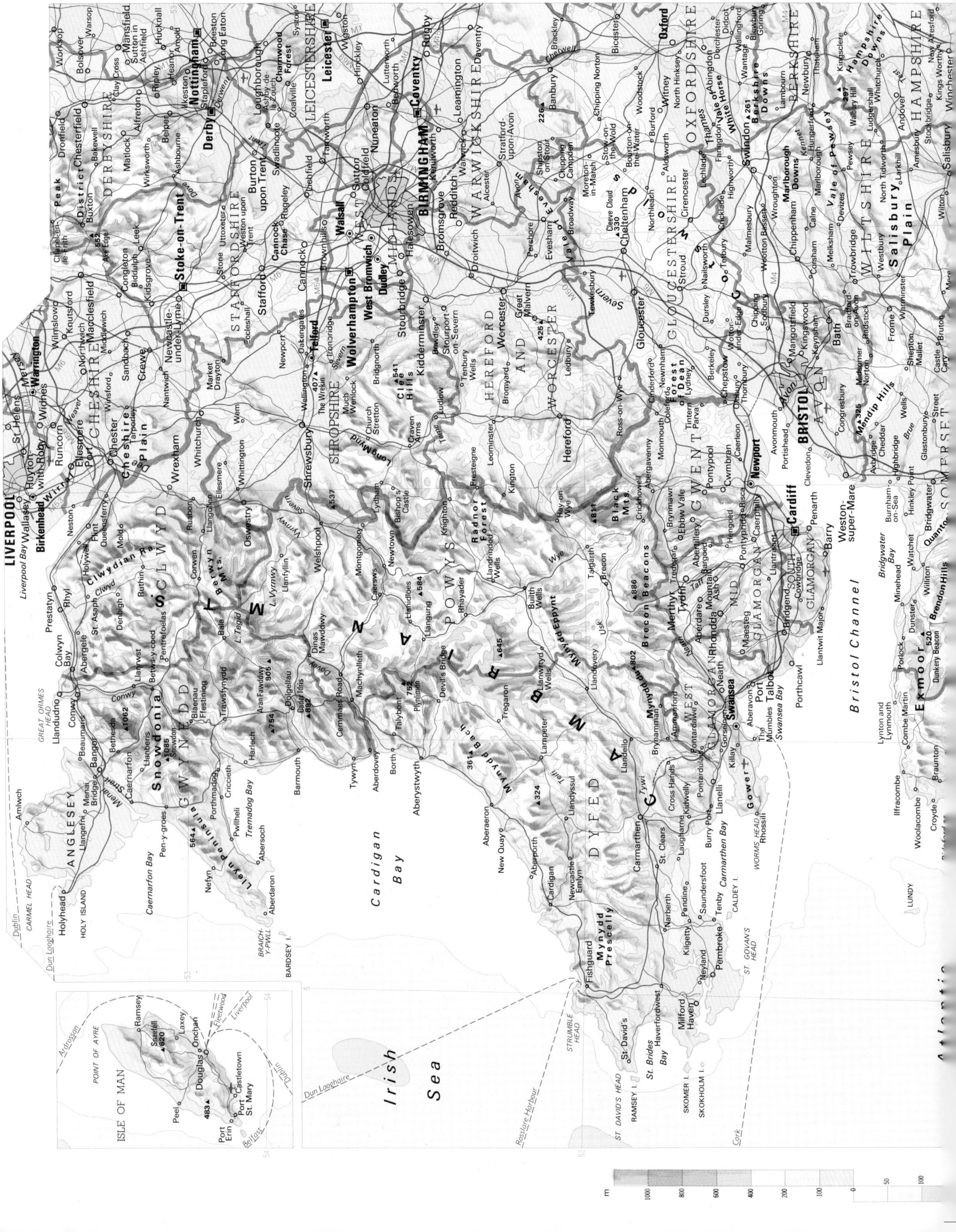

Counties and regions

DERBYSHIRE CHESHIRE STAFFORDSHIRE SHROPSHIRE LEICESTERSHIRE WARWICKSHIRE WEST MIDLANDS HEREFORD AND WORCESTER GLOUCESTERSHIRE OXFORDSHIRE BERKSHIRE WILTSHIRE HAMPSHIRE AVON SOMERSET CLWYD GWYNEDD POWYS DYFED GWENT MID GLAMORGAN SOUTH GLAMORGAN WEST GLAMORGAN ANGLESEY

Major towns and cities

LIVERPOOL BIRMINGHAM Coventry Leicester Nottingham Derby Stoke-on-Trent Wolverhampton Walsall West Bromwich Dudley Telford Shrewsbury Oxford Swindon BRISTOL Bath Cardiff Newport Swansea Hereford Worcester Gloucester Cheltenham Aberystwyth Carmarthen Wrexham Warrington Birkenhead Wallasey

Physical features

Snowdonia Cheshire Plain Cambrian Mtns Radnor Forest Brecon Beacons Black Mts Clee Hills Great Malvern Cotswolds Mendip Hills Exmoor Salisbury Plain Marlborough Downs Hampshire Downs Berkshire Downs Vale of Pewsey Clwydian Range Llyn Peninsula Lleyn Peninsula Peak District Charnwood Forest Forest of Dean Cannock Chase

Cardigan Bay Bristol Channel Swansea Bay Carmarthen Bay Tremadog Bay Caernarfon Bay Liverpool Bay Menai Strait Irish Sea

Snowdon 1085 Cadair Idris 892 Plynlimon 752

Isle of Man (inset)

ISLE OF MAN Ramsey Douglas Peel Port Erin Castletown Laxey Onchan Snaefell 620 POINT OF AYRE

Holyhead HOLY ISLAND BARDSEY I. Ardrossan Fleetwood Liverpool Dublin Dun Laoghaire Rosslare Harbour Cork Belfast

ST. DAVID'S HEAD STRUMBLE HEAD ST. GOVAN'S HEAD WORM'S HEAD GREAT ORMES HEAD CARMEL HEAD BRAICH-Y-PWLL LUNDY CALDEY I. RAMSEY I. SKOMER I. SKOKHOLM I.

Atlantic

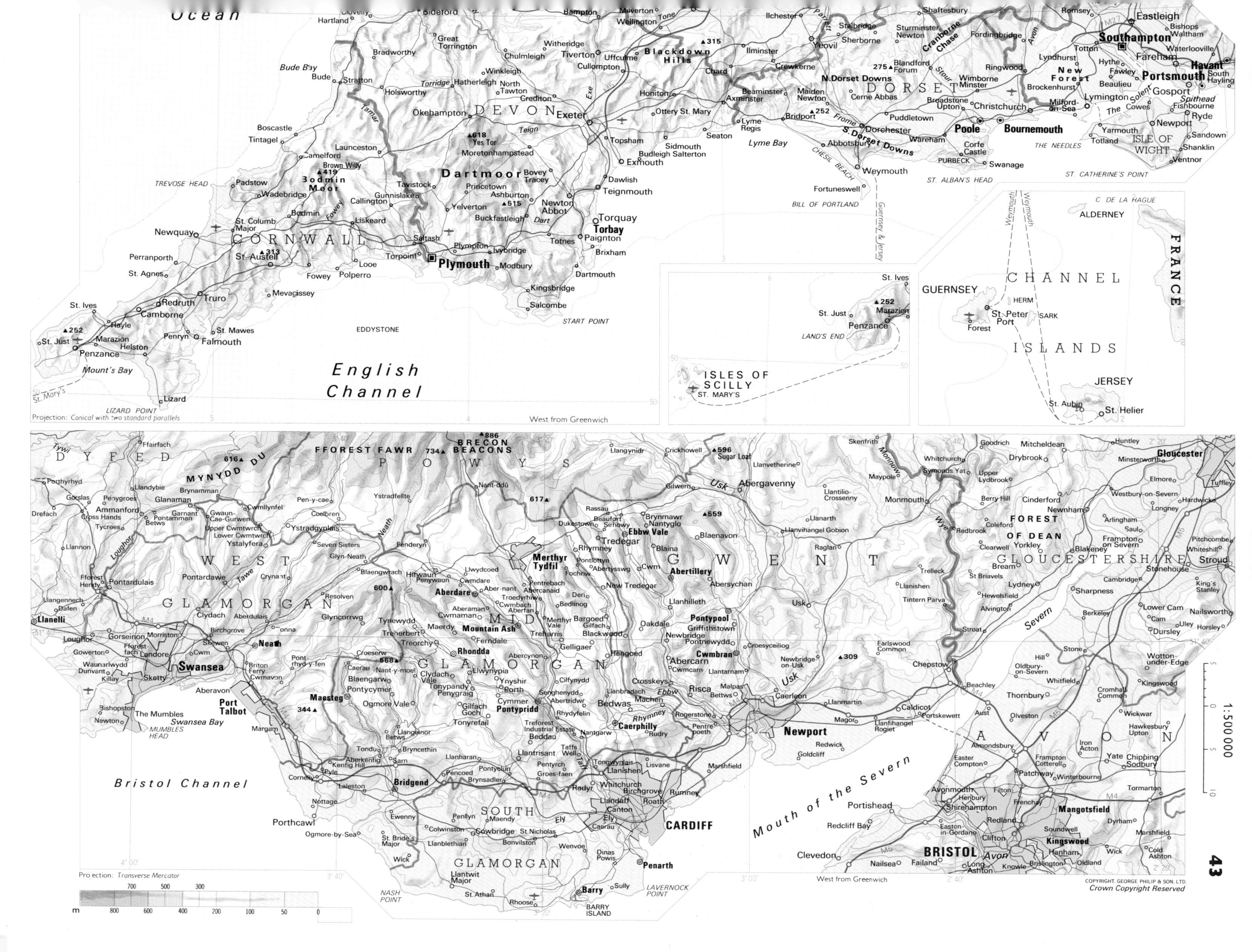
Ocean
English Channel
Bristol Channel
Mouth of the Severn
CORNWALL
DEVON
DORSET
Dartmoor
Bodmin Moor
Blackdown Hills
N.Dorset Downs
S.Dorset Downs
New Forest
ISLE OF WIGHT
Southampton
Portsmouth
Havant
Eastleigh
Fareham
Gosport
Bournemouth
Poole
Exeter
Plymouth
Torbay
Torquay
Paignton
Weymouth
Dorchester
Penzance
Truro
Falmouth
Newquay
EDDYSTONE
START POINT
LIZARD POINT
TREVOSE HEAD
ST. ALBAN'S HEAD
ST. CATHERINE'S POINT
THE NEEDLES
BILL OF PORTLAND
Projection: Conical with two standard parallels
West from Greenwich
ISLES OF SCILLY
ST. MARY'S
LAND'S END
CHANNEL ISLANDS
FRANCE
C DE LA HAGUE
ALDERNEY
GUERNSEY
HERM
SARK
St. Peter Port
JERSEY
St. Aubin
St. Helier
DYFED
POWYS
GWENT
WEST GLAMORGAN
MID GLAMORGAN
SOUTH GLAMORGAN
GLOUCESTERSHIRE
AVON
FFOREST FAWR
BRECON BEACONS
MYNYDD DU
FOREST OF DEAN
Gloucester
BRISTOL
Newport
CARDIFF
Swansea
Llanelli
Port Talbot
Neath
Bridgend
Merthyr Tydfil
Ebbw Vale
Abertillery
Pontypool
Cwmbran
Caerphilly
Pontypridd
Rhondda
Aberdare
Mountain Ash
Maesteg
Penarth
Barry
Mangotsfield
Kingswood
Abergavenny
Chepstow
Monmouth
Stroud
NASH POINT
LAVERNOCK POINT
BARRY ISLAND
MUMBLES HEAD
Projection: Transverse Mercator
1:500 000
COPYRIGHT GEORGE PHILIP & SON, LTD
Crown Copyright Reserved

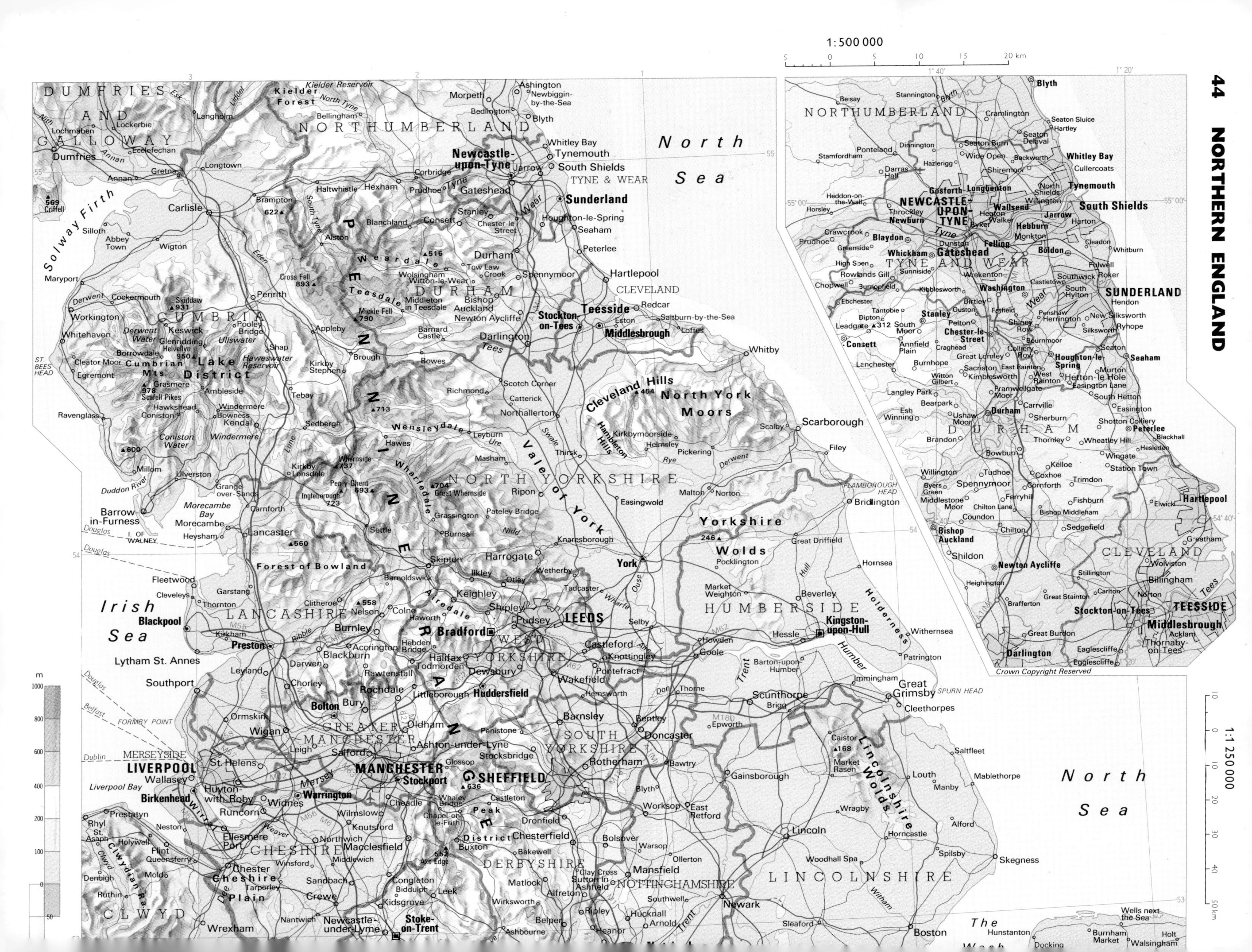

1:1 250 000

10   0   10   20   30   40   50 km

1:500 000

Inset map (upper left)

North Sea

NORTHUMBERLAND

Blyth
Whitley Bay
Tynemouth
South Shields
Cullercoats
Seaton Sluice
Seaton Delaval
Hartley
Cramlington

SUNDERLAND
Sunderland
Seaham
Hendon
Ryhope
Roker
Silksworth

NEWCASTLE-UPON-TYNE
Gateshead
Whickham
Blaydon
Newburn
Throckley
Gosforth
Longbenton
Wallsend
Walker
Byker
Jarrow
Hebburn
Boldon
Felling
Dunston
Washington
Wrekenton
Birtley

TYNE AND WEAR

Consett
Stanley
Chester-le-Street
Leadgate 312

DURHAM
Durham
Bishop Auckland
Newton Aycliffe
Shildon
Spennymoor
Brandon
Langley Park
Bearpark
Coundon
Chilton
Ferryhill

CLEVELAND
Hartlepool
Peterlee
TEESSIDE
Middlesbrough
Stockton-on-Tees
Thornaby-on-Tees
Billingham

Main map

North Sea

DUMFRIES AND GALLOWAY

Dumfries
Annan
Lockerbie
Langholm

Solway Firth

ST. BEES HEAD

CUMBRIA

Carlisle
Wigton
Maryport
Workington
Whitehaven
Egremont
Cleator Moor
Keswick
Skiddaw 931
Scafell Pikes 978
Helvellyn 950
Lake District
Ullswater
Windermere
Kendal
Penrith

PENNINE

NORTHUMBERLAND
Kielder Forest
Kielder Reservoir
Morpeth
Ashington
Blyth
Newcastle-upon-Tyne
Gateshead
Hexham
Consett

TYNE & WEAR
Sunderland
South Shields
Tynemouth
Whitley Bay
Jarrow

DURHAM
Durham
Stanley
Bishop Auckland
Darlington
Barnard Castle

CLEVELAND
Hartlepool
Peterlee
Teesside
Stockton-on-Tees
Middlesbrough
Redcar

North York Moors
Cleveland Hills
Hambleton Hills
Whitby
Scarborough
Filey

NORTH YORKSHIRE
Northallerton
Thirsk
Richmond
Catterick
Ripon
Harrogate
York
Selby
Vale of York
Wensleydale
Wharfedale

Irish Sea

Barrow-in-Furness
I. OF WALNEY
Morecambe Bay
Morecambe
Lancaster
Fleetwood
Cleveleys
Blackpool
Lytham St. Annes
Southport
FORMBY POINT

LANCASHIRE
Preston
Blackburn
Burnley
Nelson
Colne
Accrington
Chorley
Leyland

Forest of Bowland

WEST YORKSHIRE
Leeds
Bradford
Halifax
Huddersfield
Dewsbury
Wakefield
Keighley
Skipton

SOUTH YORKSHIRE
Sheffield
Rotherham
Barnsley
Doncaster

HUMBERSIDE
Kingston-upon-Hull
Beverley
Goole
Scunthorpe
Grimsby
Cleethorpes
Bridlington
Holderness
Withernsea
Yorkshire Wolds

LINCOLNSHIRE
Lincoln
Gainsborough
Market Rasen
Louth
Mablethorpe
Skegness
Boston
Sleaford
Newark
Lincolnshire Wolds

NOTTINGHAMSHIRE
Worksop
Mansfield
East Retford

GREATER MANCHESTER
Manchester
Salford
Bolton
Bury
Rochdale
Oldham
Stockport
Wigan
Leigh
Ashton-under-Lyne

MERSEYSIDE
Liverpool
Birkenhead
Wallasey
Bootle
St. Helens
Southport

CHESHIRE
Chester
Crewe
Macclesfield
Northwich
Winsford
Warrington
Wirral
Ellesmere Port
Cheshire Plain

DERBYSHIRE
Chesterfield
Buxton
Matlock
Glossop
Peak District

Stoke-on-Trent
Newcastle-under-Lyme

CLWYD
Wrexham
Rhyl
Flint
Mold
Clwydian Ra.

The Wash
Wells next the Sea
Hunstanton
Burnham Market

Crown Copyright Reserved

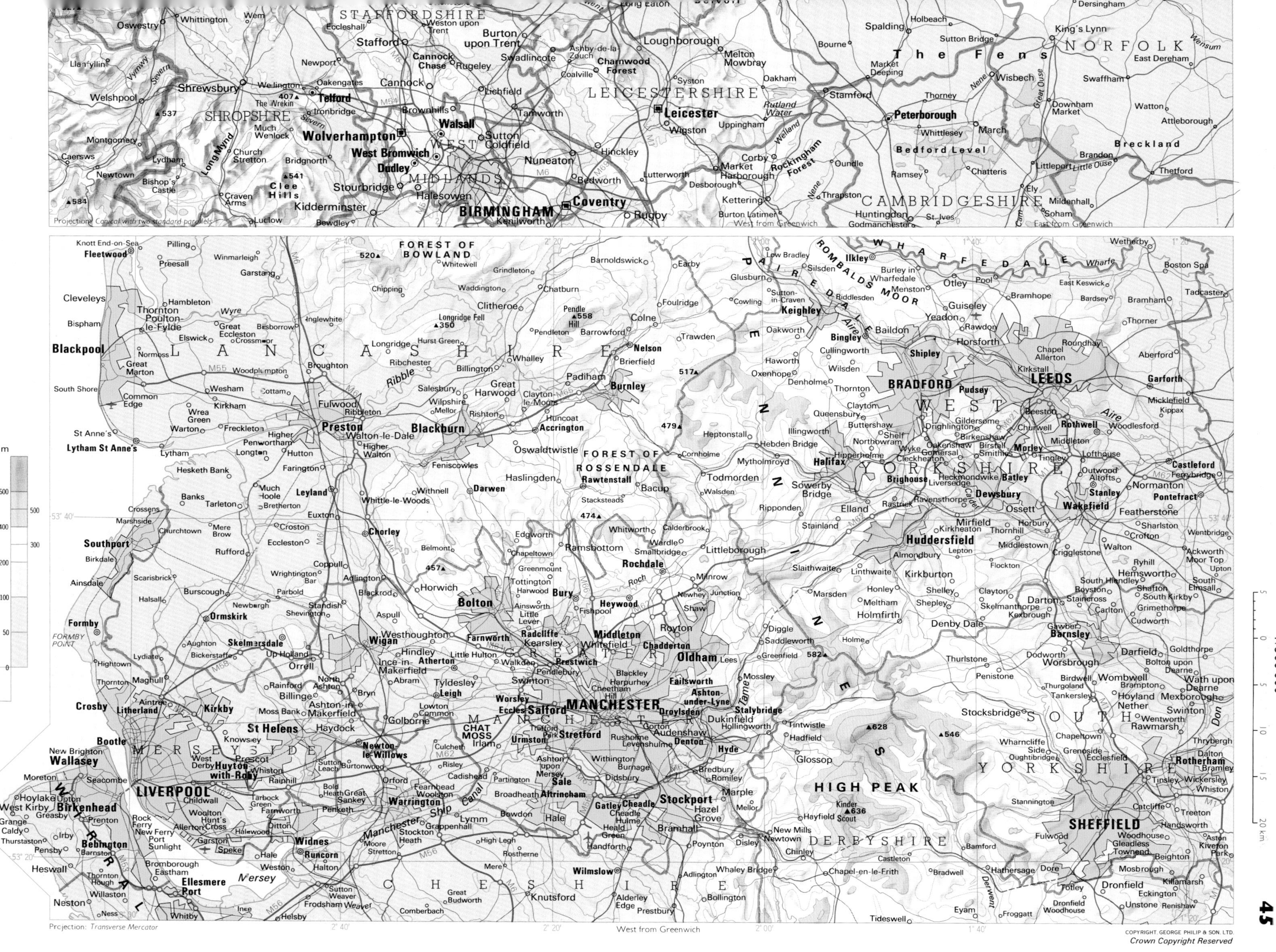

1:500 000

5   0   5   10   15   20 km

NORFOLK

The Fens

CAMBRIDGESHIRE

WHARFEDALE

AIREDALE

AIRE

LEEDS

BRADFORD

WEST YORKSHIRE

SOUTH YORKSHIRE

SHEFFIELD

Wakefield

Huddersfield

HIGH PEAK

DERBYSHIRE

PENNINE

Peterborough

Stamford

LEICESTERSHIRE

Leicester

Charnwood Forest

Loughborough

Coventry

BIRMINGHAM

WEST MIDLANDS

Walsall

West Bromwich

Dudley

Wolverhampton

STAFFORDSHIRE

SHROPSHIRE

Shrewsbury

Cannock Chase

Kidderminster

Clee Hills

FOREST OF BOWLAND

FOREST OF ROSSENDALE

Ribble

LANCASHIRE

Preston

Blackburn

Blackpool

Fleetwood

Southport

LIVERPOOL

MERSEYSIDE

Bootle

WIRRAL

Birkenhead

Wallasey

Ellesmere Port

CHESHIRE

Warrington

Widnes

St Helens

Kirkby

Wigan

Bolton

Bury

MANCHESTER

GREATER MANCHESTER

Salford

Oldham

Stockport

Rochdale

Stalybridge

Sale

Altrincham

Burnley

Nelson

Colne

Accrington

Darwen

Mersey

West from Greenwich

Projection: Transverse Mercator

m 600 500 400 300 200 100 50 0

1:1 250 000

10   0   10   20   30   40   50 km

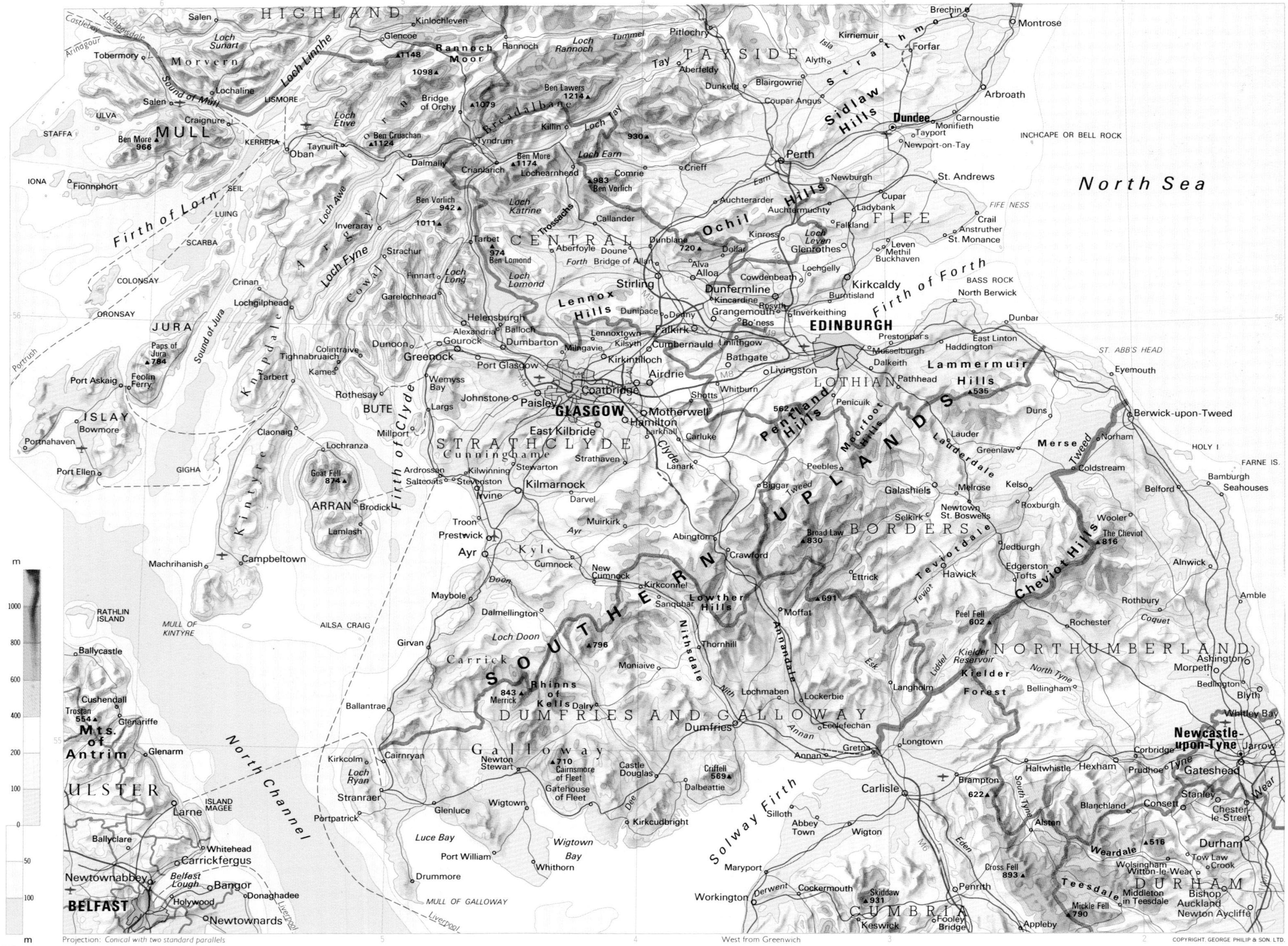

North Sea

INCHCAPE OR BELL ROCK

ST. ABB'S HEAD

FARNE IS.

HOLY I.

BASS ROCK

FIFE NESS

Firth of Forth

COPYRIGHT GEORGE PHILIP & SON LTD.

West from Greenwich

Projection: Conical with two standard parallels

Montrose
Brechin
Arbroath
Carnoustie
Monifieth
Dundee
Forfar
Kirriemuir
Newport-on-Tay
Tayport
Blairgowrie
Coupar Angus
Alyth
Isla
Pitlochry
Tummel
Kinloch Rannoch
Loch Rannoch
Rannoch Moor
△1148
1098△
Glencoe
Kinlochleven
Bridge of Orchy
△1124 Ben Cruachan
Loch Etive
Taynuilt
Oban
Connel
Dalmally
Loch Awe
△1079
Crianlarich
Tyndrum
△974 Ben Lomond
Killin
Ben More △1174
△1214 Ben Lawers
Loch Tay
Loch Earn
Comrie
△882 Ben Vorlich
Lochearnhead
Callander
Loch Katrine
Loch Lubnaig
Aberfoyle
Doune
Bridge of Allan
Dunblane
△720
Alva
Alloa
Dollar
Auchterarder
Crieff
△930
Dunkeld
Aberfeldy
Perth
Scone
Abernethy
Auchtermuchty
Newburgh
Cupar
Ladybank
Falkland
Glenrothes
Loch Leven
Kinross
Lochgelly
Cowdenbeath
Kinross
Leslie
St. Andrews
Anstruther
Crail
St. Monance
Leven
Methil
Buchhaven
Kirkcaldy
Burntisland
Inverkeithing
Dunfermline
Rosyth
Kincardine
Grangemouth
Bo'ness
Falkirk
Denny
Dumbace
Larbert
Stirling
Gargunnock
Kippen
Loch Lomond
Balloch
Dumbarton
Helensburgh
Garelochhead
Loch Long
Ben Vorlich △942
△1011
Arrochar
Loch Fyne
Strachur
Inveraray
Lochgoilhead
Cairndow
Colintraive
Tighnabruaich
Kames
Dunoon
Gourock
Greenock
Port Glasgow
Paisley
Johnstone
Kilmacolm
Bridge of Weir
Barrhead
Renfrew
Clydebank
GLASGOW
Milngavie
Kirkintilloch
Lennoxtown
Kilsyth
Cumbernauld
Coatbridge
Airdrie
Motherwell
Hamilton
Wishaw
Shotts
Whitburn
Bathgate
Armadale
Bo'ness
Linlithgow
Livingston
Broxburn
Whitburn
EDINBURGH
Leith
Musselburgh
Prestonpans
Dalkeith
Penicuik
Pathhead
Haddington
East Linton
Dunbar
North Berwick
Gullane

Firth of Clyde

Wemyss Bay
Largs
Millport
BUTE
Rothesay
ARRAN
Brodick
Lamlash
Goat Fell △874
Lochranza
Blackwaterfoot
KINTYRE
Campbeltown
Machrihanish
Claonaig
Tarbert
Knapdale
Cowal
Crinan
Lochgilphead
Ardrishaig
Gigha
Sound of Jura
Paps of Jura △784
JURA
Feolin Ferry
SCARBA
COLONSAY
ORONSAY
ISLAY
Bowmore
Port Askaig
Port Ellen
Portnahaven

North Channel

RATHLIN ISLAND
MULL OF KINTYRE
AILSA CRAIG

Ardrossan
Saltcoats
Stevenston
Kilwinning
Irvine
Troon
Prestwick
Ayr
Maybole
Girvan
Ballantrae
Dalmellington
New Cumnock
Cumnock
Mauchline
Catrine
Kilmarnock
Stewarton
Darvel
Newmilns
Galston
STRATHCLYDE
CUNNINGHAME
KYLE
CARRICK
Doon
Loch Doon
Dee
Loch Ken
Merrick △843
RHINNS OF KELLS
△710
△796
CARSPHAIRN
New Galloway
St. John's Town of Dalry
Monaiave
Moniaive
DUMFRIES AND GALLOWAY
GALLOWAY
Newton Stewart
Gatehouse of Fleet
Creetown
Castle Douglas
Kirkcudbright
Dalbeattie
Dumfries
Lochmaben
Lockerbie
Annan
Gretna
Ecclefechan
Kirtlebridge
Annandale
Nithsdale
Nithhill
Thornhill
Sanquhar
Kirkconnel
Abington
Crawford
LOWTHER HILLS
△691
Broad Law △830
Moffat
Crawfordjohn
Kirkcolm
Loch Ryan
Stranraer
Cairnryan
Portpatrick
Drummore
Port William
Whithorn
Wigtown
Glenluce
Luce Bay
Wigtown Bay
MULL OF GALLOWAY
RHINNS OF GALLOWAY

SOUTHERN UPLANDS

Biggar
Lanark
Carluke
Carstairs
Strathaven
East Kilbride
Clyde
Peebles
Broughton
Galashiels
Selkirk
Melrose
St. Boswells
Newtown
Ettrick
Lauderdale
Lauder
Greenlaw
Duns
Coldstream
Kelso
Jedburgh
Hawick
Roxburgh
Edgerston
LAMMERMUIR HILLS △535
MOORFOOT HILLS
PENTLAND HILLS △562
LOTHIAN
BORDERS
Merse
Tweed
Teviot
Ettrick
Liddel
Esk
Langholm
Newcastleton
Peel Fell △602
Kielder Forest
Kielder Reservoir
Bellingham
CHEVIOT HILLS
The Cheviot △816
Wooler
Berwick-upon-Tweed
Eyemouth
Norham
Belford
Bamburgh
Seahouses
Amble
Alnwick
Rothbury
Coquet
Morpeth
Ashington
Blyth
Whitley Bay
Newcastle-upon-Tyne
Gateshead
Jarrow
Wallsend
Stanley
Chester-le-Street
Consett
Blanchland
Hexham
Corbridge
Prudhoe
Haltwhistle
North Tyne
South Tyne
Brampton
△622
NORTHUMBERLAND
Rochester
Otterburn
Bedlington
Wear
Weardale
△516
Durham
Bishop Auckland
Newton Aycliffe
West Auckland
Witton-le-Wear
Willington
Crook
Low Low
TEESDALE
Middleton in Teesdale
Mickle Fell △790
DURHAM
Cross Fell △893
Alston
Appleby
Brough
Eden
CUMBRIA
Carlisle
Longtown
Wigton
Silloth
Abbey Town
Maryport
Cockermouth
Workington
Keswick
Skiddaw △931
Penrith
Derwent
Solway Firth

Sidlaw Hills
Lomond Hills
Ochil Hills
CENTRAL
HIGHLAND
TAYSIDE
FIFE
Lennox Hills
Campsie Fells
Firth of Tay
Firth of Forth

Liverpool (dashed line)

Larne
ISLAND MAGEE
Whitehead
Carrickfergus
Bangor
Donaghadee
Holywood
Newtownabbey
BELFAST
Belfast Lough
Newtownards
ULSTER
Glenarm
Glenariffe
Cushendall
Cushendun
Ballycastle
Ballyclare
Trostan △554
MTS. OF ANTRIM

m 1000 800 600 400 200 100 0 50 100

MULL
Tobermory
Salen
Ben More △966
Craignure
Lochaline
Loch Sunart
Morvern
Sound of Mull
Firth of Lorne
Lismore
KERRERA
SEIL
LUING
LORNE
IONA
STAFFA
ULVA
FIONNPHORT
Loch Linnhe
Loch Etive

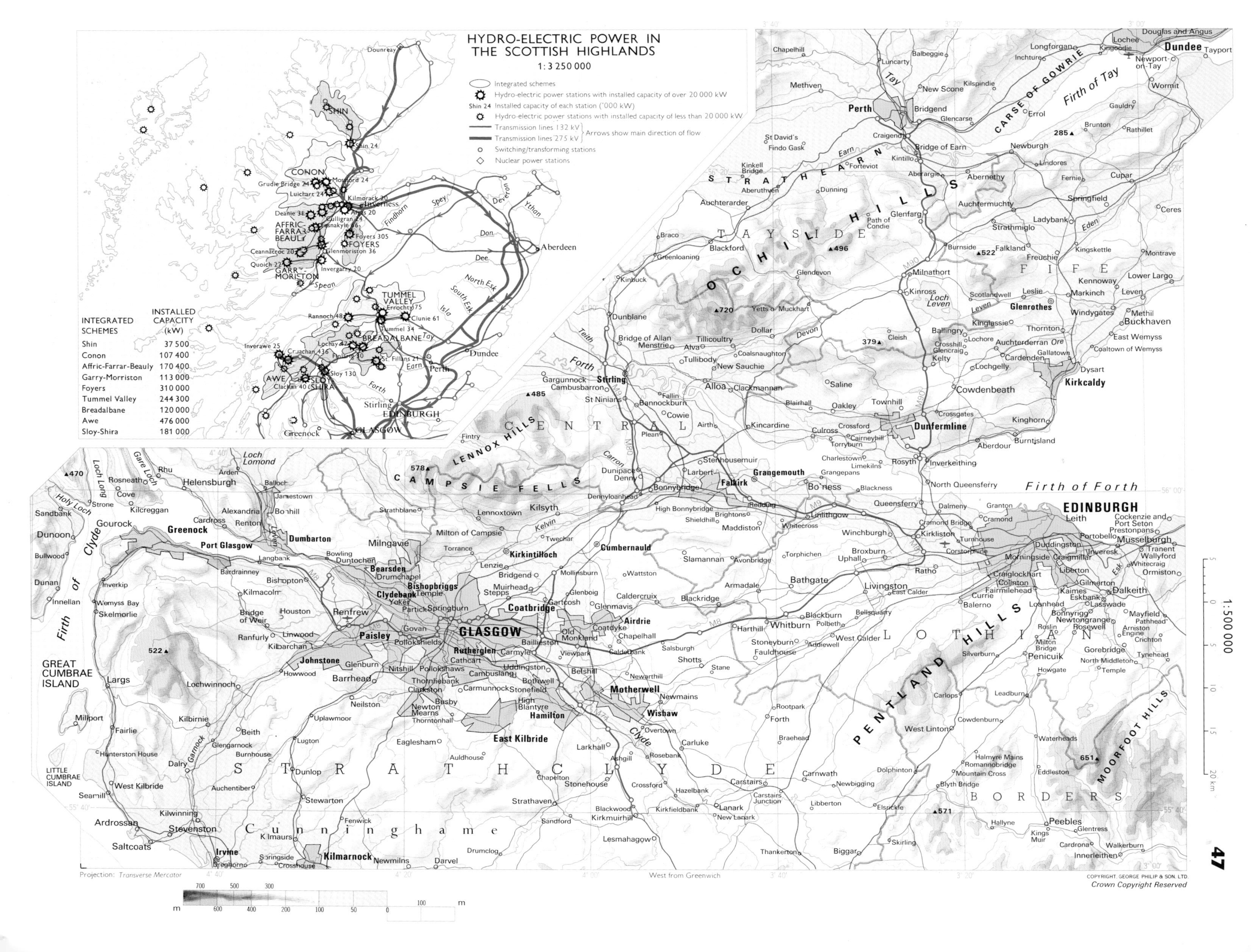

HYDRO-ELECTRIC POWER IN
THE SCOTTISH HIGHLANDS
1:3 250 000

Integrated schemes
Shin 24 Installed capacity of each station ('000 kW)
Hydro-electric power stations with installed capacity of over 20 000 kW
Hydro-electric power stations with installed capacity of less than 20 000 kW
Arrows show main direction of flow
Transmission lines 132 kV
Transmission lines 275 kV
Switching/transforming stations
Nuclear power stations

INTEGRATED SCHEMES	INSTALLED CAPACITY (kW)
Shin	37 500
Conon	107 400
Affric-Farrar-Beauly	170 400
Garry-Morriston	113 000
Foyers	310 000
Tummel Valley	244 300
Breadalbane	120 000
Awe	476 000
Sloy-Shira	181 000

1:500 000

47

COPYRIGHT GEORGE PHILIP & SON LTD.
Crown Copyright Reserved

Projection: Transverse Mercator
West from Greenwich

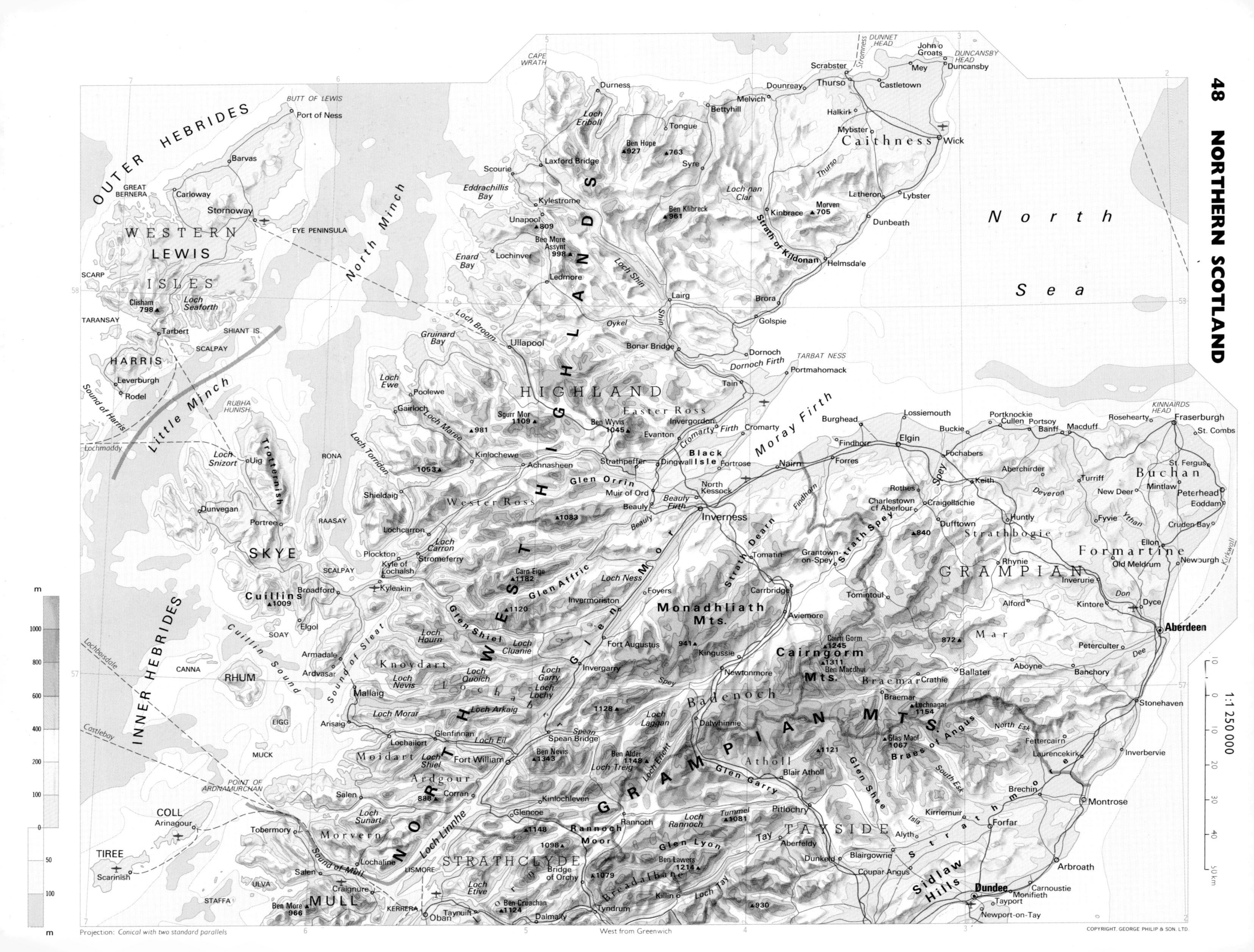

1:1 250 000

Projection: Conical with two standard parallels

West from Greenwich

COPYRIGHT GEORGE PHILIP & SON LTD.

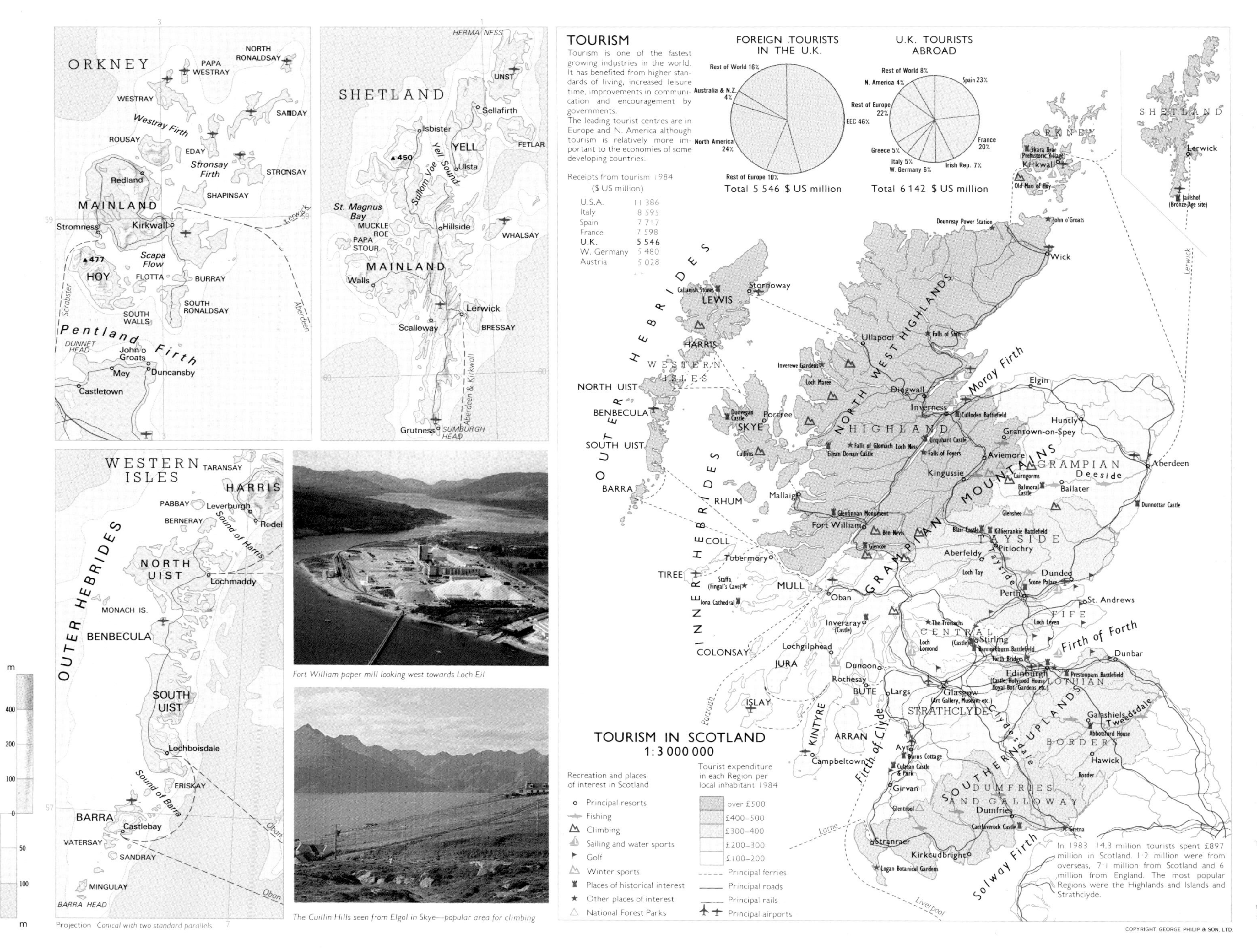

Tourism

Tourism is one of the fastest growing industries in the world. It has benefited from higher standards of living, increased leisure time, improvements in communication and encouragement by governments.

The leading tourist centres are in Europe and N. America although tourism is relatively more important to the economies of some developing countries.

Receipts from tourism 1984 ($ US million)

U.S.A.	11 386
Italy	8 595
Spain	7 717
France	7 598
U.K.	5 546
W. Germany	5 480
Austria	5 028

U.K. TOURISTS ABROAD
Total 6142 $ US million
Spain 23%, France 20%, Irish Rep. 7%, W. Germany 6%, Italy 5%, Greece 5%, N. America 4%, Rest of World 8%, Rest of Europe 22%

FOREIGN TOURISTS IN THE U.K.
Total 5 546 $ US million
EEC 46%, North America 24%, Rest of Europe 22%, Australia & N.Z. 4%, Rest of World 4%

In 1983 14.3 million tourists spent £897 million in Scotland. 1.2 million were from overseas, 7.1 million from Scotland and 6 million from England. The most popular Regions were the Highlands and Islands and Strathclyde.

TOURISM IN SCOTLAND 1:3 000 000

Tourist expenditure in each Region per local inhabitant 1984
over £500
£400–500
£300–400
£200–300
£100–200

Recreation and places of interest in Scotland
- Principal resorts
- Fishing
- Climbing
- Sailing and water sports
- Winter sports
- Golf
- Places of historical interest
- Other places of interest
- National Forest Parks

Principal ferries
Principal roads
Principal rails
Principal airports

Fort William paper mill looking west towards Loch Eil

The Cuillin Hills seen from Elgol in Skye—popular area for climbing

Projection Conical with two standard parallels

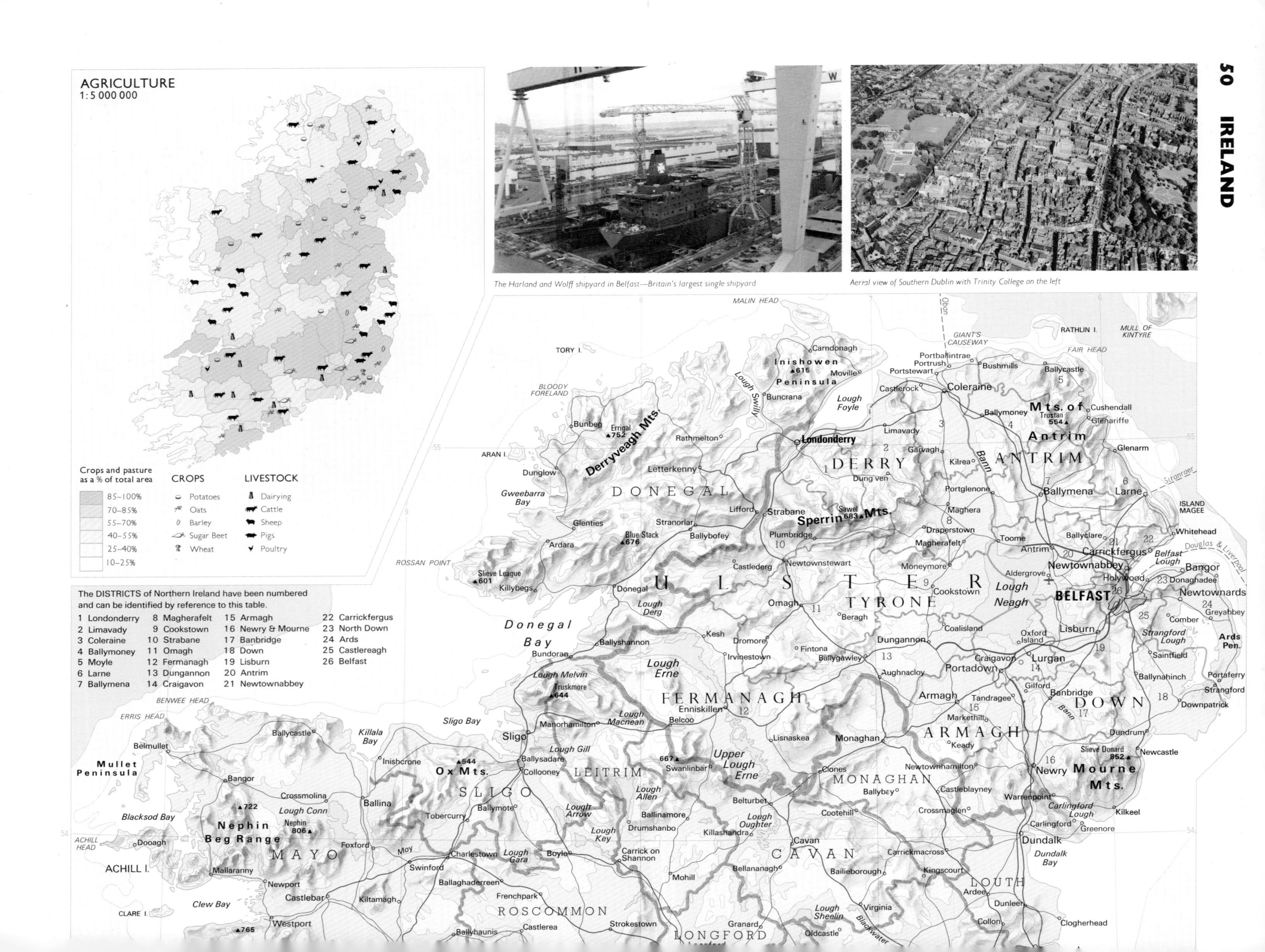

Aerial view of Southern Dublin with Trinity College on the left

The Harland and Wolff shipyard in Belfast—Britain's largest single shipyard

AGRICULTURE
1:5 000 000

Crops and pasture as a % of total area

- 85-100%
- 70-85%
- 55-70%
- 40-55%
- 25-40%
- 10-25%

CROPS
- Potatoes
- Oats
- Barley
- Sugar Beet
- Wheat

LIVESTOCK
- Dairying
- Cattle
- Sheep
- Pigs
- Poultry

The DISTRICTS of Northern Ireland have been numbered and can be identified by reference to this table.

1 Londonderry	15 Armagh
2 Limavady	16 Newry & Mourne
3 Coleraine	17 Banbridge
4 Ballymoney	18 Down
5 Moyle	19 Lisburn
6 Larne	20 Antrim
7 Ballymena	21 Newtownabbey
8 Magherafelt	22 Carrickfergus
9 Cookstown	23 North Down
10 Strabane	24 Ards
11 Omagh	25 Castlereagh
12 Fermanagh	26 Belfast
13 Dungannon	
14 Craigavon	

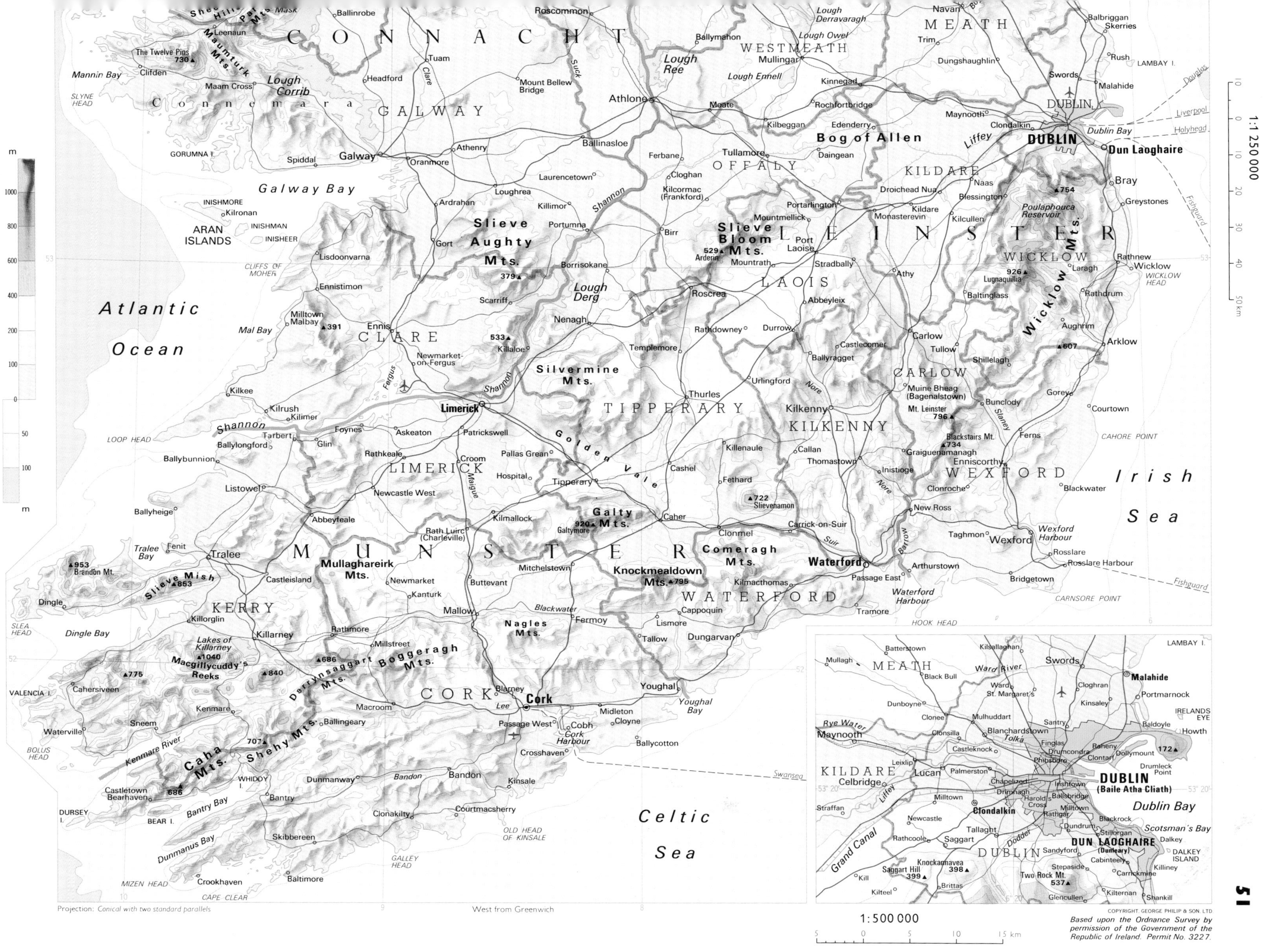
CONNACHT
LEINSTER
MUNSTER
MEATH
WESTMEATH
OFFALY
KILDARE
WICKLOW
LAOIS
CARLOW
KILKENNY
WEXFORD
TIPPERARY
WATERFORD
LIMERICK
CLARE
GALWAY
CORK
KERRY
DUBLIN
Atlantic Ocean
Irish Sea
Celtic Sea
Galway Bay
Dublin Bay
Dingle Bay
Bantry Bay
Dunmanus Bay
Tralee Bay
Youghal Bay
Mal Bay
Mannin Bay
Connemara
Lough Corrib
Lough Ree
Lough Derg
Lough Derravaragh
Lough Owel
Lough Ennell
Bog of Allen
Golden Vale
Wicklow Mts.
Slieve Bloom Mts.
Slieve Aughty Mts.
Silvermine Mts.
Galty Mts.
Knockmealdown Mts.
Comeragh Mts.
Nagles Mts.
Boggeragh Mts.
Mullaghareirk Mts.
Derrynsaggart Mts.
Shehy Mts.
Caha Mts.
Slieve Mish
Macgillycuddy's Reeks
Lakes of Killarney
Maumturk Mts.
The Twelve Pins 730
ARAN ISLANDS
INISHMORE
INISHMAN
INISHEER
GORUMNA I.
CLIFFS OF MOHER
LOOP HEAD
SLYNE HEAD
SLEA HEAD
BOLUS HEAD
MIZEN HEAD
CAPE CLEAR
GALLEY HEAD
OLD HEAD OF KINSALE
HOOK HEAD
CARNSORE POINT
CAHORE POINT
WICKLOW HEAD
VALENCIA I.
DURSEY I.
BEAR I.
WHIDDY I.
LAMBAY I.
Shannon
Suck
Clare
Fergus
Maigue
Blackwater
Lee
Bandon
Suir
Nore
Barrow
Slaney
Liffey
Kenmare River
DUBLIN
Dun Laoghaire
Limerick
Cork
Waterford
Galway
Athlone
Wexford
Tralee
Killarney
Kilkenny
Ennis
Tipperary
Clonmel
Carlow
Wicklow
Arklow
Bray
Drogheda
Navan
Mullingar
Tullamore
Port Laoise
Birr
Roscrea
Thurles
Cashel
Caher
Mallow
Fermoy
Youghal
Cobh
Kinsale
Bandon
Macroom
Bantry
Skibbereen
Listowel
Dingle
Clifden
Tuam
Ballinasloe
Loughrea
Gort
Athenry
Roscommon
Fishguard
Liverpool
Holyhead
Douglas
Swansea
1:500 000
DUBLIN (Baile Atha Cliath)
DUN LAOGHAIRE (Dunleary)
Grand Canal
Rye Water
Ward River
Tolka
Dodder
Howth
Malahide
Portmarnock
Swords
Lucan
Clondalkin
Tallaght
Maynooth
Celbridge
DALKEY ISLAND
IRELANDS EYE
Scotsman's Bay
Two Rock Mt. 537
Knockannavea 398
Saggart Hill 399
Projection: Conical with two standard parallels
West from Greenwich
COPYRIGHT GEORGE PHILIP & SON, LTD
Based upon the Ordnance Survey by permission of the Government of the Republic of Ireland. Permit No. 3227.
1:1 250 000
10 0 10 20 30 40 50 km
1:500 000
5 0 5 10 15 km
m
1000
800
600
400
200
100
0
50
100

1:2 000 000

10 0 10 20 30 40 50 60 70 80 km

Dairy farming on Holland's polder lands

NORTH SEA

WESTFRIESCHE EILANDEN

Wadden Zee

NETHERLANDS

FRIESLAND

DRENTHE

OVERIJSSEL

GELDERLAND

NOORD BRABANT

Ostfriesland

NIEDER SACHSEN

NORD RHEIN WESTFALEN

AMSTERDAM
THE HAGUE ('s Gravenhage)
ROTTERDAM
Utrecht
Haarlem
Leiden
Delft
Dordrecht
Breda
Tilburg
Eindhoven
Nijmegen
Arnhem
's Hertogenbosch
Zwolle
Groningen
Leeuwarden
Assen
Enschede
Deventer
Apeldoorn
Amersfoort
Den Helder
Alkmaar

Oldenburg
Emden
Osnabrück
Münster
Dortmund
ESSEN
DÜSSELDORF
Duisburg
Bochum
Wuppertal
Köln (COLOGNE)
Krefeld
Mönchen-Gladbach
Bonn
Koblenz
Aachen
Maastricht
Venlo
Roermond
Solingen
Remscheid
Hagen
Oberhausen
Mülheim
Gelsenkirchen
Recklinghausen
Bottrop
Herne

BELGIUM

BRUSSELS (Brussel Bruxelles)
Antwerp (Antwerpen)
Gent
Brugge (Bruges)
Ostend (Oostende)
Liège
Charleroi
Namur
Mons
Hasselt
Mechelen
Leuven
Roubaix
Lille
Tournai
Verviers

BRABANT
HAINAUT
FLANDRE ORIENTALE

LUXEMBOURG
Luxembourg
Esch
Differdange

RHEINLAND PFALZ
Trier
Wiesbaden
Mainz
Bad Kreuznach
Saarbrücken
Kaiserslautern

FRANCE
PICARDIE
ARDENNES
PAS-DE-CALAIS
SOMME
OISE
AISNE
MEURTHE
MOSELLE
MEUSE
Reims
Charleville-Mézières
Sedan
Metz
Thionville
Verdun
Cambrai
Valenciennes
Maubeuge
Soissons
Laon
Compiègne
St. Quentin
Arras
Douai

m
400
200
0

Projection : Conical with two standard parallels

East from Greenwich

COPYRIGHT. GEORGE PHILIP & SON. LTD.

INDUSTRY ON THE LOWER RHINE
1:2 000 000

— Motorways
— Railways
◆ Built up areas

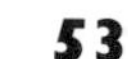

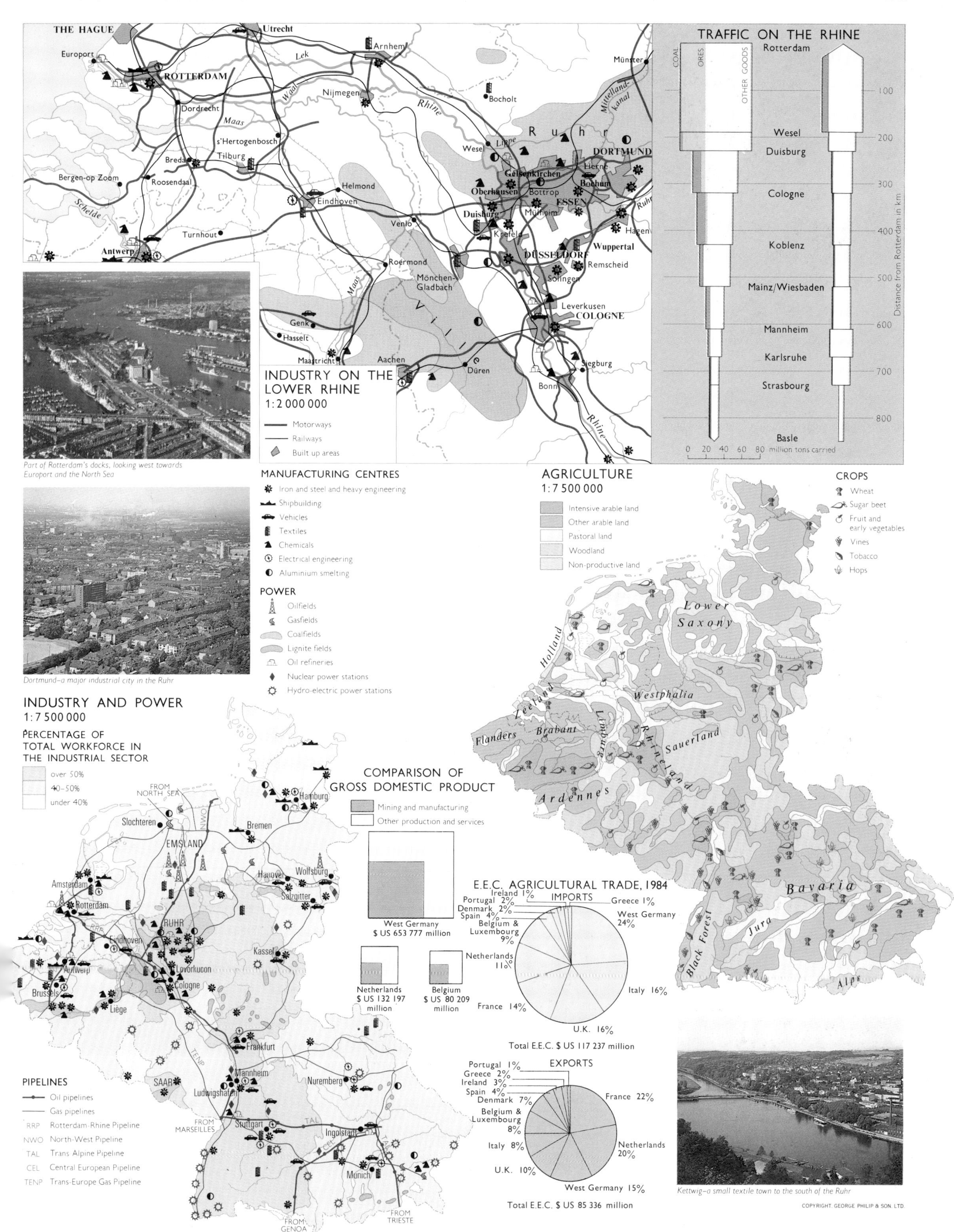

Part of Rotterdam's docks, looking west towards Europort and the North Sea

Dortmund—a major industrial city in the Ruhr

TRAFFIC ON THE RHINE

COAL / ORES / OTHER GOODS

Rotterdam
Wesel
Duisburg
Cologne
Koblenz
Mainz/Wiesbaden
Mannheim
Karlsruhe
Strasbourg
Basle

Distance from Rotterdam in km

0 20 40 60 80 million tons carried

MANUFACTURING CENTRES

✳ Iron and steel and heavy engineering
⚓ Shipbuilding
🚗 Vehicles
🏭 Textiles
▲ Chemicals
◑ Electrical engineering
◐ Aluminium smelting

POWER

⚒ Oilfields
⚐ Gasfields
▨ Coalfields
▨ Lignite fields
⌂ Oil refineries
◆ Nuclear power stations
✺ Hydro-electric power stations

AGRICULTURE
1:7 500 000

Intensive arable land
Other arable land
Pastoral land
Woodland
Non-productive land

CROPS

🌾 Wheat
🍃 Sugar beet
🍒 Fruit and early vegetables
🍇 Vines
🍂 Tobacco
🌿 Hops

INDUSTRY AND POWER
1:7 500 000

PERCENTAGE OF TOTAL WORKFORCE IN THE INDUSTRIAL SECTOR

over 50%
40–50%
under 40%

COMPARISON OF GROSS DOMESTIC PRODUCT

Mining and manufacturing
Other production and services

West Germany
$ US 653 777 million

Netherlands
$ US 132 197 million

Belgium
$ US 80 209 million

E.E.C. AGRICULTURAL TRADE, 1984

IMPORTS

Ireland 1%
Portugal 2%
Denmark 2%
Spain 4%
Greece 1%
West Germany 24%
Belgium & Luxembourg 9%
Netherlands 11%
Italy 16%
France 14%
U.K. 16%

Total E.E.C. $ US 117 237 million

EXPORTS

Portugal 1%
Greece 2%
Ireland 3%
Spain 4%
Denmark 7%
France 22%
Belgium & Luxembourg 8%
Netherlands 20%
Italy 8%
U.K. 10%
West Germany 15%

Total E.E.C. $ US 85 336 million

PIPELINES

◆— Oil pipelines
— Gas pipelines

RRP   Rotterdam-Rhine Pipeline
NWO North-West Pipeline
TAL   Trans Alpine Pipeline
CEL   Central European Pipeline
TENP Trans-Europe Gas Pipeline

Kettwig—a small textile town to the south of the Ruhr

NORTH SEA

BALTIC

NETHERLANDS
BELGIUM
LUX.
FRANCE
GERMANY
WEST GERMANY
EAST GERMANY
CZECHOSLOVAKIA
SWITZERLAND
AUSTRIA
ITALY

FLANDERS
LORRAINE
CHAMPAGNE
FRANCHE COMTÉ
SAVOY
DAUPHINÉ
PROVENCE
PIEDMONT
LOMBARDY
VENETO
EMILIA ROMAGNA
ALTO-ADIGE
TRENTINO
FRIULI-VENEZIA GIULIA
PALATINATE
SAAR
RHINELAND
NORTH RHINE-WESTPHALIA
LOWER SAXONY
SCHLESWIG-HOLSTEIN
HESSE
BAVARIA
BADEN-WÜRTTEMBERG
UPPER AUSTRIA
LOWER AUSTRIA
SALZBURG
TYROL
VORARLBERG
CARINTHIA
STYRIA
BURGENLAND
LIECHTENSTEIN

SCHLESWIG
HOLSTEIN

Flensburg
Kiel
Kiel Bay
Lübeck
Lübeck Bay
Fehmarn
Lolland
Falster
Gedser
Rügen
Sassnitz
Stralsund
Rostock
Warnemünde
Wismar
Schwerin
Güstrow
Neu Brandenburg
Szczecin (Stettin)
Goleniów
Kołobrzeg
Koszalin
Słupsk
Darłowo
Haff
Świnoujście
Wolin

Heligoland
Wilhelmshaven
Bremerhaven
Bremen
Hamburg
Altona
Harburg
Lüneburg
Cuxhaven
Emden
Oldenburg
Groningen
Leeuwarden
Den Helder
Alkmaar
Haarlem
Amsterdam
Hilversum
Leiden
The Hague
Rotterdam
Dordrecht
Utrecht
Apeldoorn
Deventer
Enschede
Arnhem
Nijmegen
Breda
Tilburg
Eindhoven
Antwerp
Ghent
Bruges
Ostend
Zeebrugge
Flushing
Brussels
Mechelen
Leuven
Aalst
Tournai
Lille
Roubaix
Tourcoing
Douai
Valenciennes
Cambrai
Charleroi
Namur
Liège
Maastricht
Aachen
Bonn
Cologne (Köln)
Düsseldorf
M.Gladbach
Krefeld
Duisburg
Oberhausen
Essen
Bochum
Dortmund
Hagen
Wuppertal
Remscheid
Hamm
Gelsenkirchen
Münster
Osnabrück
Bielefeld
Herford
Hanover (Hannover)
Hildesheim
Brunswick
Salzgitter
Goslar
Kassel
Paderborn
Detmold
Hamelin
Celle
Lüneburg
Uelzen
Stendal
Magdeburg
Brandenburg
Potsdam
BERLIN
Charlottenburg
Spandau
Frankfurt
Cottbus
Gubin
Zielona Góra
Głogów
Leszno
Poznań
Gorzów Wielkopolski
Kostrzyn
Oranienburg
Eberswalde
Neuruppin
Rathenow
Zerbst
Dessau
Bernburg
Halle
Merseburg
Leipzig
Halberstadt
Mühlhausen
Nordhausen
Erfurt
Weimar
Jena
Gera
Zeitz
Torgau
Meissen
Dresden
Görlitz
Bautzen
Legnica
Liberec
Usti nad Labem
Litoměřice
Chomutov
Most
Teplice
Karlovy Vary
Cheb
Plauen
Zwickau
Karl Marx Stadt (Chemnitz)
Reichenbach
Glauchau
Bayreuth
Hof
Prague (Praha)
Kladno
Beroun
Příbram
Plzeň (Pilsen)
Klatovy
Písek
České Budějovice
Třeboň
Jihlava
Třebíč
Brno (Brünn)
Znojmo
Gmünd
Zwettl
Krems
St. Pölten
Vienna (Wien)
Baden
Wiener Neustadt
Stockerau
Linz
Melk
Steyr
Wels
Salzburg
Bad Ischl
Gmunden
Bad Gastein
Innsbruck
Kitzbühel
Kufstein
Graz
Bruck
Leoben
Eisenerz
Mürzzuschlag
Semmering
Kapfenberg
Klagenfurt
Villach
Lienz
Bolzano
Trento
Merano
Maribor
Ljubljana
Zagreb
Trieste
Udine
Gorizia
Vittorio Veneto
Treviso
Venice (Venezia)
Padua (Padova)
Vicenza
Verona
Mantua (Mantova)
Rovigo
Ferrara
Modena
Bologna
Reggio
Parma
Piacenza
Cremona
Pavia
Milan (Milano)
Brescia
Bergamo
Como
Lecco
Novara
Vercelli
Turin
Alessandria
Asti
Cuneo
Savona
Genoa (Genova)
Spezia
Carrara
Pisa
Lucca
Florence (Firenze)
Prato
Pistoia
Ravenna
Forlì
Cesena
Rimini
Pesaro
San Marino

Münster
Westphalia
Wesel
Koblenz
Trier
Luxembourg
Thionville
Metz
Nancy
Lunéville
Épinal
Strasbourg
Haguenau
Colmar
Mulhouse
Basle
Freiburg
Karlsruhe
Pforzheim
Stuttgart
Ludwigsburg
Esslingen
Heilbronn
Tübingen
Reutlingen
Ulm
Augsburg
Munich (München)
Freising
Landshut
Ingolstadt
Regensburg
Nuremberg (Nürnberg)
Fürth
Erlangen
Bamberg
Würzburg
Schweinfurt
Aschaffenburg
Frankfurt
Offenbach
Hanau
Darmstadt
Mannheim
Heidelberg
Worms
Mainz
Wiesbaden
Ludwigshafen
Kaiserslautern
Saarbrücken
Neunkirchen
Speyer
Giessen
Fulda
Coburg
Plauen

NORTH SEA
ADRIATIC SEA
Gulf of Venice
Gulf of Genoa

Thuringian Forest
Bohemian Forest
Harz Mts.
Erz Gebirge
Swabian Jura
Black Forest
Vosges
Jura
Ardennes
Eifel
Westerwald
Taunus
Rhön
Dolomites
Carnic Alps
Karawanken
Brenner P.
St. Gotthard
Great St. Bernard P.
Mont Blanc 4807
Monte Rosa
Matterhorn
Maritime Alps

FRANCE
Reims
Soissons
Laon
St. Quentin
Charleville-Mézières
Verdun
Châlons-sur-Marne
Troyes
Langres
Dijon
Beaune
Chalon-sur-Saône
Mâcon
Villefranche
Lyons
Villeurbanne
St. Étienne
Valence
Montélimar
Nyons
Orange
Avignon
Nîmes
Arles
Aix
Marseilles
Cannes
Nice
Monaco & Monte Carlo
Grenoble
Chambéry
Annecy
Geneva
Lausanne
Montreux
Sion
Brig
Bern
Fribourg
Neuchâtel
Biel
Solothurn
Aarau
Zurich
Winterthur
St. Gallen
Luzern
Schwyz
Chur
Davos
St. Moritz
Bellinzona
Locarno
Lugano
L. Maggiore
L. Como
L. Garda
L. Geneva
L. Constance
Po
Rhine
Rhône
Danube
Elbe
Oder (Odra)
Neisse (Nysa)
Weser
Ems
Main
Rhine

East from Greenwich

Projection: Conical with two standard parallels

m
4000
3000
2000
1500
1000
400
200
0
200
m

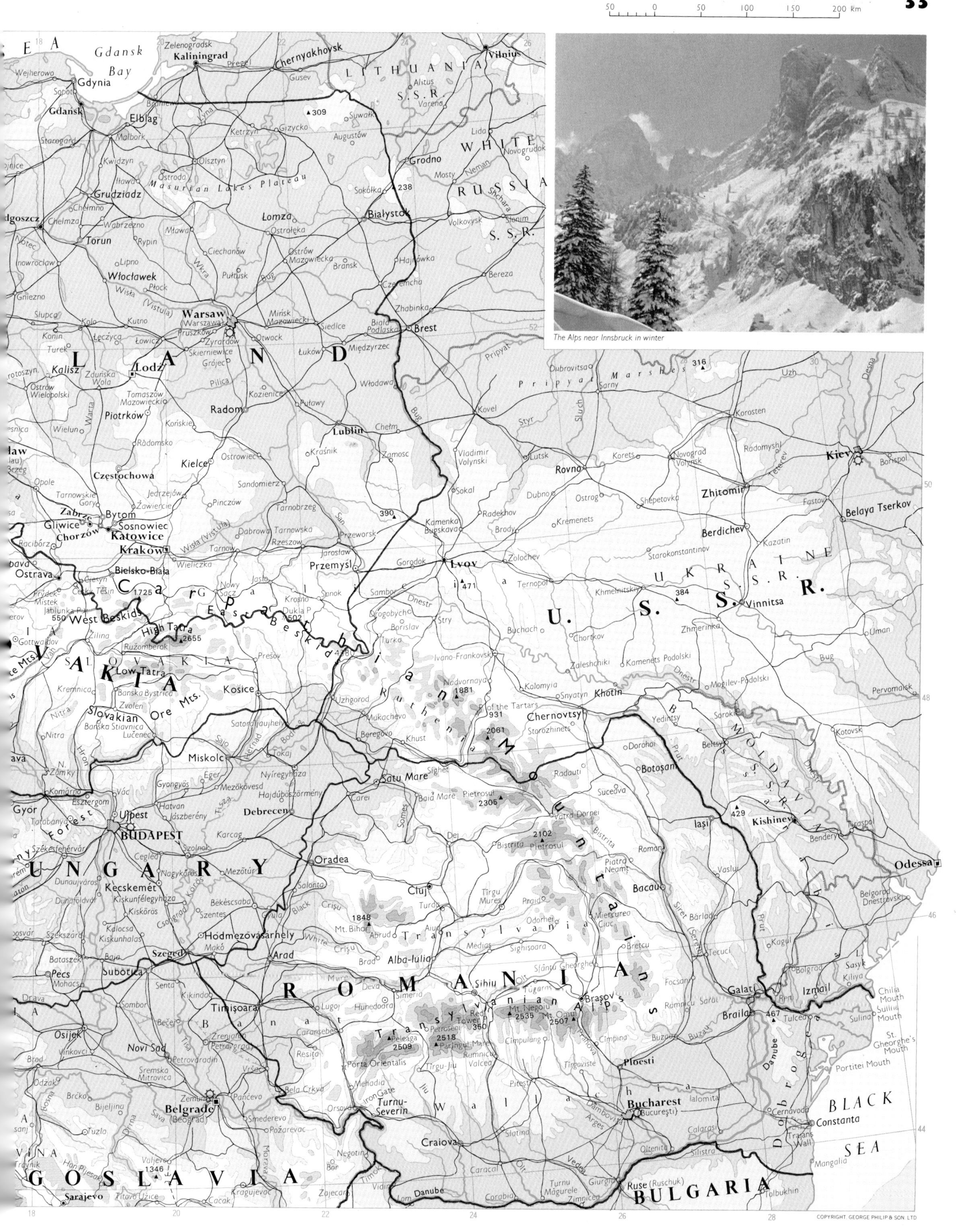

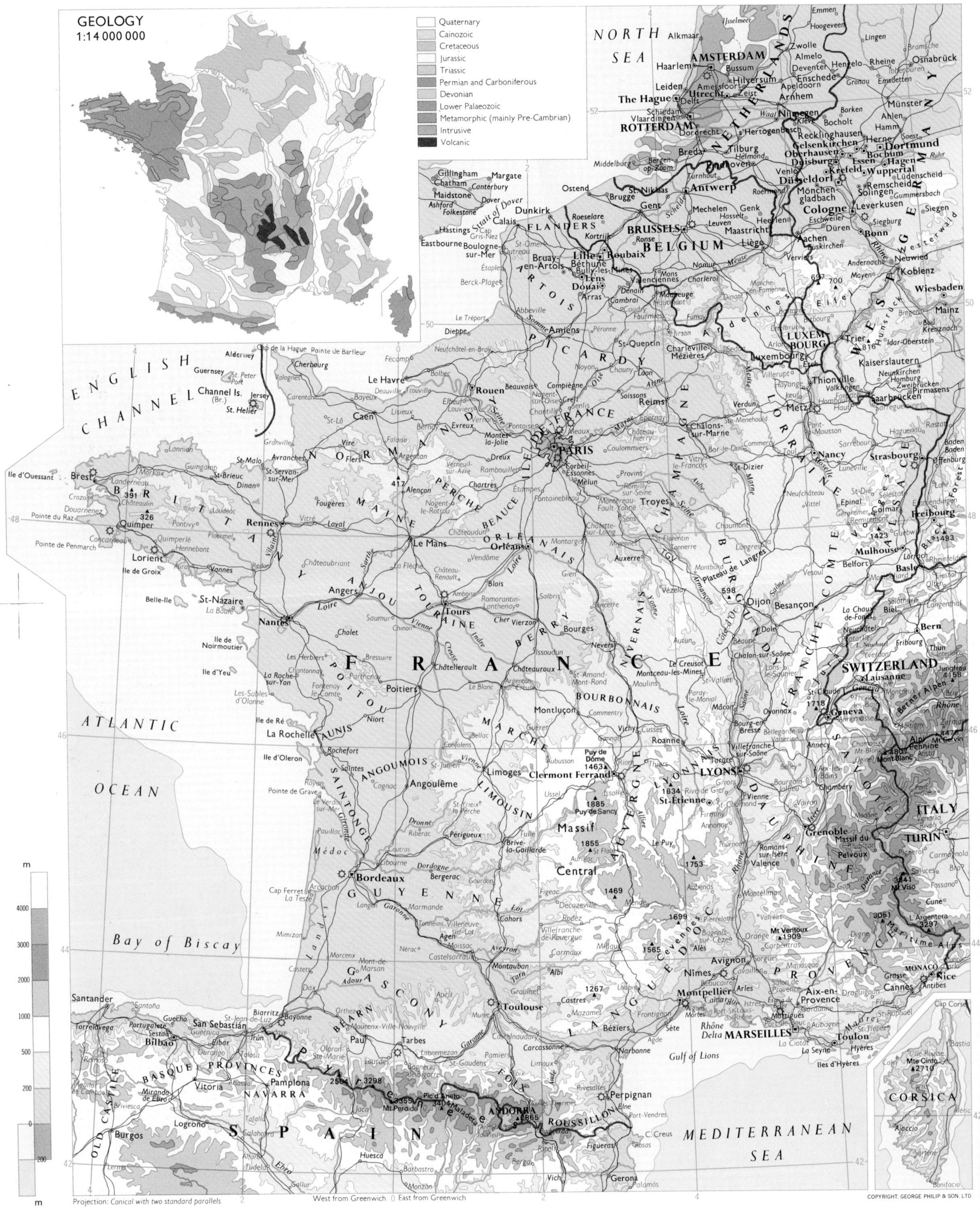

1:5 000 000

50   0   50   100   150   200 km

GEOLOGY
1:14 000 000

Quaternary
Cainozoic
Cretaceous
Jurassic
Triassic
Permian and Carboniferous
Devonian
Lower Palaeozoic
Metamorphic (mainly Pre-Cambrian)
Intrusive
Volcanic

Projection: Conical with two standard parallels

West from Greenwich 0 East from Greenwich

COPYRIGHT. GEORGE PHILIP & SON, LTD.

1 : 7 500 000

50 0 50 100 150 200 250 300 km

AGRICULTURE

Artois
Picardy
Normandy
Ile de France
Champagne
Lorraine
Brittany
Maine
Beauce
Alsace
Anjou
Touraine
Sologne
Burgundy
Franche Comté
Vendée
Poitou
Morvan
Savoie
Marche
Dauphiné
Limousin
Auvergne
Périgord
Guyenne
Landes
Languedoc
Provence
Gascony
Roussillon
Corsica

FRENCH PRODUCTION AS A PERCENTAGE OF E.E.C. PRODUCTION FOR SELECTED AGRICULTURAL GOODS, (1984)

Wheat
Barley
Maize
Sugar beet
Wine
Beef

0   20%   40%   60%   80%   100%

LAND USE
Arable land
Permanent pasture
Vineyards
Woodland
Rough grazing land
Non-productive land
—— Northern limit of the vine
- - - Northern limit of the olive

CROPS
Wheat
Oats
Maize
Sugar beet
Potatoes
Tobacco
Fruit and early vegetables

Vineyard near Avignon in Provence

Paris looking eastwards from the roof of Notre Dame

INDUSTRY AND POWER

PERCENTAGE OF TOTAL WORKFORCE IN THE INDUSTRIAL SECTOR
over 50%   40–50%   30–40%   under 30%

Dunkirk
Lille   FROM THE NETHERLANDS
Valenciennes
Maubeuge
NORD
Amiens
St. Quentin
Le Havre
Rouen
Longwy
Thionville
Hagondange
TO WEST GERMANY
Caen
Paris
Nancy
LORRAINE
Brest
Strasbourg
Rance
Brennilis
Troyes
Rennes
Le Mans
Mulhouse
Orléans
Besançon
St. Nazaire
Dijon
Montbéliard
Nantes
Chinon
TO SWITZERLAND
Le Creusot
Roanne
La Rochelle
Limoges
Lyons
Angoulème
Clermont Ferrand
St. Étienne
Grenoble
Bordeaux
Massif Central
Rhône
Parentis
Fos
Lacq
Toulouse
Marseilles
CORSICA
FROM ALGERIA
Toulon
Pyrénées

INDUSTRY
▲ Iron and steel
✳ Heavy engineering
🚗 Vehicles
⛴ Shipbuilding
■ Textiles

▲ Chemicals
⊕ Electrical engineering
▤ Paper
■ Iron ore

POWER
⛏ Oilfields
Gasfields
Major coalfields
Oil pipelines
SEPL South European Pipeline

—— Gas pipelines
Oil refineries
◆ Nuclear power stations
✿ Hydro-electric power stations

FRENCH PRODUCTION AS A PERCENTAGE OF E.E.C. PRODUCTION FOR SELECTED INDUSTRIAL GOODS, (1983)

Steel
Cars
Cotton
Coal
Iron ore
Electricity production

0   20%   40%   60%   80%   100%

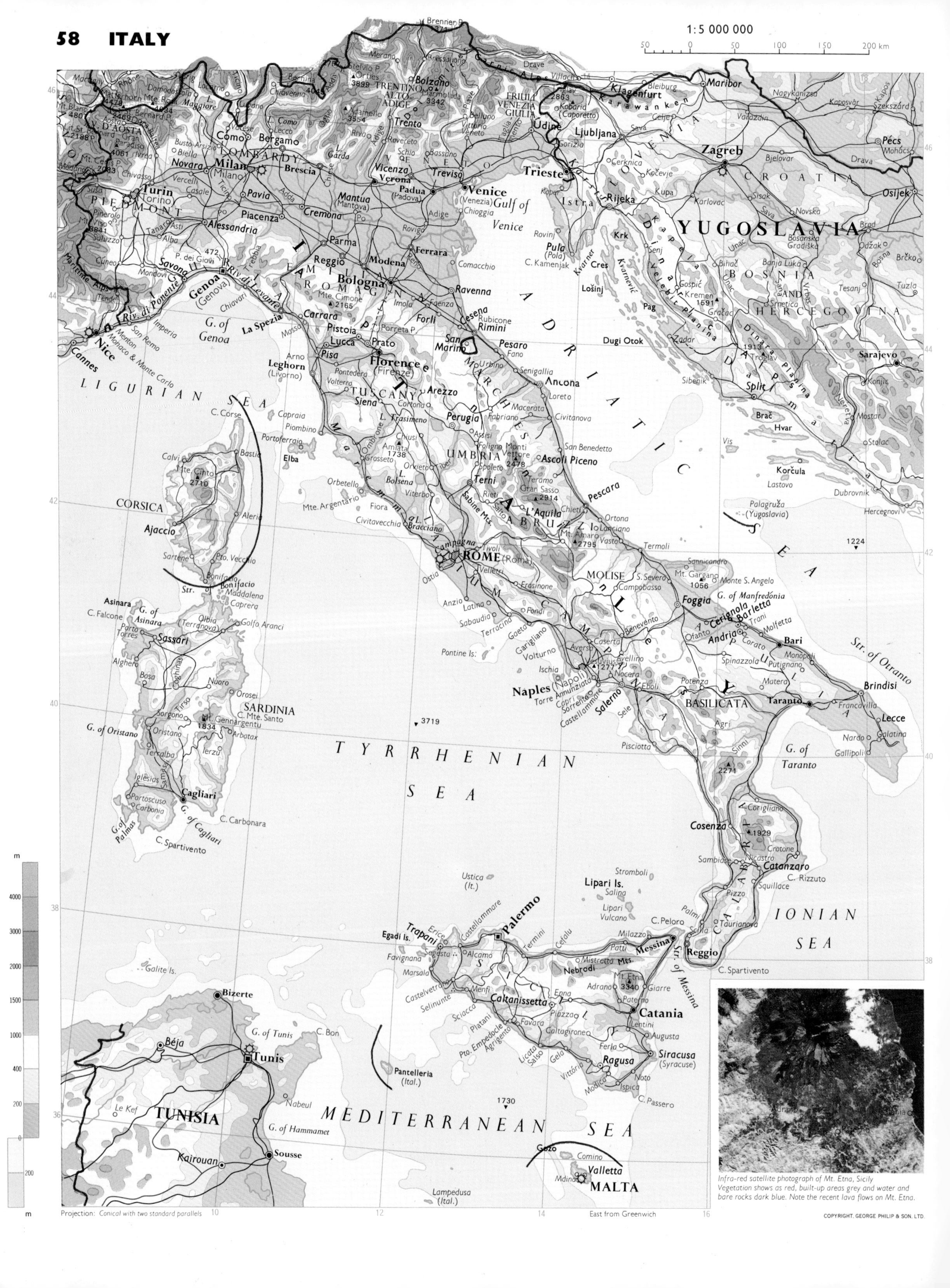

1:5 000 000

50 0 50 100 150 200 km

58 ITALY

MERANO
Brenner P.
Bolzano
TRENTINO
ALTO
ADIGE
Trento
Como
Bergamo
Novara
Milan
(Milano)
Brescia
Verona
Vicenza
Treviso
Venice
(Venezia)
Gulf of
Venice
Trieste
Klagenfurt
Maribor
Zagreb
CROATIA
Osijek
Pécs
YUGOSLAVIA

LOMBARDY
Turin
Torino
PIEMONT
Pavia
Piacenza
Cremona
Mantua
(Mantova)
Po
Adige
Rovigo
Ljubljana
Rijeka
SLOVENIA
BOSNIA
AND
HERCEGOVINA
Sarajevo

Genoa
(Genova)
La Spezia
EMILIA
ROMAGNA
Reggio
Modena
Bologna
Ferrara
Ravenna
Forlì
Rimini
San
Marino
Pesaro
Ancona
ADRIATIC
Split
Dubrovnik

Nice
Cannes
LIGURIAN
SEA
G. of
Genoa
Carrara
Pisa
Leghorn
(Livorno)
Florence
(Firenze)
TUSCANY
Siena
Arezzo
Perugia
MARCHE
UMBRIA
Ascoli Piceno
Pescara
SEA

CORSICA
Ajaccio
Bastia
Elba
Orbetello
ROME
(Roma)
L'Aquila
Gran Sasso
2914
ABRUZZI
Termoli

SARDINIA
Sassari
Oristano
Cagliari
TYRRHENIAN
SEA
Naples
(Napoli)
Salerno
MOLISE
Campobasso
Foggia
Barletta
Andria
Bari
Brindisi
Lecce
Taranto
G. of
Taranto
BASILICATA
Potenza
IONIAN
SEA
Cosenza
Catanzaro
Reggio

TUNISIA
Tunis
Bizerte
MEDITERRANEAN
SEA
Palermo
Trapani
Marsala
SICILY
Caltanissetta
Mt. Etna
3340
Catania
Siracusa
(Syracuse)
Ragusa
Messina
MALTA
Valletta

Infra-red satellite photograph of Mt. Etna, Sicily
Vegetation shows as red, built-up areas grey and water and
bare rocks dark blue. Note the recent lava flows on Mt. Etna.

Projection: Conical with two standard parallels

East from Greenwich

COPYRIGHT. GEORGE PHILIP & SON. LTD.

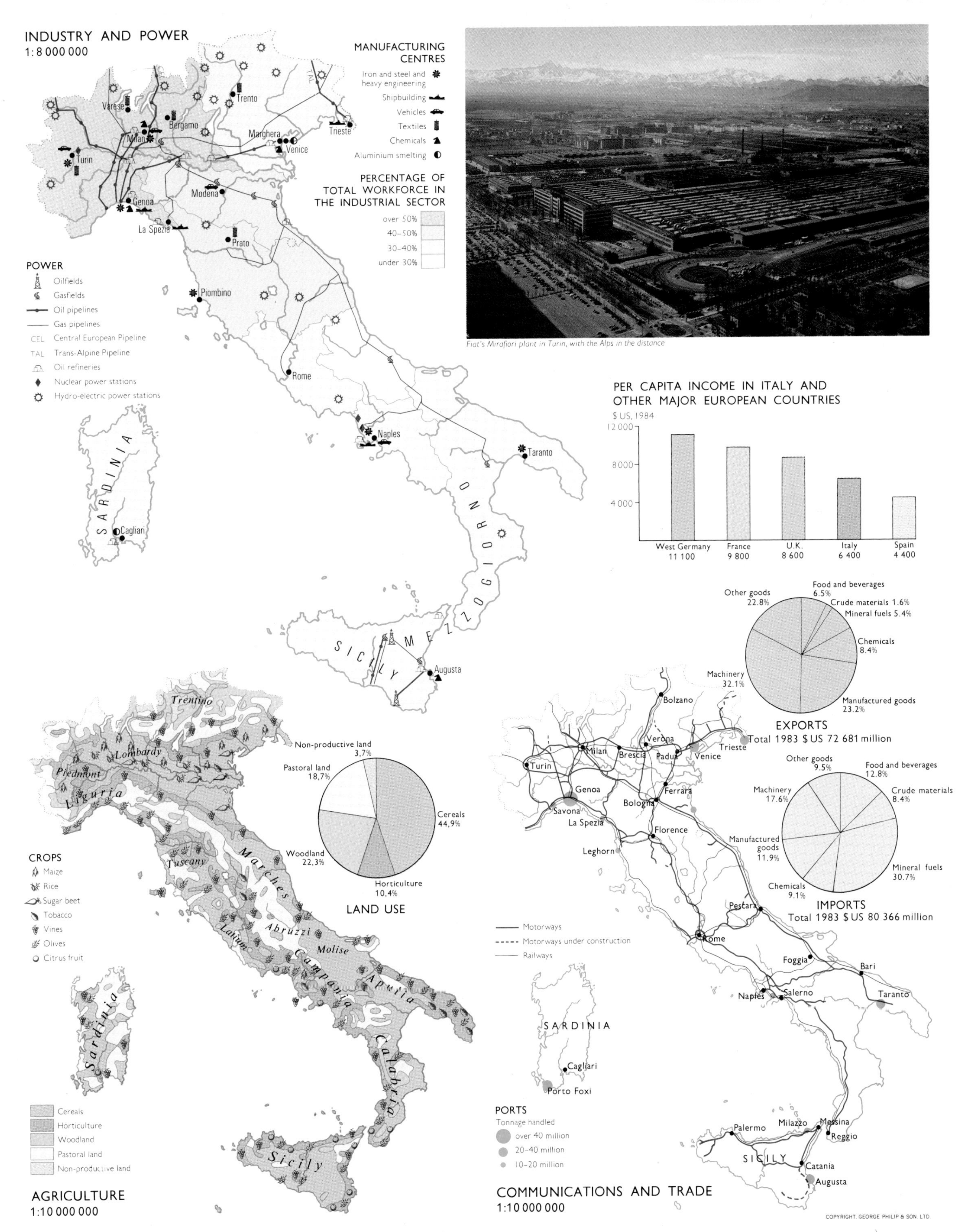

INDUSTRY AND POWER
1:8 000 000

MANUFACTURING CENTRES

✳	Iron and steel and heavy engineering
⛴	Shipbuilding
🚗	Vehicles
▮	Textiles
▲	Chemicals
◑	Aluminium smelting

PERCENTAGE OF TOTAL WORKFORCE IN THE INDUSTRIAL SECTOR

over 50%	
40–50%	
30–40%	
under 30%	

POWER

⛏	Oilfields
	Gasfields
⊶	Oil pipelines
	Gas pipelines
CEL	Central European Pipeline
TAL	Trans-Alpine Pipeline
	Oil refineries
◆	Nuclear power stations
✲	Hydro-electric power stations

Varese, Trento, Bergamo, Milan, Turin, Marghera, Venice, Trieste, Genoa, Modena, La Spezia, Prato, Piombino, Rome, Naples, Taranto, Augusta, Cagliari, SARDINIA, SICILY, MEZZOGIORNO

Fiat's Mirafiori plant in Turin, with the Alps in the distance

PER CAPITA INCOME IN ITALY AND OTHER MAJOR EUROPEAN COUNTRIES
$ US, 1984

West Germany	11 100
France	9 800
U.K.	8 600
Italy	6 400
Spain	4 400

EXPORTS
Total 1983 $ US 72 681 million

- Other goods 22.8%
- Food and beverages 6.5%
- Crude materials 1.6%
- Mineral fuels 5.4%
- Chemicals 8.4%
- Manufactured goods 23.2%
- Machinery 32.1%

IMPORTS
Total 1983 $ US 80 366 million

- Other goods 9.5%
- Food and beverages 12.8%
- Crude materials 8.4%
- Mineral fuels 30.7%
- Machinery 17.6%
- Manufactured goods 11.9%
- Chemicals 9.1%

AGRICULTURE
1:10 000 000

CROPS

🌽	Maize
	Rice
	Sugar beet
	Tobacco
🍇	Vines
🫒	Olives
○	Citrus fruit

	Cereals
	Horticulture
	Woodland
	Pastoral land
	Non-productive land

Trentino, Lombardy, Piedmont, Liguria, Tuscany, Marches, Latium, Abruzzi, Molise, Campania, Apulia, Calabria, Sardinia, Sicily

LAND USE

- Non-productive land 3.7%
- Pastoral land 18.7%
- Cereals 44.9%
- Woodland 22.3%
- Horticulture 10.4%

COMMUNICATIONS AND TRADE
1:10 000 000

——	Motorways
- - -	Motorways under construction
——	Railways

Bolzano, Verona, Milan, Brescia, Padua, Venice, Trieste, Turin, Genoa, Ferrara, Savona, La Spezia, Bologna, Florence, Leghorn, Pescara, Rome, Foggia, Bari, Naples, Salerno, Taranto, Palermo, Milazzo, Messina, Reggio, Catania, Augusta, SARDINIA, Cagliari, Porto Foxi, SICILY

PORTS
Tonnage handled

●	over 40 million
●	20–40 million
•	10–20 million

COPYRIGHT. GEORGE PHILIP & SON LTD

1 : 5 000 000

50   0   50   100   150   200 km

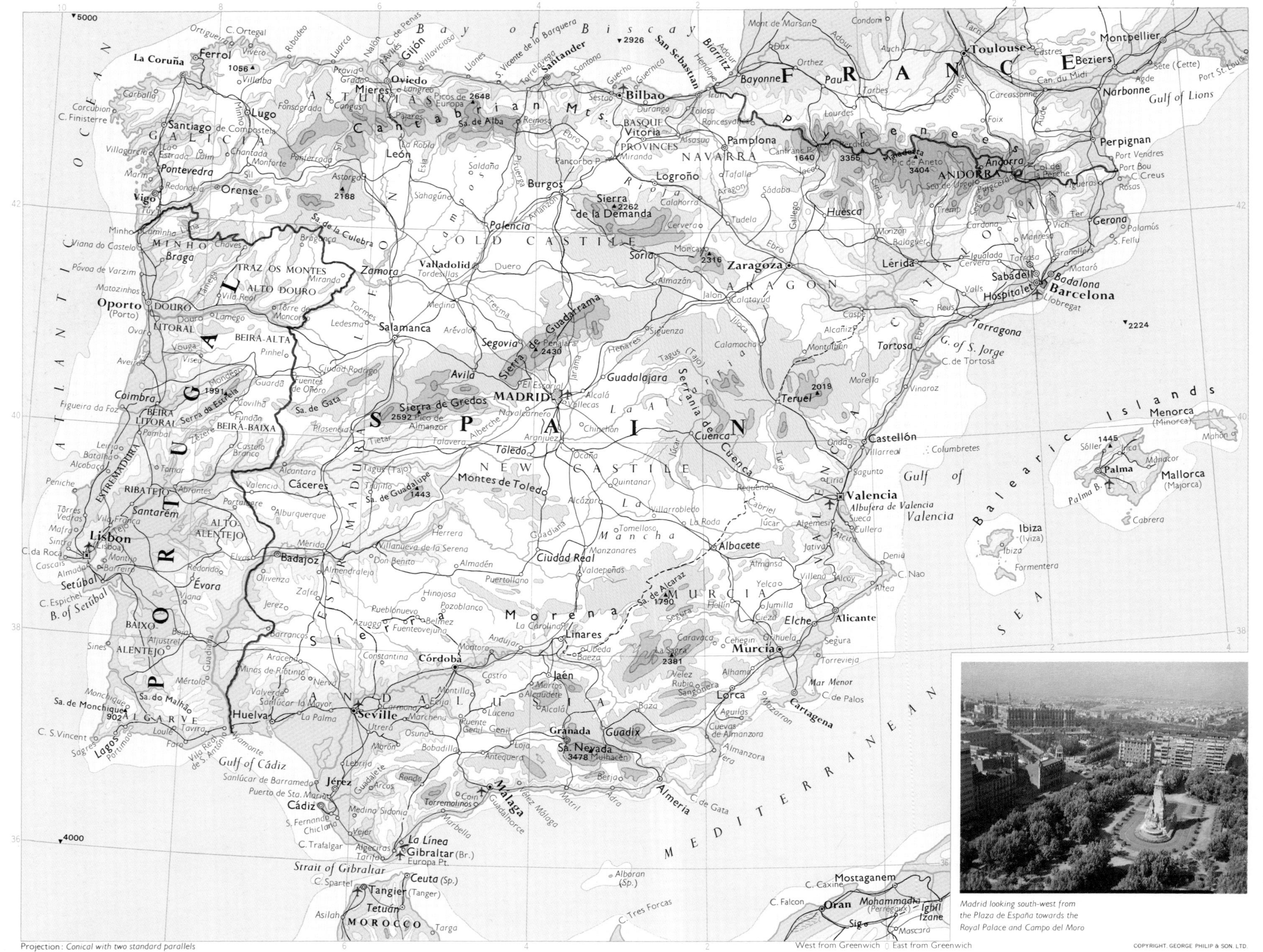

Madrid looking south-west from the
Plaza de España towards the
Royal Palace and Campo del Moro

COPYRIGHT GEORGE PHILIP & SON LTD.

West from Greenwich East from Greenwich

Projection: Conical with two standard parallels

m   3000   2000   1500   1000   400   200   0   200   m

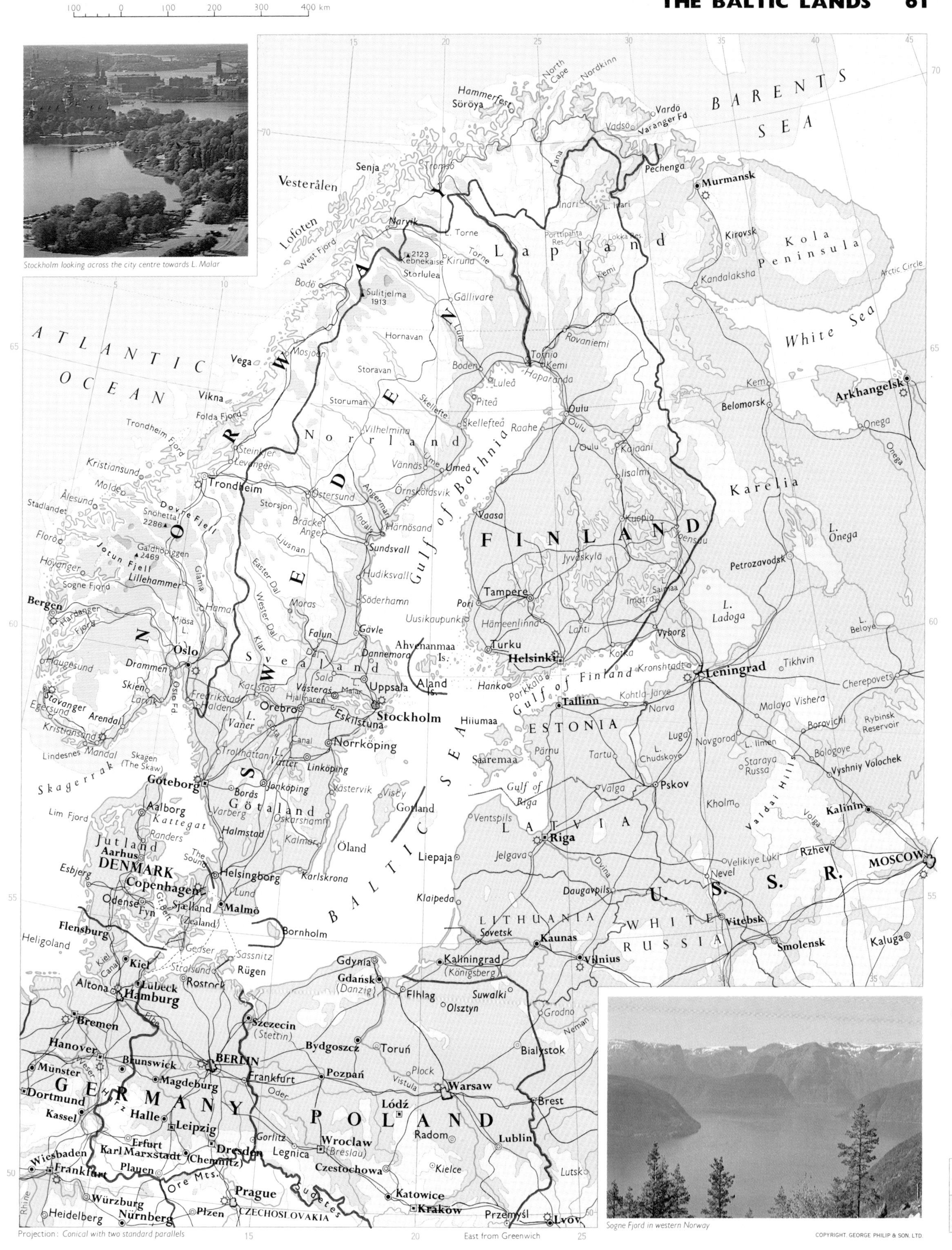

1:10 000 000

100   0   100   200   300   400 km

Stockholm looking across the city centre towards L. Malar

BARENTS SEA

North Cape   Nordkinn

Hammerfest   Söröya   Vardö   Vardø   Varanger Fd
Senja   Vadsø
Tromsö   Pechenga   **Murmansk**

Vesterålen

Lofoten   Narvik   L. Torne   Inari   L. Ivari   Kirovsk   Kola Peninsula

West Fjord   2123   Torne   Porttipahta Res.   Kandalaksha   Arctic Circle
Kebnekaise   Kiruna   Lokka Res.
Bodö   Storlulea

Sulitjelma   Gällivare   Rovaniemi   White Sea
1913

ATLANTIC   Vega   Mosjöen   Hornavan   Lule   Boden   Tornio   Kemi   Kem   **Arkhangelsk**
OCEAN   Storavan   Luleå   Haparanda   Belomorsk
Vikna   Storuman   Piteå
Folda Fjord   Vilhelmina   Skellefte   Oulu   Onega   Onega
Trondheim Fjord   Norrland   Skellefteå   Raahe   L. Oulu   Kajaani
Kristiansund   Steinkjer   Ume   Umeå   Iisalmi   Karelia
Molde   Levanger   Vännäs   L. Onega
Stadlandet   Ålesund   **Trondheim**   Östersund   Örnsköldsvik   **FINLAND**
Florö   Dovre Fjell   Storsjon   Ångerman   Härnösand   Vaasa   Petrozavodsk
2286   Bräcke   Kuopio   Saimaa
Höyanger   Snöhetta   Ange   Sundsvall   Jyväskylä   Pensau   L. Onega
Sogne Fjord   2469   Ljusnan   Indal   Hudiksvall   L. Ladoga   L. Beloye
Galdhöppigen   Jotun Fjell   Lillehammer   Moras   Söderhamn   Pori   Tampere   Imatra
Bergen   Hamar   Uusikaupunki   Hämeenlinna   Lahti   Vyborg
Hardanger   Mjösa   Falun   Gävle   **Helsinki**   Kotka   **Leningrad**   Tikhvin
Fjord   L.   Ahvenanmaa   Turku   Kronshtadt   Cherepovets
Haugesund   **Oslo**   Dannemora   Is.   Gulf of Finland   Malaya Vishera
Stavanger   Drammen   Sala   Åland   Hanko   Porkkala   **Tallinn**   Narva   Borovichi   Rybinsk
Egersund   Arendal   Sveg   Västeras   Mälar   Uppsala   Hiiumaa   Kohtla-Järve   Luga   Novgorod   L. Ilmen   Reservoir
Kristiansand   Skien   Klar   Örebro   Eskilstuna   **ESTONIA**   Pärnu   Tartu   L. Chudskoye   Staraya   Vyshniy Volochek
Lindesnes   Mandal   Larv   Karlstad   **Stockholm**   Saaremaa   Russa   Valdai Hills
Skagen   Vättern   Norrköping   Gulf of   Valga   Pskov   Kholm   Kalinin
Skagerrak   (The Skaw)   Trollhättan   Linköping   Riga   Volga
Lim Fjord   **Göteborg**   Borås   Jönköping   Västervik   Visby   **LATVIA**   Velikie Luki   Rzhev   **MOSCOW**
Aalborg   Götaland   Varberg   Gotland   Ventspils   **Riga**   Dvina
Jutland   Kattegat   Randers   Halmstad   Kalmar   Öland   Liepaja   Jelgava   Daugavpils   Nevel   U.S.S.R.
Aarhus   Helsingborg   Karlskrona   Klaipeda   **U.S.S.R.**   Kaluga
DENMARK   The   Lund   Vitebsk
Esbjerg   Sound   **Malmö**   Bornholm   **LITHUANIA**   **WHITE**   Smolensk
Copenhagen   Sovetsk   **Kaunas**   **RUSSIA**
Odense   Fyn   Sjælland   **Vilnius**
Flensburg   (Zealand)   Gdynia   **Kaliningrad**   Grodna
Heligoland   Kiel   Gedser   Sassnitz   **Gdańsk**   (Königsberg)   Suwalki   Neman
Altona   Kiel   Stralsund   Rügen   (Danzig)   Elblag   Olsztyn   Bialystok
Hamburg   Canal   Rostock   **Szczecin**   Grodno
Bremen   Elbe   (Stettin)   **Bydgoszcz**   Toruń   **Warsaw**   Brest
Hanover   **BERLIN**   **Poznań**   Plock   Lublin
Münster   Brunswick   Magdeburg   Frankfurt   Vistula   Lutsk
Dortmund   Weser   Halle   Oder   **POLAND**   Radom   Lvov
Kassel   **GERMANY**   **Leipzig**   **Lódź**
Wiesbaden   Erfurt   Gorlitz   **Wroclaw**   Czestochowa   Kielce
Frankfurt   Plauen   Karl Marxstadt   Legnica   (Breslau)   Katowice
Heidelberg   Würzburg   **Dresden**   **Krakow**   Przemyśl
Nürnberg   Prague   Plzen   Ore Mts.   **CZECHOSLOVAKIA**

BALTIC SEA

Gulf of Bothnia

NORWAY   **SWEDEN**   **Svealand**

Lapland

Sogne Fjord in western Norway

Projection : Conical with two standard parallels   East from Greenwich   COPYRIGHT GEORGE PHILIP & SON LTD.

m   2000   1000   400   200   0   200   m

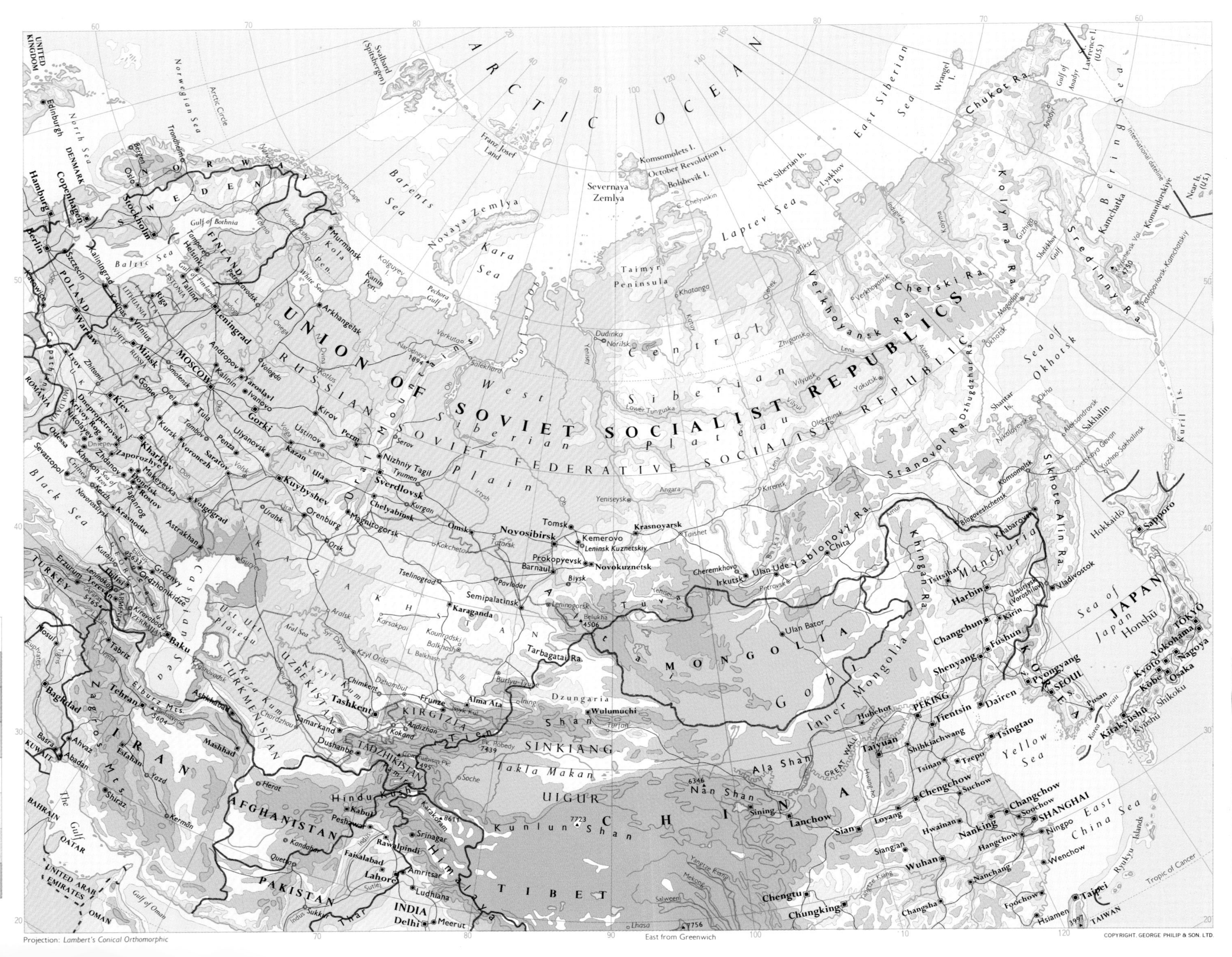

1:35 000 000

200 0 200 400 600 800 1000 1200 1400 km

1 : 40 000 000

400 0 400 800 1200 1600 km

Moscow, The Red Square

NATIONALITIES OF THE U.S.S.R., 1979
(Population figures in millions)

Russians 137

Other nationalities 13

Chuvash 1
Latvians 1
Kirgiz 2
Germans 2
Jews 2
Lithuanians 2
Turkmen 2
Georgians 3
Moldavians 3
Tadzhiks 3
Azerbaijans 4
Armenians 4
Tartars 5
Kazakhs 6
White Russians 7
Uzbeks 9

Ukrainians 42

URBAN AND RURAL POPULATION STRUCTURE

URBAN POPULATION

FEMALES

Age
80+
70–79
60–69
50–59
40–49
30–39
20–29
10–19
0–9

MALES

Population in millions 10 5 5 10

RURAL POPULATION

FEMALES

Age
80+
70–79
60–69
50–59
40–49
30–39
20–29
10–19
0–9

MALES

Population in millions 10 5 5 10

THE GROWTH OF POPULATION
within present boundaries
(Population figures in millions)

	1913	1940	1950	1959	1970	1980
URBAN	159,2 18%	194,1 29%	178,5 38%	208,8 47%	241,7 56%	264,5 63%
RURAL	82%	71%	62%	53%	44%	37%

POPULATION OF CITIES AND TOWNS

■ Over 1 million inhabitants

● 500 000–1 million inhabitants

● 250 000–500 000 inhabitants

• 100 000–250 000 inhabitants

DENSITY OF POPULATION
inhabitants per km²

| over 100 |
| 50–100 |
| 25–50 |
| 10–25 |
| 1–10 |
| under 1 |
| uninhabited |

COPYRIGHT GEORGE PHILIP & SON LTD.

Leningrad
Minsk
Moscow
Gorki
Kazan
Kharkov
Dnepropetrovsk
Donetsk
Odessa
Kuybyshev
Perm
Sverdlovsk
Chelyabinsk
Ufa
Omsk
Novosibirsk
Tashkent
Baku
Tbilisi

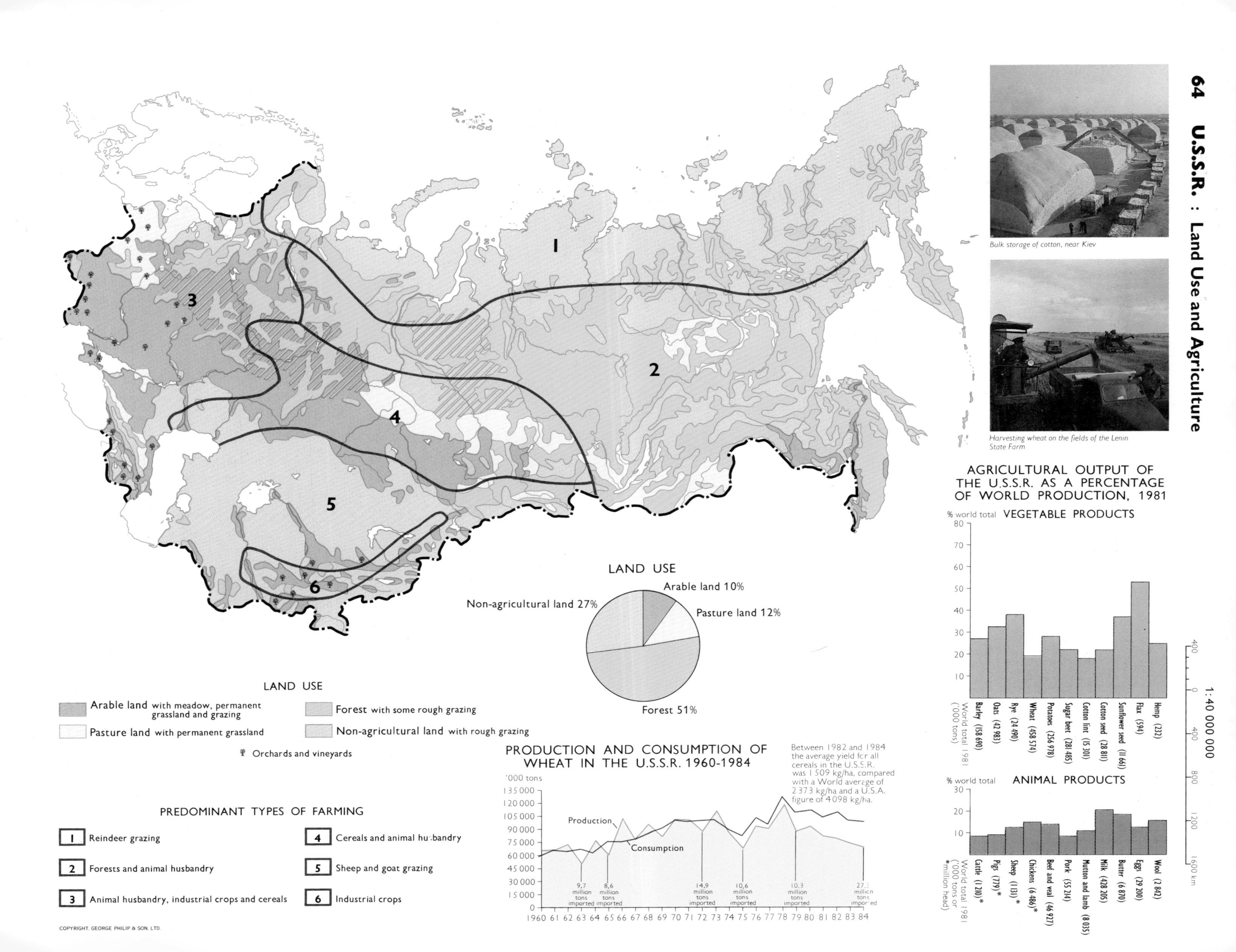

Bulk storage of cotton, near Kiev

Harvesting wheat on the fields of the Lenin State Farm

1 : 40 000 000

400   0   400   800   1200   1600 km

AGRICULTURAL OUTPUT OF THE U.S.S.R. AS A PERCENTAGE OF WORLD PRODUCTION, 1981 VEGETABLE PRODUCTS

Hemp (232)
Flax (594)
Sunflower seed (11 661)
Cotton seed (28 811)
Cotton lint (15 301)
Sugar beet (281 485)
Potatoes (256 978)
Wheat (458 574)
Rye (24 490)
Oats (42 983)
Barley (158 690)

% world total
80 70 60 50 40 30 20 10

World total 1981 ('000 tons)

ANIMAL PRODUCTS

Wool (2 842)
Eggs (29 200)
Butter (6 870)
Milk (428 205)
Mutton and lamb (8 035)
Pork (55 214)
Beef and veal (46 927)
Chickens (6 486)*
Sheep (1 131)*
Pigs (779)*
Cattle (1 210)*

% world total
30 20 10

World total 1981 ('000 tons or *million head)

LAND USE

Arable land 10%
Pasture land 12%
Forest 51%
Non-agricultural land 27%

LAND USE

Arable land with meadow, permanent grassland and grazing

Pasture land with permanent grassland

Forest with some rough grazing

Non-agricultural land with rough grazing

❊ Orchards and vineyards

PREDOMINANT TYPES OF FARMING

1 Reindeer grazing

2 Forests and animal husbandry

3 Animal husbandry, industrial crops and cereals

4 Cereals and animal husbandry

5 Sheep and goat grazing

6 Industrial crops

PRODUCTION AND CONSUMPTION OF WHEAT IN THE U.S.S.R. 1960-1984

Between 1982 and 1984 the average yield for all cereals in the U.S.S.R. was 1509 kg/ha, compared with a World average of 2373 kg/ha and a U.S.A. figure of 4098 kg/ha.

'000 tons

135 000
120 000
105 000
90 000
75 000
60 000
45 000
30 000
15 000
0

Production

Consumption

2.7 million tons imported
10.3 million tons imported
10.6 million tons imported
14.9 million tons imported
8.6 million tons imported
9.7 million tons imported

1960 61 62 63 64 65 66 67 68 69 70 71 72 73 74 75 76 77 78 79 80 81 82 83 84

COPYRIGHT GEORGE PHILIP & SON, LTD.

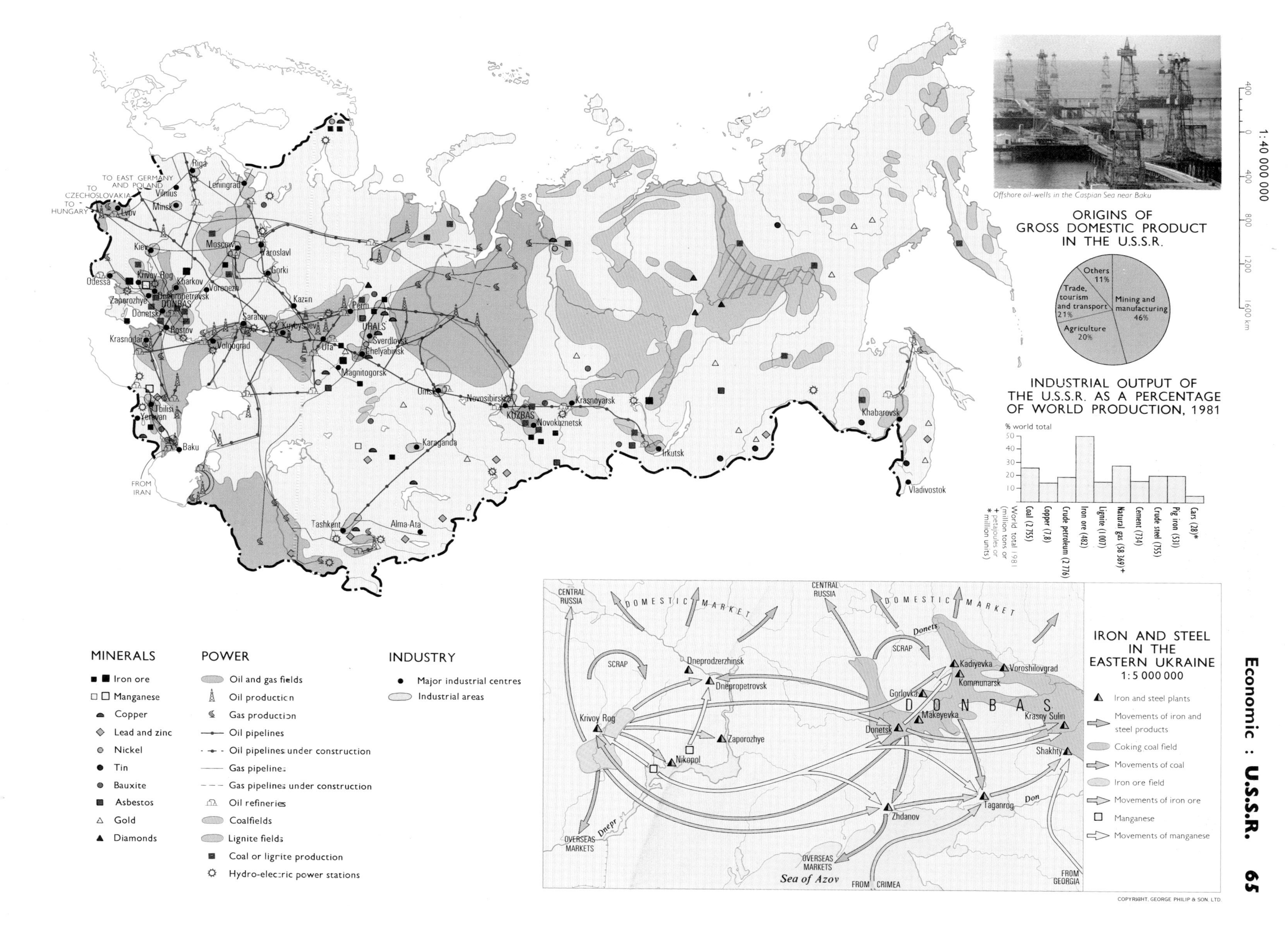

SCALE 1:40 000 000

Offshore oil-wells in the Caspian Sea near Baku

ORIGINS OF GROSS DOMESTIC PRODUCT IN THE U.S.S.R.

Mining and manufacturing 46%

Agriculture 20%

Trade, tourism and transport 21%

Others 11%

INDUSTRIAL OUTPUT OF THE U.S.S.R. AS A PERCENTAGE OF WORLD PRODUCTION, 1981

Cars (28)*
Pig iron (531)
Crude steel (755)
Cement (734)
Natural gas (58 369)+
Lignite (1 007)
Iron ore (482)
Crude petroleum (2 776)
Copper (7,8)
Coal (2 755)

% world total

World total 1981 (million tons or + petajoules or * million units)

IRON AND STEEL IN THE EASTERN UKRAINE
1:5 000 000

Iron and steel plants
Movements of iron and steel products
Coking coal field
Movements of coal
Iron ore field
Movements of iron ore
Manganese
Movements of manganese

DONBAS

Kadiyevka · Voroshilovgrad · Krasny Sulin · Shakhty
Kommunarsk
Makeyevka
Gorlovka
Donetsk · Taganrog · Zhdanov

Dnepropetrovsk · Zaporozhye
Dneprodzerzhinsk
Nikopol
Krivoy Rog

DOMESTIC MARKET

CENTRAL RUSSIA

SCRAP

Sea of Azov

FROM GEORGIA

FROM CRIMEA

OVERSEAS MARKETS

Don

Dnepr

INDUSTRY
● Major industrial centres
Industrial areas

POWER
● Oil and gas fields
⚒ Oil production
⚒ Gas production
●―― Oil pipelines
●-‐- Oil pipelines under construction
―― Gas pipelines
‐-‐ Gas pipelines under construction
Oil refineries
Coalfields
Lignite fields
■ Coal or lignite production
✿ Hydro-electric power stations

MINERALS
■ Iron ore
□ Manganese
● Copper
◆ Lead and zinc
● Nickel
● Tin
● Bauxite
■ Asbestos
△ Gold
▲ Diamonds

Vladivostok
Khabarovsk
Irkutsk
Krasnoyarsk
Novokuznetsk
KUZBAS
Novosibirsk
Karaganda
Alma Ata
Tashkent
Sverdlovsk
Chelyabinsk
Magnitogorsk
URALS
Perm
Ufa
Kazan
Kuybyshev
Saratov
Volgograd
Yaroslavl
Gorki
Moscow
Leningrad
Riga
Vilnius
Minsk
Kiev
Krivoy Rog
Kharkov
Dnepropetrovsk
Zaporozhye
DONBAS
Donetsk
Rostov
Odessa
Krasnodar
Tbilisi
Baku

TO EAST GERMANY AND POLAND
TO CZECHOSLOVAKIA
TO HUNGARY
FROM IRAN

1 : 50 000 000

500   0   500   1000   1500   2000 km

COPYRIGHT GEORGE PHILIP & SON, LTD

East from Greenwich

Projection: Bonne

BHU – Bhutan
B'DESH – Bangladesh
LEB – Lebanon

Gan (Addu Atoll)

Oceans and Seas: PACIFIC OCEAN, ARCTIC OCEAN, INDIAN OCEAN, Bering Sea, Sea of Okhotsk, Japan Sea, Yellow Sea, East China Sea, South China Sea, Philippine Sea, Celebes Sea, Banda Sea, Arafura Sea, Sulu Sea, Bay of Bengal, Arabian Sea, Caspian Sea, Aral Sea, Black Sea, Mediterranean Sea, Red Sea, Baltic Sea, North Sea, Kara Sea, Laptev Sea, Barents Sea, Timor Sea, Flores Sea, Ceram Sea, Molucca Sea, Makasar Strait, Gulf of Thailand, Gulf of Oman, The Gulf (Persian Gulf), Gulf of Aden

Countries / Regions: U.S.S.R., CHINA, MONGOLIA, INDIA, PAKISTAN, AFGHANISTAN, IRAN (PERSIA), IRAQ, SAUDI ARABIA, TURKEY, SYRIA, ISRAEL, JORDAN, LEB, CYPRUS, KUWAIT, QATAR, UNITED ARAB EMIRATES, OMAN, YEMEN, SOUTH YEMEN, BAHRAIN, BURMA, THAILAND (SIAM), VIETNAM, MALAYSIA, BRUNEI, INDONESIA, PHILIPPINES, SRI LANKA (Ceylon), NEPAL, BHU, B'DESH, KOREA, SOMALI REP., ETHIOPIA, SUDAN, EGYPT, LIBYA, KENYA, TANZANIA, UGANDA, RWANDA, BURUNDI, ZAIRE, ZAMBIA, MALAWI, SEYCHELLES, EUROPE, AFRICA, AUSTRALIA, New Guinea, ICELAND, BRITISH ISLES, Greenland, Scandinavia, Finland

Cities: Tokyo, Osaka, Kyoto, Yokohama, Nagoya, Nagasaki, Peking, Shanghai, Nanking, Wuhan, Hong Kong, Macau, Canton (Kwangchow), Foochow, Chungking, Chengtu, Kunming, Sian, Tientsin, Tsingtao, Dairen, Shenyang, Harbin, Changchun, Soochow, Hankow, Lanchow, Urumchi, Kashgar, Yarkand, Lhasa, Delhi, Bombay, Calcutta, Madras, Hyderabad, Ahmadabad, Karachi, Lahore, Kanpur, Lucknow, Varanasi, Allahabad, Agra, Nagpur, Kozhikode (Calicut), Colombo, Kabul, Kandahar, Herat, Mashhad, Tehran, Esfahan, Shiraz, Zahedan, Tabriz, Baghdad, Basra, Kuwait, Riyadh, Mecca, Medina, Abu Dhabi, Muscat, Aden, Jerusalem, Damascus, Beirut, Aleppo, Ankara, Istanbul, Izmir, Erzurum, Tbilisi, Yerevan, Baku, Moscow, Leningrad, Kiev, Warsaw, Berlin, Vienna, Belgrade, Athens, Thessaloniki, Rome, Paris, London, Stockholm, Odessa, Rostov, Astrakhan, Volgograd, Sverdlovsk, Chelyabinsk, Magnitogorsk, Omsk, Novosibirsk, Tomsk, Krasnoyarsk, Irkutsk, Chita, Khabarovsk, Vladivostok, Tashkent, Samarkand, Bukhara, Ashkhabad, Khiva, Mary, Alma Ata, Semipalatinsk, Barnaul, Kemerovo, Tobolsk, Tyumen, Orenburg, Ufa, Guryev, Arkhangelsk, Murmansk, Ulan Bator, Seoul, Pusan, Pyongyang, Manila, Davao, Zamboanga, Kuala Lumpur, Singapore, Penang, Jakarta, Ujung Pandang, Kuching, Rangoon, Mandalay, Bangkok, Hanoi, Ho Chi Minh City, Phnom Penh, Haiphong, Cairo, Alexandria, Aswan, Khartoum, Addis Ababa, Nairobi, Mombasa, Mogadishu, Dar es Salaam, El Obeid, Port Sudan, Massawa, Djibouti, Pondicherry

Physical features: Himalaya, Karakoram Ra., Kunlun, Tien Shan, Plateau of Tibet, Altyn Tagh, Taklamakan, Tarim Basin, Dzungaria, Plateau of Mongolia, Gobi, Manchuria, Plain of North China, Great Khingan, Sikhote Alin, Verkhoyansk Range, Kolyma Range, Stanovoy Ra., Yablonovy Ra., Sayan Mts., Altai, Central Siberian Plateau, West Siberian Plain, Ural Mts., Kirghiz Steppe, Turan Plain, Hindu Kush, Plateau of Iran, Elburz Mts., Caucasus, Zagros Mts., Taurus Mts., Anatolia, Syrian Desert, Mesopotamia, Rub' al Khali, An Nafud, Asir, Deccan, Western Ghats, Eastern Ghats, Malay Peninsula, Kamchatka, Sakhalin, Hokkaido, Honshu, Kyushu, Shikoku, Kuril Is., Aleutian Is., Bonin Is., Ryukyu Is., Andaman Is. (India), Nicobar Is. (India), Lakshadweep (India), Maldives, Socotra, Kuria Muria Is., Bahrain, Hainan, Taiwan, Luzon, Mindanao, Palawan, Borneo, Sumatera, Sulawesi, Halmahera, Ceram, Timor, Java, Irian Jaya, Somali Peninsula, Sinai, Libyan Desert, Scandinavia, Svalbard, Novaya Zemlya, Severnaya Zemlya, Taimyr Peninsula, New Siberian Is., Wrangel I., Greenland, Ethiopian Highlands

Rivers: Ob, Yenisei, Lena, Amur, Indus, Ganges, Brahmaputra, Huang Ho (Yellow R.), Yangtze (Si-kiang), Mekong, Irrawaddy, Salween, Tigris, Euphrates, Nile, Syr Darya, Amu Darya, Volga, Don, Dnepr, Danube, Rhine, Elbe, Vistula, Ural, Angara, Kolyma, Indigirka, Olenek, Kotuy, Lower Tunguska, Khatanga, Yana, Dvina, Godavari, Narmada, Kabul, Helmand

Other labels: Tropic of Cancer, Arctic Circle, Equator, East C. (C. Dezhnev), Bering Str., North Cape, Kola, C. Comorin, Polk Strait, Gulf of Tongking, Strait of Malaka, Sunda Str., Palk Strait, Gan (Addu Atoll), Amirantes, Seychelles

10497 Deep, 4506, 5604, 1022, 10542, 10594

Depth/Height scale (m):
6000 4000 2000 1000 400 200 0 200 2000

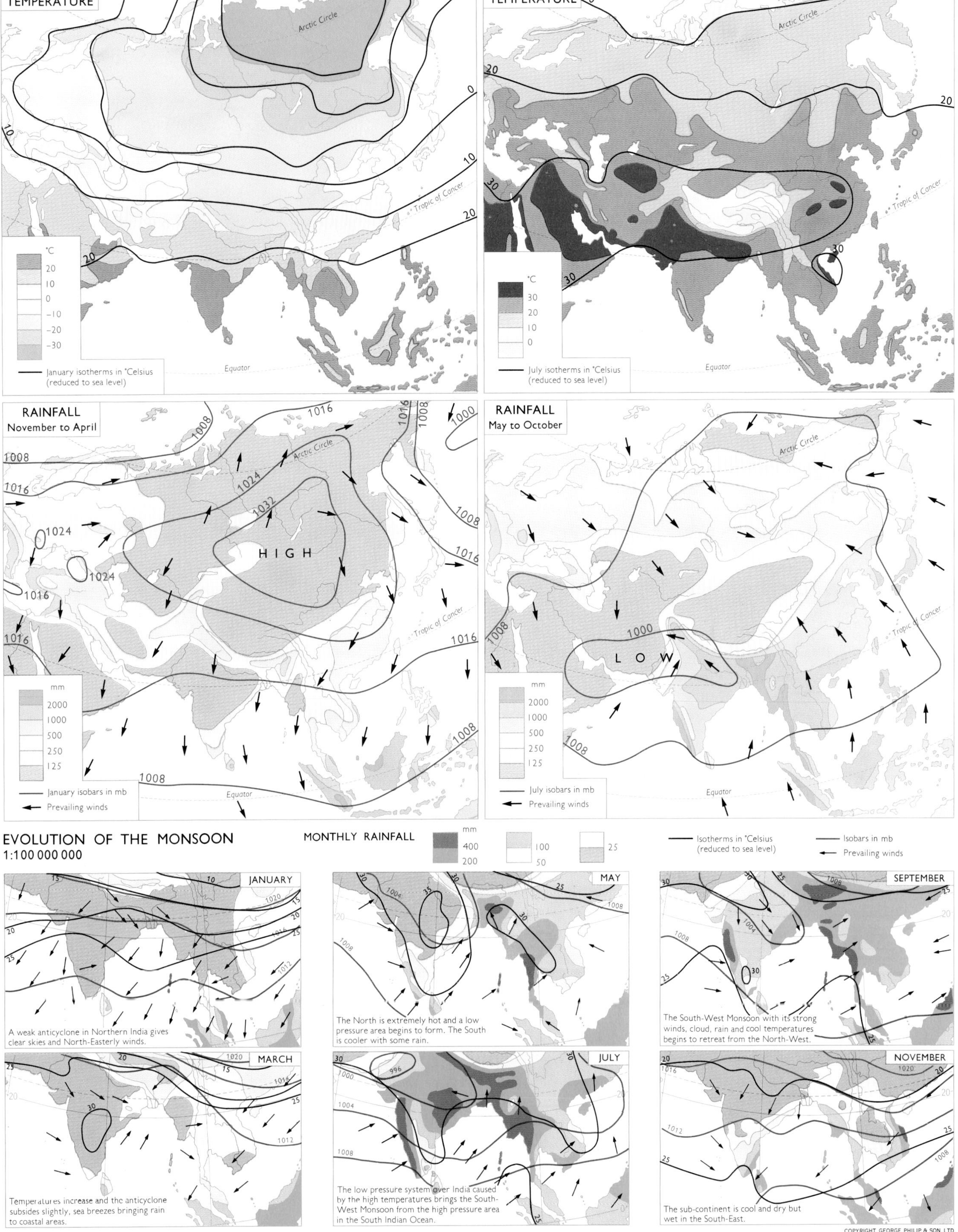

1:110 000 000

1000   0   1000   2000   3000   4000 km

JANUARY TEMPERATURE

Arctic Circle

Tropic of Cancer

°C
20
10
0
−10
−20
−30

— January isotherms in °Celsius (reduced to sea level)

Equator

JULY TEMPERATURE

Arctic Circle

Tropic of Cancer

°C
30
20
10
0

— July isotherms in °Celsius (reduced to sea level)

Equator

RAINFALL
November to April

Arctic Circle

HIGH

Tropic of Cancer

mm
2000
1000
500
250
125

— January isobars in mb
← Prevailing winds

Equator

RAINFALL
May to October

Arctic Circle

LOW

Tropic of Cancer

mm
2000
1000
500
250
125

— July isobars in mb
← Prevailing winds

Equator

EVOLUTION OF THE MONSOON
1:100 000 000

MONTHLY RAINFALL

mm
400
200
100
50
25

— Isotherms in °Celsius (reduced to sea level)
— Isobars in mb
← Prevailing winds

JANUARY

A weak anticyclone in Northern India gives clear skies and North-Easterly winds.

MARCH

Temperatures increase and the anticyclone subsides slightly, sea breezes bringing rain to coastal areas.

MAY

The North is extremely hot and a low pressure area begins to form. The South is cooler with some rain.

JULY

The low pressure system over India caused by the high temperatures brings the South-West Monsoon from the high pressure area in the South Indian Ocean.

SEPTEMBER

The South-West Monsoon with its strong winds, cloud, rain and cool temperatures begins to retreat from the North-West.

NOVEMBER

The sub-continent is cool and dry but wet in the South-East.

COPYRIGHT. GEORGE PHILIP & SON. LTD.

LANGUAGES
1:110 000 000

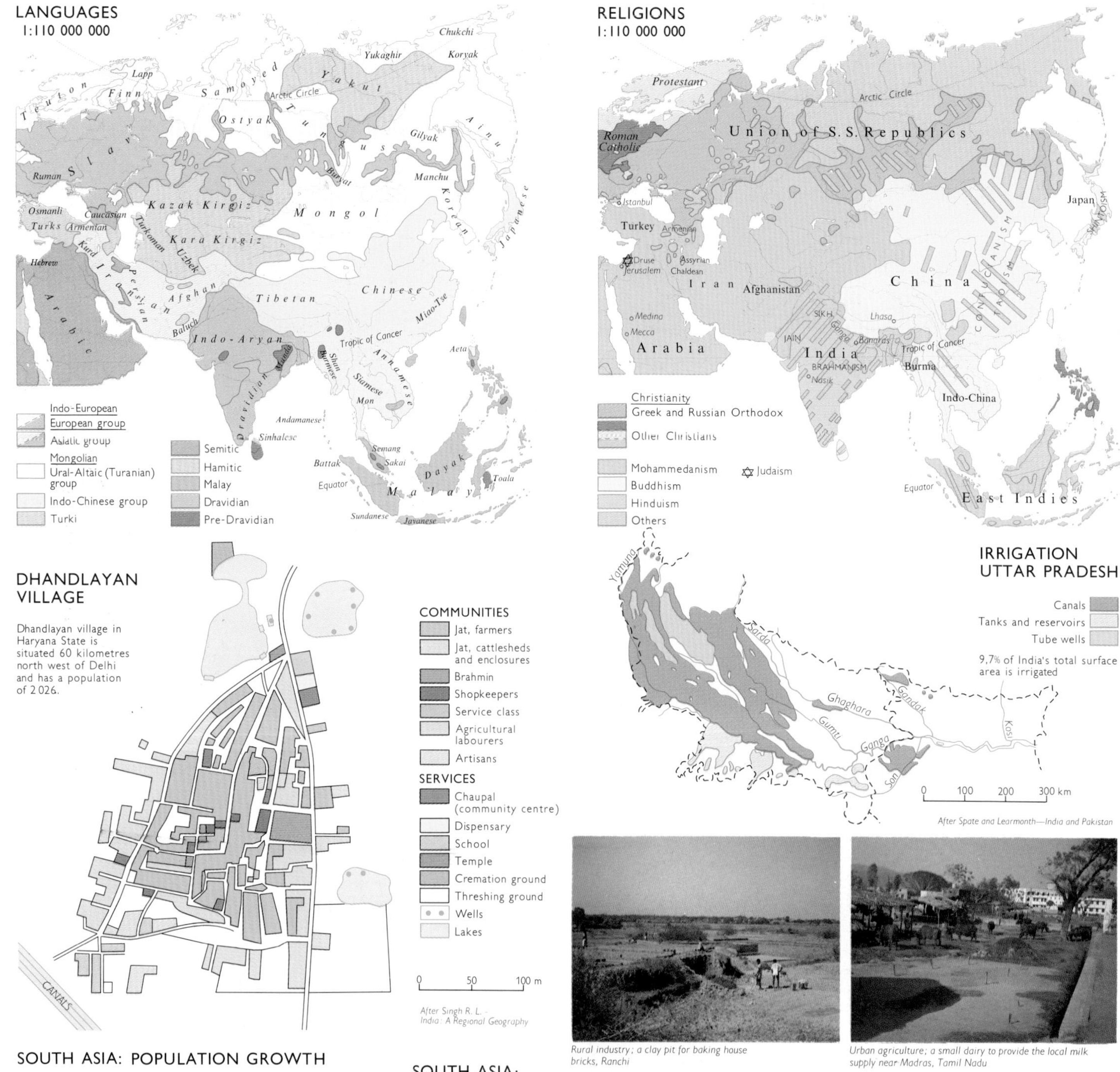

Teuton
Lapp
Finn
Samoyed
Yakut
Yukaghir
Chukchi
Koryak
Arctic Circle
Ostyak
Tungus
Gilyak
Ainu
Slav
Ruman
Buryat
Manchu
Osmanli
Turks
Caucasian
Armenian
Turkoman
Kazak Kirgiz
Uzbek
Mongol
Korean
Japanese
Kara Kirgiz
Hebrew
Kurd
Persian
Afghan
Tibetan
Chinese
Miao-Tse
Arabic
Baluch
Indo-Aryan
Tropic of Cancer
Annamese
Aeta
Dravidian
Shan
Burmese
Siamese
Mon
Andamanese
Sinhalese
Battak
Semang
Sakai
Toala
Equator
Dayak
Malay
Sundanese
Javanese

Indo-European
European group
Asiatic group
Mongolian
Ural-Altaic (Turanian) group
Indo-Chinese group
Turki

Semitic
Hamitic
Malay
Dravidian
Pre-Dravidian

RELIGIONS
1:110 000 000

Protestant
Roman Catholic
Arctic Circle
Union of S.S. Republics
Japan
Istanbul
Turkey
Armenian
Druse
Jerusalem
Assyrian
Chaldean
Iran
Afghanistan
China
Medina
Mecca
Arabia
SIKH.
Lhasa
Ganga
Banaras
JAIN
India
Tropic of Cancer
BRAHMANISM
Nasik
Burma
Indo-China
Equator
East Indies

Christianity
Greek and Russian Orthodox
Other Christians
Mohammedanism
Buddhism
Hinduism
Others
Judaism

DHANDLAYAN VILLAGE

Dhandlayan village in Haryana State is situated 60 kilometres north west of Delhi and has a population of 2 026.

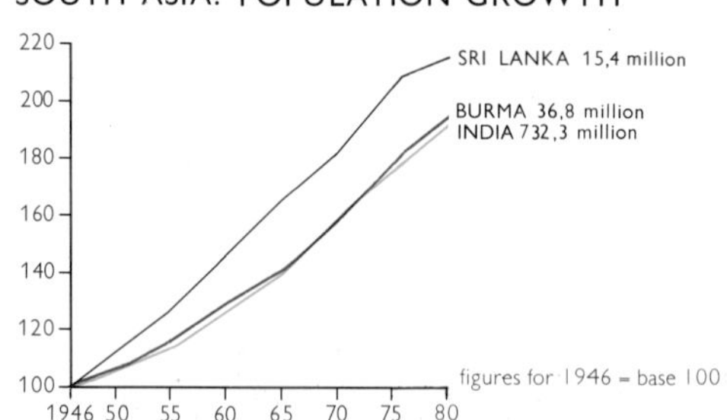

CANALS

COMMUNITIES
Jat, farmers
Jat, cattlesheds and enclosures
Brahmin
Shopkeepers
Service class
Agricultural labourers
Artisans

SERVICES
Chaupal (community centre)
Dispensary
School
Temple
Cremation ground
Threshing ground
Wells
Lakes

0   50   100 m

After Singh R. L.
India : A Regional Geography

IRRIGATION UTTAR PRADESH

Yamuna
Sarda
Ghaghara
Gumti
Ganga
Gandak
Kosi
Son

Canals
Tanks and reservoirs
Tube wells

9,7% of India's total surface area is irrigated

0   100   200   300 km

After Spate and Learmonth—India and Pakistan

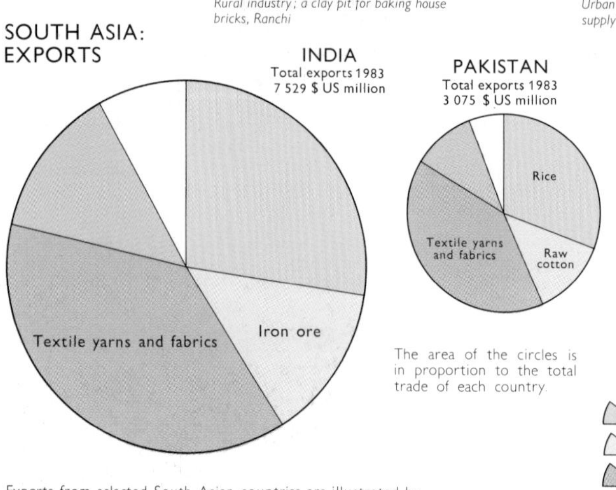

Rural industry; a clay pit for baking house bricks, Ranchi

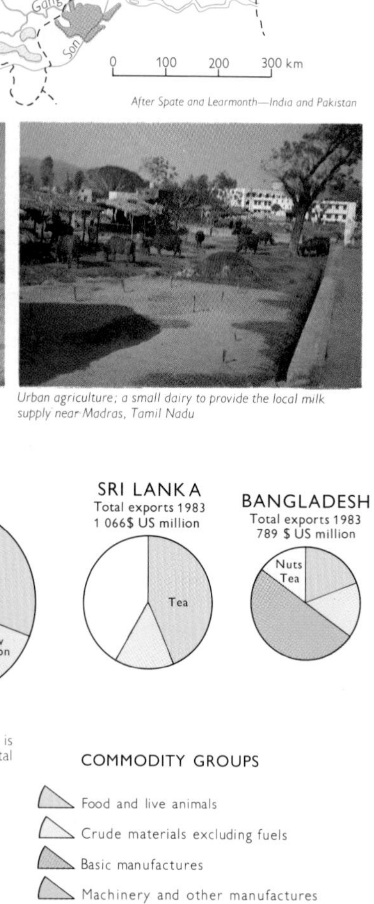

Urban agriculture; a small dairy to provide the local milk supply near Madras, Tamil Nadu

SOUTH ASIA: POPULATION GROWTH

220
200
180
160
140
120
100
1946 50 55 60 65 70 75 80

SRI LANKA 15,4 million
BURMA 36,8 million
INDIA 732,3 million

figures for 1946 = base 100

SOUTH ASIA: DEMOGRAPHY

	Population million	Birth rate per '000 population	Death rate per '000 population	Expectation of life at birth	Population density per km²
India	732.3	33.8	11.9	55	223
Bangladesh	94.7	44.8	17.5	50	657
Pakistan	89.7	44.4	9.6	50	112
Burma	36.8	30.5	11.0	55	54
Nepal	15.7	41.7	18.4	46	112
Sri Lanka	15.4	26.2	6.1	69	234

SOUTH ASIA: EXPORTS

INDIA
Total exports 1983
7 529 $ US million

Textile yarns and fabrics
Iron ore

PAKISTAN
Total exports 1983
3 075 $ US million

Rice
Textile yarns and fabrics
Raw cotton

SRI LANKA
Total exports 1983
1 066$ US million

Tea

BANGLADESH
Total exports 1983
789 $ US million

Nuts
Tea

The area of the circles is in proportion to the total trade of each country.

COMMODITY GROUPS
Food and live animals
Crude materials excluding fuels
Basic manufactures
Machinery and other manufactures
Others : beverages, mineral fuels, chemicals, edible oils and fats

Exports from selected South Asian countries are illustrated by proportional circles divided into principal commodity groups: the main goods in each group are shown.

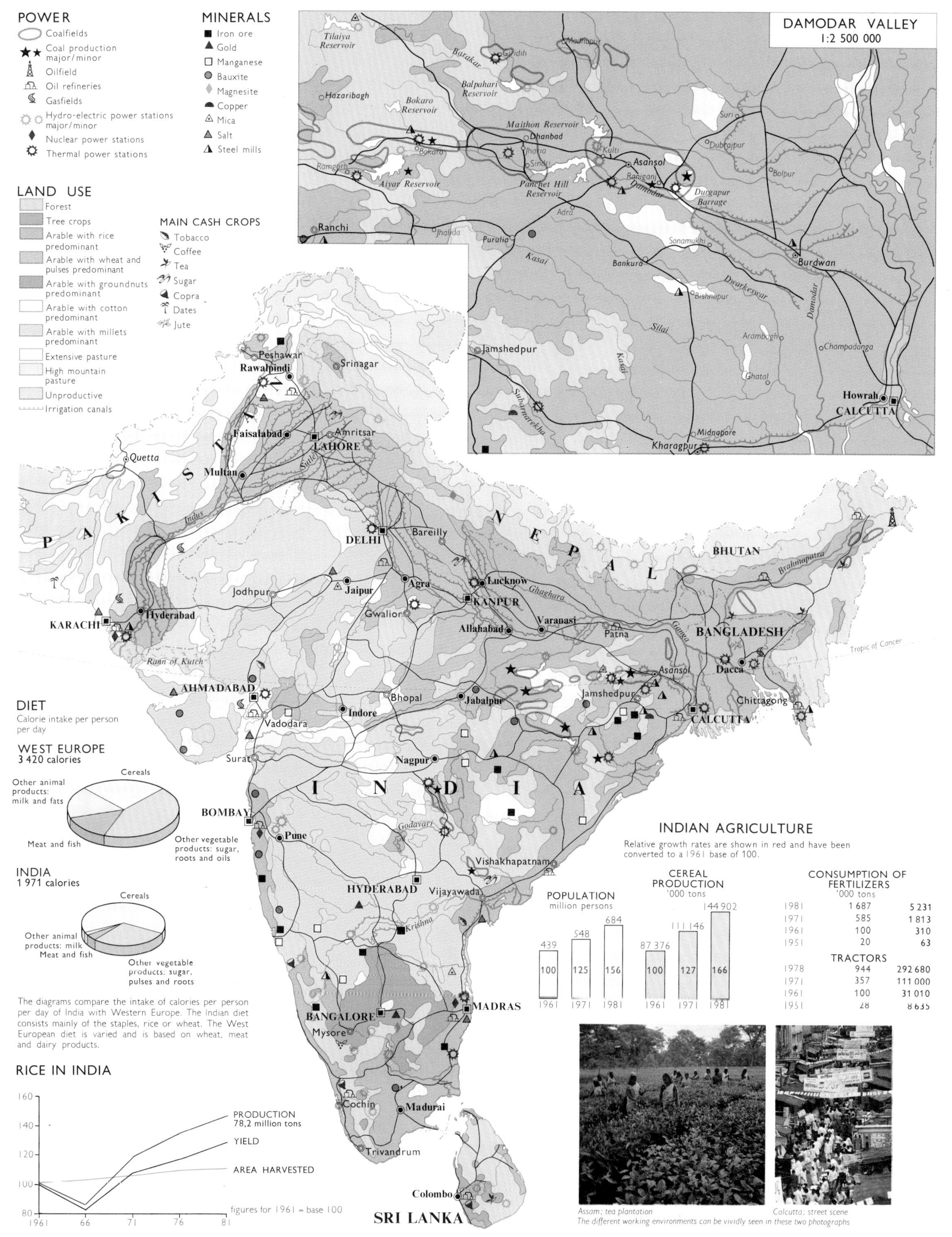

Economic : SOUTH ASIA 69

1:17 500 000

100 0 100 200 300 400 500 600 700 km

POWER
- ⬭ Coalfields
- ★★ Coal production major/minor
- 🗼 Oilfield
- 🏭 Oil refineries
- 🗲 Gasfields
- ⚙⚙ Hydro-electric power stations major/minor
- ◆ Nuclear power stations
- ⚙ Thermal power stations

MINERALS
- ■ Iron ore
- ▲ Gold
- □ Manganese
- ● Bauxite
- ◆ Magnesite
- ● Copper
- △ Mica
- ▲ Salt
- △ Steel mills

LAND USE
- Forest
- Tree crops
- Arable with rice predominant
- Arable with wheat and pulses predominant
- Arable with groundnuts predominant
- Arable with cotton predominant
- Arable with millets predominant
- Extensive pasture
- High mountain pasture
- Unproductive
- Irrigation canals

MAIN CASH CROPS
- 🌿 Tobacco
- ☕ Coffee
- 🍃 Tea
- 🌾 Sugar
- 🌴 Copra
- 🌴 Dates
- 🌿 Jute

DAMODAR VALLEY
1:2 500 000

DIET
Calorie intake per person per day

WEST EUROPE
3 420 calories

- Cereals
- Other animal products: milk and fats
- Meat and fish
- Other vegetable products: sugar, roots and oils

INDIA
1 971 calories

- Cereals
- Other animal products: milk
- Meat and fish
- Other vegetable products: sugar, pulses and roots

The diagrams compare the intake of calories per person per day of India with Western Europe. The Indian diet consists mainly of the staples, rice or wheat. The West European diet is varied and is based on wheat, meat and dairy products.

RICE IN INDIA

PRODUCTION 78,2 million tons
YIELD
AREA HARVESTED

figures for 1961 = base 100

INDIAN AGRICULTURE
Relative growth rates are shown in red and have been converted to a 1961 base of 100.

POPULATION
million persons

	1961	1971	1981
persons	439	548	684
index	100	125	156

CEREAL PRODUCTION
'000 tons

	1961	1971	1981
tons	87 376	111 146	144 902
index	100	127	166

CONSUMPTION OF FERTILIZERS
'000 tons

1981	1 687	5 231
1971	585	1 813
1961	100	310
1951	20	63

TRACTORS

1978	944	292 680
1971	357	111 000
1961	100	31 010
1951	28	8 635

Assam; tea plantation
Calcutta; street scene
The different working environments can be vividly seen in these two photographs

COPYRIGHT. GEORGE PHILIP & SON. LTD.

1:20 000 000

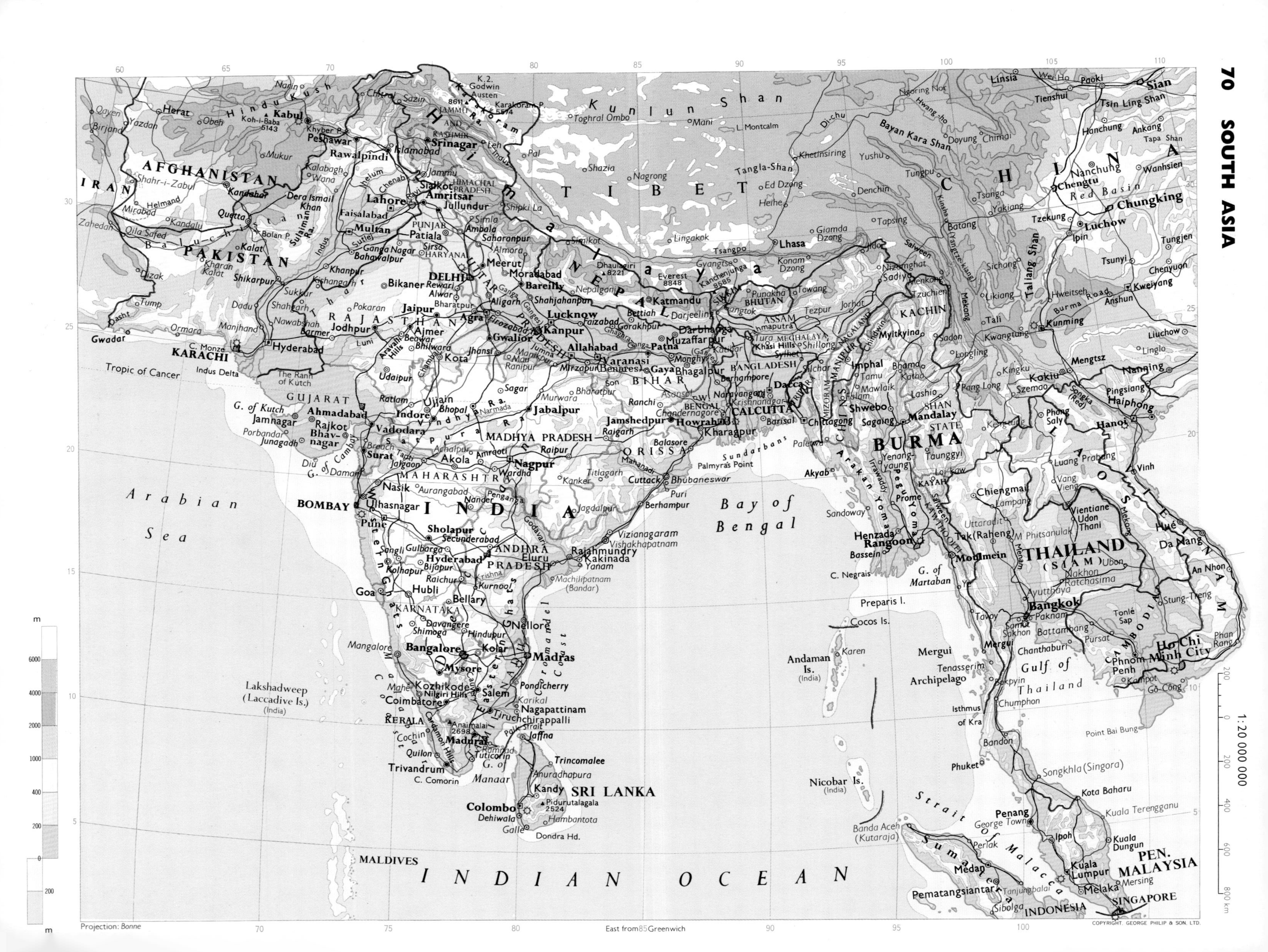
200 0 200 400 600 800 km
COPYRIGHT. GEORGE PHILIP & SON. LTD.
Projection: Bonne
East from 85 Greenwich
Tropic of Cancer
AFGHANISTAN
PAKISTAN
INDIA
NEPAL
BHUTAN
BANGLADESH
BURMA
THAILAND (SIAM)
SRI LANKA
PEN. MALAYSIA
SINGAPORE
INDONESIA
IRAN
TIBET
CHINA
LAOS
CAMBODIA
VIETNAM
MALDIVES
Arabian Sea
Bay of Bengal
INDIAN OCEAN
Gulf of Thailand
Strait of Malacca
Andaman Is. (India)
Nicobar Is. (India)
Lakshadweep (Laccadive Is.) (India)
Kunlun Shan
Himalaya
Karakoram
Hindu Kush
KARACHI
BOMBAY
DELHI
CALCUTTA
Madras
Kabul
Lahore
Katmandu
Lhasa
Dacca
Rangoon
Bangkok
Hanoi
Colombo
Mandalay
Kuala Lumpur
Everest 8848
K.2 Godwin Austen 8611
m
6000
4000
2000
1000
400
200
0
200
m

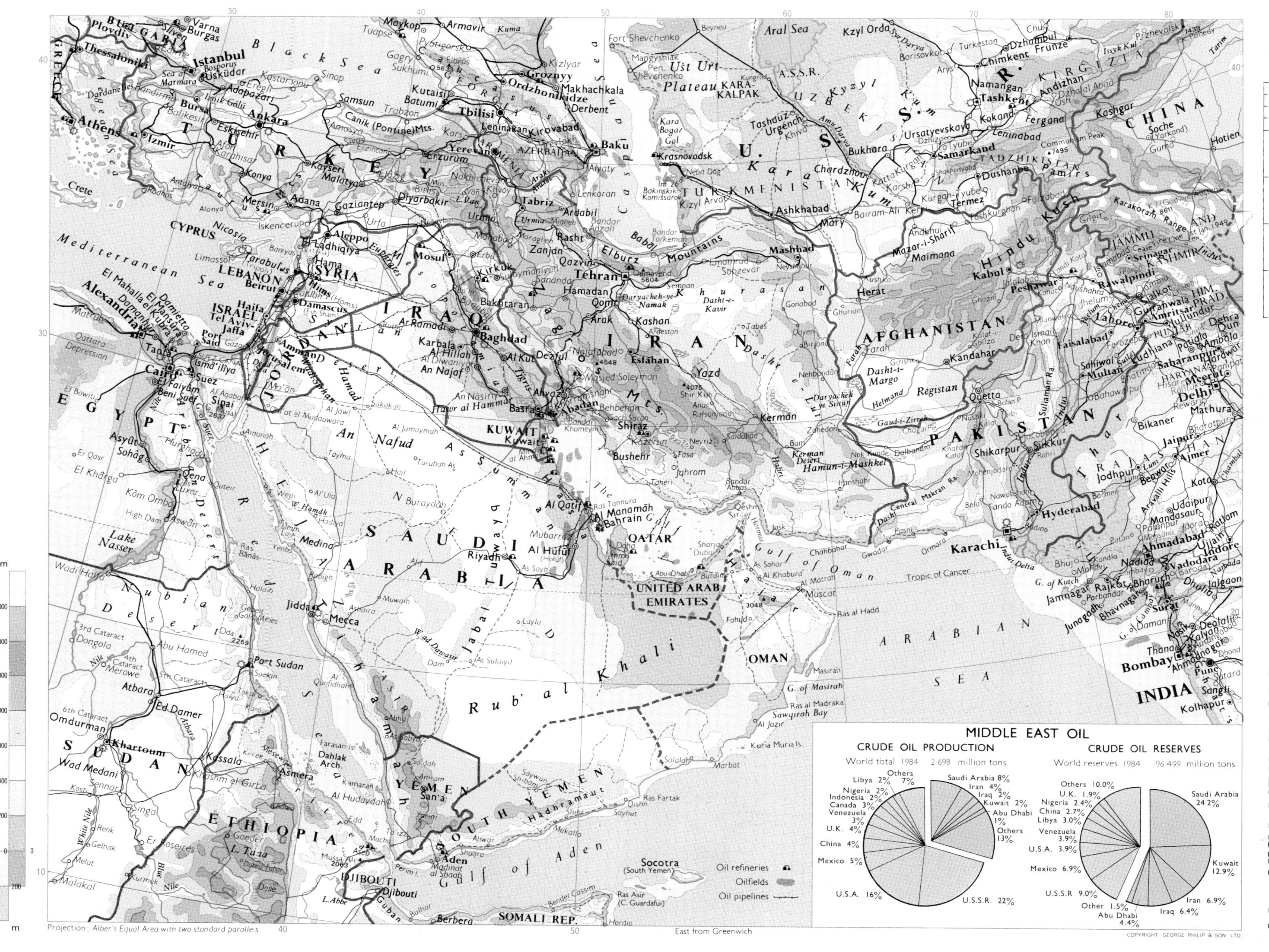
1:20 000 000
200 0 200 400 600 800 km
MIDDLE EAST OIL
CRUDE OIL PRODUCTION
World total 1984 2 698 million tons
Saudi Arabia 8%
Iran 4%
Iraq 2%
Kuwait 2%
Abu Dhabi 1%
Others 13%
U.S.S.R. 22%
U.S.A. 16%
Mexico 5%
China 4%
U.K. 4%
Venezuela 3%
Canada 3%
Indonesia 2%
Nigeria 2%
Libya 2%
Others 7%
CRUDE OIL RESERVES
World reserves 1984 96 499 million tons
Saudi Arabia 24.2%
Kuwait 12.9%
Iran 6.9%
Iraq 6.4%
Abu Dhabi 4.4%
Other 1.5%
U.S.S.R 9.0%
Mexico 6.9%
U.S.A. 3.9%
Venezuela 3.9%
Libya 3.0%
China 2.7%
Nigeria 2.4%
U.K. 1.9%
Others 10.0%
Oil refineries
Oilfields
Oil pipelines
East from Greenwich
Projection: Alber's Equal Area with two standard paralles
m 6000 4000 3000 2000 1000 400 200 0 200 m

1 : 20 000 000

200 0 200 400 600 800 km

POPULATION
1 : 60 000 000

inhabitants per km²
over 600
200–600
100–200
50–100
10–50
1–10
under 1
■ Towns of over 1 million inhabitants

COPYRIGHT GEORGE PHILIP & SON LTD.

Projection: Bonne

East from Greenwich

PACIFIC OCEAN

INDIAN OCEAN

SOUTH CHINA SEA

CELEBES SEA

SULU SEA

BANDA SEA

CERAM SEA

MOLUCCA SEA

JAVA SEA

FLORES SEA

ARAFURA SEA

TIMOR SEA

ANDAMAN SEA

CHINA
BURMA
LAOS
VIET-NAM
THAILAND (SIAM)
CAMBODIA
MALAYSIA
PHILIPPINES
BORNEO
KALIMANTAN
SUMATRA
SARAWAK
SABAH
BRUNEI
SULAWESI (CELEBES)
IRIAN JAYA
INDONESIA
AUSTRALIA
TAIWAN (FORMOSA)
HONG KONG
LUZON
MINDANAO
JAVA
BALI
MADURA

RANGOON
BANGKOK
PHNOM PENH
HO CHI MINH CITY (SAIGON)
HANOI
HAIPHONG
KUALA LUMPUR
SINGAPORE
JAKARTA
BANDUNG
SURABAYA
MANILA
KUCHING
BANJARMASIN
PALEMBANG
MEDAN
PADANG

m
4000
2000
1000
400
200
0
200

Manila
Hong Kong
Jakarta
Bandung
Surabaya
Singapore
Bangkok
Rangoon
Ho Chi Minh City
Equator

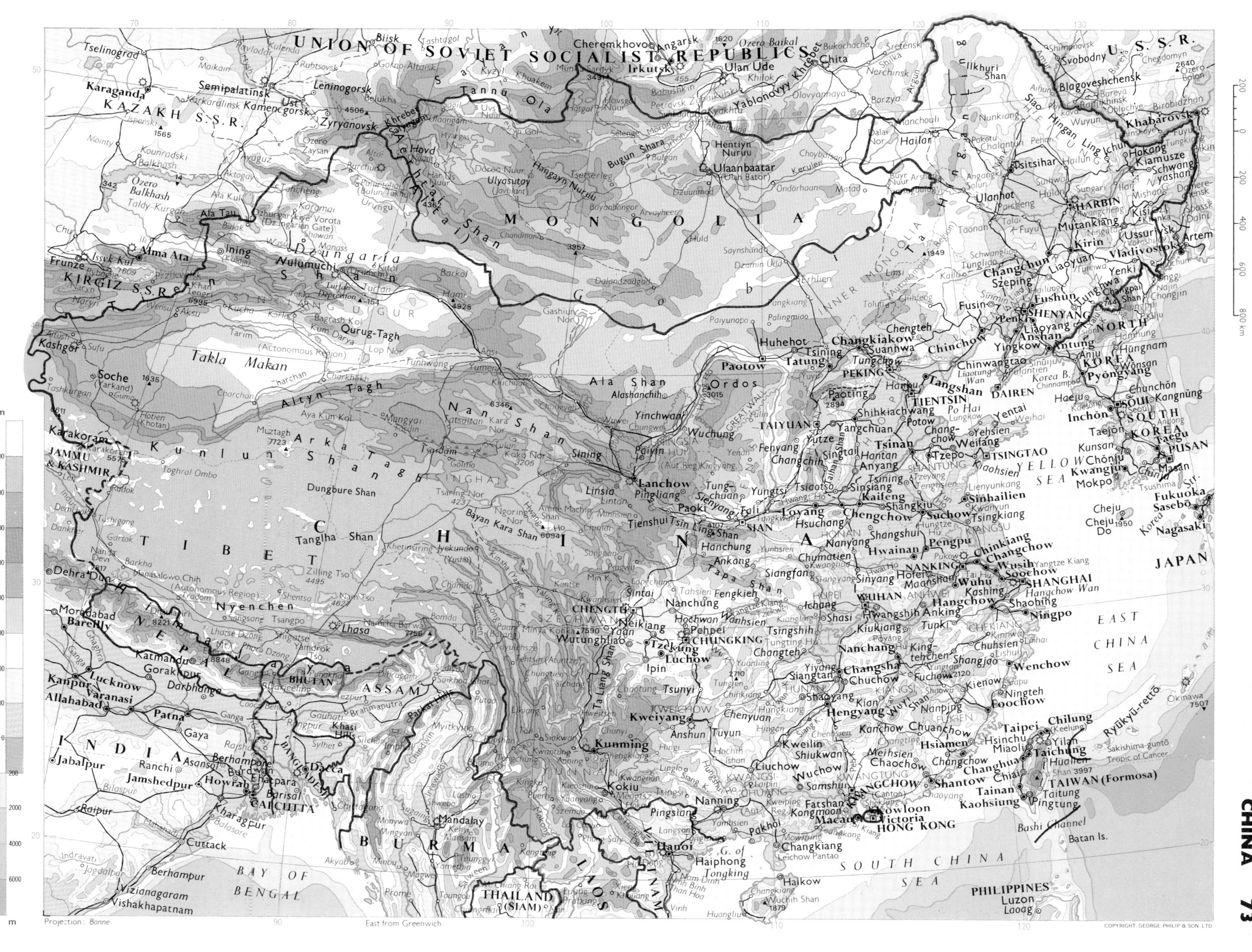

1:20 000 000

East from Greenwich

Projection: Bonne

1:35 000 000

200 0 200 400 600 800 1000 1200 1400 km

DENSITY OF POPULATION

POPULATION OF CITIES AND TOWNS

■ Over 2 million inhabitants
● 1-2 million inhabitants
● 250 000 – 1 million inhabitants
• 100 000 – 250 000 inhabitants

DENSITY OF POPULATION
inhabitants per km²

over 200
100–200
50–100
10–50
1–10
under 1
uninhabited

Harbin
Changchun
Fushun
Shenyang
Anshan
Peking
Tientsin
Dairen
Taiyuan
Tsinan
Lanchow
Tsingtao
Sian
Chengchow
Chengtu
Nanking
Shanghai
Wuhan
Chungking
Kunming
Canton
Hong Kong
Taipei
Kaohsiung

THE GROWTH OF POPULATION

millions
5 000
4 000
3 000
2 000
1 000
0

WORLD
ASIA (excl. U.S.S.R.)
CHINA

1800 1850 1900 1950 1985

In the latest census, taken on 1.7.82
China's population was reported to be 1 008 175 288

LAND USE AND AGRICULTURE

Over 50% of the total area of land is cultivated
Over 10% of the total area of land is cultivated
Under 10% of the total area of land is cultivated
Non-cultivated with grazing land
Forest
Oasis cultivation

PREDOMINANT TYPES OF FARMING

1	Corn, kaoliang, soybeans
2	Spring wheat
3	Corn, winter wheat, barley
4	Millet, corn, winter wheat
5	Rice, wheat, rape, peas
6	Rice, winter wheat
7	Rice, wheat, beans
8	Rice, wheat, peas, tea
9	Two rice crops, wheat, sugar cane
10	Grassland with diverse agriculture in irrigated areas
11	Non-agricultural land with rough grazing

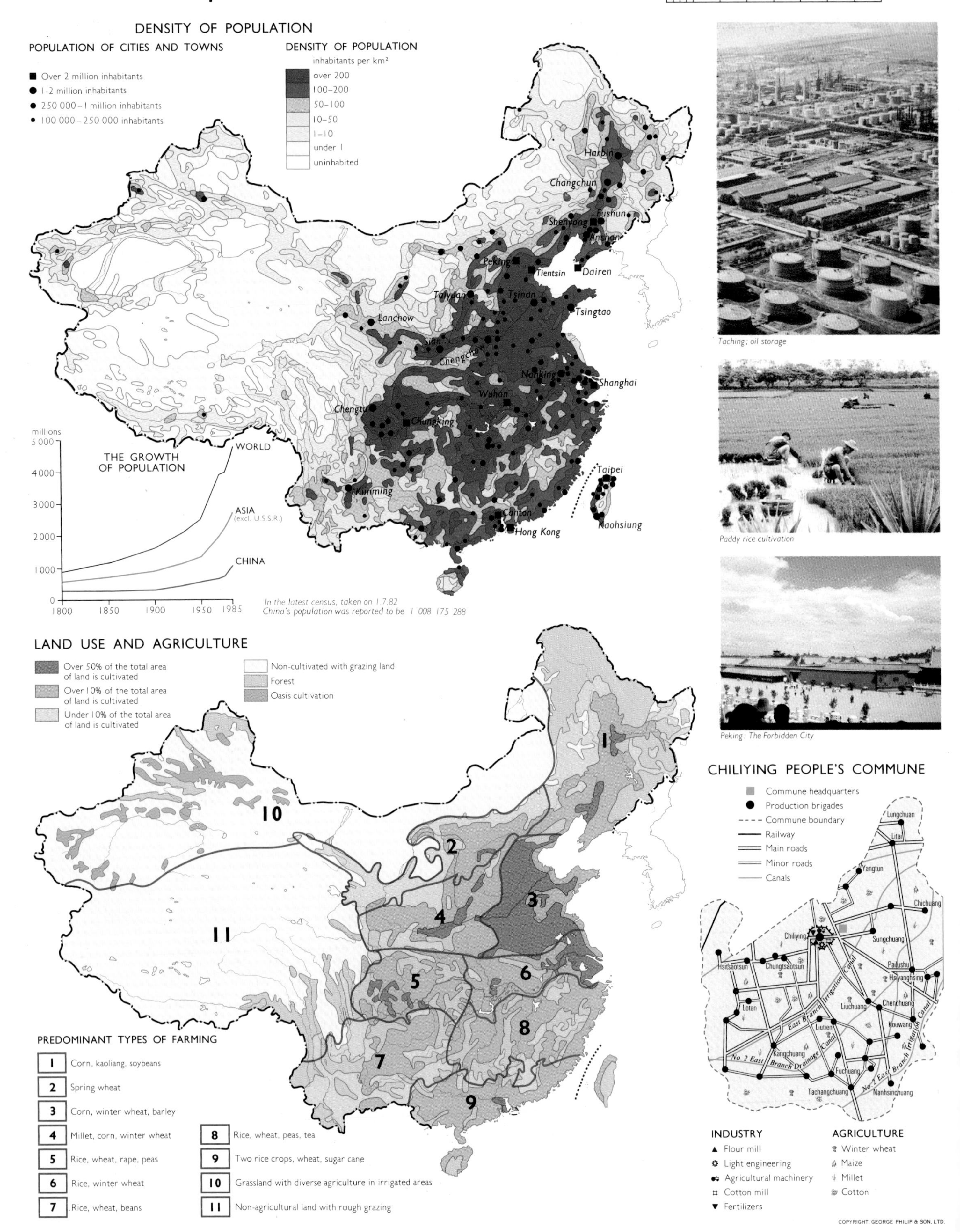

Taching; oil storage

Paddy rice cultivation

Peking : The Forbidden City

CHILIYING PEOPLE'S COMMUNE

■ Commune headquarters
● Production brigades
--- Commune boundary
━ Railway
Main roads
Minor roads
Canals

Lungchuan
Litai
Yangtun
Chichuang
Chiliying
Sungchuang
Hsiaotsun
Chungtsaotsun
Pailushu
Hsiyanghsing
Lotan
Liuchuang
Chenchuang
Liutien
Kouwang
East Branch Irrigation Canal
Kangchuang
Fuchuang
No. 2 East Branch Drainage Canal
Tachangchuang
Nanhsinchuang

INDUSTRY

▲ Flour mill
✿ Light engineering
⚙ Agricultural machinery
⊞ Cotton mill
▼ Fertilizers

AGRICULTURE

🌾 Winter wheat
Maize
Millet
Cotton

COPYRIGHT. GEORGE PHILIP & SON. LTD.

1:50 000 000

500   0   500   1000   1500   2000 km

COAL

▬ Coalfields, near the surface
▬ Coalfields, deeply buried
★ Important production centres
⭑ Other production centres

Hokang ★
Fusin ★ Fushun
Tatung
Hwainan ⭑

COAL PRODUCTION
million tons

1 000
800
600
400
200

1960 1965 1970 1975 1980 1985

PETROLEUM

⛭ Oilfields
⛭ Oil refineries
— Oil pipelines
◈ Gasfields
◆ Oil shale
⛭ Oil shale refinery
◆ Uranium

Karamai
Tushantze
Urumchi
Lenghu
Mangyai
Yumen
Lanchow
Yanchang
Taching
Fusin
Fushun
Dairen
Shengli
Nanking
Shanghai
Nanchung
Tzekung
Miaoli
Kaohsiung

CRUDE PETROLEUM
million tons

150
120
90
60
30

1960 1965 1970 1975 1980 1985

IRON AND STEEL

▲ Iron and steel plants (major centres underlined)
△ Iron plant
▲ Steel plant
■ Iron ore
□ Coking coal

Urumchi
Kiuchuan
Paotow
Suanhwa
Anshan
Penki
Fushun
Shihkingshan
Taiyuan
Hantan Fengfeng
Pingtingshan
Hwainan
Nanking
Shanghai
Wuhan
Maanshan
Chungking
Hwangshih
Tsunyi
Siangtan
Anning
Shiukwan
Canton
Taipei

IRON AND STEEL PRODUCTION
million tons

60
45
30
15

1960 1965 1970 1975 1980 1985

Iron   Steel

NON-FERROUS MINERALS

Sikwangshan
Fenghwang
Tungchwan
Tayu
Kokiu

PRODUCTION '000 tons

	1949	1956	1965	1974	1983	% world production
+ Tungsten	7	11	10	11	12.5	32
▲ Antimony	3	13	15	12	13.0	27
● Tin	5	14	24	23	17.5	8
● Bauxite	—	150	400	600	1 900	2
□ Manganese	51	158	300	300	1 800	7
▲ Copper	5	12	90	150	175	2
● Mercury	—	0.6	0.9	1	0.85	14

	1949	1956	1965	1974	1983	% world production
◆ Lead	2	35	100	140	160	5
◆ Zinc	0	35	100	135	160	2
□ Molybdenum	—	1.2	1.5	1.5	2.0	3
◆ Magnesite	—	—	1000	1000	2 000	18

MACHINERY

⚙ Railway equipment
★ Mining machinery
⊟ Textile machinery
⚙ Agricultural machinery
◉ Diversified
✿ Machine tools
⚓ Shipbuilding
⚙ Motor vehicles

Urumchi
Tsitsihar
Harbin
Changchun
Shenyang
Fushun
Changkiakow
Shanhaikwan
Peking
Dairen
Tientsin
Tangshan
Taiyuan
Yutze
Tsingtao
Lanchow
Tsinan
Loyang
Chengchow
Sian
Nanking
Shanghai
Chengtu
Wuhan
Chungking
Changsha
Nanchang
Siangtan
Chuchow
Hengyang
Kanchow
Taipei
Kunming
Keelung
Canton
Kaohsiung

MACHINE TOOLS
thousand units

200
160
120
80
40

1960 1965 1970 1975 1980 1985

TEXTILES

⊓ Cotton
⊟ Linen
★ Wool
◊ Silk

Harbin
Antung
Khotan
Peking
Lanchow
Chengchow
Sian
Wusih Nantung
Nanking Shanghai
Chungking
Hangchow
Canton
Shuntak Hong Kong

COTTON CLOTH PRODUCTION
thousand million linear metres

16
12
8
4

1960 1965 1970 1975 1980 1985

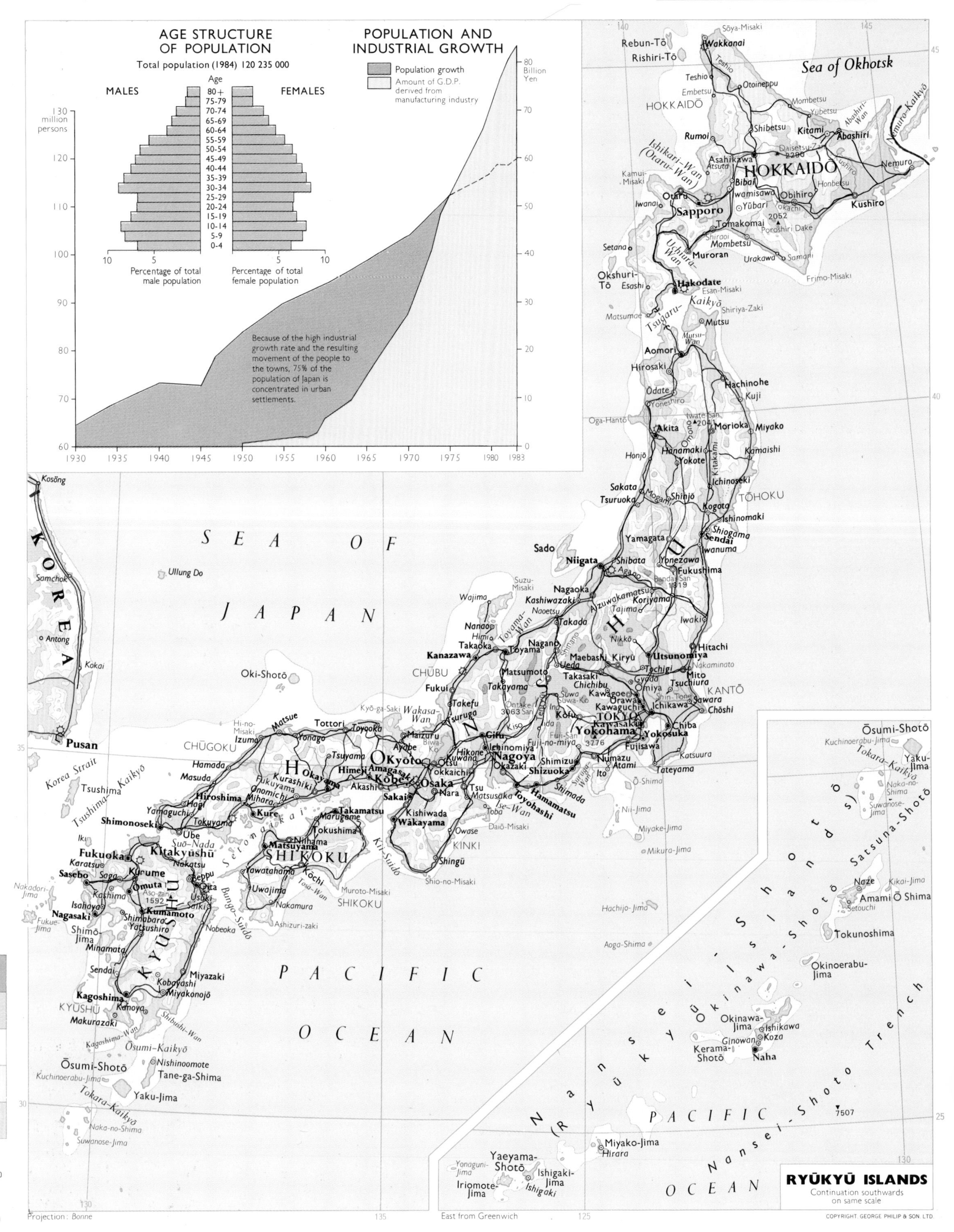

1 : 7 500 000

AGE STRUCTURE
OF POPULATION
Total population (1984) 120 235 000

POPULATION AND
INDUSTRIAL GROWTH

■ Population growth
□ Amount of G.D.P. derived from manufacturing industry

MALES   Age   FEMALES
80+
75-79
70-74
65-69
60-64
55-59
50-54
45-49
40-44
35-39
30-34
25-29
20-24
15-19
10-14
5-9
0-4

Percentage of total male population
Percentage of total female population

Because of the high industrial growth rate and the resulting movement of the people to the towns, 75% of the population of Japan is concentrated in urban settlements.

Sea of Okhotsk

HOKKAIDŌ

Sapporo

KOREA

SEA OF JAPAN

TŌHOKU

CHŪBU

KANTŌ

CHŪGOKU

Kyōto
Ōsaka
TOKYO
Yokohama

KINKI

Hiroshima

SHIKOKU

KYŪSHŪ

Nagasaki

Kagoshima

PACIFIC OCEAN

RYŪKYŪ ISLANDS
Continuation southwards
on same scale

Ōsumi-Shotō

Projection: Bonne

East from Greenwich

COPYRIGHT GEORGE PHILIP & SON. LTD.

1:18 750 000

100 0 100 200 300 400 500 600 700 km

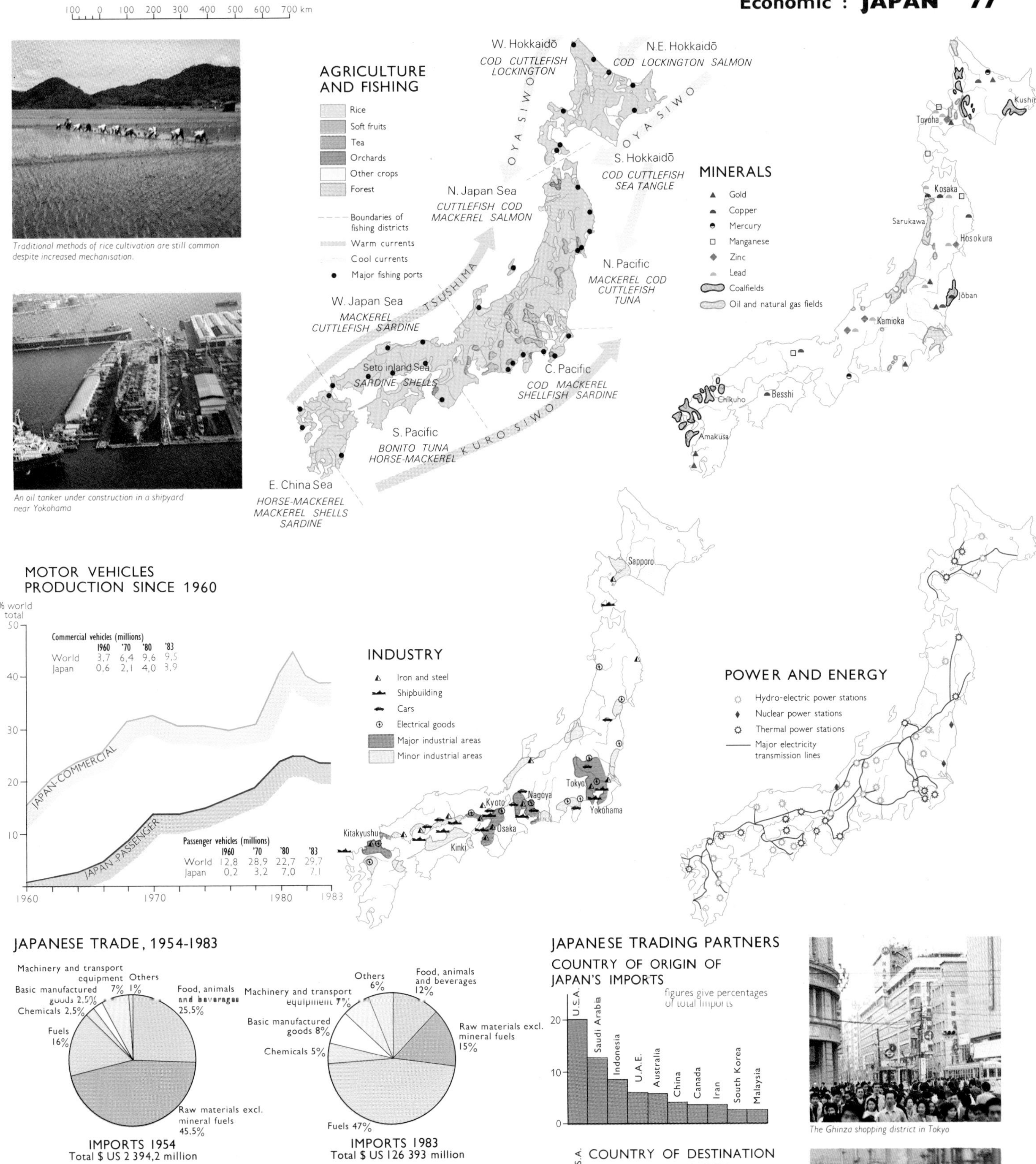

Traditional methods of rice cultivation are still common despite increased mechanisation.

An oil tanker under construction in a shipyard near Yokohama

AGRICULTURE AND FISHING

- Rice
- Soft fruits
- Tea
- Orchards
- Other crops
- Forest

-- -- Boundaries of fishing districts
≈≈≈ Warm currents
~~~ Cool currents
● Major fishing ports

W. Hokkaidō
COD CUTTLEFISH LOCKINGTON

N.E. Hokkaidō
COD LOCKINGTON SALMON

S. Hokkaidō
COD CUTTLEFISH SEA TANGLE

N. Japan Sea
CUTTLEFISH COD MACKEREL SALMON

N. Pacific
MACKEREL COD CUTTLEFISH TUNA

W. Japan Sea
MACKEREL CUTTLEFISH SARDINE

C. Pacific
COD MACKEREL SHELLFISH SARDINE

Seto Inland Sea
SARDINE SHELLS

S. Pacific
BONITO TUNA HORSE-MACKEREL

E. China Sea
HORSE-MACKEREL MACKEREL SHELLS SARDINE

OYA SIWO
TSUSHIMA
KURO SIWO

MINERALS

- ▲ Gold
- ⬠ Copper
- ● Mercury
- ☐ Manganese
- ◆ Zinc
- ▲ Lead
- Coalfields
- Oil and natural gas fields

Kushiro
Toyoha
Kosaka
Sarukawa
Hosokura
Jōban
Kamioka
Chikuho
Besshi
Amakusa

MOTOR VEHICLES PRODUCTION SINCE 1960

% world total

Commercial vehicles (millions)	1960	'70	'80	'83
World	3,7	6,4	9,6	9,5
Japan	0,6	2,1	4,0	3,9

JAPAN-COMMERCIAL

JAPAN PASSENGER

Passenger vehicles (millions)	1960	'70	'80	'83
World	12,8	28,9	22,7	29,7
Japan	0,2	3,2	7,0	7,1

1960   1970   1980   1983

INDUSTRY

- ▲ Iron and steel
- ⬇ Shipbuilding
- Cars
- ① Electrical goods
- Major industrial areas
- Minor industrial areas

Tokyo
Kyoto
Nagoya
Osaka
Yokohama
Kitakyushu
Kinki

POWER AND ENERGY

- ○ Hydro-electric power stations
- ◆ Nuclear power stations
- ✿ Thermal power stations
- — Major electricity transmission lines

Sapporo

JAPANESE TRADE, 1954-1983

IMPORTS 1954
Total $ US 2 394,2 million

- Machinery and transport equipment 7%
- Others 1%
- Food, animals and beverages 25.5%
- Basic manufactured goods 2.5%
- Chemicals 2.5%
- Fuels 16%
- Raw materials excl. mineral fuels 45.5%

IMPORTS 1983
Total $ US 126 393 million

- Others 6%
- Food, animals and beverages 12%
- Machinery and transport equipment 7%
- Basic manufactured goods 8%
- Chemicals 5%
- Raw materials excl. mineral fuels 15%
- Fuels 47%

EXPORTS 1954
Total $ US 1 629,5 million

- Others 10%
- Food, raw materials and mineral fuels 14%
- Chemicals 5%
- Machinery, cars and ships 13%
- Iron and steel 10%
- Manufactured goods 48%

EXPORTS 1983
Total $ US 146 927 million

- Others 9%
- Food, raw materials and mineral fuels 2%
- Chemicals 7%
- Manufactured goods 11%
- Iron and steel 9%
- Machinery cars and ships 64%

JAPANESE TRADING PARTNERS

COUNTRY OF ORIGIN OF JAPAN'S IMPORTS

figures give percentages of total imports

U.S.A., Saudi Arabia, Indonesia, U.A.E., Australia, China, Canada, Iran, South Korea, Malaysia

COUNTRY OF DESTINATION OF JAPAN'S EXPORTS

figures give percentages of total exports

U.S.A., Saudi Arabia, South Korea, W. Germany, Hong Kong, U.K., China, Singapore, Australia, Indonesia

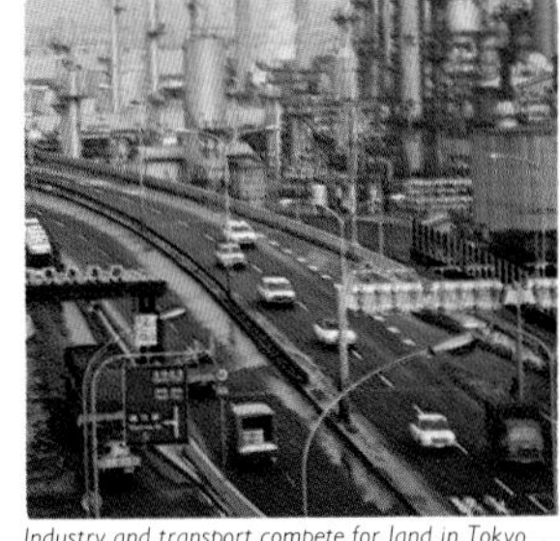

The Ghinza shopping district in Tokyo

Industry and transport compete for land in Tokyo.

COPYRIGHT. GEORGE PHILIP & SON LTD.

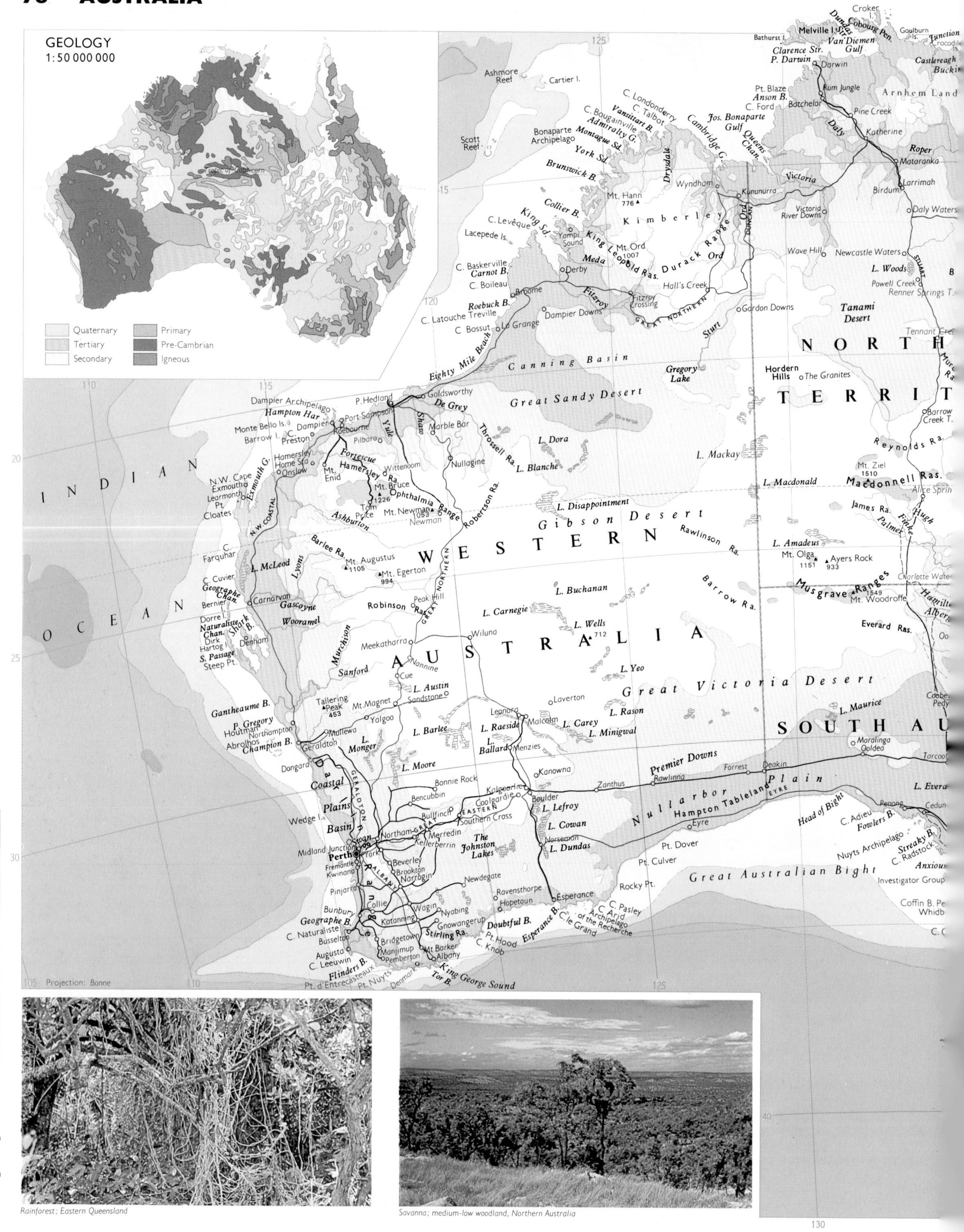

GEOLOGY
1:50 000 000

Legend:
- Quaternary
- Tertiary
- Secondary
- Primary
- Pre-Cambrian
- Igneous

Projection: Bonne

m
2000
1500
1000
400
200
0
600
6000
12 000
18 000

Rainforest; Eastern Queensland

Savanna; medium-low woodland, Northern Australia

Map labels (selected):

INDIAN OCEAN

WESTERN AUSTRALIA

SOUTH AU...

NORTH TERRIT...

Croker
Dundas
Cobourg Pen.
Goulburn Is.
Junction
Crocodile
Melville I.
Van Diemen Gulf
Castlereagh
Buckin...
Bathurst I.
Clarence Str.
P. Darwin
Darwin
C. Ford
C. Talbot
Pt. Blaze
Anson B.
Batchelor
Rum Jungle
Arnhem Land
Ashmore Reef
Cartier I.
C. Londonderry
C. Bougainville
Jos. Bonaparte Gulf
Cambridge G.
Queens Chan.
Pine Creek
Scott Reef
Bonaparte Archipelago
Montague Sd.
Admiralty G.
Vansittart B.
Katherine
Roper
Matarinka
York Sd.
Brunswick B.
Wyndham
Kununurra
Victoria
Birdum
Larrimah
Collier B.
Mt. Hann 776
Kimberley
Ord
Duncan
Daly Waters
C. Lévêque
King Sd.
King Leopold Ras.
Mt. Ord 1007
Durack Ord
Victoria River Downs
Lacepede Is.
Meda
Hall's Creek
Wave Hill
Newcastle Waters
C. Baskerville
Carnot B.
Derby
L. Woods
C. Boileau
Broome
Fitzroy Crossing
Gordon Downs
Powell Creek
Renner Springs T.
Roebuck B.
Fitzroy
Great Northern
Sturt
Tanami Desert
C. Latouche Treville
Dampier Downs
Tennant Cre...
C. Bossut
Le Grange
Eighty Mile Beach
Canning Basin
Gregory Lake
Hordern Hills
The Granites
Barrow Creek T.
Dampier Archipelago
P. Hedland
Goldsworthy
De Grey
Great Sandy Desert
Hampton Har
Port Sampson
Monte Bello Is.
Dampier
Roebourne
Yule
Shaw
Marble Bar
L. Dora
Reynolds Ra.
Barrow I.
Pilbara
Throssell Ra.
Nullagine
L. Blanche
Mt. Ziel 1510
Hamersley
Home Sta.
Onslow
Fortescue
Wittenoom
Hamersley Ra.
L. Mackay
Macdonnell Ras.
Alice Sprin...
N.W. Cape
Exmouth
Learmonth
Pt. Cloates
Mt. Enid
Mt. Bruce 1226
Ophthalmia Range
Robertson Ra.
L. Macdonald
James Ra.
Hugh
Palmer
Ashburton
Tom Price
Mt. Newman 1053
Newman
Gibson Desert
Rawlinson Ra.
L. Disappointment
L. Amadeus
C. Farquhar
Barlee Ra.
Mt. Augustus 1105
Mt. Egerton 994
Peak Hill
L. Buchanan
Mt. Olga 1151
Ayers Rock 933
Charlotte Water
C. Cuvier
L. McLeod
Lyons
Geographe Chan.
Bernier I.
Carnarvon
Gascoyne
Wooramel
Robinson
Peak Hill
L. Carnegie
Barrow Ra.
Musgrave Ranges
Mt. Woodroffe 1549
Hamersl...
Albers...
Dorre I.
Naturaliste Chan.
Dirk Hartog I.
Denham
Wiluna
L. Wells 712
Everard Ras.
S. Passage
Steep Pt.
Meekatharra
Nannine
L. Yeo
Gantheaume B.
Sanford
Cue
L. Austin
Sandstone
Great Victoria Desert
L. Maurice
Cooba Pedy
P. Gregory
Houtman
Abrolhos
Northampton
Tallering Peak 453
Mt. Magnet
Yalgoo
Laverton
L. Rason
Maralinga
Ooldea
Tarcoo...
Champion B.
Geraldton
L. Monger
Leonora
Malcolm
L. Carey
L. Minigwal
Dongara
L. Moore
L. Raeside
Menzies
Premier Downs
Forrest
Deakin
L. Everar...
Wedge I.
Bonnie Rock
Ballard
Kanowna
Rawlinna
Hampton Tableland
Head of Bight
C. Adieu
Fowlers B.
Penong
Cedun...
Midland Junction
Swan
Northam
Bencubbin
Kalgoorlie
Zanthus
Nullarbor Plain
Eyre
Streaky B.
C. Radstock
Perth
York
Merredin
Coolgardie
Boulder
L. Lefroy
Eyre
Pt. Dover
Nuyts Archipelago
Anxious
Fremantle
Kwinana
Beverley
Kellerberrin
Southern Cross
The Johnston Lakes
L. Cowan
Norseman
Pt. Culver
Great Australian Bight
Investigator Group
Pinjarra
Brookton
Norrogin
Newdegate
L. Dundas
Rocky Pt.
Coffin B. Pe...
Whidb...
Bunbury
Collie
Wagin
Nyabing
Ravensthorpe
Hopetoun
C. Pasley
Geographe B.
C. Naturaliste
Busselton
Katanning
Gnowangerup
Doubtful B.
C. Arid
Archipelago of the Recherche
Augusta
C. Leeuwin
Bridgetown
Manjimup
Pemberton
Mt. Barker
Stirling Ra.
Pt. Hood
Esperance B.
C. le Grand
Esperance
Flinders B.
Pt. d'Entrecasteaux
Pt. Nuyts
Denmark
Albany
C. Knob
Tor B.
King George Sound

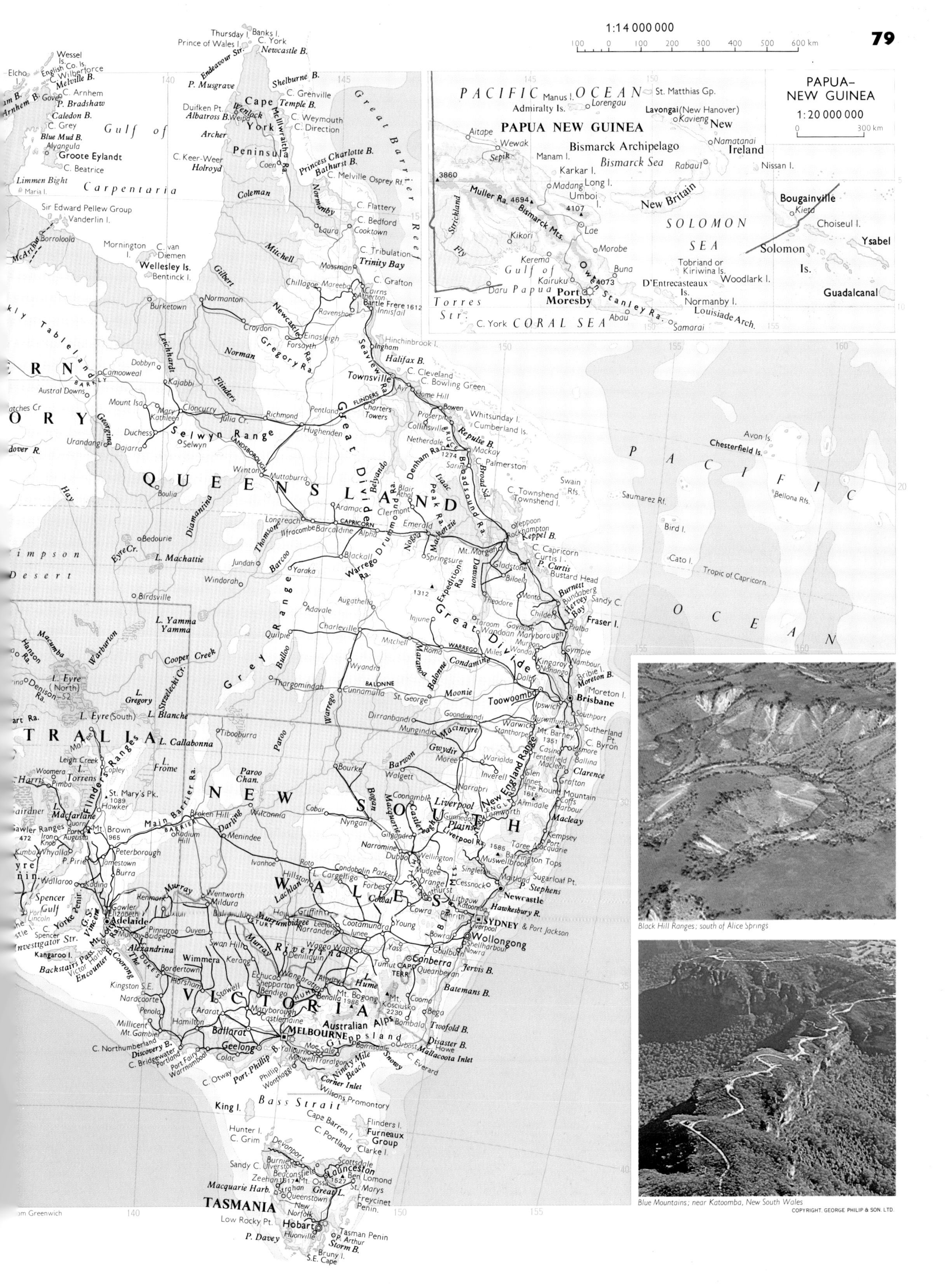

Black Hill Ranges; south of Alice Springs

Blue Mountains; near Katoomba, New South Wales

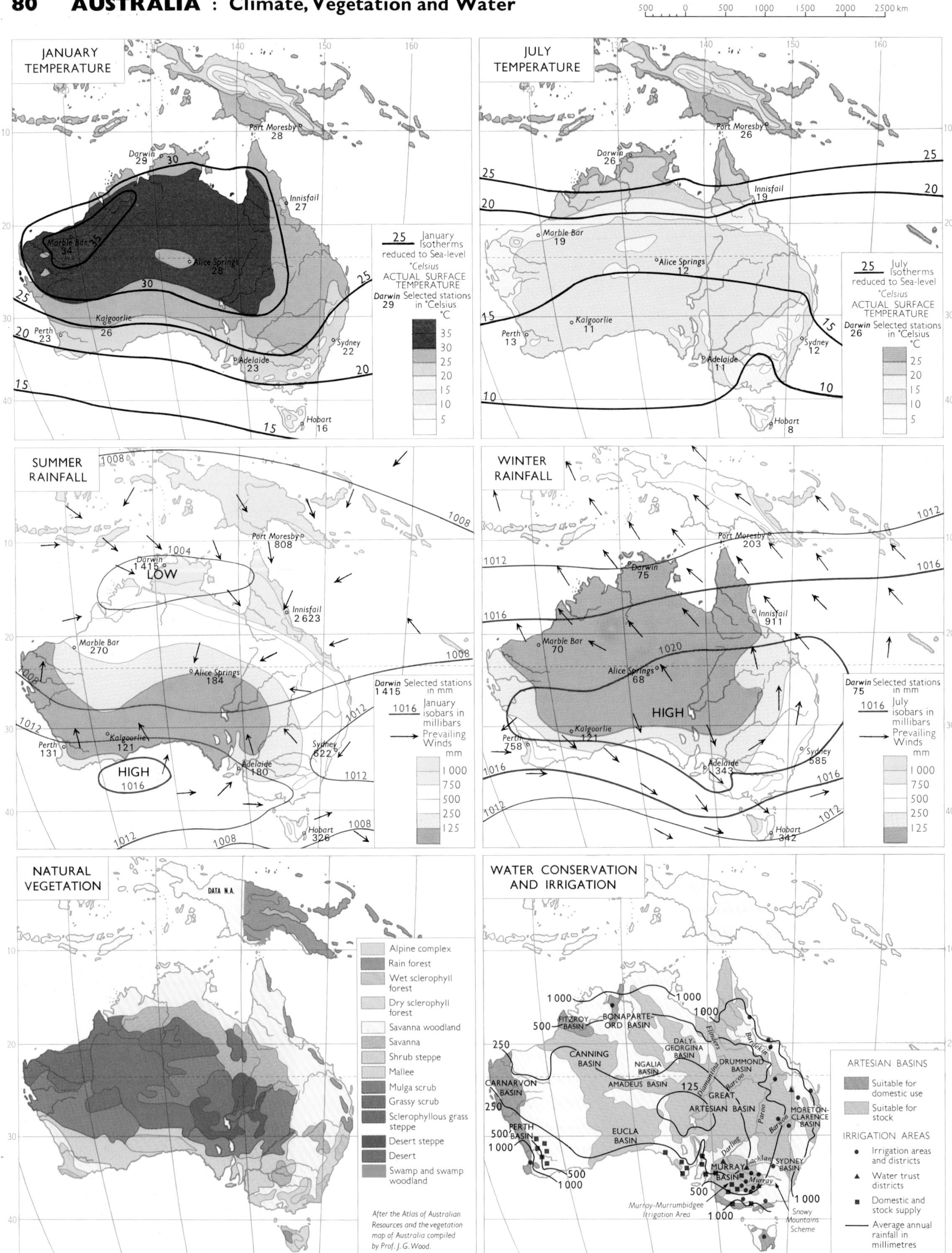

1:60 000 000

JANUARY TEMPERATURE

25 January Isotherms reduced to Sea-level °Celsius
ACTUAL SURFACE TEMPERATURE
Darwin Selected stations in °Celsius
°C
35
30
25
20
15
10
5

Port Moresby 28
Darwin 29 30
Innisfail 27
Marble Bar 34
Alice Springs 28
30
25
Kalgoorlie 26
Perth 23
Sydney 22
Adelaide 23
20
15
Hobart 16
15

JULY TEMPERATURE

25 July Isotherms reduced to Sea-level °Celsius
ACTUAL SURFACE TEMPERATURE
Darwin Selected stations in °Celsius
°C
25
20
15
10
5

Port Moresby 26
Darwin 26
Innisfail 19
25
20
Marble Bar 19
Alice Springs 12
15
Perth 13
Kalgoorlie 11
Sydney 12
Adelaide 11
10
Hobart 8
10

SUMMER RAINFALL

1008
1008
1004
LOW
Darwin 1415
Port Moresby 808
Innisfail 2623
Marble Bar 270
Alice Springs 184
1008
1008
1012
Perth 131
Kalgoorlie 121
Sydney 622
HIGH 1016
Adelaide 180
1012
1012
Hobart 326
1008
1008
1012

Darwin Selected stations in mm
1415
1016 January isobars in millibars
→ Prevailing Winds
mm
1 000
750
500
250
125

WINTER RAINFALL

1012
1016
1012
1016
Port Moresby 203
Darwin 75
Innisfail 911
Marble Bar 70
1020
Alice Springs 68
HIGH
Perth 758
Kalgoorlie 121
Sydney 585
Adelaide 343
1016
1016
1012
Hobart 342
1016

Darwin Selected stations in mm
75
1016 July isobars in millibars
→ Prevailing Winds
mm
1 000
750
500
250
125

NATURAL VEGETATION

DATA N.A.

Alpine complex
Rain forest
Wet sclerophyll forest
Dry sclerophyll forest
Savanna woodland
Savanna
Shrub steppe
Mallee
Mulga scrub
Grassy scrub
Sclerophyllous grass steppe
Desert steppe
Desert
Swamp and swamp woodland

After the Atlas of Australian Resources and the vegetation map of Australia compiled by Prof. J. G. Wood.

Projection: *Mollweide's Homolographic* 130 East from 140 Greenwich 150

WATER CONSERVATION AND IRRIGATION

1000
1000
1000
FITZROY BASIN
BONAPARTE-ORD BASIN
500
250
DALY-GEORGINA BASIN
CANNING BASIN
NGALIA BASIN
DRUMMOND BASIN
CARNARVON BASIN
AMADEUS BASIN
125
GREAT ARTESIAN BASIN
250
MORETON-CLARENCE BASIN
500
PERTH BASIN
1000
EUCLA BASIN
500
SYDNEY BASIN
500
MURRAY BASIN
1000
Murray-Murrumbidgee Irrigation Area
Murray
Snowy Mountains Scheme
1000
1000

ARTESIAN BASINS
Suitable for domestic use
Suitable for stock
IRRIGATION AREAS
● Irrigation areas and districts
▲ Water trust districts
■ Domestic and stock supply
— Average annual rainfall in millimetres

130 East from 140 Greenwich 150

COPYRIGHT. GEORGE PHILIP & SON. LTD.

LAND USE

1:30 000 000

200 0 200 400 600 800 1000 1200 km

Wool

Wool and fat lamb production

Intensive beef cattle rearing

Extensive beef cattle rearing

Dairy farming

Grain production (principally wheat)

Cash crops (fruit, vegetables, cotton, sugar cane etc.)

Areas of scattered subsistence farming

Forests and timber reserves

Non-agricultural land

■ Major urban areas

PAPUA NEW GUINEA
same scale

Tropic of Capricorn

Perth

Brisbane

Adelaide

Sydney

Melbourne

MEAT PRODUCTION

'000 tonnes

2000

1500

1000

500

1960 1965 1970 1975 1980 1985

Beef

Mutton

AREA OF CROPS

Others
Vegetables
Fruit
Sunflower
Sugar Cane
Hay Crops
Legumes
Silage Crops
Oats
Barley
Wheat

DIRECTION OF WOOL EXPORTS

Others
Belgium and Luxembourg
Netherlands
China
U.S.A.
India
S. Korea
Poland
West Germany
Taiwan
France
Italy
U.S.S.R.
Japan

THE RIVER MURRAY SCHEME

Water conservation
■ Dams
Weirs
Locks and weirs
Reservoirs
Watershed of Murray Basin
Catchment area

Water consumption
Heavily irrigated areas
Irrigated areas, all uses
Irrigated areas, livestock and domestic areas

0 50 100 150 km

PIPELINE TO WHYALLA AND WOOMERA

SOUTH

L. Ana Branch
L. Victoria
Darling

NEW SOUTH

WALES

L. Brewster

Lachlan

Morgan
Murray
Renmark
Wentworth
Mildura
Loxton

Balranald

Hay

Murrumbidgee

AUSTRALIA

Billabong Creek
Yanko Creek
Edward
Wakool
Deniliquin

Wagga Wagga
Tumut
Burrinjuck Res.
L. George

PROPOSED PIPELINE

Adelaide

Gulf St. Vincent

Mannum
Tailem Bend
Wellington
L. Alexandrina
L. Albert

L. Tyrrell

L. Hindmarsh

MURRAY MOUTH BARRAGES

The Coorong

Avoca
Loddon
Avoca

VICTORIA

Echuca
Yarrawonga
L. Mulwala
Shepparton
Waranga Res.
Albury
Murray
Hume Res.
Geehi Dam
Ovens
Mitta Mitta
L. Eucumbene
Guthega Dam

Canberra A.C.T.
SNOWY MOUNTAINS SCHEME

Snowy

PACIFIC OCEAN

L. Lonsdale Res.
Wimmera
Laanecoorie Res.
Cairn Curran Res.
Eppalock Res.
Tullaroop Res.
Coliban Storages
Bendigo
Seymour
Goulburn
Eildon Res.

Ballarat

Intensive grazing and mixed farming near Orange, N.S.W.

SNOWY-MURRAY DEVELOPMENT

■ Pumping stations
Direction of flow
Power stations

The Snowy-Murray Development diverts the flow of the Snowy River to the Upper Murray Basin. The water is used for electricity generation and irrigation.

metres
2000

1500

1000

500

Great Dividing Range

Guthega Pondage

Geehi Res.

Island Bend Pondage

GUTHEGA 60 000 kW

SNOWY-GEEHI TUNNEL
14 km

EUCUMBENE-SNOWY TUNNEL
24 km

Lake Eucumbene

Lake Jindabyne

Murray 2 Pondage

MURRAY 1
950 000 kW

Swampy Plain River

Khancoban Pondage

MURRAY 2
550 000 kW

metres
2000

1500

1000

500

Lake Eucumbene, N.S.W.

1:40 000 000

400   0   400   800   1200   1600 km

MINERALS, INDUSTRY AND POWER

FUEL AND POWER

Coal basins
- Black (bituminous) coal
- Lignite (brown coal)

- ♦ Oilfields
- ◊ Gasfields
- —— Oil pipelines
- —— Gas pipelines
- ★ Main mining centres
- ◆ Uranium
- ○ Hydro-electric power stations
- ◎ Thermal power stations

MAJOR METALS
- ● Bauxite
- ● Copper
- ■ Iron ore
- ▲ Lead
- ● Tin
- ◆ Zinc

OTHER METALS
- ▲ Antimony
- ▲ Gold
- □ Manganese
- ● Nickel
- ▽ Silver
- + Tungsten

OTHER MINERALS
- ■ Asbestos
- ▽ Mineral sands (rutile, ilmenite and zircon)
- △ Opals
- ▲ Salt
- ● Gypsum

Map labels: Darwin, Mount Bundey, Frances Creek, Gove, Rum Jungle, Weipa, Groote Eylandt, McArthur River, Wolfram Camp, Yampi Sound, Pompey's Pillar, Mount Garnet, Port Hedland, Tennant Creek, Mount Isa, Collinsville, Barrow I., Dampier, Pilbara, Mount Nicholas, Duchess, Mount Morgan, Rockhampton, Rough Range, Mount Tom Price, Mount Newman, Tropic of Capricorn, Mereenie, Palm Valley, Arcturus, Moura, Peak Hill, Warburton Range, Blackstone Range, Gilmore, Meekatharra, Coober Pedy, Gidgealpa-Moomba, Roma, Brisbane, Mount Magnet, Moonie, Grafton, Dongara, Broken Hill, Cobar, Kalgoorlie, Whyalla, Port Pirie, Wallerawang, Newcastle, Perth, Norseman, Adelaide, Sydney, Kwinana, Ravensthorpe, Kangaroo I., Port Kembla, Snowy Mountains Scheme, Geelong, Melbourne, Gippsland Shelf, King I., Read-Rosebery, Rossarden, Bell Bay, Hobart

50% of potential Australian H.E.P. is in Tasmania

EMPLOYMENT IN MANUFACTURING INDUSTRY
'000 people

	N.S.W.	Victoria	Queensland	South Australia	West Australia	Tasmania	Australia
Food, drink, tobacco	54,9	51,4	32,3	16,8	12,0	5,9	174,5
Textiles, clothing, footwear	28,9	58,6	4,3	6,8	2,1	0,2	102,1
Wood, paper, printing	59,5	51,8	21,8	14,5	13,8	7,9	171,7
Machinery, metal goods	143,5	91,5	30,3	29,2	23,0	2,1	324,9
Transport equipment	31,7	54,4	12,0	17,4	4,8	0,6	121,0
Others	66,1	55,3	13,9	11,2	9,3	7,4	158,7
Total manufacturing	384,6	363,0	114,6	95,9	65,0	24,1	1052,9

POPULATION

Map labels: Darwin, Cairns, Mount Isa, Townsville, Mackay, Tropic of Capricorn, Rockhampton, Bundaberg, Maryb, Redcli, Toowoomba, Brisb, Ipswich, Gold, Kalgoorlie, Whyalla, Broken Hill, Tamworth, Perth, Orange, Newcastle, Gosford, Sydney, Adelaide, Wagga Wagga, Goulburn, Wollongong, Bendigo, Albury, Canberra, Ballarat, Melbourne, Geelong, Moe, Burnie, Launceston, Hobart

COMMUNICATIONS

AIRPORTS
Total passenger traffic, (excluding international flights)
- ● Over 1 000 000
- ● 250 000–1 000 000
- ● 100 000–250 000
- • under 100 000

Sydney   International airport

PORTS
Volume of shipping entering ports '000 tonnes
- ■ Over 10 000
- ■ 2 000–10 000
- ■ 1 000–2 000
- ▪ under 1 000

Map labels: Darwin, Weipa, Groote Eylandt, Yampi Sound, Cairns, Port Hedland, Townsville, Bowen, Dampier, Mount Isa, Mackay, Tropic of Capricorn, Alice Springs, Rockhampton, Gladstone, Geraldton, Brisbane, Coolangatta, Tamworth, Perth, Whyalla, Dubbo, Fremantle, Port Pirie, Newcastle, Bunbury, Wallaroo, Adelaide, Sydney, Port Lincoln, Wagga Wagga, Port Kembla, Albany, Kangaroo I., Canberra, Geelong, Melbourne, Burnie, Port Latta, Devonport, Wynard, Launceston, Hobart

DENSITY OF POPULATION
inhabitants per km²
- 100–200
- 50–100
- 25–50
- 10–25
- 1–10
- under 1

URBAN POPULATION
- ■ Over 1 million inhabitants
- ◉ 500 000–1 million inhabitants
- ◉ 100 000–500 000 inhabitants
- ◉ 20 000–100 000 inhabitants
- —— Aboriginal reserves

—— Major roads   —— Railways
262 897 kilometres sealed roads   ---- Railways under construction
801 589 kilometres total roads   39 065 route kilometres of railways

In Australia 90% of passenger travel and 20% of freight transport is by road.

GROWTH OF POPULATION IN AUSTRALIA
million people

Total population 1984 15.5 million
Total population of Australia

15
10
5

1881 91 1901 11 21 31 41 51 61 71 81 84

- Rest of Australia
- Victoria
- New South Wales

COUNTRY OF ORIGIN OF FOREIGN BORN
'000 people

- Other 214.9
- Africa 90.2
- America 96.2
- Other Europe 205.0
- Other Asia 321.9
- Lebanon 49.6
- Malta 57.0
- Poland 59.4
- Netherlands 96.0
- Germany 110.8
- Greece 146.6
- Yugoslavia 149.3
- U.K. and Ireland 1132.6
- Italy 275.9
- New Zealand 176.7

Immigration has been a major factor in the development of Australia. Over 3 million (or 22%) of Australia's population are immigrants.

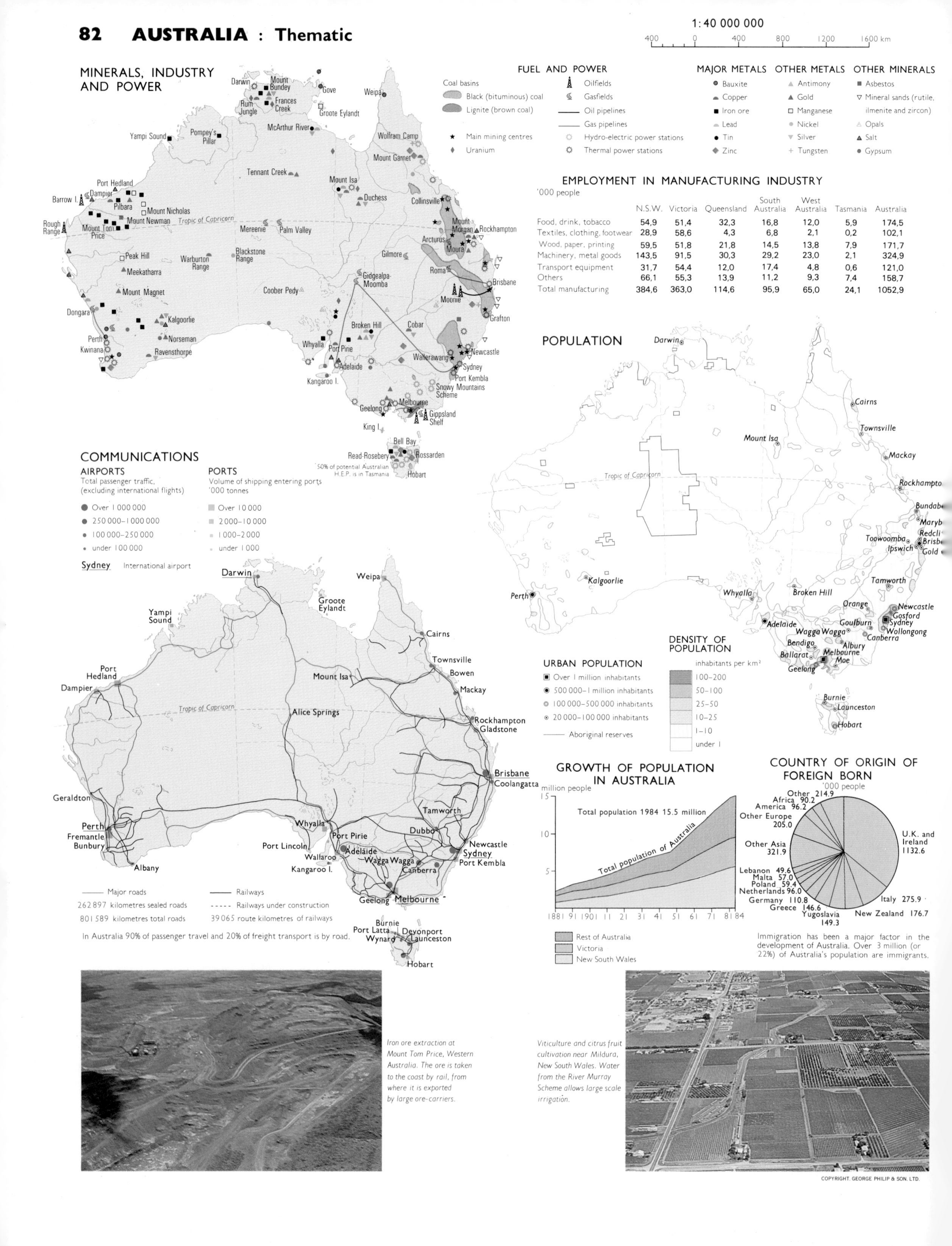

Iron ore extraction at Mount Tom Price, Western Australia. The ore is taken to the coast by rail, from where it is exported by large ore-carriers.

Viticulture and citrus fruit cultivation near Mildura, New South Wales. Water from the River Murray Scheme allows large scale irrigation.

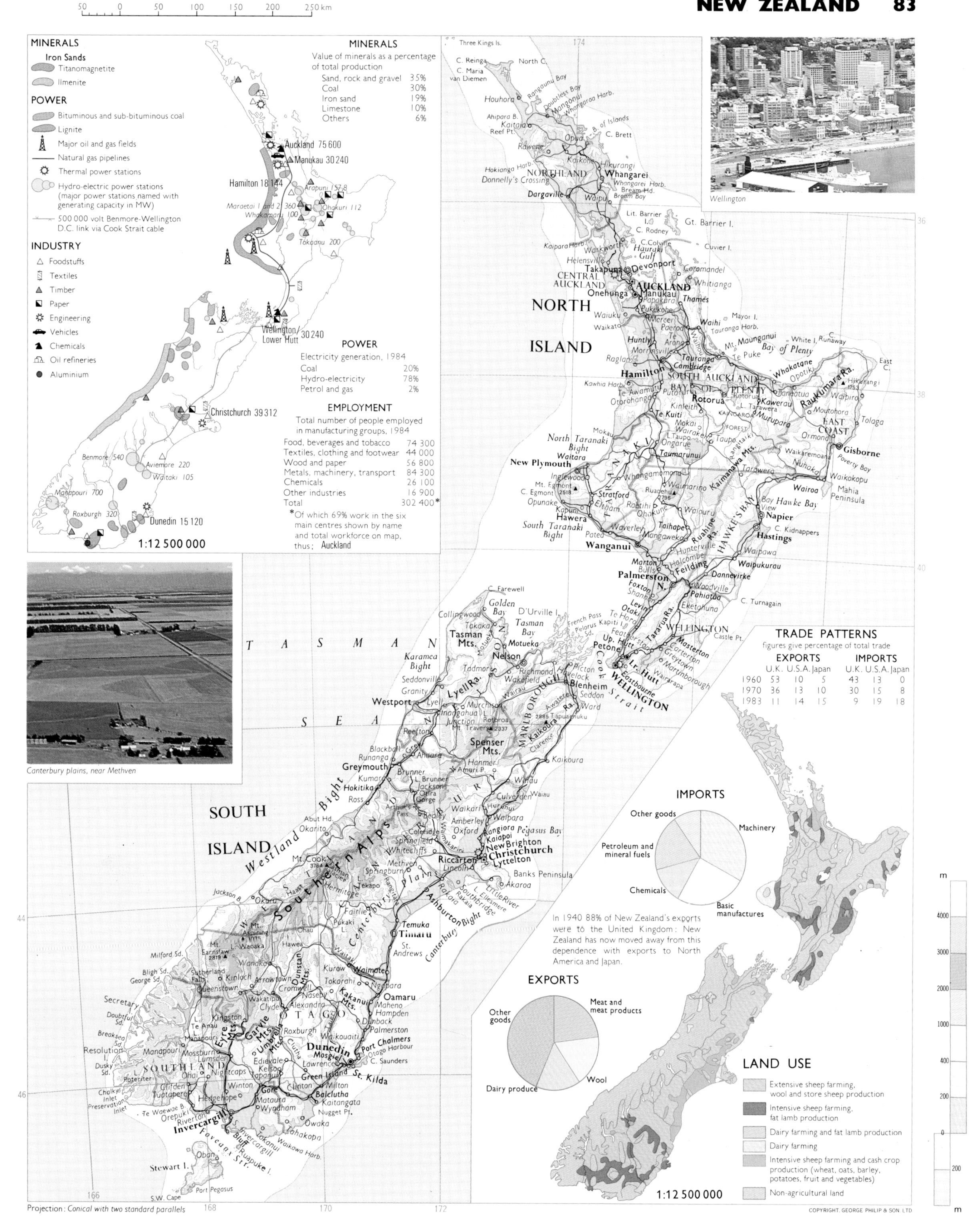
1:6 000 000
50 0 50 100 150 200 250 km
MINERALS
Iron Sands
Titanomagnetite
Ilmenite
POWER
Bituminous and sub-bituminous coal
Lignite
Major oil and gas fields
Natural gas pipelines
Thermal power stations
Hydro-electric power stations (major power stations named with generating capacity in MW)
500 000 volt Benmore-Wellington D.C. link via Cook Strait cable
INDUSTRY
Foodstuffs
Textiles
Timber
Paper
Engineering
Vehicles
Chemicals
Oil refineries
Aluminium
MINERALS
Value of minerals as a percentage of total production
Sand, rock and gravel 35%
Coal 30%
Iron sand 19%
Limestone 10%
Others 6%
Auckland 75 600
Manukau 30 240
Hamilton 18 144
Wellington/Lower Hutt 30 240
Christchurch 39 312
Dunedin 15 120
POWER
Electricity generation, 1984
Coal 20%
Hydro-electricity 78%
Petrol and gas 2%
EMPLOYMENT
Total number of people employed in manufacturing groups, 1984
Food, beverages and tobacco 74 300
Textiles, clothing and footwear 44 000
Wood and paper 56 800
Metals, machinery, transport 84 300
Chemicals 26 100
Other industries 16 900
Total 302 400*
*Of which 69% work in the six main centres shown by name and total workforce on map, thus: Auckland
1:12 500 000
Wellington
Canterbury plains, near Methven
NORTH ISLAND
SOUTH ISLAND
TASMAN SEA
AUCKLAND
WELLINGTON
Christchurch
Dunedin
Invercargill
Hamilton
Rotorua
Gisborne
Napier
Hastings
New Plymouth
Wanganui
Palmerston N.
Nelson
Blenheim
Greymouth
Timaru
Cook Strait
Southern Alps
TRADE PATTERNS
figures give percentage of total trade
EXPORTS: U.K. U.S.A. Japan | IMPORTS: U.K. U.S.A. Japan
1960 53 10 5 | 43 13 0
1970 36 13 10 | 30 15 8
1983 11 14 15 | 9 19 18
IMPORTS
Other goods
Machinery
Petroleum and mineral fuels
Chemicals
Basic manufactures
In 1940 88% of New Zealand's exports were to the United Kingdom: New Zealand has now moved away from this dependence with exports to North America and Japan.
EXPORTS
Other goods
Meat and meat products
Dairy produce
Wool
LAND USE
Extensive sheep farming, wool and store sheep production
Intensive sheep farming, fat lamb production
Dairy farming and fat lamb production
Dairy farming
Intensive sheep farming and cash crop production (wheat, oats, barley, potatoes, fruit and vegetables)
Non-agricultural land
1:12 500 000
Projection: Conical with two standard parallels
COPYRIGHT GEORGE PHILIP & SON LTD

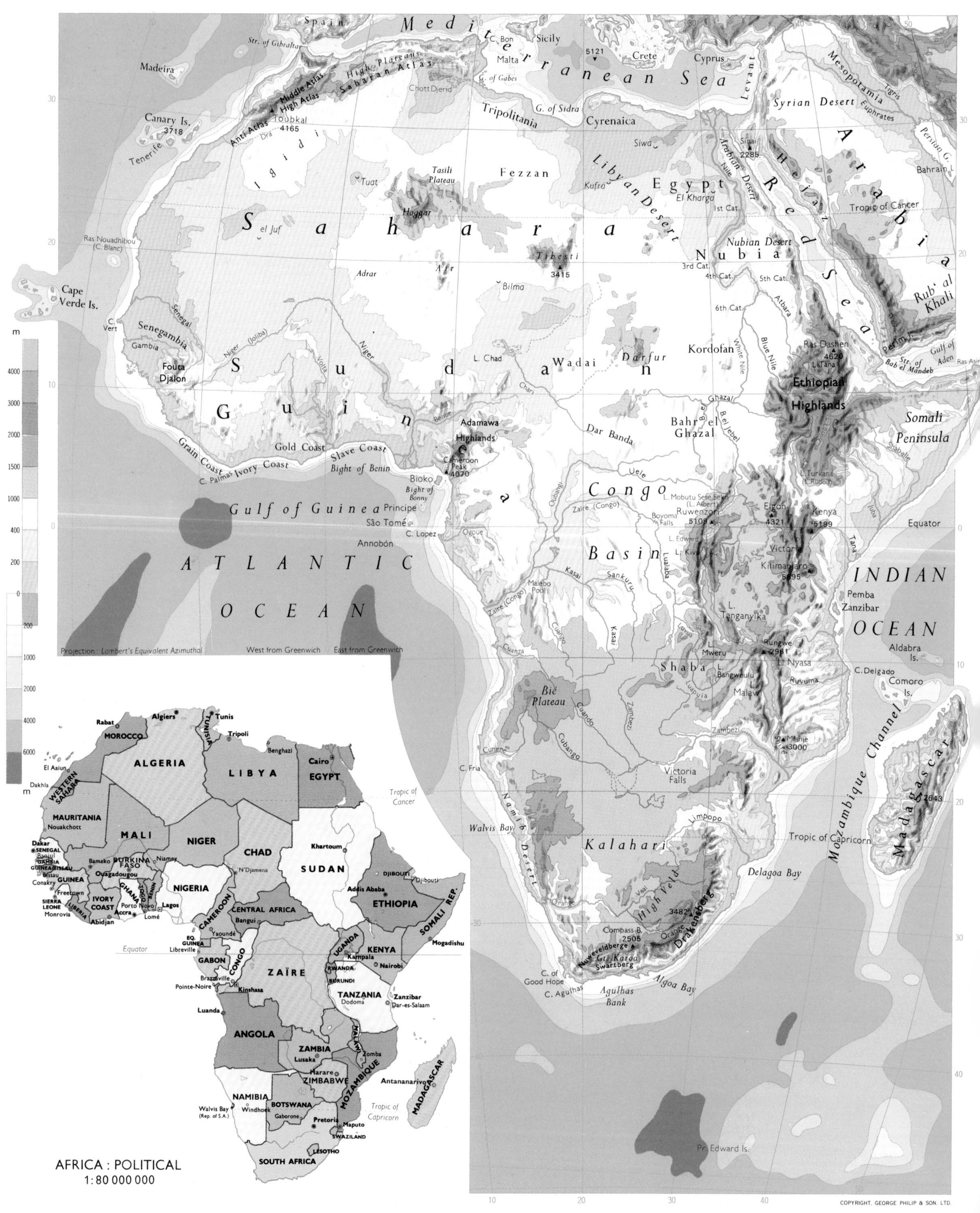

1 : 40 000 000

400   0   400   800   1200   1600 km

AFRICA : POLITICAL
1 : 80 000 000

COPYRIGHT. GEORGE PHILIP & SON. LTD.

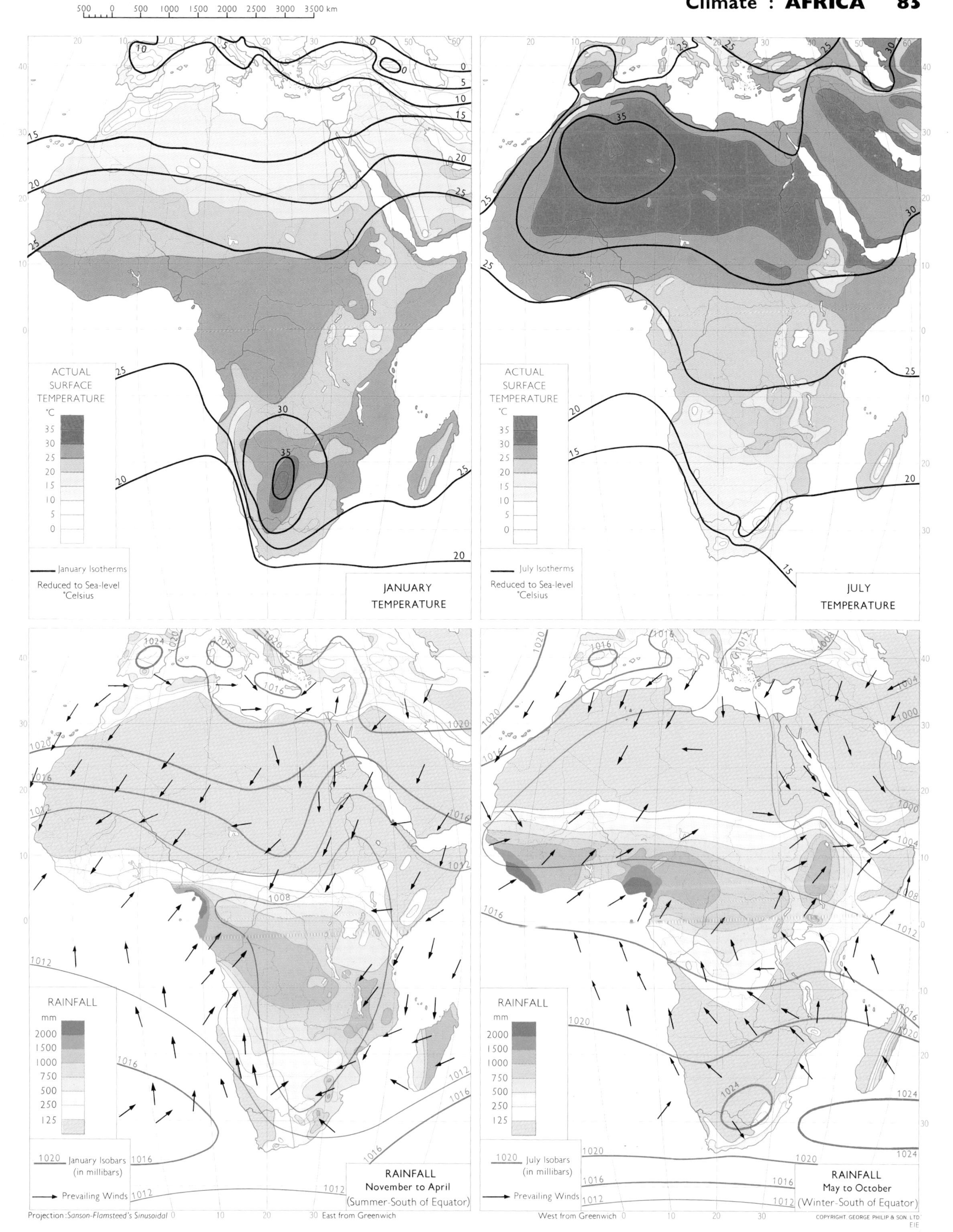

1 : 80 000 000

500   0   500 1000 1500 2000 2500 3000 3500 km

ACTUAL
SURFACE
TEMPERATURE
°C
35
30
25
20
15
10
5
0

—— January Isotherms
Reduced to Sea-level
°Celsius

JANUARY
TEMPERATURE

ACTUAL
SURFACE
TEMPERATURE
°C
35
30
25
20
15
10
5
0

—— July Isotherms
Reduced to Sea-level
°Celsius

JULY
TEMPERATURE

RAINFALL

mm
2000
1500
1000
750
500
250
125

1020 January Isobars
       (in millibars)
——➤ Prevailing Winds

RAINFALL
November to April
(Summer-South of Equator)

RAINFALL

mm
2000
1500
1000
750
500
250
125

1020 July Isobars
       (in millibars)
——➤ Prevailing Winds

RAINFALL
May to October
(Winter-South of Equator)

Projection: Sanson-Flamsteed's Sinusoidal   0   30   East from Greenwich

West from Greenwich   0   10   20   30

COPYRIGHT GEORGE PHILIP & SON LTD
E/E

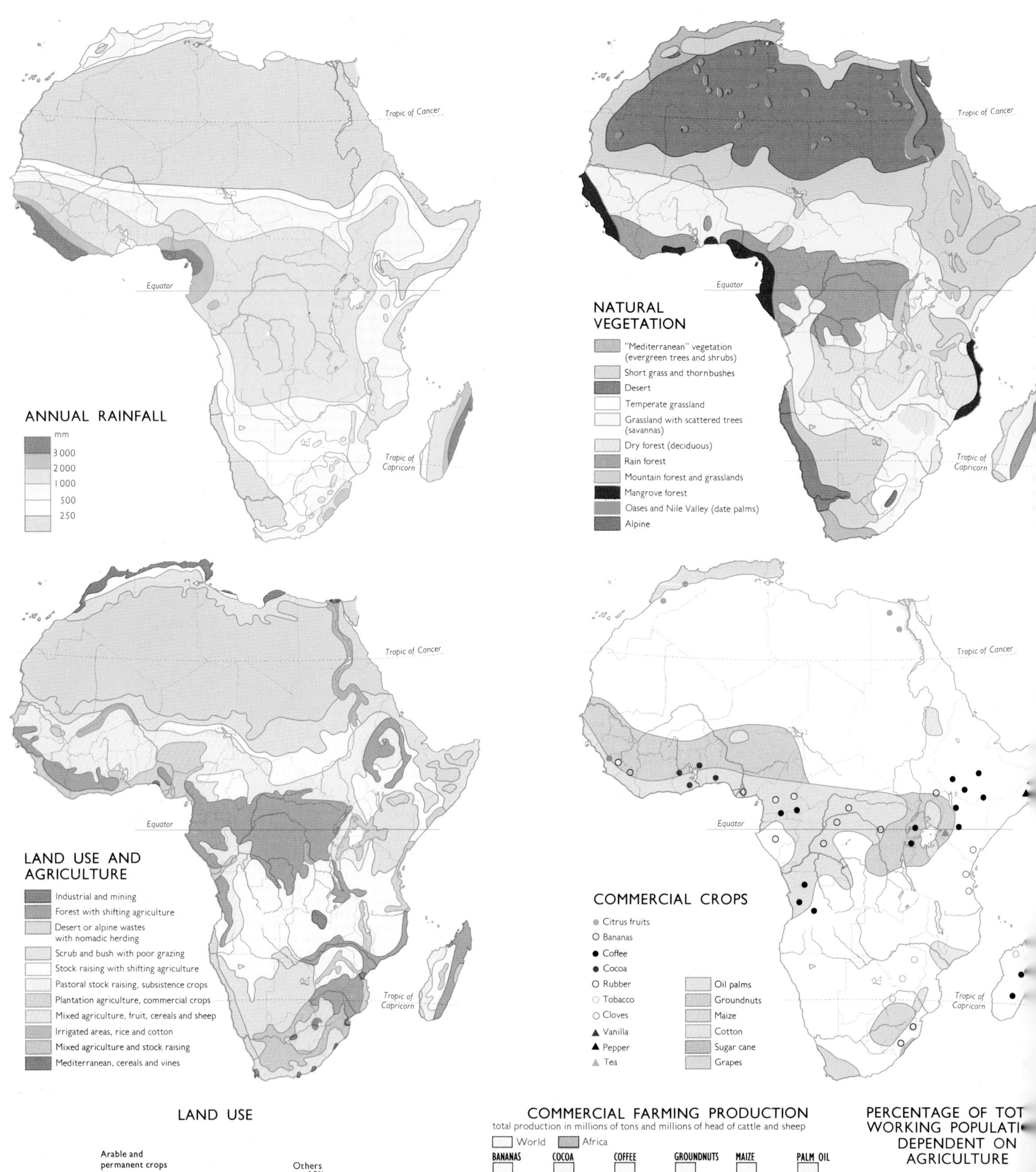

ANNUAL RAINFALL

mm
3 000
2 000
1 000
500
250

NATURAL
VEGETATION

"Mediterranean" vegetation
(evergreen trees and shrubs)

Short grass and thornbushes

Desert

Temperate grassland

Grassland with scattered trees
(savannas)

Dry forest (deciduous)

Rain forest

Mountain forest and grasslands

Mangrove forest

Oases and Nile Valley (date palms)

Alpine

LAND USE AND
AGRICULTURE

Industrial and mining

Forest with shifting agriculture

Desert or alpine wastes
with nomadic herding

Scrub and bush with poor grazing

Stock raising with shifting agriculture

Pastoral stock raising, subsistence crops

Plantation agriculture, commercial crops

Mixed agriculture, fruit, cereals and sheep

Irrigated areas, rice and cotton

Mixed agriculture and stock raising

Mediterranean, cereals and vines

COMMERCIAL CROPS

● Citrus fruits
○ Bananas
● Coffee
● Cocoa
○ Rubber
○ Tobacco
○ Cloves
▲ Vanilla
▲ Pepper
▲ Tea

Oil palms
Groundnuts
Maize
Cotton
Sugar cane
Grapes

LAND USE

Arable and
permanent crops
6%

Others
44%

Permanent pasture
and crops
27%

Forest and woodland
23%

AFRICA

Others
10%

Arable
35%

Forest
30%

Permanent pasture 25%

E.E.C.

COMMERCIAL FARMING PRODUCTION
total production in millions of tons and millions of head of cattle and sheep

☐ World   ▨ Africa

	World	Africa
BANANAS	41,3	4,9
COCOA	1,8	1,0
COFFEE	5,2	1,1
GROUNDNUTS	20,7	3,9
MAIZE	450,7	22,2
PALM OIL	6,9	1,4
CATTLE	1272,5	175,6
SHEEP	1139,5	188,6
SUGAR CANE	935,8	68,5
WINE	32,8	1,2
WOOL	2,9	0,2
CITRUS FRUITS	56,1	4,6

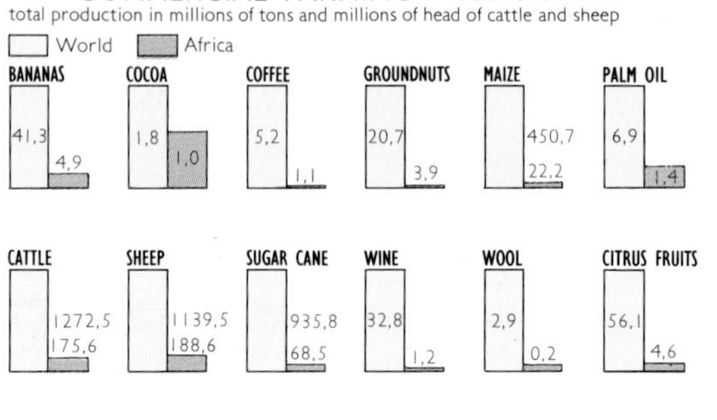

PERCENTAGE OF TOT
WORKING POPULATI
DEPENDENT ON
AGRICULTURE

63%
AFRICA

27%
SOUTH AF

73%
ZAÏRE

46%
ALGER

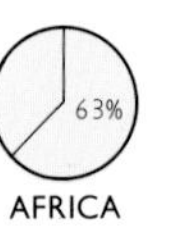

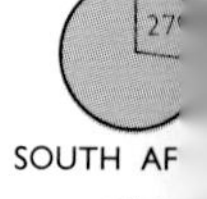

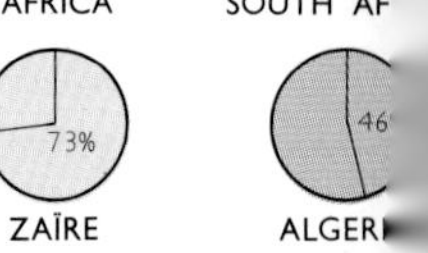

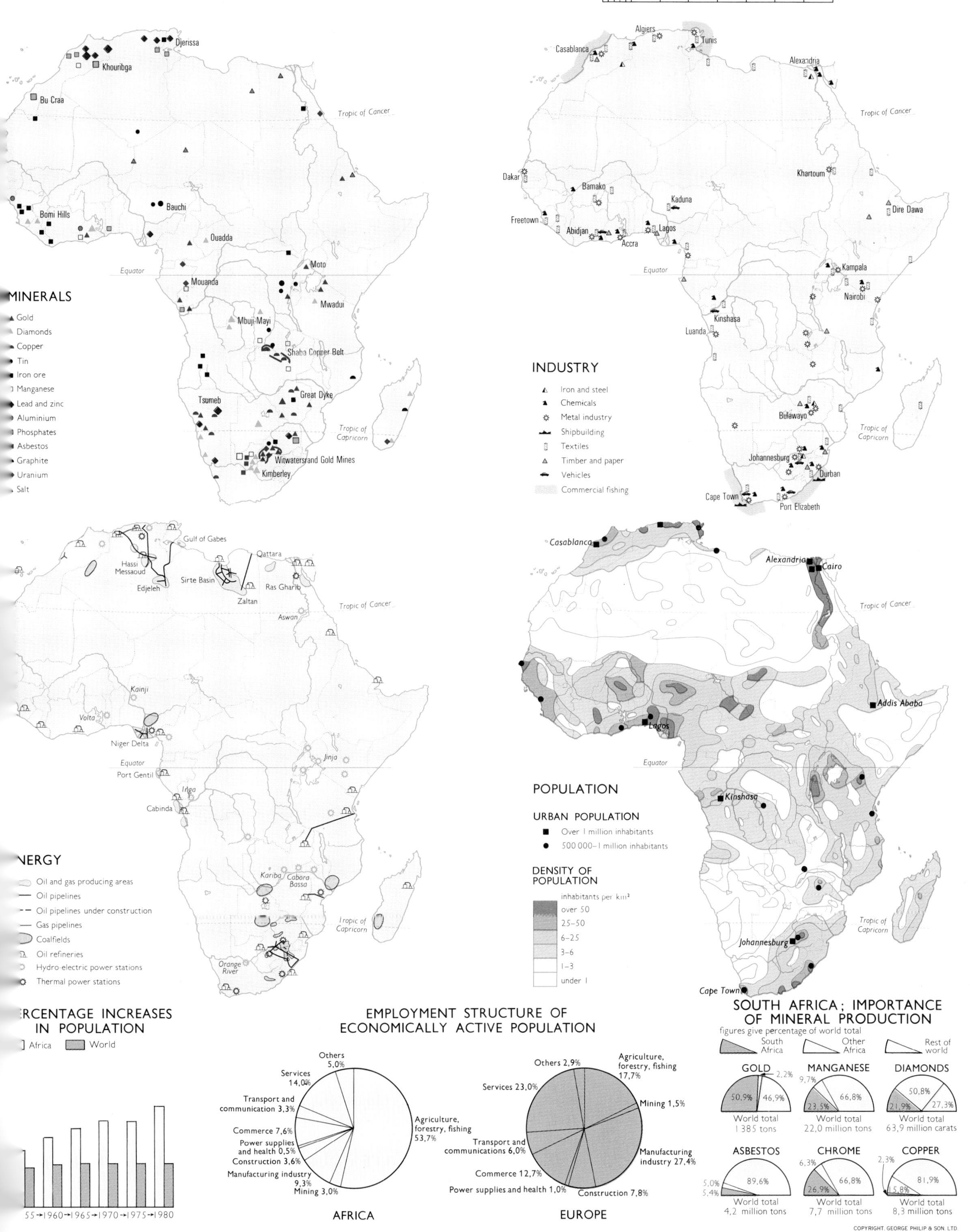

MINERALS

- ▲ Gold
- ▲ Diamonds
- ■ Copper
- ■ Tin
- ■ Iron ore
- ◻ Manganese
- ■ Lead and zinc
- ▲ Aluminium
- ◼ Phosphates
- ▲ Asbestos
- ◆ Graphite
- ▲ Uranium
- ▲ Salt

Bu Craa
Djerissa
Khouribga
Bomi Hills
Bauchi
Ouadda
Moto
Mouanda
Mwadui
Mbuji-Mayi
Shaba Copper Belt
Tsumeb
Great Dyke
Witwatersrand Gold Mines
Kimberley
Tropic of Cancer
Equator
Tropic of Capricorn

INDUSTRY

- ▲ Iron and steel
- ▲ Chemicals
- ✿ Metal industry
- ⛴ Shipbuilding
- ▯ Textiles
- △ Timber and paper
- 🚗 Vehicles
- ▨ Commercial fishing

Algiers
Tunis
Casablanca
Alexandria
Dakar
Khartoum
Bamako
Kaduna
Dire Dawa
Freetown
Abidjan
Lagos
Accra
Kampala
Nairobi
Kinshasa
Luanda
Bulawayo
Johannesburg
Durban
Cape Town
Port Elizabeth

ENERGY

- ◠ Oil and gas producing areas
- — Oil pipelines
- ‑‑ Oil pipelines under construction
- — Gas pipelines
- ◠ Coalfields
- ⚒ Oil refineries
- ⚙ Hydro-electric power stations
- ✿ Thermal power stations

Gulf of Gabes
Qattara
Hassi Messaoud
Sirte Basin
Edjeleh
Ras Gharib
Zaltan
Aswan
Kainji
Volta
Niger Delta
Jinja
Port Gentil
Inga
Cabinda
Kariba
Cabora Bassa
Orange River

POPULATION

URBAN POPULATION

- ■ Over 1 million inhabitants
- ● 500 000–1 million inhabitants

DENSITY OF POPULATION

inhabitants per km²
- over 50
- 25–50
- 6–25
- 3–6
- 1–3
- under 1

Casablanca
Alexandria
Cairo
Addis Ababa
Lagos
Kinshasa
Johannesburg
Cape Town

PERCENTAGE INCREASES IN POPULATION

- ◻ Africa
- ▨ World

55→1960→1965→1970→1975→1980

EMPLOYMENT STRUCTURE OF ECONOMICALLY ACTIVE POPULATION

AFRICA

- Others 5,0%
- Services 14,0%
- Transport and communication 3,3%
- Commerce 7,6%
- Power supplies and health 0,5%
- Construction 3,6%
- Manufacturing industry 9,3%
- Mining 3,0%
- Agriculture, forestry, fishing 53,7%

EUROPE

- Others 2,9%
- Services 23,0%
- Transport and communications 6,0%
- Commerce 12,7%
- Power supplies and health 1,0%
- Construction 7,8%
- Manufacturing industry 27,4%
- Mining 1,5%
- Agriculture, forestry, fishing 17,7%

SOUTH AFRICA: IMPORTANCE OF MINERAL PRODUCTION

figures give percentage of world total

- South Africa
- Other Africa
- Rest of world

GOLD 2,2%
50,9% 46,9%
World total 1 385 tons

MANGANESE 9,7%
23,5% 66,8%
World total 22,0 million tons

DIAMONDS
50,8% 27,3%
21,9%
World total 63,9 million carats

ASBESTOS
5,0% 89,6%
5,4%
World total 4,2 million tons

CHROME 6,3%
26,9% 66,8%
World total 7,7 million tons

COPPER 2,3%
15,8% 81,9%
World total 8,3 million tons

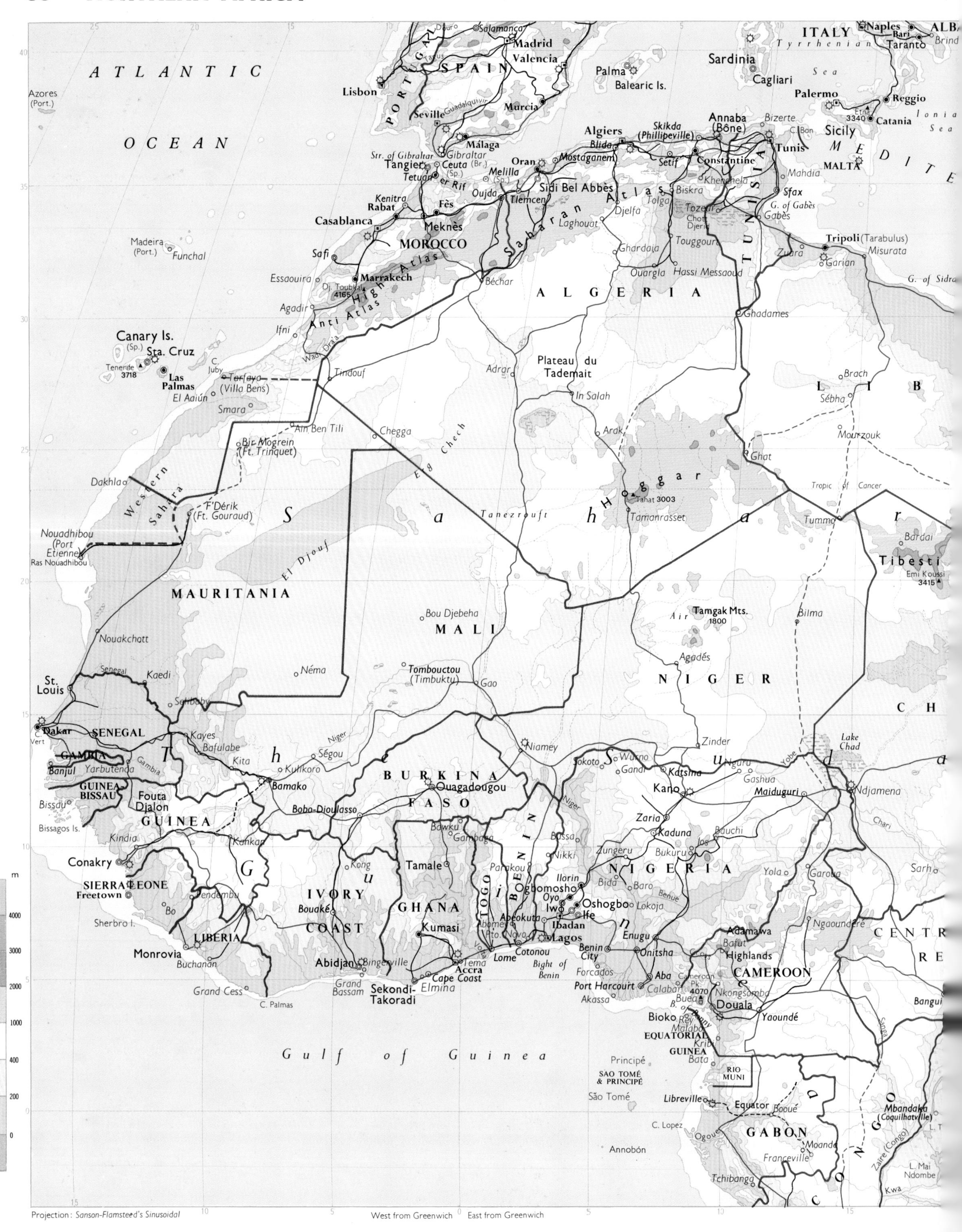

ATLANTIC

OCEAN

Azores
(Port.)

SPAIN

Madrid
Valencia

ITALY

Naples
Bari
Taranto
Brind

Tyrrhenian

Sardinia

Sea

Cagliari

Palermo

Sicily

Reggio
Etna 3340
Catania

MEDITE

Ionia
Sea

MALTA

PORTUGAL

Lisbon

Salamanca

Seville
Guadalquivir

Murcia

Valencia

Palma

Balearic Is.

Algiers
Skikda (Phillipeville)

Annaba
(Bône)

Bizerte

Tunis

Str. of Gibraltar
Gibraltar
Tangier
Tetuán
er Rif
Ceuta (Sp.)

Málaga

Melilla
(Sp.)

Oran

Blida
Mostaganem

Setif

Constantine

C. Bon

Khenchela

Mahdia

Kenitra
Rabat
Fès
Oujda
Tlemcen

Sidi Bel Abbès

Tolga
Biskra

Sfax

G. of Gabès

Casablanca
Meknès

Laghouat

Djelfa

Touggourt

Tozeur
Chott
Djerid

Gabès

Zuara

Safi

MOROCCO

Ghardaia

Ouargla

Hassi Messaoud

Tripoli (Tarabulus)
Misurata

Essaouira

Dj. Toubkal
4165

Marrakech

Béchar

ALGERIA

Ghadames

Gârian

G. of Sidra

Agadir

Anti Atlas

High Atlas

Saharan Atlas

TUNISIA

Madeira
(Port.)
Funchal

Ifni

Wad Dra'a

Tindouf

Adrar

Plateau du
Tademait

In Salah

Ghadames

Brach

Sébha

LIB

B

Canary Is.
(Sp.)
Sta. Cruz

C. Juby
Terfaya
(Villa Bens)

Arak

Ghat

Tropic of Cancer

Mourzouk

Tenerife
3718

Las
Palmas

El Aaiún

Smara

Aïn Ben Tili

Chegga

Eg Chech

Haggar

Tahat 3003
Tamanrasset

Tummo

Bardai

Tibesti

Dakhla

Western Sahara

F'Dérik
(Ft. Gouraud)

S

a

Tanezrouft

h

a

r

Emi Koussi
3415

Bir Mogrein
(Ft. Trinquet)

Nouadhibou
(Port Etienne)
Ras Nouadhibou

El Djouf

Bou Djebeha

Air

Tamgak Mts.
1800

Bilma

Nouakchott

MAURITANIA

MALI

Agadès

NIGER

Kaedi

Néma

Tombouctou
(Timbuktu)

Gao

CH

St.
Louis

Senegal

Selibaby

SENEGAL

Kayes
Bafulabe
Kita

Niger

Ségou

Niamey

Sokoto
Wurno
Gandi

Zinder

Nguru
Yobe

Lake
Chad

Ndjamena

a

Dakar
C. Vert

GAMBIA
Banjul
Yarbutenda

Gambia

Bamako

Kulikoro

h

BURKINA

Ouagadougou

e

Bawku
Gambaga

Katsina

Kano

Gashua

Maiduguri

Chari

Bissau

GUINEA
BISSAU

Fouta
Djalon

GUINEA

Bissagos Is.

Kindia

Kankan

Bafulabe

FASO

Bobo-Dioulasso

Bassa

Zaria

Kaduna

Zungeru

Bukuru
Bauchi

Conakry

G

Kong

Tamale

Parakou

Ilorin

Bida
Baro

NIGERIA

Yola

Garoua

SIERRA LEONE
Freetown

u

IVORY
COAST

Bouaké

GHANA

Kumasi

TOGO

BENIN

Ogbomosho
Oyo
Iwo
Oshogbo
Ife

Lokoja

Benue

Sendembu

Bo

Sherbro I.

Kong

Abeokuta
Abo. Novo
Ato-Novo

Ibadan

Adamawa

Bafut
Highlands

CENTRE
RE

LIBERIA

Grand Cess

Monrovia
Buchanan

Bingerville

Abidjan

Grand
Bassam

Sekondi-
Takoradi

Bouaké

C. Palmas

Tema
Accra
Cape Coast
Elmina

Lome
Cotonou

Lagos

Bight of
Benin

Benin
City

Forçados

Onitsha

Enugu

Aba

Ngaoundéré

Cameroon Pk.
4070

Nkongsamba

Kribi

Douala

CAMEROON

Bangui

Port Harcourt

Calabar
Buea

Bioko
Malabo

Yaoundé

Libreville

Equator

EQUATORIAL
GUINEA

Bata

RIO
MUNI

Booué

GABON

C. Lopez

Francville

Mbandaka
(Coquilhatville)

Principé

SAO TOMÉ
& PRINCIPE

São Tomé

Annobón

Moanda

CONGO

Zaire (Congo)

L. Mai
Ndombe

Kwa

Tchibanga

Ogoué

Gulf of Guinea

m

4000

3000

2000

1000

400

200

0

200

m

West from Greenwich   East from Greenwich

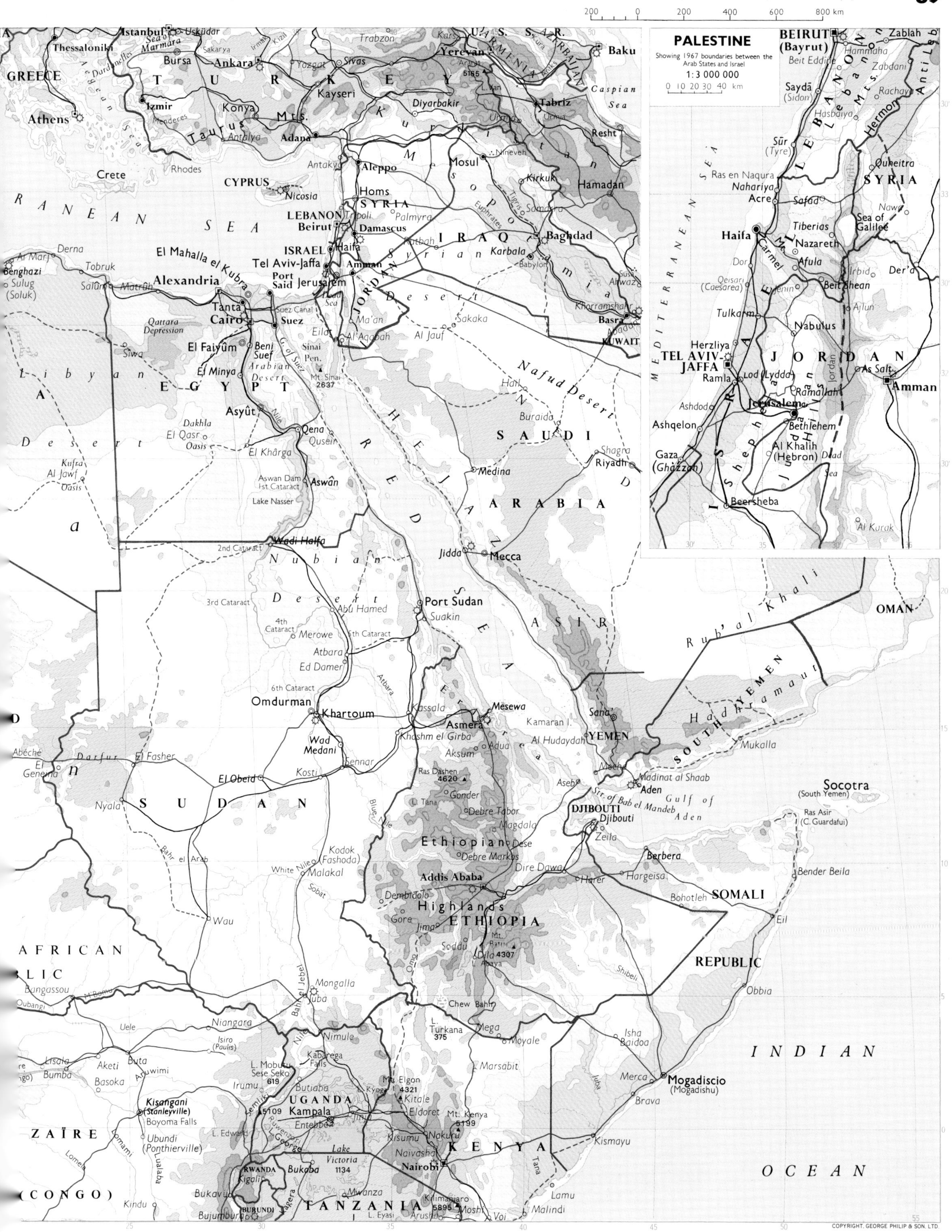

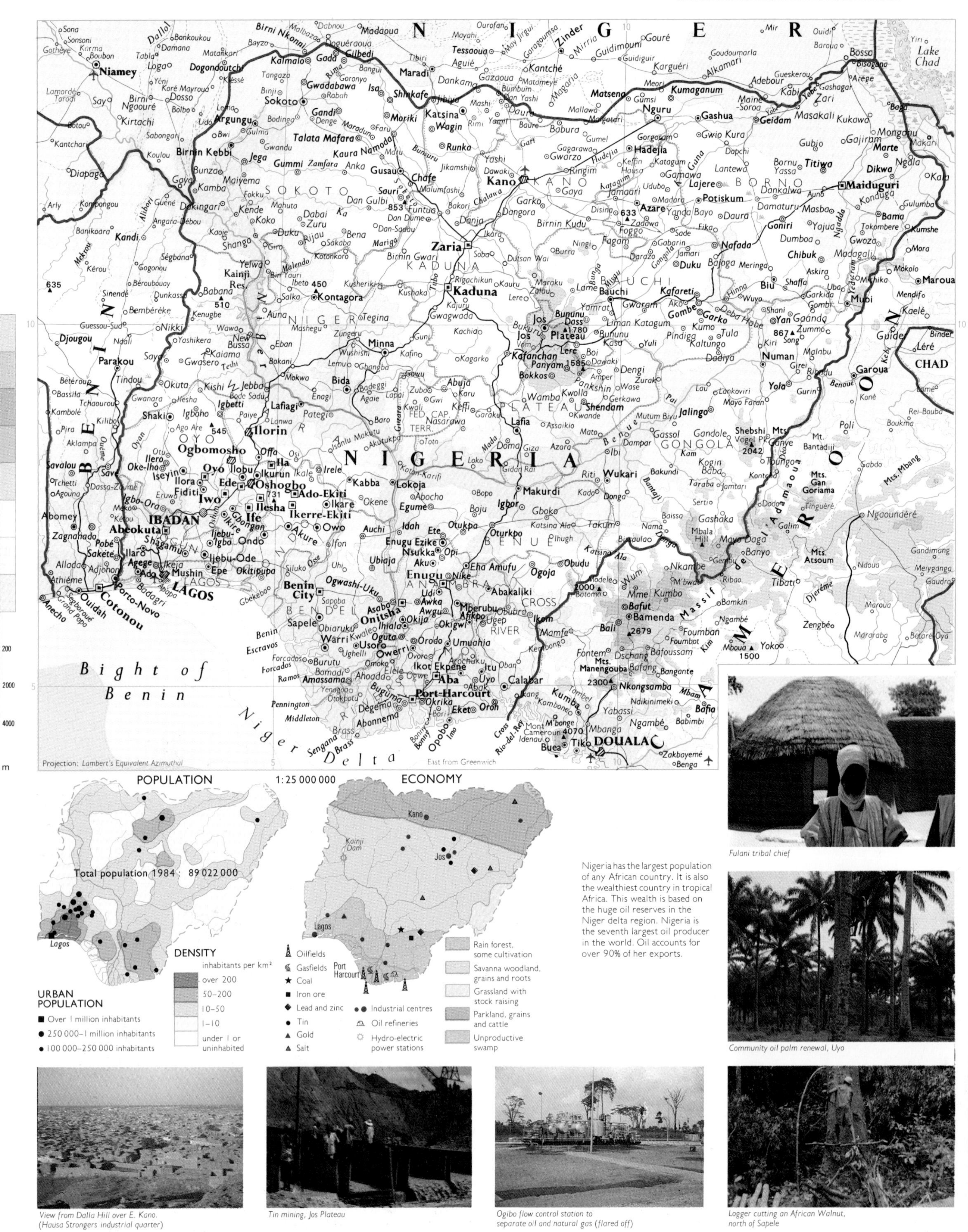

1:7 500 000

POPULATION
1:25 000 000
ECONOMY

Total population 1984 : 89 022 000

Nigeria has the largest population of any African country. It is also the wealthiest country in tropical Africa. This wealth is based on the huge oil reserves in the Niger delta region. Nigeria is the seventh largest oil producer in the world. Oil accounts for over 90% of her exports.

DENSITY
inhabitants per km²
over 200
50–200
10–50
1–10
under 1 or uninhabited

URBAN POPULATION
- Over 1 million inhabitants
- 250 000–1 million inhabitants
- 100 000–250 000 inhabitants

Oilfields
Gasfields
Coal
Iron ore
Lead and zinc
Tin
Gold
Salt
Industrial centres
Oil refineries
Hydro-electric power stations

Rain forest, some cultivation
Savanna woodland, grains and roots
Grassland with stock raising
Parkland, grains and cattle
Unproductive swamp

Fulani tribal chief

Community oil palm renewal, Uyo

View from Dalla Hill over E. Kano. (Hausa Strongers industrial quarter)

Tin mining, Jos Plateau

Ogibo flow control station to separate oil and natural gas (flared off)

Logger cutting an African Walnut, north of Sapele

COPYRIGHT. GEORGE PHILIP & SON. LTD.

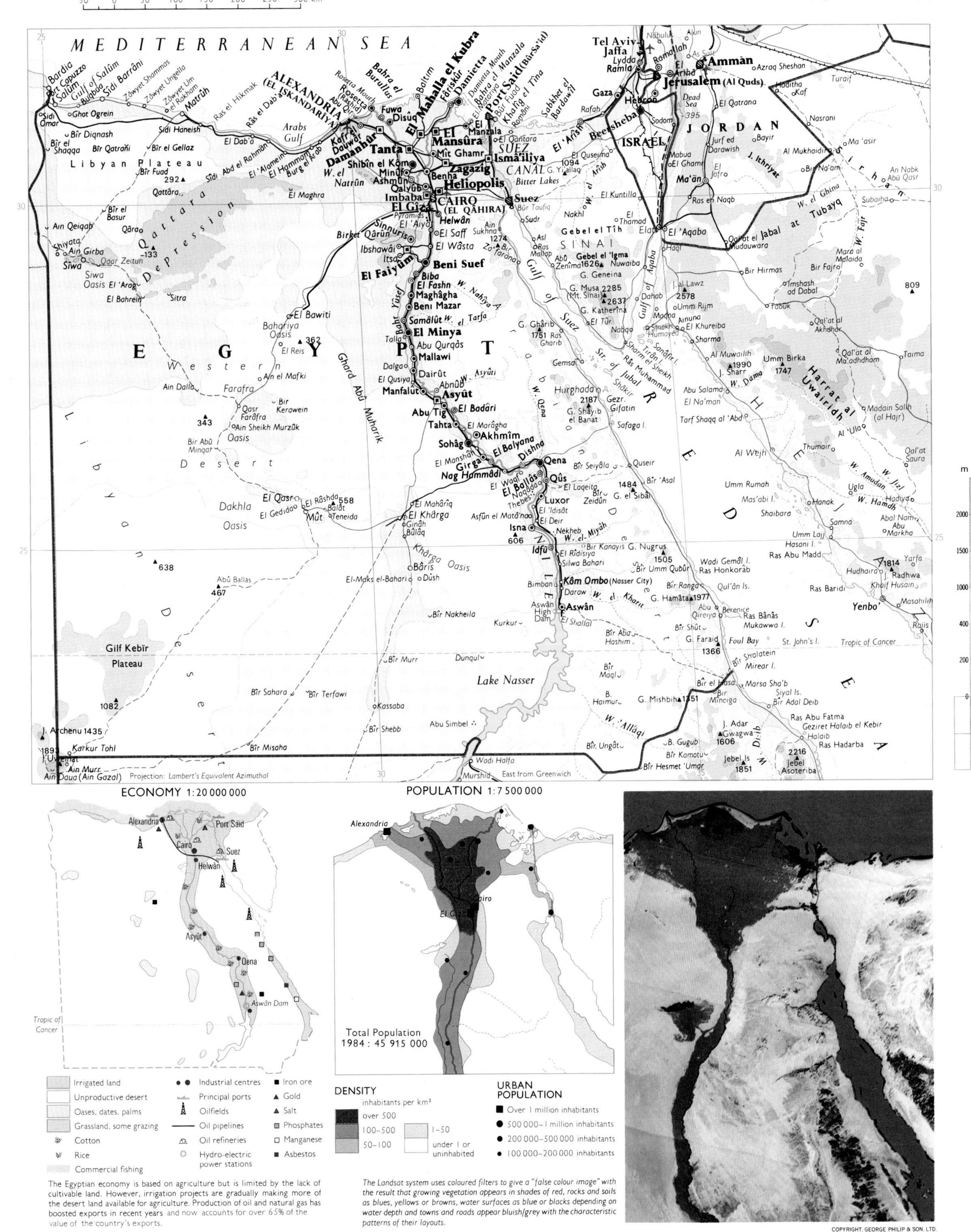

1:7 500 000

50 0 50 100 150 200 250 300 km

MEDITERRANEAN SEA

ALEXANDRIA (EL ISKANDARIYA)

Port Saïd (Bûr Saïd)

Tel Aviv-Jaffa

Amman

Jerusalem (Al Quds)

ISRAEL

JORDAN

SUEZ CANAL

CAIRO (EL QÂHIRA)

El Gîza

Heliopolis

Helwân

Suez

SINAI

Gebel el Tîh

Gebel el Igma 1626

G. Musa 2285 (Mt. Sinai)

2637

El Faiyûm

Beni Suef

El Minya

Mallawi

Asyût

Sohâg

Girga

Nag Hammâdi

El Balyana

Qena

Luxor

Thebes

Isna

Idfû

Kôm Ombo (Nasser City)

Aswân

Aswân High Dam

Lake Nasser

EGYPT

Libyan Plateau

Qattara Depression -133

Siwa Oasis

Bahariya Oasis

Farafra Oasis

Western Desert

Libyan Desert

Gilf Kebîr Plateau

Dakhla Oasis

Kharga Oasis

RED SEA

Tropic of Cancer

Abu Simbel

Wadi Halfa

Projection: Lambert's Equivalent Azimuthal

East from Greenwich

ECONOMY 1:20 000 000

Alexandria

Port Said

Cairo

Suez

Helwân

Asyût

Qena

Aswân Dam

Tropic of Cancer

Irrigated land	●● Industrial centres	■ Iron ore
Unproductive desert	⚓ Principal ports	▲ Gold
Oases, dates, palms	⛏ Oilfields	■ Salt
Grassland, some grazing	— Oil pipelines	▣ Phosphates
⚘ Cotton	⚒ Oil refineries	□ Manganese
⚘ Rice	⊚ Hydro-electric power stations	■ Asbestos
Commercial fishing		

The Egyptian economy is based on agriculture but is limited by the lack of cultivable land. However, irrigation projects are gradually making more of the desert land available for agriculture. Production of oil and natural gas has boosted exports in recent years and now accounts for over 65% of the value of the country's exports.

POPULATION 1:7 500 000

Alexandria

Cairo

El Gîza

Total Population
1984 : 45 915 000

DENSITY
inhabitants per km²

over 500
100–500
50–100
1–50
under 1 or uninhabited

URBAN POPULATION

■ Over 1 million inhabitants
● 500 000–1 million inhabitants
● 200 000–500 000 inhabitants
• 100 000–200 000 inhabitants

The Landsat system uses coloured filters to give a "false colour image" with the result that growing vegetation appears in shades of red, rocks and soils as blues, yellows or browns, water surfaces as blue or blacks depending on water depth and towns and roads appear bluish/grey with the characteristic patterns of their layouts.

1:20 000 000

200   0   200   400   600   800 km

NIGERIA

CAMEROON

CENTRAL AFRICAN REPUBLIC

SUDAN

ETHIOPIA

SOMALI REP.

Cross
Cameroon Pk. 4070
Douala
Yaounde
B. of Bonny
Bioko
Bata
EQUATORIAL GUINEA
RIO MUNI
Libreville
C. Lopez
Ogowe
Franceville
GABON
Equator

Bangui
Ubangi
Bangassou
M. Bomu
Uele
Niangara
Wadelai
Nile
Nimule
Juba
Bahr el Jebel
Mongalla
Omo
Chew Bahir (L. Stefanie)
L. Turkana
Marsabit
Juba

Sanga
Zaire (Congo)
Lisala
Aketi
Buta
Aruwimi
Isiro
Irumu
L. Mobutu Sese Seko
Kabarega Falls
Butiaba
L. Kyoga
Mt. Elgon 4321
Kitale
Eldoret
Mt. Kenya 5199
Lamu
Malindi

CONGO
Mbandaka
Congo
Kwa
Basoko
Kisangani
Stanley Falls
Ubundi
Lomami
Lualaba
Semliki
Ruwenzori 5109
L. Edward
L. George
Kampala
Entebbe
Jinja
Kisumu
UGANDA
Naivasha
Nairobi
KENYA
Nakuru
Kilimanjaro 5895
Moshi
Voi
Tana
Kismayu

Brazzaville
Pool Malebo
Kinshasa
Cabinda
Boma
Muanda
Matadi
Ilebo
Sankuru
Kindu
Kasongo
Kwilu
Bukavu
RWANDA
Kigali
BURUNDI
Bujumbura
Kigoma
Mwanza
Bukoba
Lake Victoria
Mwanza
Arusha
L. Manyara
Tabora
Mombasa and Kilindini
Tanga
Pemba
Zanzibar
Bagamoyo
Dar-es-Salaam

Pointe Noire
Pte Noire
ZAÏRE (CONGO)
Kananga
Lusambo
Kongolo
Kabalo
Kalemie
L. Eyasi
Dodoma
Kongwa
Mpwapwa
Morogoro
Rufiji
Mafia

Ambriz
Luanda
Cuanza
Malanje
Lulua
Kasai
Kamina
Bukama
Shaba
L. Mweru
Mpanda
L. Rukwa
Kasanga
Mbala
TANZANIA
Mbeya
Tukuyu
Tringa
Kilwa

Lobito
Benguela
Bié
Huambo
Plateau
ANGOLA
Luau
Likasi
Lubumbashi
Kitwe
Ndola
Karonga
Manda
Lindi
Mikindani
C. Delgado
Ruvuma

Namibe
Gt. Fish Bay
Cubango
Cunene
Seshek
Lealui
Mongu
Barotseland
Zambezi
Kafue
Kabwe
ZAMBIA
Lusaka
Chipata
Lilongwe
MALAWI
Salima
Blantyre
Zomba
Limbe
Shirwa
Nampula
Mozambique

Owambo
Etosha Pan
Otavi
Grootfontein
Botlete
Sesheke
Victoria Falls
Livingstone
Kafue
L. Kariba
Zumbo
Zambezi
Cabora Bassa Dam
Tete
Sena
Quelimane
Chinde

C. Frio
Damaraland
Makgadikgadi Salt pan
Hwange
Wankie
Harare
ZIMBABWE
Gweru
Mutare
Sabi
Beira
Sofala

Swakopmund
Walvis Bay
Windhoek
BOTSWANA
Serowe
Shoshong
Palapye
Matabeleland
Bulawayo
Matopo Hills
Gt. Zimbabwe
Masvingo
West Nicholson
Gwanda
Sabi
MOZAMBIQUE
Inhambane

NAMIBIA
Nossob
Kalahari
Tropic of Capricorn
Gaborone
Messina
Limpopo
Pietersburg
Olifants
TRANSVAAL

Hardap Dam
Namaland
Molopo
Mmabatho
Pretoria
Lydenburg
Barberton
Springs
Maputo
Delagoa Bay

Lüderitz
Possession I.
Karas Mts.
Keetmanshoop
Vryburg
Soweto
Johannesburg
Germiston
Vereeniging
Kroonstad
Kroonstad
SWAZILAND

Orange
Upington
Kimberley
ORANGE FREE STATE
Vaal
Chirundu
Mt. aux Sources 3298
Ladysmith
St. Lucia Bay
NATAL
Maseru
LESOTHO 3482
Pietermaritzburg

Port Nolloth
Bushmanland
Bloemfontein
Stormberg
Dr.
Umzi
Durban
Umtata

De Aar
Calvinia
Kompasberg 2504
Graaff-Reinet
Nuweveldberge
Karoo
Swartberg
Oudtshoorn
CAPE PROVINCE
East London
Mdantsane
Grahamstown
INDIAN OCEAN

SOUTH AFRICA
St. Helena Bay
Cape Town
Table Mountain
C. of Good Hope
Paarl
Mosselbaai
Algoa Bay
Pt. Elizabeth
C. Agulhas

ATLANTIC OCEAN

MADAGASCAR
On the same scale as the main map

C. Bobraomby
Antsiranana
Nossi-Bé
Andoany
Vohimarina
Tsaratanana 2876
Andapa
Mahajanga
Maroantsetra
Marovoay
Besalampy
Fenoarivo
Alaotra
Ambatondrazaka
Toamasina
Maintirano
Mdevatanana
Belo-Tsiribihina
Antananarivo 2643
Antsirabe
Mahanoro
Morondava
Mananjary
Morombé
Fianarantsoa
Mangoky
Ihosy
Manakara
Ankazoabo
Betroka
Farafangana
Toliara
Bekily
1956
Ambovombé
Tropic of Capricorn
C. Vohimena
Taolanaro

m
3000
2000
1000
400
200
0
200
m

Projection: Sanson-Flamsteed's Sinusoidal

East from Greenwich

COPYRIGHT. GEORGE PHILIP & SON, LTD.

1 : 7 500 000

50   0   50   100 150 200 250   300 km

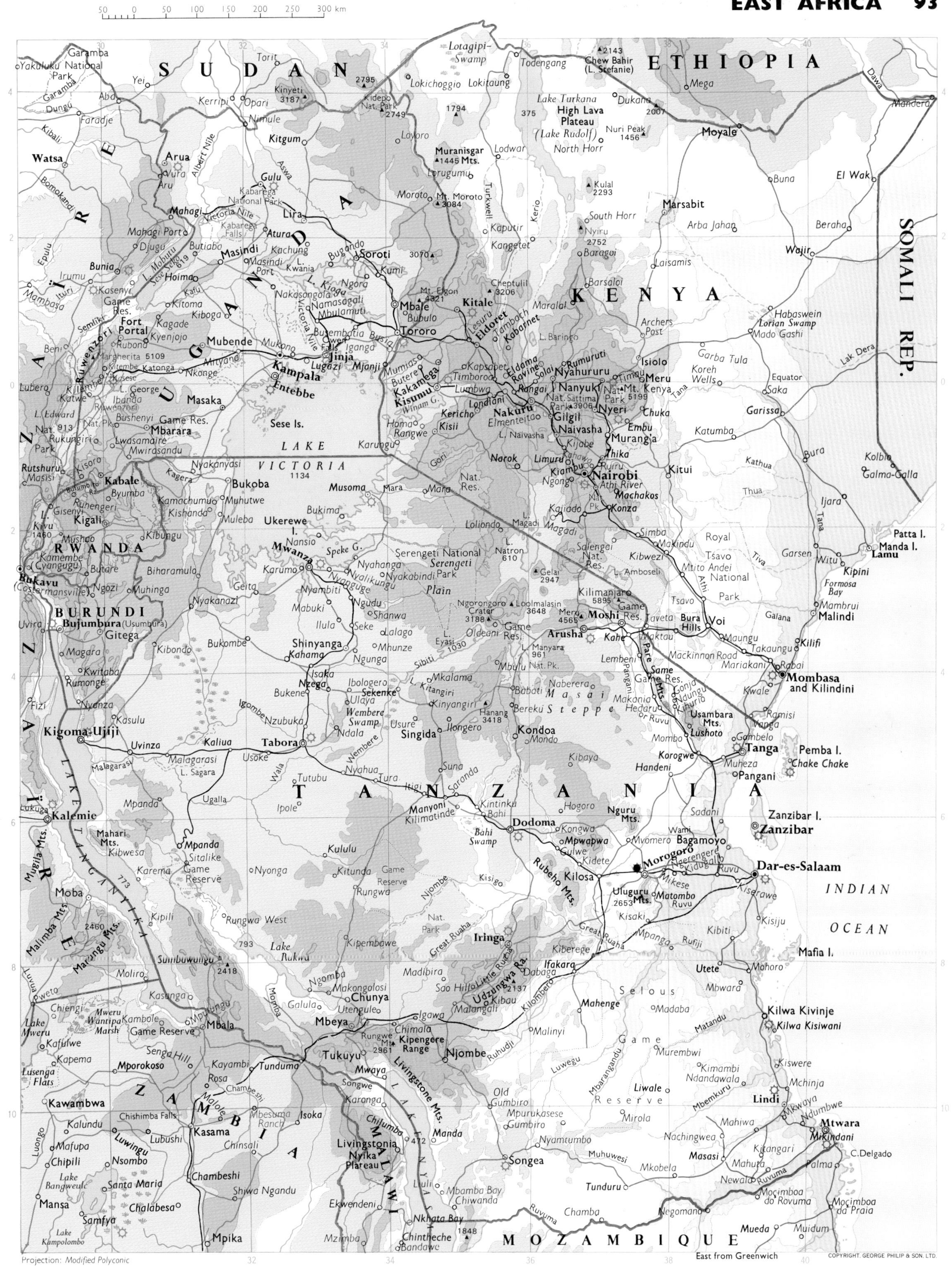

SUDAN

ETHIOPIA

SOMALI REP.

Garamba
National
Park
Yakuluku
Garamba
Dunga
Kibali
Faradje
Watsa
Bomokandi
Aba
Yei
Torit
Kerripi
Opari
Nimule
Kidepo
Nat. Park
2749
Kitgum
2795
Lotagipi-
Swamp
Todengang
Lokichoggio
Lokitaung
Chew Bahir
(L. Stefanie)
2143
Mega
Dawa
Mandera

Kinyeti
3187
Layoro
Muranisgar
1445 Mts.
1794
375
Lake Turkana
High Lava
Plateau
(Lake Rudolf)
North Horr
Dukana
2007
Moyale
Mega

Arua
Vura
Aru
Gulu
Aswa
Kabarega
National Park
Moroto
Mt. Moroto
3084
Turkwell
Lodwar
Nuri Peak
1456
South Horr
Nyiru
2752
Arba Jahan
Beraha

Mahagi
Mahagi Port
Victoria Nile
Atura
Djugu
Butiaba
Kabarega
Falls
Masindi
Masindi
Port
Kachung
Kwania
Lira
3070
Soroti
Kumi
Ngora
Kaputir
Kangetet
Baragoi
Laisamis
Wajir
El Wak

Epulu
Irumu
Bunia
Kasenyi
Game
Res.
Kitoma
Kiboga
Mbale
4321
Mt. Elgon
Kitale
Cheptulil
3206
Maralal
Barsaloi
Archers
Post
Marsabit
Buna
Habaswein
Lorian Swamp
Mado Gashi

Ituri
Mambasa
Semliki
Beni
Fort
Portal
Kyenjojo
Mubende
Mukono
Jinja
Tororo
Bubulo
Eldoret
Kabarnet
L. Baringo
Isiolo
Meru
Koreh
Wells
Garba Tula
Lak Dera

Margherita 5109
Ruwenzori
Katonga
Nkonge
Kampala
Entebbe
Butere
Mumias
Kakamega
Kapsabet
Timboroa
Eldama
Ravine
Solai
Rumuruti
Nyahururu
Mt. Kenya
5199
Nat.
Park
Nanyuki
Chuka
Garissa

L. George
L. Edward
Nat. 913
Bushenyi
Masaka
Kisumu
Lumbwa
Kericho
Londiani
Rangai
Nakuru
Naivasha
Gilgil
Nyeri
Embu
Murang'a
Katumba
Bura

RWANDA
Mbarara
Sese Is.
Winam G.
Homa
Kisii
Rangwe
L. Naivasha
Elmenteita
Kijabe
Limuru
Thika
Kiambu
Ngong
Nairobi
Kitui
Thua
Kathua
Kolbia
Galma-Galla

Kigali
Kabale
Bukoba
LAKE
VICTORIA
1134
Musoma
Mara
Nat.
Res.
Narok
Athi River
Machakos
Konza
Kiamba
Ijara

BURUNDI
Bukavu
(Costermansville)
Mushoro
L. Kivu
1460
Ukerewe
Nansio
Mwanza
Speke G.
Karumo
Serengeti
Plain
Lolgorien
Magadi
Salengai
Nat.
Res.
Kibwezi
Simba
Makindu
Royal
Tsavo
Mtito Andei
National
Garsen
Witu
Patta I.
Manda I.
Lamu
Kipini

Bujumbura
(Usumbura)
Gitega
Biharamulo
Nyakanazi
Geita
Nyambiti
Bukima
Nyahanga
Valikungu
Nyanguge
Nyamuguge
Ngudu
Shanwa
Serengeti National
Park
Ngorongoro
Crater
3188
Gelai
2947
Natron
610
Kilimanjaro
5895
Moshi
Taveta
Game
Res.
Voi
Bura
Hills
Galana
Mambrui
Malindi

Ngozi
Muhinga
Kibondo
Mabuki
Seke
Lalago
Mhunze
Oldeani
L.
Eyasi
1030
Merv
4565
Arusha
Kahe
Mackinnon Road
Mariakani
Takaungu
Kilifi

Uvira
Mugara
Kwitaba
Rumonge
Nyanza
Kasulu
Kombe
Shinyanga
Kahama
Ngunga
Isaka
Ulaya
Singida
Hanang
3418
Babati
Mbulu
Nat. Pk.
Manyara
961
Lembeni
Same
Korogwe
Handeni
Usambara
Mts.
Lushoto
Mombo
Gonja
Naberera
Masai
Steppe
Makania
Hedaru
Ruvu
Usambara
Mts.
Gambelo
Wanga
Ramisi
Kwale
Mombasa
and Kilindini

Kigoma-Ujiji
Uvinza
Malagarasi
L. Sagara
Nzega
Bukene
Nzubuka
Igombe
Ibologero
Kitangiri
Kinyangiri
Usure
Hongero
Kondoa
Mondo
Kibaya
Handeni
Pangani
Tanga
Pemba I.
Chake Chake

Kalemie
Mpanda
Usoke
Tabora
Kaliua
Wala
Tutubu
Tura
Wembere
Swamp
Ndala
Nyahua
Igiri
Caronda
Suna
Manyoni
Kilimatinde
Kintinku
Bahi
Hogoro
Nguru
Mts.
Sadani
Muheza
Pangani
Zanzibar I.
Zanzibar

Moba
Mahari
Mts.
Kibwesa
Karema
Sitalike
Game
Reserve
Nyonga
Kululu
Itunda
Game
Reserve
Rungwa
Bahi
Swamp
Dodoma
Kongwa
Mpwapwa
Gulwe
Kidete
Mvomero
Bagamoyo
Wami
Dar-es-Salaam

Kipili
Mpanda
Mahenge
Njombe
Ipole
Kisigo
Njombe
Kibiti
Mpanga
Morogoro
Kilosa
Mikese
Kidugalo
Ruvu
Kisarawe
Kisiju
INDIAN
OCEAN

Sumbawanga
2418
Rungwa West
793
Lake
Rukwa
Kipembawe
Nat.
Park
Great Ruaha
Iringa
Kiberege
Great
Ruaha
Uluguru
2653
Mts.
Matombo
Ruvu
Kisaki
Mafia I.

Moliro
Kasanga
Mbala
Senga
Hill
Mporokoso
Kayambi
Mbeya
Tukuyu
Mwaya
Songwe
Chimala
Kipengere
Range
2961
Madibira
Sao Hill
Ifakara
Dabaga
Udzungwa Ra.
Kibau
2137
Malangali
Mahenge
Madaba
Mbwara
Selous
Utete
Mohoro
Utengule
Kilwa Kivinje
Kilwa Kisiwani

Kawambwa
Kalundu
Luongo
Mafupa
Chipili
Nsombo
Lake
Bangweulu
Lusenga
Flats
Mwanza
Wantipa
Marsh
Kambole
Mbeya
Chunya
Makongolosi
Ngomba
Galula
Igawa
Ruhudji
Njombe
Luwegu
Game
Muremwbi
Matandu
Liwale
Ndandawala
Kimambi
Mchinja
Lindi

Luwingu
Lubushi
Kasama
Chinsali
Rosa
Chambeni
Tunduma
Sangwa
Rungwe M.
Livingstone Mts.
Livingstonia
Nyika
Plateau
412
Manda
Old
Gumbiro
Mpurukasese
Gumbiro
Nyamtumbo
Mirola
Masasi
Kitangari
Mahua
Mahiwa
Mtwara
Mikindani
C. Delgado

Mansa
Samfya
Lake
Kampolombo
Chalabesa
Mpika
Santa Maria
Shiwa Ngandu
Chambeshi
Chishimba Falls
Mbesuma
Ranch
Isoka
Chilumba
Chambeshi
1848
Chintheche
Bandawe
Mbamba Bay
Chiwanda
Ruvuma
Chamba
Songea
Tunduru
Muhuwesi
Mkobela
Newala
Masasi
Nachingwea
Masasi
Ruvuma
Negomano
Mueda
Muidum
Moçimboa
do Rovuma
Moçimboa
da Praia

ZAMBIA

MALAWI

TANZANIA

MOZAMBIQUE

Projection: Modified Polyconic

East from Greenwich

COPYRIGHT. GEORGE PHILIP & SON. LTD

m
4000
3000
2000
1500
1000
400
200
0
200
m

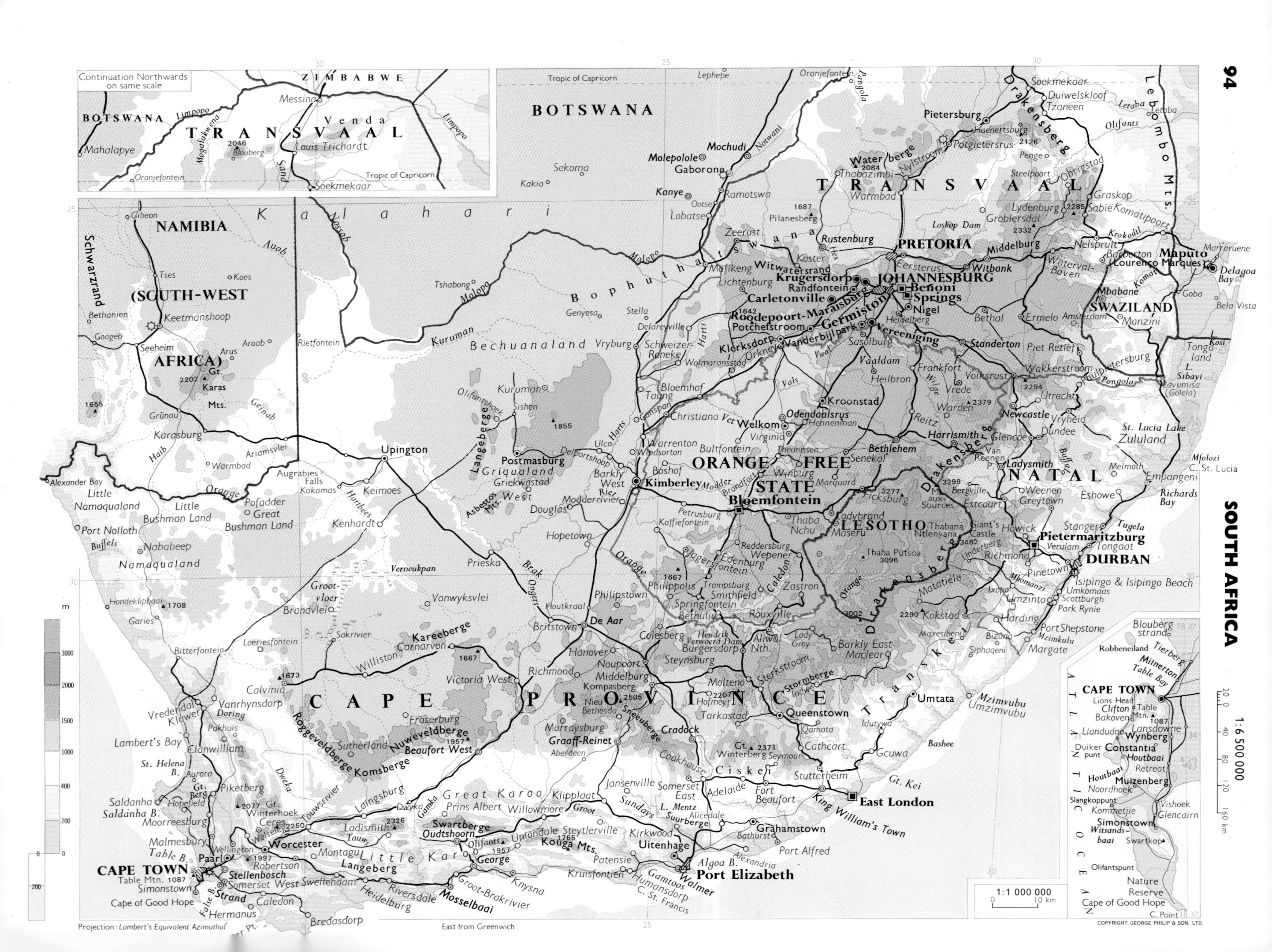

SOUTH AFRICA

1 : 6 500 000

20   0   40   80   120   160 km

CAPE TOWN

1 : 1 000 000

10   0   10 km

BOTSWANA

TRANSVAAL

SWAZILAND

NATAL

ORANGE FREE STATE

CAPE PROVINCE

LESOTHO

NAMIBIA

SOUTH-WEST AFRICA

ZIMBABWE

Continuation Northwards on same scale

Projection: Lambert's Equivalent Azimuthal

East from Greenwich

m   3000   2000   1500   1000   400   200   0

1:10 500 000

100   0   100   200   300   400 km

NAMIBIA

- East Caprivi
- Kavango
- Bushmanland
- Tswanaland
- Rehoboth Area
- Namaland
- Hereroland
- Owambo
- Kaokoland
- Damaraland

SOUTH AFRICA

- Transkei
- Bophuthatswana
- Lebowa
- Ciskei
- Gazankulu
- Venda
- KwaNdebele
- KwaZulu
- KaNgwane (Swazi)
- Qwa Qwa

Selected growth points
- ■ in border areas
- ◆ in homelands
- ○ ○ Major towns

NAMIBIA

Ohopoho
Oshakati
Rundu
Katima Mulilo
Tsumkwe
Okakarara
Khorixas
Okombahe
Windhoek
Rehoboth
Walvis Bay
Tropic of Capricorn
Tses

(SOUTH WEST AFRICA)

During the 1950's, the South African government gave recognition to the development of Black Homeland States.

Black people in South Africa were to be given automatic citizenship to one of these Homelands. This would be determined by exact tribal origin. Black people, resident or not, in a Homeland would therefore lose South African citizenship and take on that of their attached Homeland.

So far, ten Homelands have been given official recognition by the Republic of South Africa. Only four Homeland states, Transkei, Ciskei, Bophuthatswana and Venda, have been granted independence by South Africa. Neither these states, nor any of the other Homelands are recognized by any other nation.

VENDA
Makwarela
Sibasa
Giyani
Pietersburg
Lebowakgomo
Nyamazane
Mmabatho
Krugersdorp
Johannesburg
Roodepoort-Maraisburg   Benoni
Germiston   Springs
Vereeniging
Pretoria
BOPHUTHATSWANA
Mbabane
SWAZI-LAND
Welkom
Phuthaditjhaba
Ulundi
Kimberley
Ladysmith
Richards Bay
Bloemfontein
Maseru
LESOTHO
Pietermaritzburg
Durban
SOUTH AFRICA
Umtata
TRANSKEI
Zwelitsha   East London
Port Elizabeth
Cape Town
Port Nolloth

RESIDENCE PATTERN OF BANTU GROUPS

percentage living outside homeland
percentage living within homeland

100 75 50 25 0

Others
Venda
Ndebele and others
Swazi
Shangaan
South Sotho
North Sotho
Tswana
Xhosa
Zulu

VALUE OF AGRICULTURAL PRODUCTION

R '000
40 000
30 000
20 000
10 000
5 000
0

Plant produce
Animal produce

Venda   Gazankulu   Ciskei   Bophuthatswana   Lebowa   KwaZulu   Transkei

POPULATION

'000 people

Urban	Population density
Rural	

2 250
2 000
1 750
1 500
1 250
1 000
750
500
250

inhabitants per km²
80
70
60
50
40
30
20
10

KwaZulu   Transkei   Lebowa   Bophuthatswana   Ciskei   Venda   Gazankulu   KaNgwane   Qwa Qwa

GROSS DOMESTIC PRODUCT

R '000
150 000
125 000
100 000
75 000
50 000
25 000
12 500
0

Agriculture, forestry, hunting, fishing
Mining, quarrying
Manufacturing, industry, services, finance, trade, transport, health, education

Qwa Qwa   KaNgwane   Venda   Gazankulu   Ciskei   Lebowa   Bophuthatswana   KwaZulu   Transkei

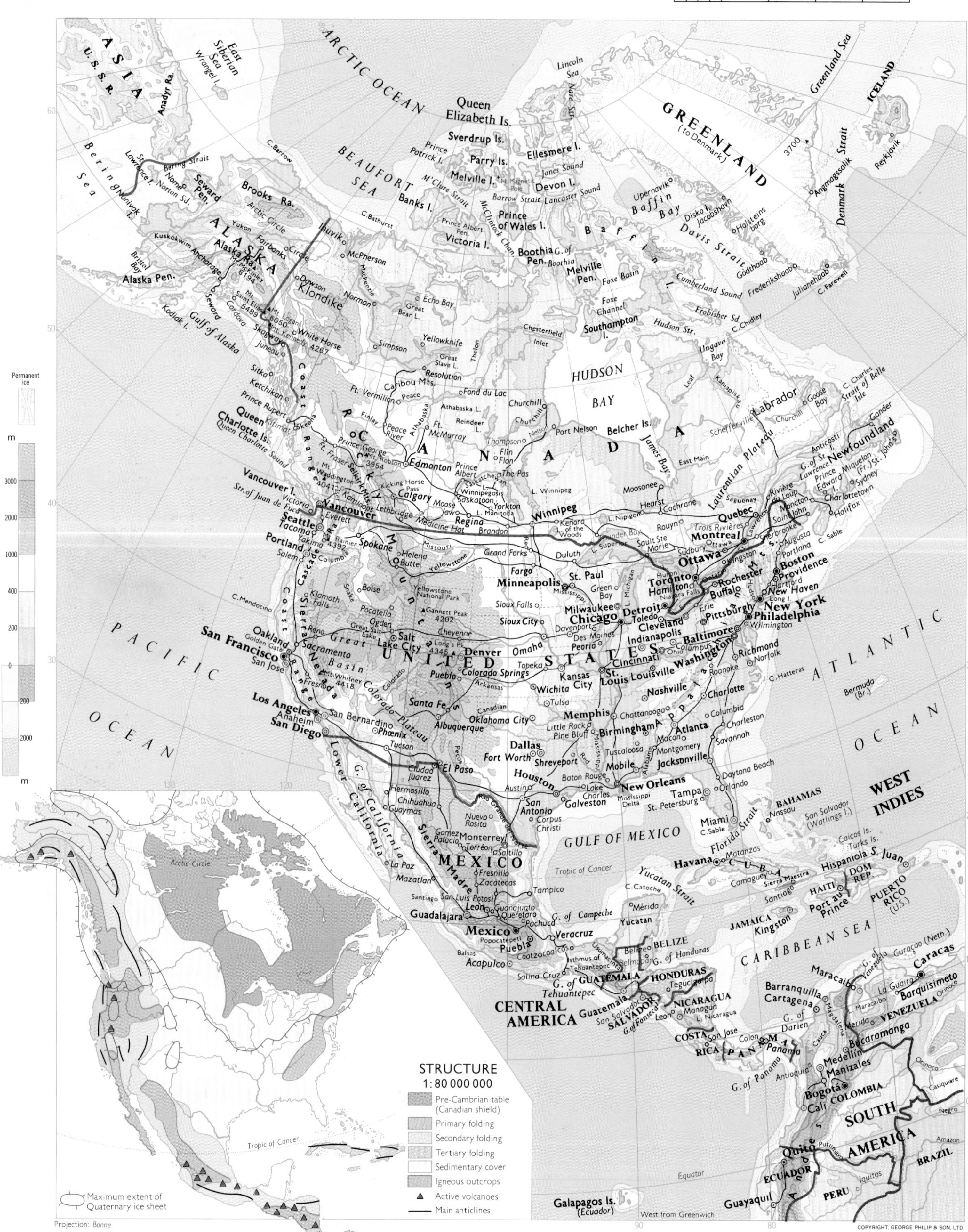

1 : 40 000 000

Permanent
ice

m

3000

2000

1000

400

200

0

200

2000

m

STRUCTURE
1 : 80 000 000

Pre-Cambrian table
(Canadian shield)

Primary folding

Secondary folding

Tertiary folding

Sedimentary cover

Igneous outcrops

▲ Active volcanoes

— Main anticlines

Maximum extent of
Quaternary ice sheet

Projection: Bonne

West from Greenwich

COPYRIGHT. GEORGE PHILIP & SON. LTD

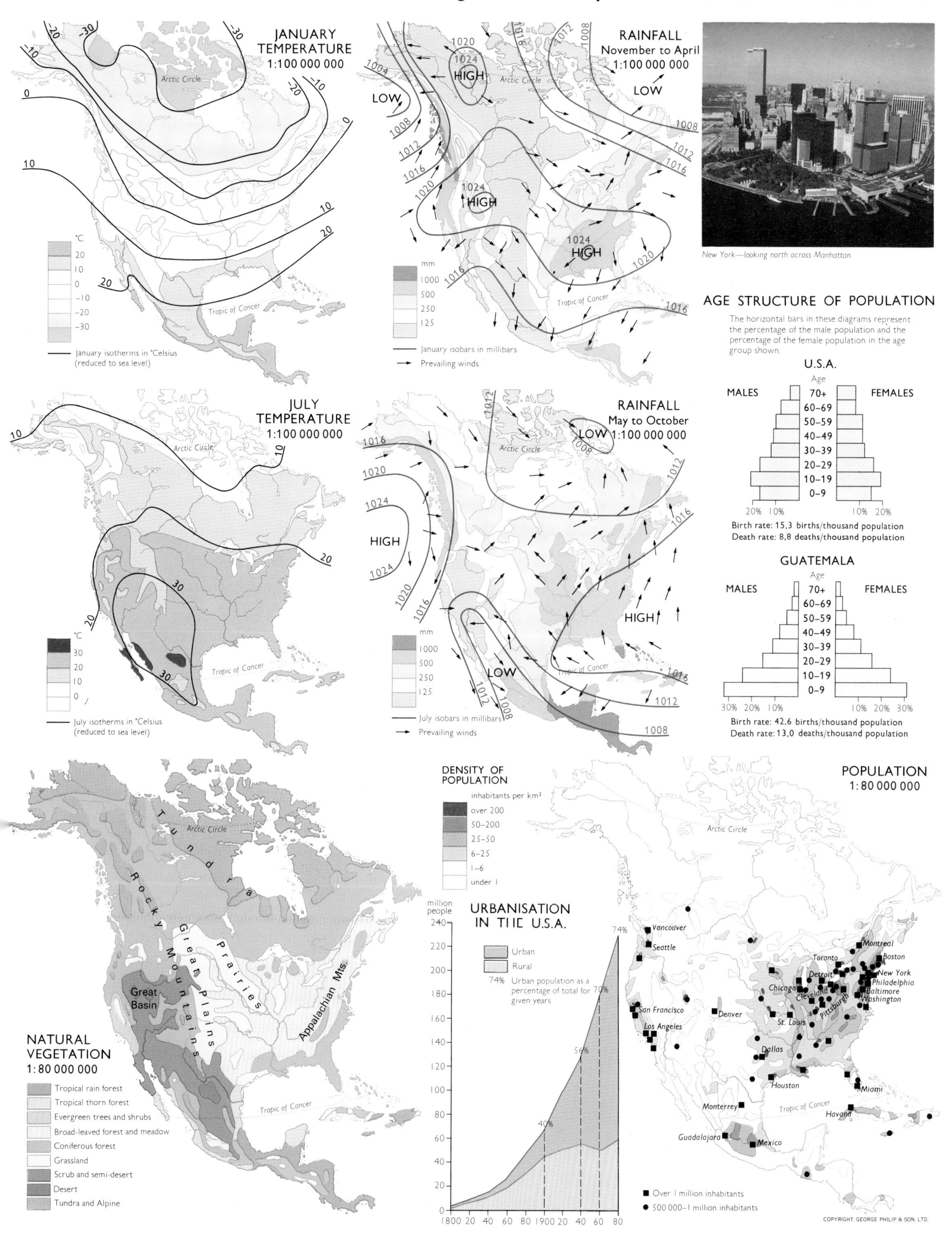

JANUARY TEMPERATURE
1:100 000 000

Arctic Circle

Tropic of Cancer

°C
20
10
0
-10
-20
-30

—— January isotherms in °Celsius (reduced to sea level)

RAINFALL
November to April
1:100 000 000

HIGH
LOW
LOW
HIGH
HIGH

Arctic Circle

Tropic of Cancer

mm
1000
500
250
125

—— January isobars in millibars
→ Prevailing winds

New York—looking north across Manhattan

AGE STRUCTURE OF POPULATION

The horizontal bars in these diagrams represent the percentage of the male population and the percentage of the female population in the age group shown.

U.S.A.

Age
MALES   70+   FEMALES
60-69
50-59
40-49
30-39
20-29
10-19
0-9

20% 10%   10% 20%

Birth rate: 15,3 births/thousand population
Death rate: 8,8 deaths/thousand population

GUATEMALA

Age
MALES   70+   FEMALES
60-69
50-59
40-49
30-39
20-29
10-19
0-9

30% 20% 10%   10% 20% 30%

Birth rate: 42,6 births/thousand population
Death rate: 13,0 deaths/thousand population

JULY TEMPERATURE
1:100 000 000

Arctic Circle

Tropic of Cancer

°C
30
20
10
0

—— July isotherms in °Celsius (reduced to sea level)

RAINFALL
May to October 1:100 000 000

LOW
HIGH
HIGH
LOW

Arctic Circle

Tropic of Cancer

mm
1000
500
250
125

—— July isobars in millibars
→ Prevailing winds

NATURAL VEGETATION
1:80 000 000

Tundra
Rocky Mountains
Great Plains
Prairies
Great Basin
Appalachian Mts.

Arctic Circle

Tropic of Cancer

Tropical rain forest
Tropical thorn forest
Evergreen trees and shrubs
Broad-leaved forest and meadow
Coniferous forest
Grassland
Scrub and semi-desert
Desert
Tundra and Alpine

DENSITY OF POPULATION

inhabitants per km²
over 200
50-200
25-50
6-25
1-6
under 1

URBANISATION IN THE U.S.A.

million people
240
220
200
180
160
140
120
100
80
60
40
20
0

Urban
Rural

74% Urban population as a percentage of total for given years

74%
56%
40%

1800 20 40 60 80 1900 20 40 60 80

POPULATION
1:80 000 000

Arctic Circle

Tropic of Cancer

Vancouver
Seattle
Montreal
Toronto
Boston
Detroit
New York
Chicago
Cleveland
Philadelphia
San Francisco
Pittsburgh
Baltimore
Denver
St. Louis
Washington
Los Angeles
Dallas
Houston
Miami
Monterrey
Havana
Guadalajara
Mexico

■ Over 1 million inhabitants
● 500 000–1 million inhabitants

COPYRIGHT. GEORGE PHILIP & SON. LTD.

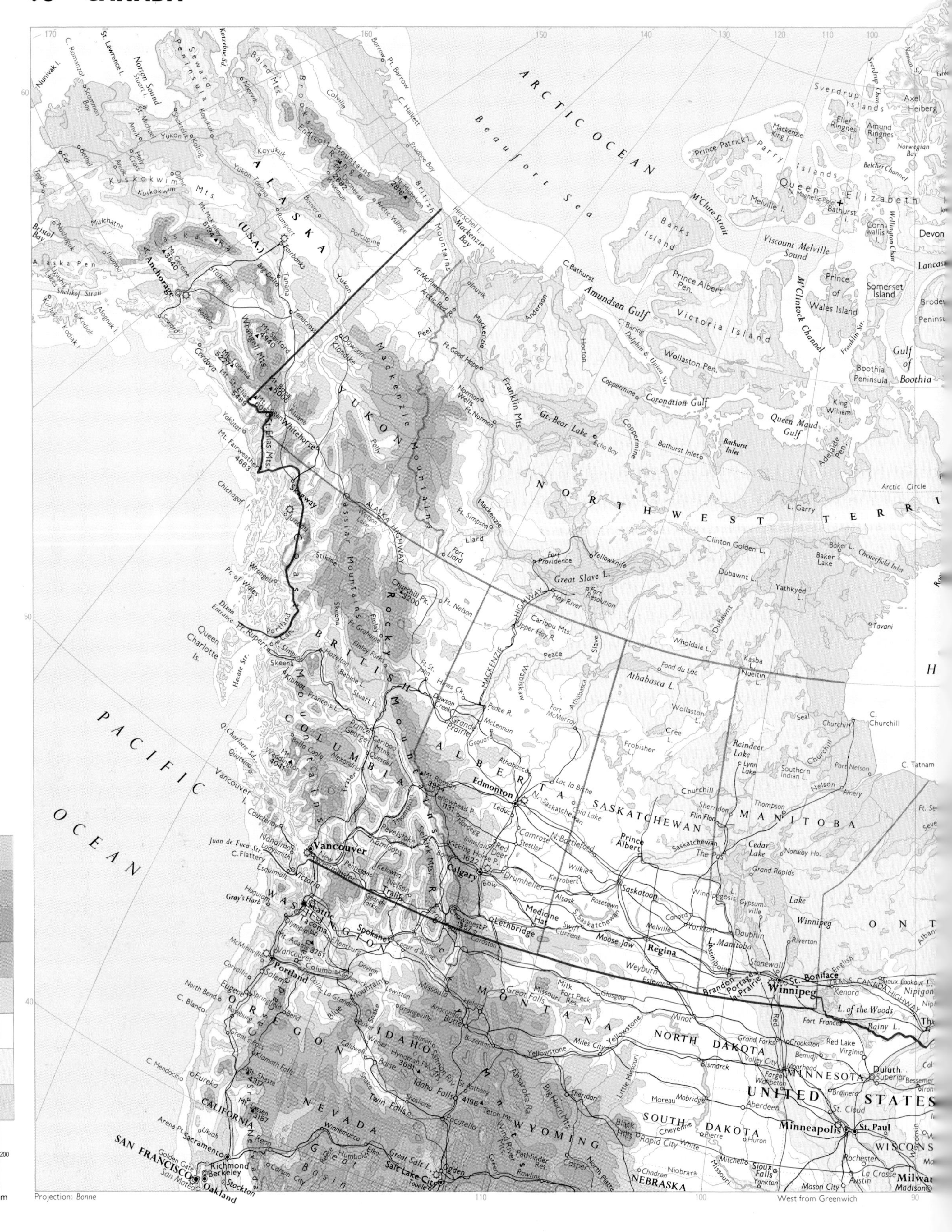
ARCTIC OCEAN
Beaufort Sea
PACIFIC OCEAN
ALASKA (U.S.A.)
YUKON
NORTHWEST TERRI
BRITISH COLUMBIA
ALBERTA
SASKATCHEWAN
MANITOBA
WASHINGTON
OREGON
IDAHO
MONTANA
WYOMING
NEVADA
CALIFORNIA
NORTH DAKOTA
SOUTH DAKOTA
NEBRASKA
MINNESOTA
UNITED STATES
Brooks Range
Endicott Mountains
Alaska Ra.
Wrangell Mts.
Mackenzie Mountains
Rocky Mountains
Coast Mountains
Cassiar Mountains
Selkirk Mts.
Franklin Mts.
Great Basin
Anchorage
Fairbanks
Whitehorse
Dawson
Juneau
Vancouver
Victoria
Seattle
Tacoma
Portland
Spokane
Edmonton
Calgary
Lethbridge
Saskatoon
Regina
Moose Jaw
Prince Albert
Winnipeg
St. Boniface
Brandon
Minneapolis
St. Paul
Duluth
Salt Lake City
Ogden
SAN FRANCISCO
Oakland
Berkeley
Richmond
Sacramento
Yellowknife
Great Slave L.
Gt. Bear Lake
Athabasca L.
Lake Winnipeg
L. of the Woods
Banks Island
Victoria Island
Prince of Wales Island
Somerset Island
Queen Elizabeth Islands
Parry Islands
Sverdrup Islands
Amundsen Gulf
Coronation Gulf
Queen Maud Gulf
Gulf of Boothia
M'Clure Strait
Viscount Melville Sound
M'Clintock Channel
Arctic Circle
ALASKA HIGHWAY
MACKENZIE HIGHWAY
TRANS-CANADA HIGHWAY
Queen Charlotte Is.
Vancouver I.
Mt. McKinley 6194
Mt. Logan 6050
Mt. St. Elias 5489
Mt. Robson 3954
Mt. Rainier
Projection: Bonne
West from Greenwich

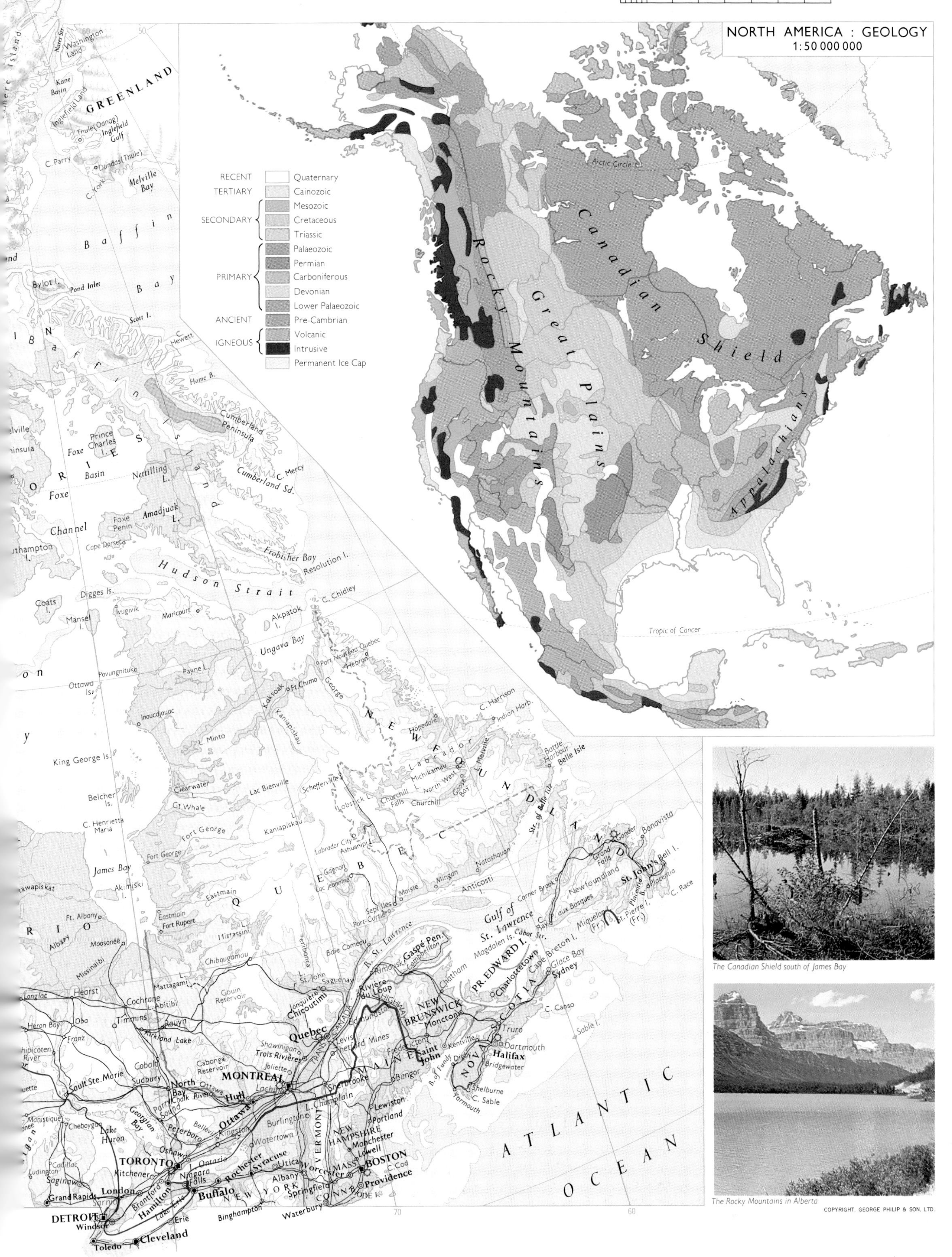

The Canadian Shield south of James Bay

The Rocky Mountains in Alberta

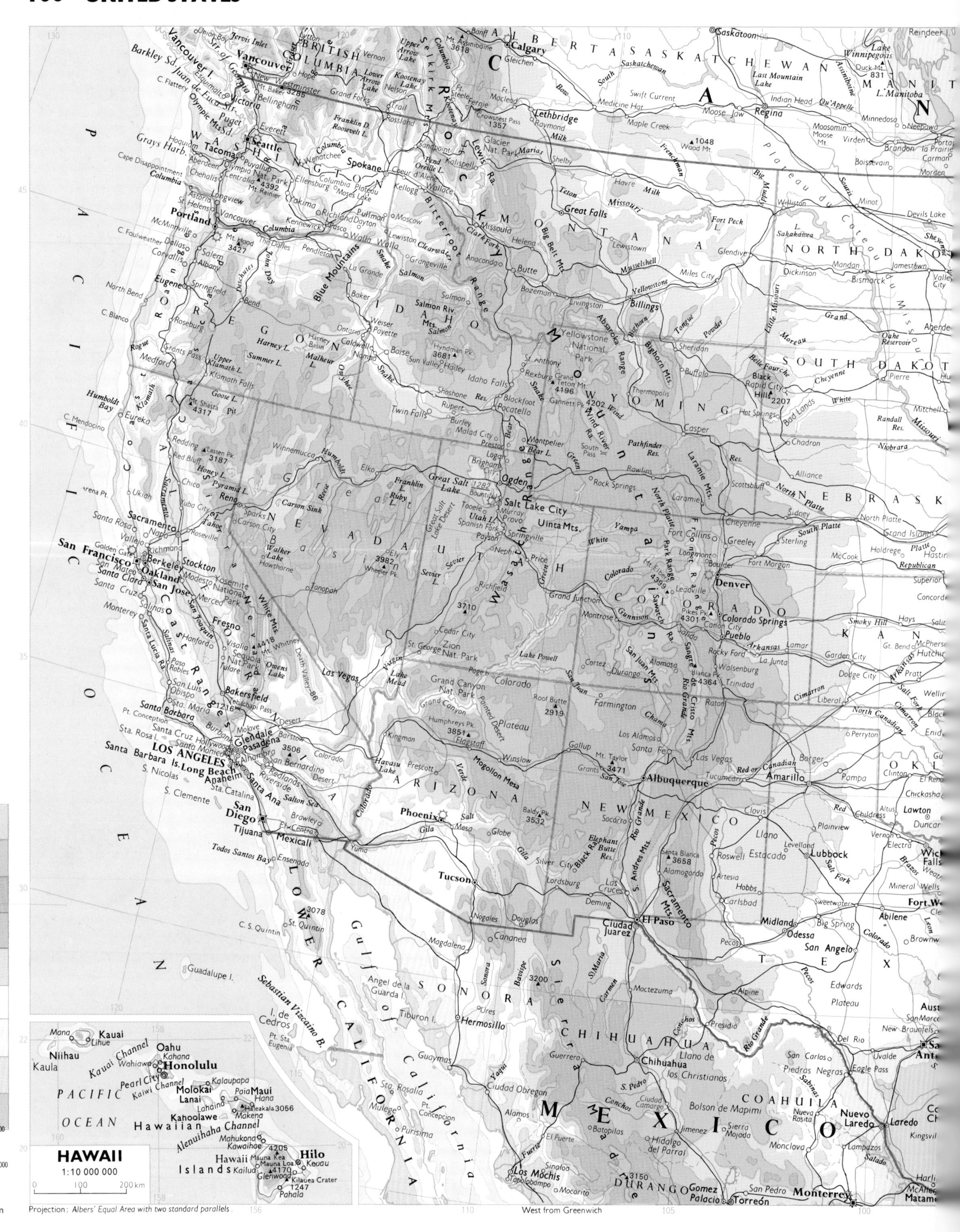

HAWAII
1:10 000 000

0   100   200km

Projection: Albers' Equal Area with two standard parallels.

West from Greenwich

GULF OF MEXICO

ATLANTIC OCEAN

BAHAMAS

MAINE

NEW BRUNSWICK

MINNESOTA

WISCONSIN

IOWA

MISSOURI

ILLINOIS

INDIANA

OHIO

KENTUCKY

TENNESSEE

ARKANSAS

OKLAHOMA

MISSISSIPPI

ALABAMA

LOUISIANA

GEORGIA

FLORIDA

SOUTH CAROLINA

NORTH CAROLINA

VIRGINIA

WEST VIRGINIA

PENNSYLVANIA

NEW JERSEY

MARYLAND

DELAWARE

NEW YORK

VERMONT

NEW HAMPSHIRE

MASS.

CONN.

Lake Winnipeg

Lake Superior

Lake Michigan

Lake Huron

Lake Erie

Lake Ontario

St. Lawrence

MONTREAL

Quebec

Ottawa

TORONTO

Buffalo

DETROIT

CHICAGO

Milwaukee

Minneapolis   St. Paul

Des Moines

Kansas City

St. Louis

Memphis

Little Rock

Dallas

Houston

New Orleans

Baton Rouge

Birmingham

Atlanta

Montgomery

Jacksonville

Tampa

Orlando

Miami

Savannah

Charleston

Columbia

Raleigh

Charlotte

Knoxville

Nashville

Chattanooga

Louisville

Cincinnati

Indianapolis

Columbus

Cleveland

PITTSBURGH

WASHINGTON D.C.

Baltimore

PHILADELPHIA

NEW YORK

Boston

Winnipeg

Duluth

Tulsa

Oklahoma City

Fort Worth

1:60 000 000

500   0   500   1000   1500   2000   2500 km

Stockyard, Phoenix

The spring wheat harvest

Aerial photograph of field patterns, Saskatchewan

Land Use; lumber mill, Washington, Seattle

MAIZE
1,5%   55,4%
43,1%
World total
450,7 million tons

SORGHUM
36,7%   63,3%
World total
78,9 million tons

:: MAIZE
:: SORGHUM
:: SUGAR CANE
one dot represents 8 000 hectares

figures give percentage of world total

U.S.A.   Canada   Rest of world

WHEAT
82,5%
4,0%
13,5%
World total
522,9 million tons

TOBACCO
85,8%
1,5%
12,7%
World total
6,2 million tons

:: SPRING WHEAT
:: WINTER WHEAT
:: TOBACCO
one dot represents 8 000 hectares

COTTON LINT
84%
16%
World total
17,7 million tons

BARLEY
86,4%
6%
7,6%
World total
171,3 million tons

:: OATS
:: BARLEY
:: COTTON
one dot represents 8 000 hectares

BEEF CATTLE
71,6%
2,3%
26,1%
World total
46,7 million head

DAIRY CATTLE
94,1%
1,0%
4,9%
World total
224,0 million head

:: DAIRY CATTLE
:: BEEF CATTLE
one dot represents 20 000 head

LAND USE

Forest and woodland
Predominantly arable land
Predominantly pastoral land
Semi-desert and steppe
Desert
Tundra and Alpine areas
Swamp and marshland

**MEAN ANNUAL
FROST-FREE PERIOD**

Killing frost liable annually
Occasional killing frost

over 300 days
240–300 days
180–240 days
120–180 days
60–120 days
under 60 days

COPYRIGHT. GEORGE PHILIP & SON. LTD.

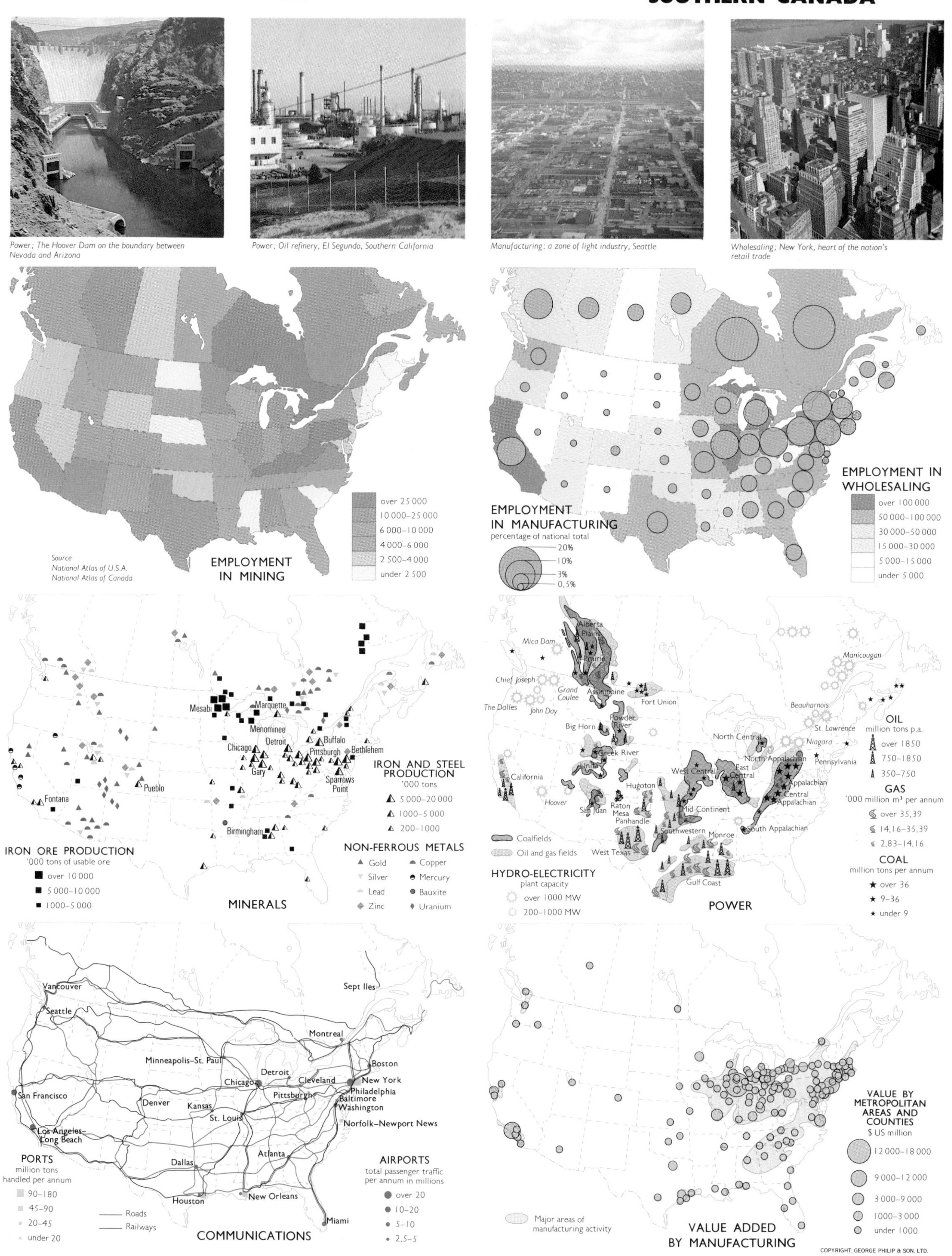

1:60 000 000

500   0   500   1000   1500   2000   2500 km

Power; The Hoover Dam on the boundary between
Nevada and Arizona

Power; Oil refinery, El Segundo, Southern California

Manufacturing; a zone of light industry, Seattle

Wholesaling; New York, heart of the nation's
retail trade

EMPLOYMENT
IN MINING

over 25 000
10 000–25 000
6 000–10 000
4 000–6 000
2 500–4 000
under 2 500

Source
National Atlas of U.S.A.
National Atlas of Canada

EMPLOYMENT
IN MANUFACTURING
percentage of national total
20%
10%
3%
0,5%

EMPLOYMENT IN
WHOLESALING
over 100 000
50 000–100 000
30 000–50 000
15 000–30 000
5 000–15 000
under 5 000

MINERALS

Mesabi
Marquette
Menominee
Detroit   Buffalo
Chicago   Pittsburgh   Bethlehem
Gary   Sparrows
                        Point
Fontana
Pueblo

Birmingham

IRON AND STEEL
PRODUCTION
'000 tons
5 000–20 000
1 000–5 000
200–1000

IRON ORE PRODUCTION
'000 tons of usable ore
over 10 000
5 000–10 000
1000–5000

NON-FERROUS METALS
Gold   Copper
Silver   Mercury
Lead   Bauxite
Zinc   Uranium

POWER

Mica Dam   Alberta
                Plains
Chief Joseph   Prairie   Manicougan
The Dalles   Grand   Assiniboine
John Day   Coulee   Fort Union
                Big Horn   Powder   Beauharnois
                           River   St. Lawrence
                Green River North Central   Niagara
California   Uinta   North Appalachian Pennsylvania
                       Hugoton   West Central East
Hoover   San Juan Raton   Central Appalachian
                     Mesa   Central
                     Panhandle Mid-Continent   Appalachian
                        Southwestern Monroe   South Appalachian
West Texas   Gulf Coast

Coalfields
Oil and gas fields

HYDRO-ELECTRICITY
plant capacity
over 1000 MW
200–1000 MW

OIL
million tons p.a.
over 1850
750–1850
350–750

GAS
'000 million m³ per annum
over 35,39
14,16–35,39
2,83–14,16

COAL
million tons per annum
over 36
9–36
under 9

COMMUNICATIONS

Vancouver
Seattle   Sept Iles

                   Montreal

Minneapolis–St. Paul
                            Boston
San Francisco   Chicago Detroit
                   Cleveland New York
         Denver Pittsburgh Philadelphia
              Kansas   Baltimore
Los Angeles–   St. Louis   Washington
Long Beach   Norfolk–Newport News

                  Atlanta
        Dallas
              Houston   New Orleans
                            Miami

PORTS
million tons
handled per annum
90–180
45–90
20–45
under 20

Roads
Railways

AIRPORTS
total passenger traffic
per annum in millions
over 20
10–20
5–10
2,5–5

VALUE ADDED
BY MANUFACTURING

Major areas of
manufacturing activity

VALUE BY
METROPOLITAN
AREAS AND
COUNTIES
$ US million
12 000–18 000
9 000–12 000
3 000–9 000
1000–3000
under 1000

1:6 000 000

50   0   50   100   150   200   250 km

THE ST. LAWRENCE SEAWAY

The St. Lawrence Seaway was opened in 1959 enabling 80% of ocean-going vessels to reach Lake Superior. The Seaway is frozen for over 90 days each year from late December to early April.

Lake Superior 177 m
Lake Michigan Lake Huron 177 m
Lake Erie 174 m
Windsor
Hamilton 75 m
Lake Ontario
Welland Canal
L. St. Laurent 73 m
L. St. Francis 46 m
L. St. Louis 20 m
Montreal 20 m
7 m
Quebec
to Atlantic Ocean

QUEBEC
Cargo loaded (million tons)
Wheat 0.6   Barley 0.5   Zinc ore 0.4   Newsprint 0.2   Others 1.2

MONTREAL
Cargo loaded (million tons)
Fuel oil 0.9   Cement 0.4   Barley 0.2   Others 1.8   Wheat 1.5

HAMILTON
Cargo unloaded (million tons)
Coal 5.1   Iron ore 3.9   Fuel oil 0.5   Others 2.6

ST. LAWRENCE SEAWAY TRAFFIC
Total cargo carried (million tons)

OTHER GOODS 13.0
MANUFACTURES 1.6
AGRICULTURAL PRODUCTS 22.5
Wheat 10.9   Others 11.6
MINERALS 31.0
Iron ore 18.1   Coal 7.3   Others 5.6

POWER AND MINERALS

Oilfields
Gasfields
Oil pipelines
Gas pipelines
Coalfields
Coal production
Oil refineries
Hydro-electric power stations
Railways

Iron
Cobalt
Nickel
Vanadium
Copper
Lead
Zinc

Silver
Titanium
Uranium
Asbestos
Mica
Potash
Salt

LAND USE

Forest and woodland, mostly ungrazed
Woodland and forest, with some arable and pasture
Arable
Arable and pasture, with some woodland and forest
Pasture with some woodland and forest
Swamp
Urban and industrial areas

COPYRIGHT GEORGE PHILIP & SON LTD.

Projection: Alber's Equal area with two standard parallels

West from Greenwich

LAKE SUPERIOR
LAKE HURON
LAKE ONTARIO
LAKE ERIE
Lake Michigan
Georgian Bay

O N T A R I O
Q U E B E C
NEW HAMPSHIRE
VERMONT
NEW YORK
MASS.
CONN.
NEW JERSEY
DELAWARE
MARYLAND
PENNSYLVANIA
WEST VIRGINIA
OHIO
INDIANA
MICHIGAN
WISCONSIN

Quebec
Trois Rivières
Shawinigan
Drummondville
Thetford Mines
St. Hyacinthe
Sherbrooke
St. Jean
MONTREAL
St. Jerome
Hull
OTTAWA
Renfrew
Pembroke
North Bay
L. Nipissing
Sudbury
Copper Cliff
Sault Ste. Marie
Bancroft
Peterborough
Barrie
Kitchener
Guelph
TORONTO
Hamilton
Brantford
London
Sarnia
Port Huron
St. Clair
Chatham
Windsor
DETROIT
Pontiac
Ann Arbor
Flint
Bay City
Saginaw
Lansing
Jackson
Battle Creek
Kalamazoo
Grand Rapids
Muskegon
Traverse City
Cadillac
Cheboygan
Manistique
Escanaba
Menominee
Marinette
Green Bay
Appleton
Sheboygan
MILWAUKEE
Racine
Kenosha
Evanston
CHICAGO
Gary
Hammond
Joliet
Kankakee
South Bend
Elkhart
Kokomo
Logansport
La Fayette
Fort Wayne
Lima
Findlay
Toledo
Sandusky
Lorain
Lakewood
CLEVELAND
Akron
Canton
Youngstown
New Castle
PITTSBURGH
McKeesport
Wheeling
Bellaire
Zanesville
Columbus
Springfield
Dayton
Middletown
CINCINNATI
Marion
Mansfield
Monroe
Clarksburg
Parkersburg
Cumberland
Johnstown
Altoona
Williamsport
Hagerstown
York
Lancaster
Harrisburg
Reading
Allentown
Bethlehem
Wilkes Barre
Scranton
Binghamton
Elmira
Ithaca
BUFFALO
Niagara Falls
St. Catherines
Oshawa
Belleville
Kingston
Watertown
Oswego
Rochester
Auburn
Syracuse
Utica
Rome
Schenectady
Amsterdam
Troy
Albany
Saratoga Sps.
Glen Falls
Montpelier
Burlington
Concord
Manchester
Lowell
Lawrence
Lynn
BOSTON
Cambridge
Quincy
Brockton
Worcester
Pawtucket
Woonsocket
Providence
Fall River
New Bedford
Portland
Hartford
New Britain
Waterbury
New Haven
Bridgeport
Springfield
Holyoke
Kingston
Poughkeepsie
Yonkers
NEW YORK
Jersey City
Newark
Elizabeth
Paterson
Trenton
PHILADELPHIA
Camden
Wilmington
Atlantic City
Dover
BALTIMORE
WASHINGTON D.C.

Adirondack Mts. 1629
Catskill Mts. 1281
Hudson
Delaware
Susquehanna
Connecticut
Champlain
Ottawa
St. Lawrence
Allegheny
Ohio
Au Sable
Manitoulin I.
Long Island
Isle Royale
Hancock
Marquette
Negaunee
Ishpeming
Iron Mt. 604
646
960
556
759
1917

1:6 000 000

50 0 50 100 150 200 250 km

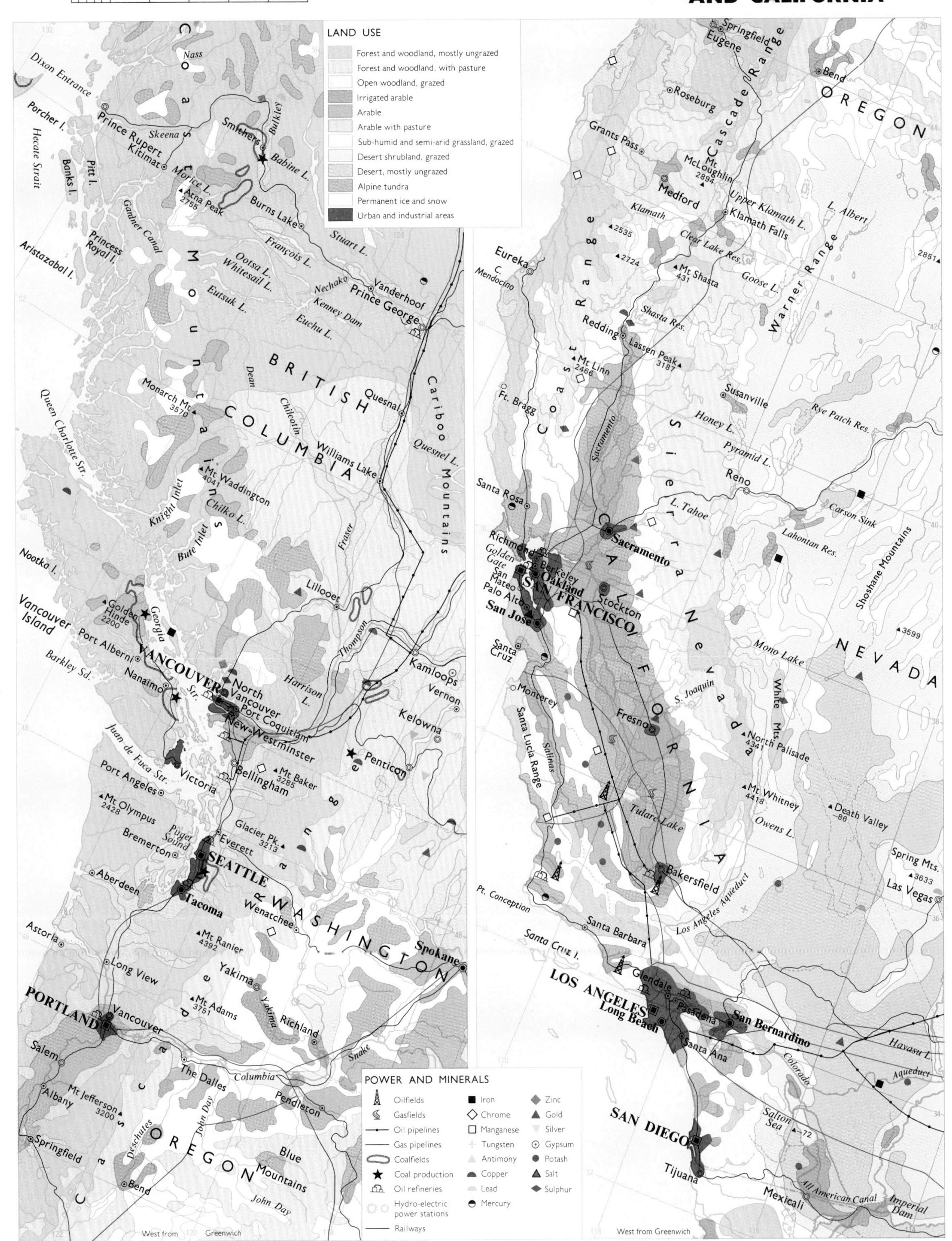

Projection: Lambert's Equivalent Azimuthal

Projection: Alber's Equal Area with two standard parallels

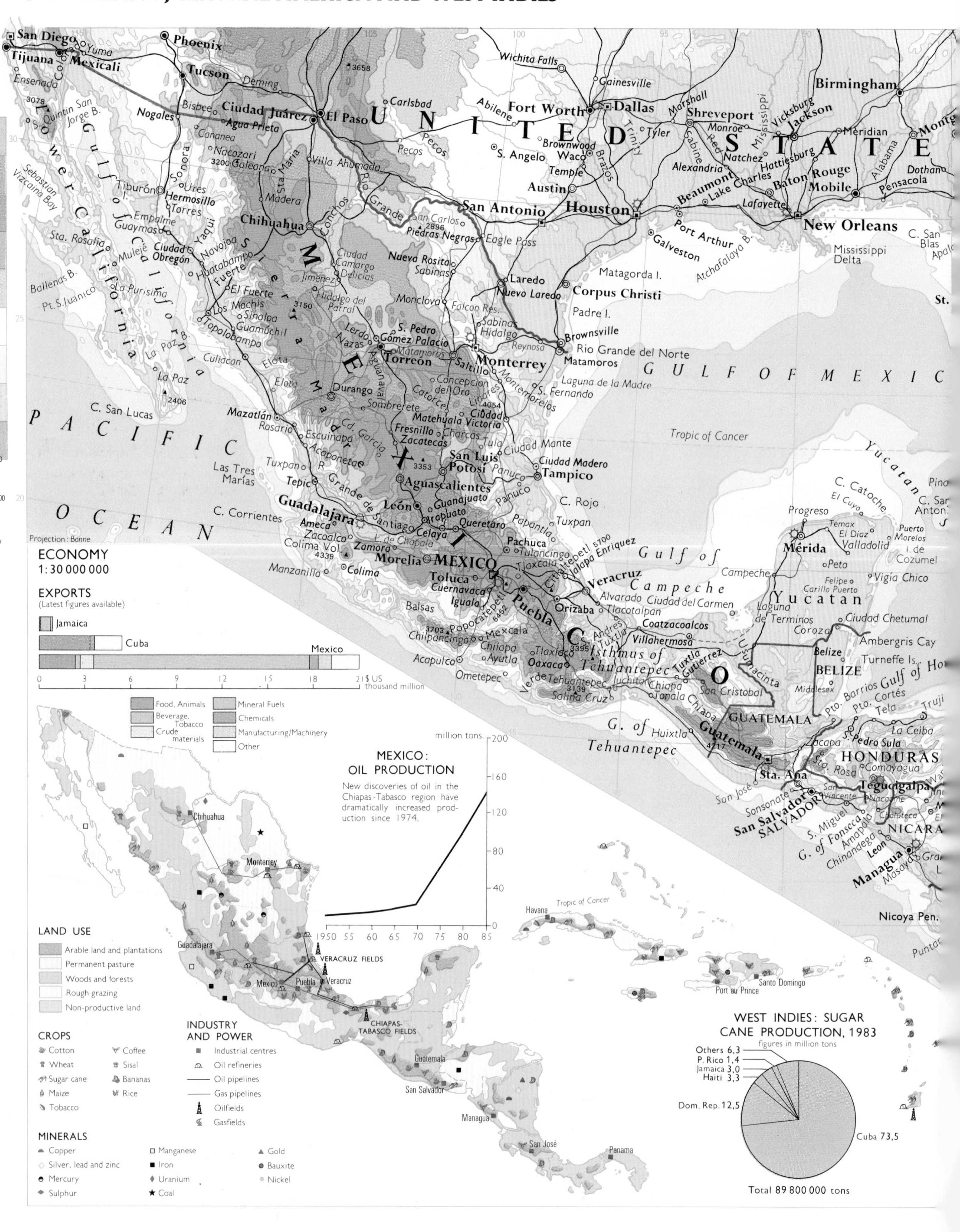

ECONOMY
1:30 000 000

EXPORTS
(Latest figures available)

Jamaica		
Cuba		
Mexico		

0   3   6   9   12   15   18   21 $ US
thousand million

Food, Animals	Mineral Fuels
Beverage, Tobacco	Chemicals
Crude materials	Manufacturing/Machinery
	Other

MEXICO: OIL PRODUCTION

New discoveries of oil in the Chiapas-Tabasco region have dramatically increased production since 1974.

million tons. 200
160
120
80
40

1950 55 60 65 70 75 80 85

LAND USE

	Arable land and plantations
	Permanent pasture
	Woods and forests
	Rough grazing
	Non-productive land

CROPS

Cotton	Coffee
Wheat	Sisal
Sugar cane	Bananas
Maize	Rice
Tobacco	

INDUSTRY AND POWER

Industrial centres	
Oil refineries	
Oil pipelines	
Gas pipelines	
Oilfields	
Gasfields	

VERACRUZ FIELDS

CHIAPAS-TABASCO FIELDS

MINERALS

Copper	Manganese	Gold
Silver, lead and zinc	Iron	Bauxite
Mercury	Uranium	Nickel
Sulphur	Coal	

WEST INDIES: SUGAR CANE PRODUCTION, 1983
figures in million tons

Others 6,3
P. Rico 1,4
Jamaica 3,0
Haiti 3,3
Dom. Rep. 12,5
Cuba 73,5

Total 89 800 000 tons

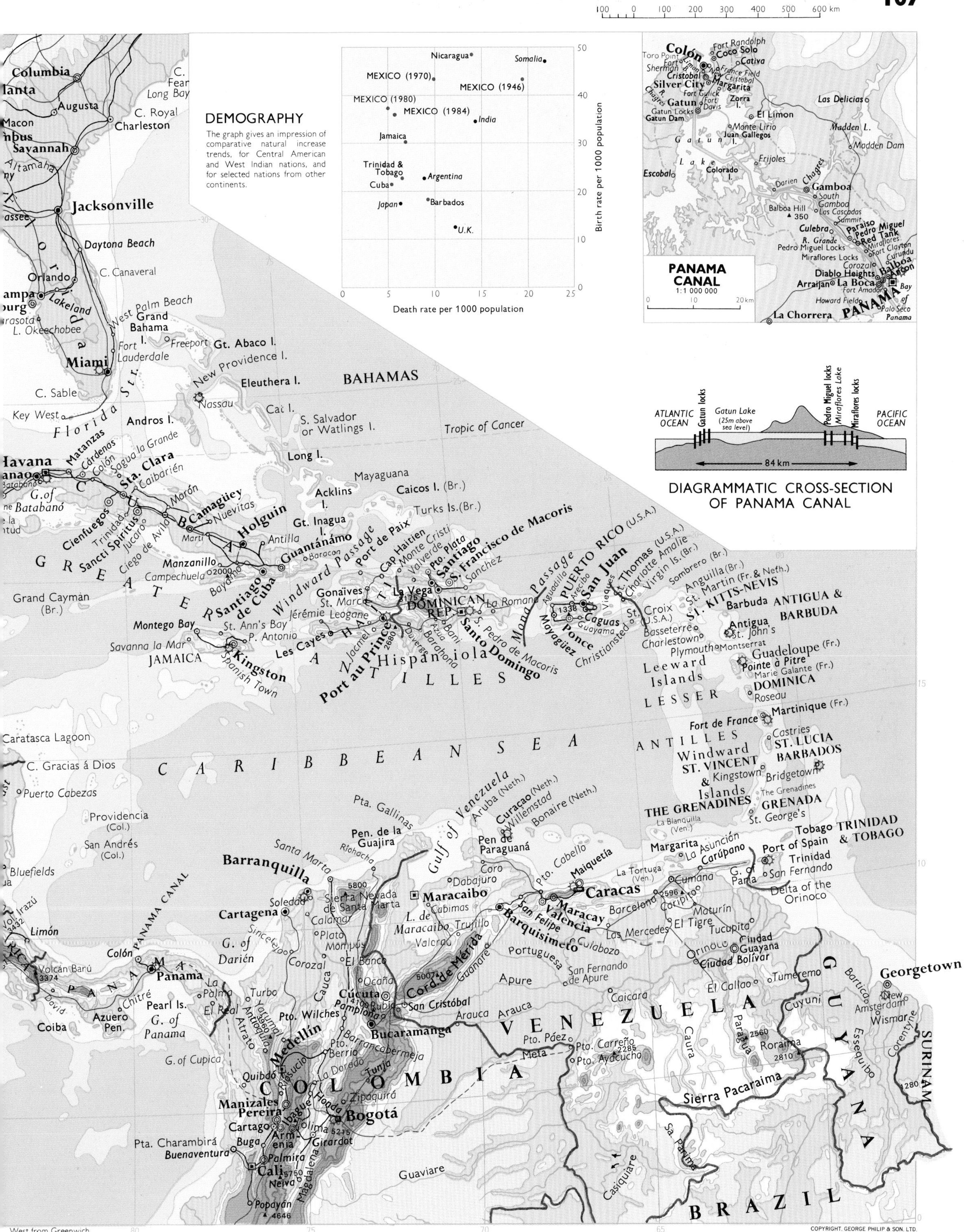

1:40 000 000

400   0   400   800   1200   1600 km

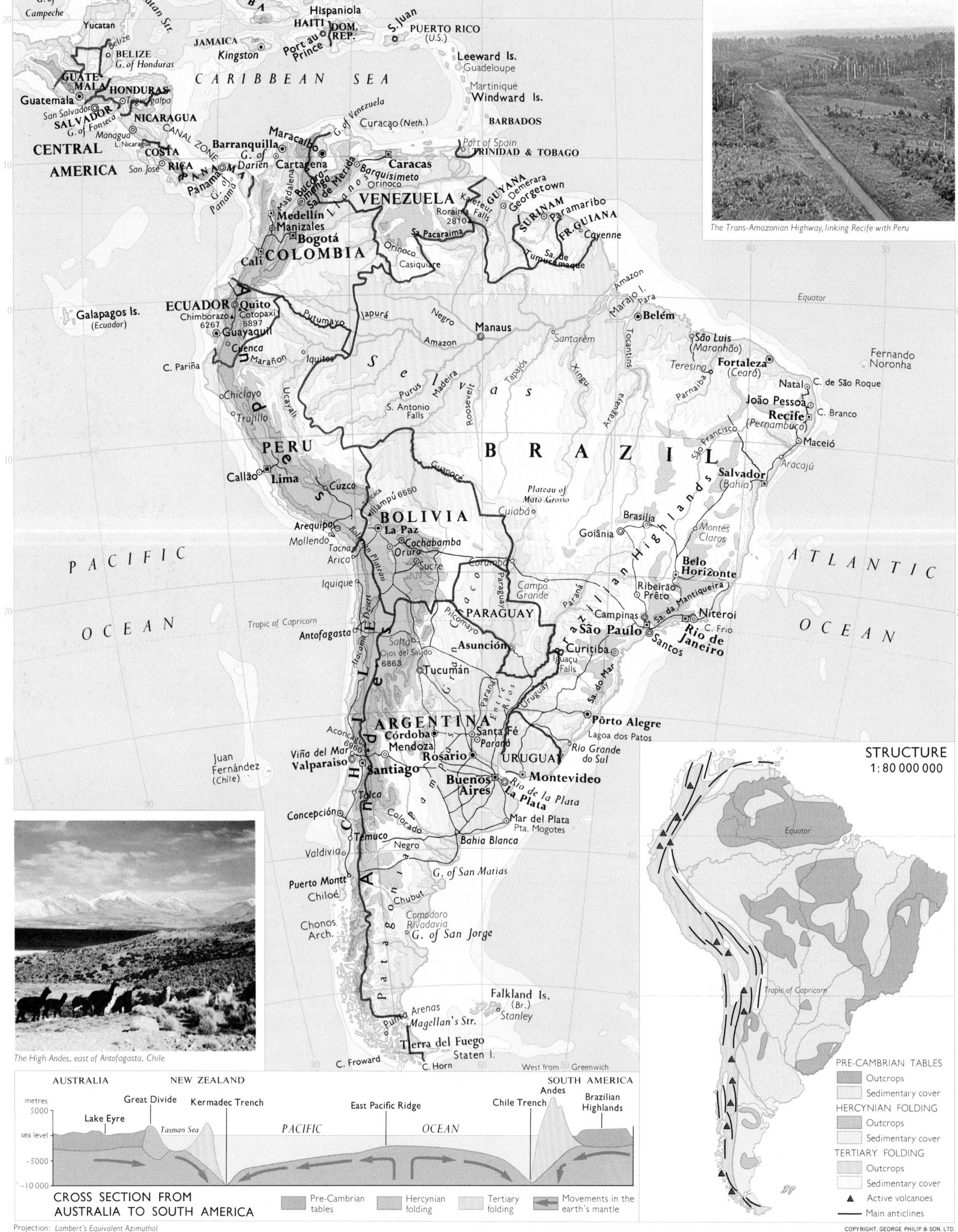

The Trans-Amazonian Highway, linking Recife with Peru

G. of Campeche
C. Catoche
Yucatan Str.
CUBA
Turks Is.
JAMAICA
Kingston
Hispaniola
HAITI
DOM. REP.
Port au Prince
S. Juan
PUERTO RICO (U.S.)

C A R I B B E A N   S E A

Leeward Is.
Guadeloupe
Martinique
Windward Is.
BARBADOS

Yucatan
BELIZE
G. of Honduras
GUATE-MALA
Guatemala
HONDURAS
Tegucigalpa
San Salvador
SALVADOR
G. of Fonseca
NICARAGUA
Managua
L. Nicaragua
COSTA RICA
San José
CENTRAL AMERICA
PANAMA
Panamá
CANAL ZONE
G. of Darien

Barranquilla
Maracaibo
G. of Venezuela
Cartagena
Curaçao (Neth.)
TRINIDAD & TOBAGO
Port of Spain
Bucaramanga
Caracas
Barquisimeto
Orinoco
VENEZUELA
Kaieteur Falls
Roraima 2810
GUYANA
Demerara
Georgetown
SURINAM
Paramaribo
FR. GUIANA
Cayenne
Sa. de Tumucumaque

Medellín
Manizales
Bogotá
Sa. de Merida
Llanos
Orinoco
Casiquiare
Sa. Pacaraima

Cali
COLOMBIA

Galapagos Is. (Ecuador)
ECUADOR
Quito
Chimborazo 6267
Cotopaxi 5897
Guayaquil
Cuenca
C. Pariña
Marañon
Putumayo
Japurá
Negro
Amazon
Marajó I.
Pará
Belém
Equator
Fernando Noronha

Chiclayo
Trujillo
Iquitos
Ucayali
Purus
Madeira
Roosevelt
Tapajós
Xingu
Tocantins
Araguaia
Parnaiba
Teresina
São Luis (Maranhão)
Fortaleza (Ceará)
C. de São Roque
Natal
João Pessoa
Recife (Pernambuco)
C. Branco

PERU
S. Antonio Falls
S e l v a s
Manaus
Santarém
Amazon
São Francisco
Maceió
Aracajú

Callão
Lima
Cuzco
B R A Z I L
Salvador (Bahia)

Titicaca
Illampú 6550
BOLIVIA
La Paz
Cochabamba
Arequipa
Mollendo
Tacna
Arica
Oruro
Sucre
Corumbá
Plateau of Mato Grosso
Cuiabá
Goiânia
Brasília
Brasília
Montes Claros
Brazilian Highlands
Ribeirão Prêto

Iquique
Campo Grande
Paraguay
Belo Horizonte
Sa. da Mantiqueira
Campinas
Niteroi
São Paulo
C. Frio
Rio de Janeiro
Santos
Sa. do Mar

PACIFIC
OCEAN
Tropic of Capricorn
Antofagasta
Atacama Desert
Bolivian Plateau
Ojos del Salado 6863
Salto
PARAGUAY
Asunción
Gran Chaco
Pilcomayo
Paraná
Iguaçu Falls
Curitiba
ATLANTIC OCEAN
Pôrto Alegre
Lagoa dos Patos
Rio Grande do Sul

Tucumán
Aconcagua 6960
Córdoba
ARGENTINA
Santa Fé
Paraná
Entre Rios
URUGUAY
Montevideo
Lagoa do Mirim

Juan Fernández (Chile)
Viña del Mar
Valparaiso
Santiago
Mendoza
Rosario
Buenos Aires
Rio de la Plata
La Plata
Mar del Plata
Pta. Mogotes

Talca
Concepción
Colorado
Negro
Bahia Blanca
Valdivia
Temuco
G. of San Matias
Chiloé
Chubut
Puerto Montt
Chonos Arch.
Comodoro Rivadavia
G. of San Jorge
A N D E S
C O R D I L L E R A
P a t a g o n i a

Falkland Is. (Br.)
Stanley

Punta Arenas
Magellan's Str.
C. Froward
Tierra del Fuego
Staten I.
C. Horn
West from Greenwich

The High Andes, east of Antofagasta, Chile

Elevation scale (m)

4000
2000
1000
400
200
0
200
2000
m

CROSS SECTION FROM AUSTRALIA TO SOUTH AMERICA

AUSTRALIA   NEW ZEALAND   SOUTH AMERICA

metres
5000
sea level
-5000
-10000

Lake Eyre
Great Divide
Kermadec Trench
Tasman Sea
PACIFIC   **OCEAN**
East Pacific Ridge
Chile Trench
Andes
Brazilian Highlands

Pre-Cambrian tables	Hercynian folding	Tertiary folding	Movements in the earth's mantle

Projection: Lambert's Equivalent Azimuthal

STRUCTURE

1:80 000 000

Equator

Tropic of Capricorn

PRE-CAMBRIAN TABLES
Outcrops
Sedimentary cover
HERCYNIAN FOLDING
Outcrops
Sedimentary cover
TERTIARY FOLDING
Outcrops
Sedimentary cover
▲ Active volcanoes
— Main anticlines

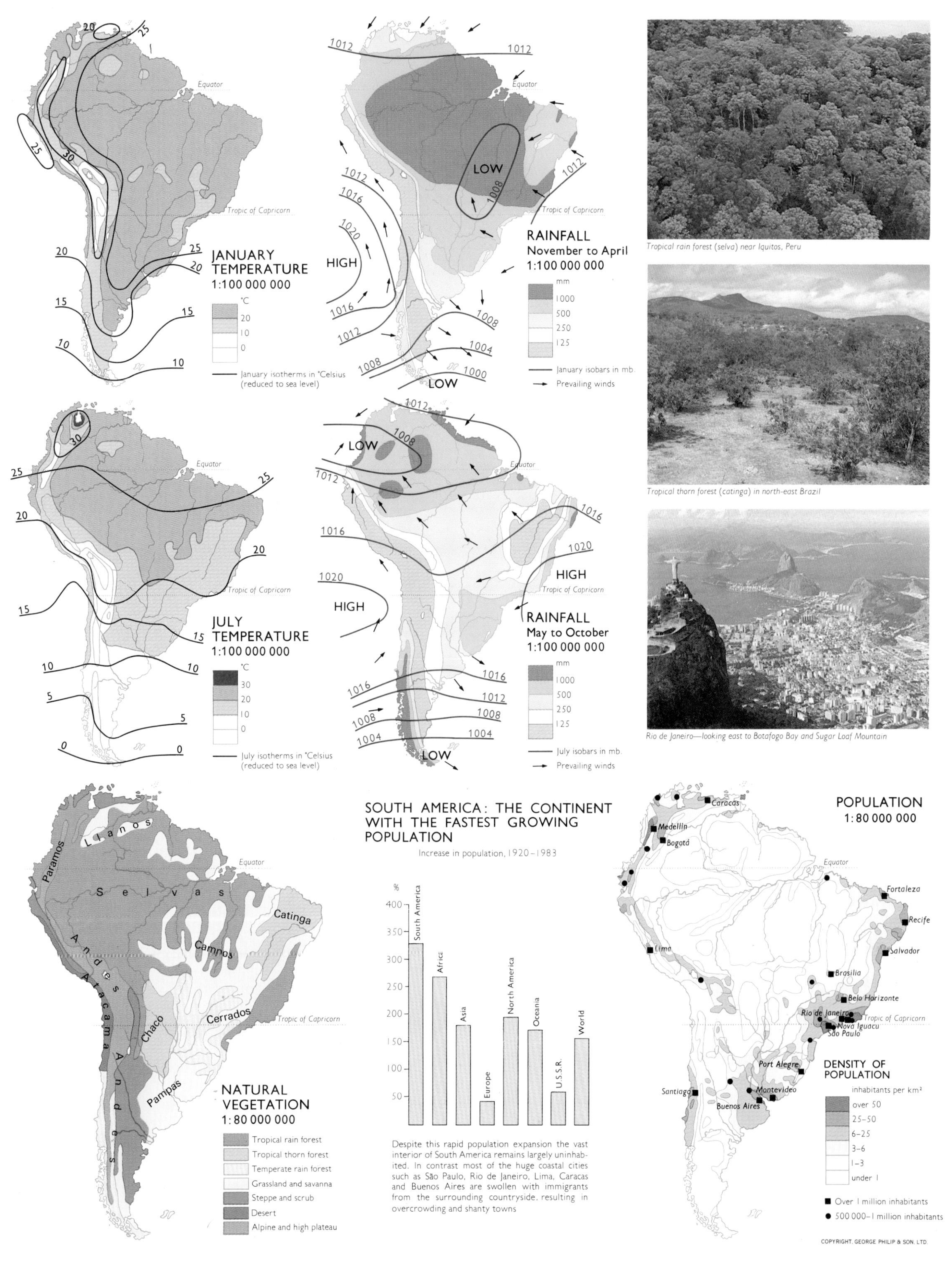

JANUARY TEMPERATURE
1:100 000 000

°C
20
10
0

January isotherms in °Celsius
(reduced to sea level)

JULY TEMPERATURE
1:100 000 000

°C
30
20
10
0

July isotherms in °Celsius
(reduced to sea level)

RAINFALL
November to April
1:100 000 000

mm
1000
500
250
125

January isobars in mb.
Prevailing winds

RAINFALL
May to October
1:100 000 000

mm
1000
500
250
125

July isobars in mb.
Prevailing winds

Tropical rain forest (selva) near Iquitos, Peru

Tropical thorn forest (catinga) in north-east Brazil

Rio de Janeiro—looking east to Botafogo Bay and Sugar Loaf Mountain

NATURAL VEGETATION
1:80 000 000

Tropical rain forest
Tropical thorn forest
Temperate rain forest
Grassland and savanna
Steppe and scrub
Desert
Alpine and high plateau

SOUTH AMERICA: THE CONTINENT WITH THE FASTEST GROWING POPULATION

Increase in population, 1920–1983

Despite this rapid population expansion the vast interior of South America remains largely uninhabited. In contrast most of the huge coastal cities such as São Paulo, Rio de Janeiro, Lima, Caracas and Buenos Aires are swollen with immigrants from the surrounding countryside, resulting in overcrowding and shanty towns

POPULATION
1:80 000 000

DENSITY OF POPULATION

inhabitants per km²
over 50
25–50
6–25
3–6
1–3
under 1

■ Over 1 million inhabitants
● 500 000–1 million inhabitants

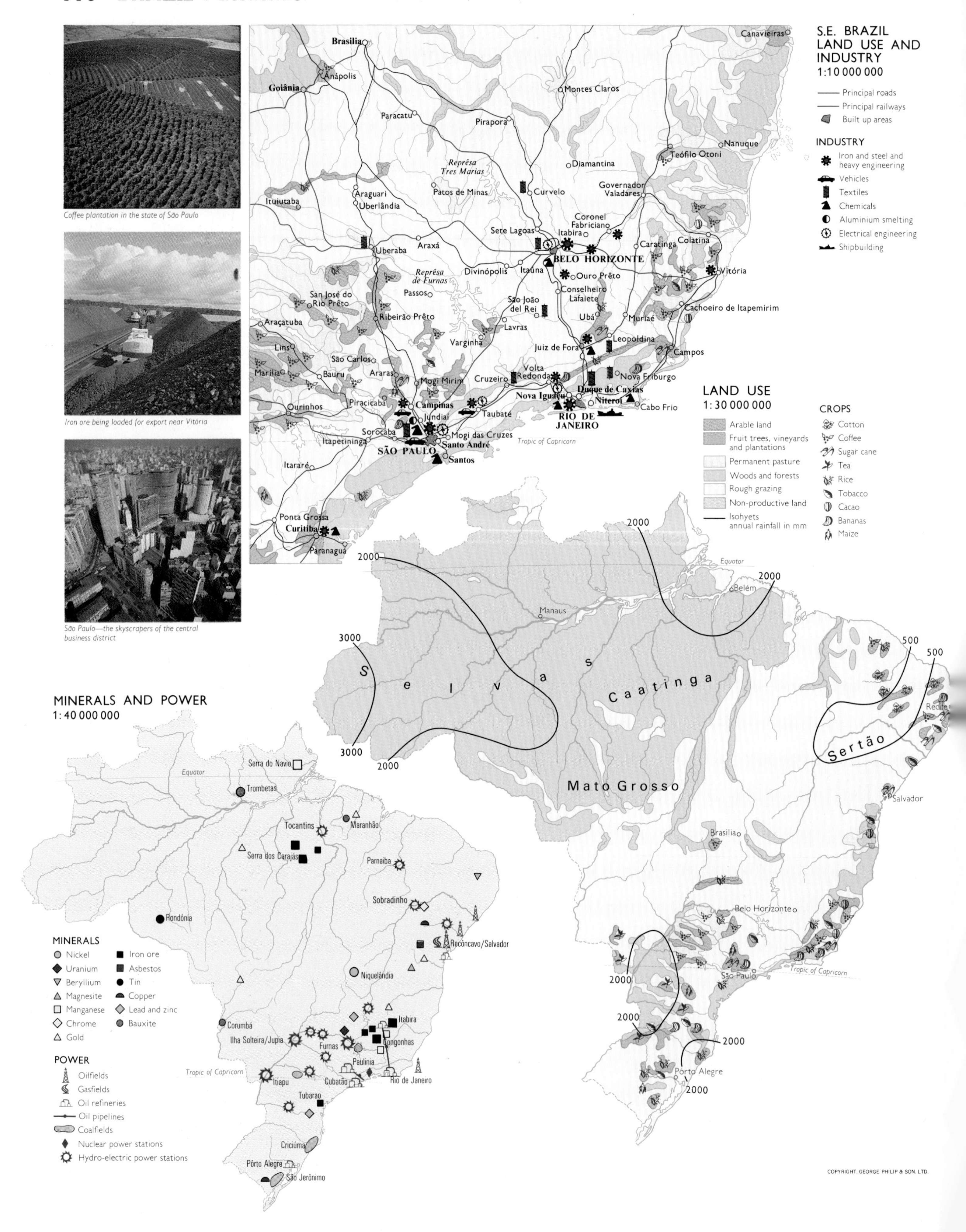

Coffee plantation in the state of São Paulo

Iron ore being loaded for export near Vitória

São Paulo—the skyscrapers of the central business district

S.E. BRAZIL LAND USE AND INDUSTRY
1:10 000 000

— Principal roads
— Principal railways
◢ Built up areas

INDUSTRY
✳ Iron and steel and heavy engineering
🚗 Vehicles
▮ Textiles
▲ Chemicals
◖ Aluminium smelting
⊕ Electrical engineering
⚓ Shipbuilding

LAND USE
1:30 000 000

Arable land
Fruit trees, vineyards and plantations
Permanent pasture
Woods and forests
Rough grazing
Non-productive land
— Isohyets annual rainfall in mm

CROPS
Cotton
Coffee
Sugar cane
Tea
Rice
Tobacco
Cacao
Bananas
Maize

MINERALS AND POWER
1:40 000 000

MINERALS
● Nickel   ■ Iron ore
▽ Uranium   ▲ Asbestos
▽ Beryllium   ● Tin
△ Magnesite   ◆ Copper
□ Manganese   ◇ Lead and zinc
◇ Chrome   ● Bauxite
△ Gold

POWER
⛏ Oilfields
⬡ Gasfields
⛭ Oil refineries
— Oil pipelines
⬭ Coalfields
◆ Nuclear power stations
✷ Hydro-electric power stations

Serra do Navio
Trombetas
Tocantins
Maranhão
Serra dos Carajás
Parnaiba
Rondônia
Sobradinho
Recôncavo/Salvador
Niquelândia
Corumbá
Itabira
Ilha Solteira/Jupia
Furnas
Congonhas
Paulinia
Itiapu
Cubatão
Rio de Janeiro
Tubarao
Criciuma
Pôrto Alegre
São Jerónimo

Equator
Tropic of Capricorn

S.E. BRAZIL LAND USE map labels
Brasília
Goiânia
Anápolis
Paracatu
Montes Claros
Pirapora
Diamantina
Teófilo Otoni
Nanuque
Ituiutaba
Araguari
Uberlândia
Patos de Minas
Curvelo
Governador Valadares
Represa Tres Marias
Coronel Fabriciano
Colatina
Sete Lagoas
Itabira
Caratinga
Uberaba
Araxá
BELO HORIZONTE
Vitória
San José do Rio Prêto
Divinópolis
Itaúna
Ouro Prêto
Represa de Furnas
Passos
Conselheiro Lafaiete
Cachoeiro de Itapemirim
Ribeirão Prêto
São João del Rei
Ubá
Muriaé
Araçatuba
Lavras
Varginha
Juiz de Fora
Leopoldina
Campos
Lins
São Carlos
Marília
Bauru
Araras
Cruzeiro
Volta Redonda
Nova Friburgo
Mogi Mirim
Piracicaba
Campinas
Taubaté
Nova Iguaçu
Duque de Caxias
Niterói
Ourinhos
Jundiai
RIO DE JANEIRO
Sorocaba
Mogi das Cruzes
Cabo Frio
Itapetininga
SÃO PAULO
Santo André
Itararé
Santos
Tropic of Capricorn
Ponta Grossa
Curitiba
Paranaguá

Physical regions (large map)
Manaus
Belém
Selvas
Caatinga
Recife
Sertão
Mato Grosso
Salvador
Brasilia
Belo Horizonte
São Paulo
Tropic of Capricorn
Pôrto Alegre
Equator

2000
3000
3000
2000
2000
500
500
2000
2000
2000
2000

Canavieiras

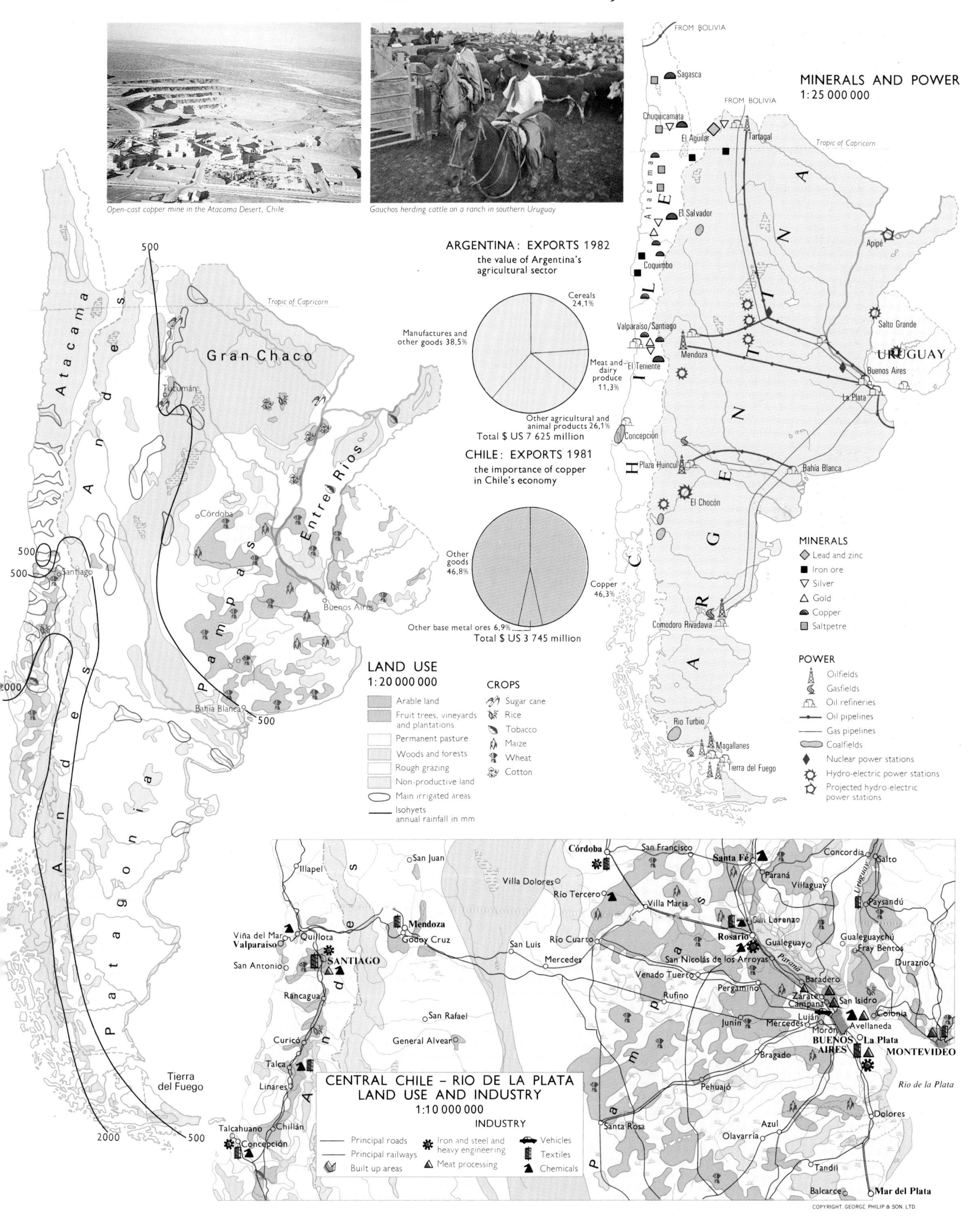

Open-cast copper mine in the Atacama Desert, Chile

Gauchos herding cattle on a ranch in southern Uruguay

FROM BOLIVIA

FROM BOLIVIA

MINERALS AND POWER
1:25 000 000

Sagasca

Chuquicamata
El Aguilar
Tartagal

Tropic of Capricorn

El Salvador

Apipé

Coquimbo

Valparaíso/Santiago

URUGUAY

El Teniente
Mendoza

Salto Grande

Concepción

Buenos Aires

La Plata

Plaza Huincul

Bahía Blanca

El Chocón

Río Turbio

Comodoro Rivadavia

Magallanes

Tierra del Fuego

MINERALS
◇ Lead and zinc
■ Iron ore
▽ Silver
△ Gold
◓ Copper
▨ Saltpetre

POWER
⛏ Oilfields
⬡ Gasfields
⛩ Oil refineries
⋯ Oil pipelines
— Gas pipelines
◠ Coalfields
◆ Nuclear power stations
✲ Hydro-electric power stations
✺ Projected hydro-electric power stations

ARGENTINA : EXPORTS 1982
the value of Argentina's agricultural sector

- Cereals 24,1%
- Manufactures and other goods 38,5%
- Meat and dairy produce 11,3%
- Other agricultural and animal products 26,1%

Total $ US 7 625 million

CHILE : EXPORTS 1981
the importance of copper in Chile's economy

- Other goods 46,8%
- Copper 46,3%
- Other base metal ores 6,9%

Total $ US 3 745 million

LAND USE
1:20 000 000

- Arable land
- Fruit trees, vineyards and plantations
- Permanent pasture
- Woods and forests
- Rough grazing
- Non-productive land
- Main irrigated areas
- Isohyets annual rainfall in mm

CROPS
- 🌾 Sugar cane
- Rice
- Tobacco
- Maize
- Wheat
- Cotton

Atacama
Andes
Gran Chaco
Tropic of Capricorn
Tucumán
Entre Ríos
Córdoba
Pampas
Buenos Aires
Bahía Blanca
Patagonia
Andes
Tierra del Fuego

500
500
500
500
2000
500
2000
500

CENTRAL CHILE – RIO DE LA PLATA
LAND USE AND INDUSTRY
1:10 000 000
INDUSTRY

— Principal roads
— Principal railways
  Built up areas
△ Iron and steel and heavy engineering
▲ Meat processing
🚚 Vehicles
▲ Textiles
◣ Chemicals

Illapel
San Juan
San Francisco
Córdoba
Santa Fé
Concordia
Salto
Villa Dolores
Paraná
Villaguay
Río Tercero
Villa María
Viña del Mar
Quillota
Mendoza
Godoy Cruz
Río Cuarto
Rosario
Gualeguay
Gualeguaychú
Fray Bentos
Valparaíso
San Luis
San Nicolás de los Arroyas
Durazno
San Antonio
SANTIAGO
Mercedes
Venado Tuerto
BarADERO
Rancagua
Rufino
Pergamino
Zárate
San Isidro
Colonia
San Rafael
Junín
Mercedes
Luján
Avellaneda
General Alvear
Morón
Curicó
Bragado
BUENOS AIRES
La Plata
MONTEVIDEO
Talca
Pehuajó
Río de la Plata
Linares
Azul
Dolores
Tierra del Fuego
Santa Rosa
Olavarría
Talcahuano
Chillán
Concepción
Tandil
Balcarce
Mar del Plata
San Lorenzo

COPYRIGHT GEORGE PHILIP & SON LTD.

1:50 000 000

500   0   500   1000   1500   2000 km

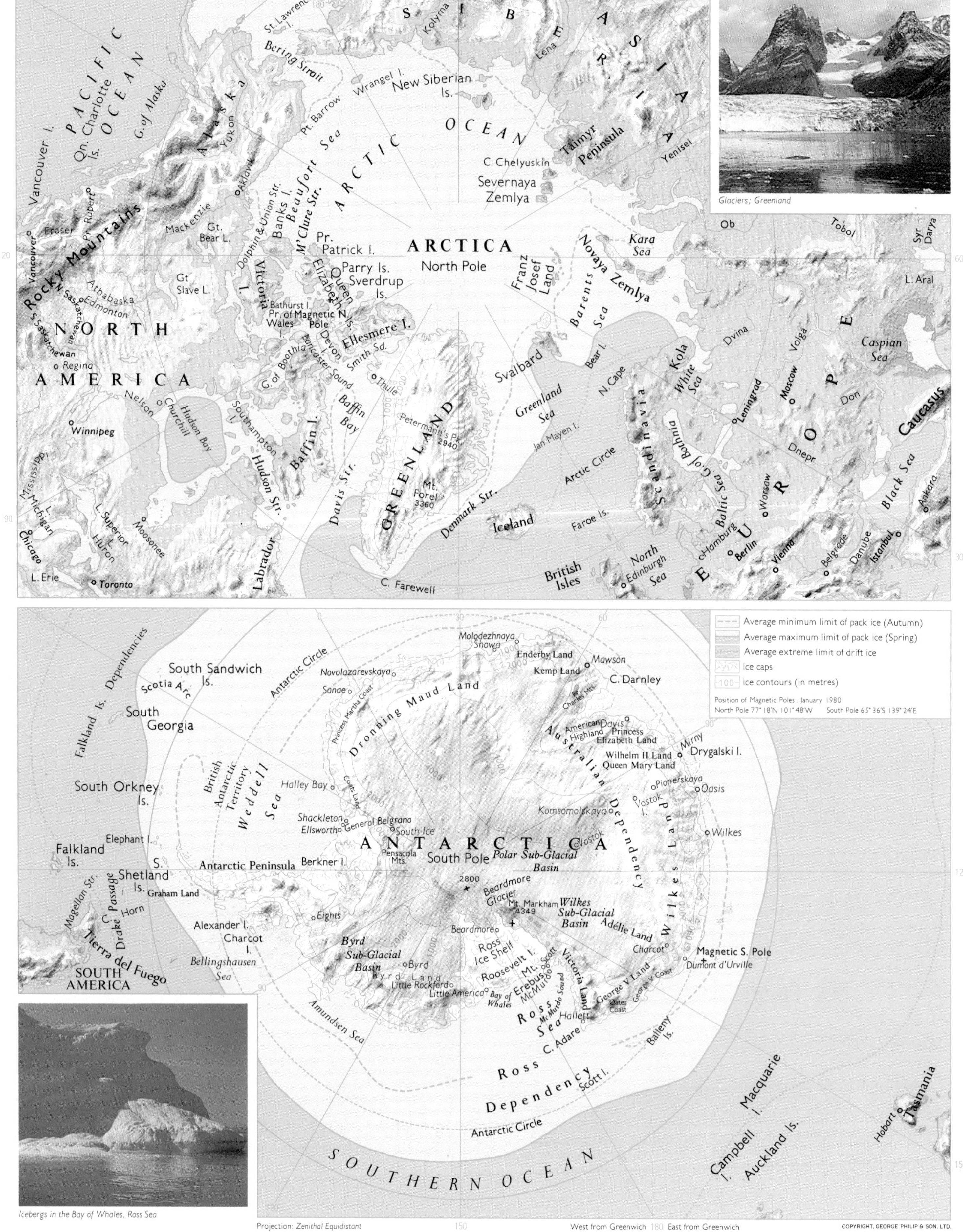

Glaciers; Greenland

Average minimum limit of pack ice (Autumn)
Average maximum limit of pack ice (Spring)
Average extreme limit of drift ice
Ice caps
100 Ice contours (in metres)

Position of Magnetic Poles, January 1980
North Pole 77° 18'N 101° 48'W   South Pole 65° 36'S 139° 24'E

Icebergs in the Bay of Whales, Ross Sea

Projection: Zenithal Equidistant

West from Greenwich 180 East from Greenwich

COPYRIGHT. GEORGE PHILIP & SON. LTD.

Country or Dependency	Total G.D.P. million $	G.D.P. per capita $	Annual average change %	Origin of G.D.P. % Agricultural	Origin of G.D.P. % Mining and Mf'g
Afghanistan	1 858	111	0.1	69	14
Algeria	44 926	2 262	8.7	6	40
Angola	2 701	432	—	—	—
Argentina	122 195	4 124	1.3	13	35
Australia	166 691	10 981	2.8	6	27
Austria	58 490	7 747	3.0	4	28
Bangladesh	15 298	158	4.4	48	9
Belgium	75 306	7 638	2.3	2	25
Belize	175	1 093	4.1	19	12
Benin	1 035	303	−4.3	44	6
Bhutan	88	76	—	—	—
Bolivia	6 238	1 054	3.2	22	23
Botswana	1 022	1 043	11.6	7	36
Brazil	283 076	2 232	6.5	10	25
Brunei	2 859	12 997	—	1	83
Bulgaria	25 415*	2 836*	6.0	18	57
Burkina Faso	1 079	163	6.6	37	11
Burma	6 176	165	6.3	50	11
Burundi	1 027	238	5.0	52	8
Cameroon	7 786	871	6.8	23	26
Canada	327 555	13 034	3.1	3	23
Central African Rep.	796	342	6.2	32	15
Chad	805	191	−4.5	41	17
Chile	17 796	1 524	7.6	6	31
China	224 263	218	6.0	45	42
Colombia	32 414	1 149	5.6	18	24
Congo	1 918	1 128	—	8	48
Costa Rica	2 911	1 193	4.2	21	22
Cuba	15 244	1 526	2.5	10	35
Cyprus	2 009	3 043	9.3	9	17
Czechoslovakia	44 112*	2 853*	3.2	8	59
Denmark	50 252	9 834	2.1	5	18
Djibouti	303	917	−1.9	4	8
Dominican Republic	8 575	1 439	4.5	17	20
Ecuador	11 684	1 283	6.0	14	36
Egypt	31 750	711	8.7	19	29
El Salvador	4 564	873	−0.2	21	16
Equatorial Guinea	73	243	—	—	—
Ethiopia	4 429	135	2.9	45	10
Fiji	1 092	1 629	7.5	17	9
Finland	47 148	9 661	3.1	7	24
France	445 913	8 115	2.8	4	26
Gabon	3 603	3 246	—	5	49
Gambia	214	369	—	56	3
Germany, East	72 820*	4 368*	4.2	9	74
Germany, West	554 511	9 064	3.5	2	31
Ghana	8 182	668	1.3	51	6
Greece	29 361	2 966	3.7	16	18
Guatemala	9 030	1 201	5.0	25	16
Guinea	910	369	3.1	43	8
Guinea-Bissau	110	166	—	—	—
Guyana	410	436	2.9	20	15
Haiti	1 029	199	4.1	32	17
Honduras	3 148	744	6.5	24	15
Hong Kong	26 786	4 997	12.0	1	21
Hungary	15 705*	1 473*	3.4	13	38
Iceland	2 636	10 983	4.1	10	20
India	173 883	233	3.7	29	17
Indonesia	79 994	500	7.9	25	30
Iran	128 921	3 161	−4.7	17	21
Iraq	48 879	3 700	2.4	7	60
Irish Republic	16 613	4 733	3.7	11	24
Israel	24 483	6 075	3.1	7	23
Italy	316 190	5 549	3.6	5	27
Ivory Coast	7 586	856	7.1	27	11
Jamaica	3 185	1 428	−2.5	7	22
Japan	1 183 975	9 928	5.0	3	31
Jordan	3 761	1 113	0.9	7	17
Kampuchea	592	83	2.2	41	17
Kenya	5 442	279	5.4	27	11
Korea, South	81 129	1 999	7.2	14	30
Kuwait	21 225	12 709	3.2	1	56
Laos	300	87	—	—	—
Lebanon	3 438	1 273	—	8	13
Lesotho	365	253	11.7	21	6
Liberia	871	456	2.7	14	23
Libya	29 885	8 974	7.7	2	52
Luxembourg	3 344	9 289	2.7	3	29
Madagascar	2 991	325	1.8	33	19
Malawi	1 334	208	5.7	37	13
Malaysia	23 796	1 566	8.2	20	31
Mali	2 179	297	—	53	7
Malta	937	2 466	10.2	4	31
Mauritania	811	483	0.7	22	19
Mauritius	913	932	6.4	12	15
Mexico	171 267	2 284	7.1	7	31
Morocco	14 697	687	4.3	14	23
Mozambique	3 272	322	—	—	—
Nepal	2 212	141	2.6	53	4
Netherlands	111 259	7 716	2.5	4	26
New Zealand	21 554	6 736	4.0	10	25
Nicaragua	2 955	998	1.7	25	21
Niger	2 513	473	—	43	16
Nigeria	64 956	730	1.7	25	25
Norway	49 149	11 872	4.2	3	32
Oman	8 826	7 480	—	3	51
Pakistan	31 138	334	6.4	22	19
Panama	4 541	2 123	6.1	9	9
Papua New Guinea	2 372	767	0.6	34	23
Paraguay	4 460	1 360	10.6	29	16
Peru	11 428	595	2.1	8	36
Philippines	27 750	520	6.0	25	27
Poland	58 867*	1 595*	−0.8	17	50
Portugal	23 811	2 344	4.9	8	30
Puerto Rico	18 671	5 710	2.8	2	39
Romania	17 915*	843*	7.2	15	58
Rwanda	1 440	250	8.3	40	19
Saudi Arabia	120 937	12 094	8.7	2	53
Senegal	2 117	331	1.1	17	3
Sierra Leone	1 297	380	3.7	30	19
Singapore	17 848	7 083	9.0	1	25
Somali Republic	1 540	303	—	—	—
South Africa	73 556	2 448	3.5	5	38
Spain	149 193	3 853	1.6	6	27
Sri Lanka	4 768	309	5.6	26	16
Sudan	6 634	345	6.3	36	8
Surinam	1 235	3 529	6.3	9	21
Swaziland	527	925	7.2	20	19
Sweden	88 048	10 570	1.2	3	21
Switzerland	97 120	15 081	1.9	2	21
Syria	19 140	1 927	5.9	20	17
Tanzania	5 127	299	2.7	46	9
Thailand	36 529	725	7.4	20	21
Togo	815	296	2.7	27	15
Trinidad and Tobago	8 115	7 056	6.2	2	29
Tunisia	8 132	1 208	6.5	13	23
Turkey	40 551	858	2.4	21	26
Uganda	3 360	238	−0.4	73	6
U.S.S.R.	711 818*	2 588*	4.1	20	46
United Kingdom	367 983	6 514	0.8	2	20
United States	3 276 000	13 968	3.1	2	24
Uruguay	3 980	1 331	3.9	12	22
Venezuela	46 461	2 757	2.8	7	40
Vietnam	4 682	98	2.8	29	7
Western Samoa	50*	320*	—	—	—
Yemen, North	2 838	477	5.4	28	7
Yemen, South	290	176	4.6	19	27
Yugoslavia	49 736*	2 196*	5.4	14	42
Zaïre	5 443	180	−3.7	26	16
Zambia	2 767	443	−1.6	14	35
Zimbabwe	6 612	876	3.2	13	27

The Gross Domestic Product (G.D.P.) is a measure of a country's total production of goods and services. For comparison national currencies have been converted to U.S. dollars. Owing to difficulties in the use of exchange rates and individual national methods of calculation of G.D.P., the figures must be used cautiously. For countries where the G.D.P. figure was not available, Gross National Product (G.N.P.) figures are given which are a measure of the total value of goods and services produced in a given country together with its imports from abroad. For communist countries the Net Material Product (N.M.P.) is given. This is not strictly comparable with the G.D.P., it is the total value of goods and services but excludes many of the latter, for example, public administration, defense costs and professional services. The figures quoted are usually for 1984 or at least for a year in the period 1975-84 and the annual rate of change is from 1975-81.

* Gross National Product Figures

International Trade

Country I = Imports E = Exports		Total trade (million U.S. $)	Primary Comms. as a % of total trade	Manuf'd Goods as a % of total trade	Fuels as a % of total trade	Value of trade per capita (U.S. $)
Afghanistan (1982)	I	695	17	74	9	41
	E	708	73	14	13	42
Algeria	I	10286	25	73	2	484
	E	11861	1	1	98	602
Angola	I	636	17	78	5	74
	E	2029	39	9	52	238
Argentina	I	4583	11	76	13	151
	E	8107	67	26	7	269
Australia*	I	23424	9	80	11	1507
	E	23998	55	22	23	1529
Austria	I	19631	12	74	14	2988
	E	15741	11	88	1	2083
Bahamas	I	3025	2	6	92	13152
	E	2346	1	4	95	10361
Bahrain	I	3530	7	40	53	9193
	E	3138	—	18	82	8172
Bangladesh	I	2042	29	60	11	21
	E	934	34	62	4	10
Barbados	I	659	19	65	16	2636
	E	391	21	61	18	1564
Belgium-Lux.	I	55247	19	64	17	5395
	E	51699	13	79	8	5049
Benin (1982)	I	476	20	70	10	97
	E	43	100	0	0	18
Bolivia	I	631	15	84	1	101
	E	773	53	33	14	124
Brazil	I	15210	10	36	54	115
	E	27005	55	39	6	204
Brunei (1983)	I	728	16	83	1	3792
	E	3386	0	1	99	17635
Bulgaria*	I	12715	5	81	14	1414
	E	12850	51	49	0	1432
Burkina Faso (1983)	I	288	27	57	16	45
	E	57	85	15	0	9
Burma	I	239	10	87	3	6
	E	310	95	4	1	8
Burundi	I	186	15	70	15	41
	E	98	98	2	0	22
Cameroon	I	1106	13	83	4	117
	E	882	44	9	47	93
Canada*	I	73999	11	82	7	2942
	E	86817	26	60	14	3455
Chad (1982)	I	109	16	70	14	23
	E	58	84	8	8	12
Chile	I	3191	14	71	15	269
	E	3657	45	53	2	308
China	I	26183	32	67	1	19
	E	24871	26	52	22	21
Colombia	I	4052	14	73	13	181
	E	3462	66	19	15	112
Congo (1983)	I	806	21	65	14	488
	E	1066	4	6	90	646
Costa Rica	I	1085	11	69	20	443
	E	978	70	29	1	372
Cuba	I	8144	19	54	27	815
	E	6197	88	12	0	618
Cyprus	I	1364	19	62	19	2067
	E	575	36	56	8	871
Czechoslovakia*	I	17080	16	54	30	1105
	E	17196	6	89	5	1112
Denmark	I	16973	17	63	20	3322
	E	16349	37	58	5	3196
Dominican Republic*	I	1257	17	47	36	215
	E	868	76	24	0	132
Ecuador	I	1716	9	89	2	188
	E	2581	32	4	64	282
Egypt (1983)	I	10274	36	60	4	224
	E	3215	22	12	66	70
El Salvador	I	970	20	55	25	171
	E	708	55	42	3	141
Ethiopia (1983)	I	875	13	62	25	27
	E	403	91	1	8	12
Fiji	I	450	19	58	23	670
	E	256	63	18	19	381
Finland	I	12443	12	61	27	2549
	E	13505	17	78	5	2768
France	I	103807	16	59	25	1888
	E	93276	20	76	4	1698
Gabon (1983)	I	853	18	81	1	755
	E	1975	12	3	85	1748
Germany, East*	I	22940	—	—	—	1376
	E	28436	—	—	—	1490
Germany, West	I	151246	19	60	21	2489
	E	169784	7	90	3	2758
Ghana	I	591	12	61	27	200
	E	571	85	15	—	160
Greece	I	9434	19	54	27	960
	E	4811	40	53	7	486
Guatemala	I	1278	8	54	38	151
	E	1129	69	29	2	154
Guyana	I	601	19	59	22	639
	E	91	94	6	0	97
Haiti (1983)	I	314	26	61	13	61
	E	187	49	51	0	36
Honduras (1983)	I	804	11	67	22	188
	E	667	91	9	0	169
Hong Kong	I	28567	17	76	7	5329
	E	28317	7	92	1	5283
Hungary	I	8091	14	63	23	760
	E	8563	27	64	9	803
Iceland	I	838	19	65	16	3417
	E	738	73	27	0	3033
India	I	13501	14	42	44	18
	E	8474	40	59	1	11
Indonesia	I	13882	11	64	25	87
	E	21888	14	10	76	137
Iran (1983)	I	18296	16	84	0	283
	E	20247	2	0	98	476
Iraq (1981)	I	7903	14	86	0	521
	E	10530	1	99	1	695
Irish Republic	I	9663	16	71	13	2730
	E	9629	32	67	1	2720
Israel	I	8411	13	69	18	2036
	E	5804	18	82	0	1190
Italy	I	84215	22	47	31	1414
	E	73303	8	87	5	1292
Ivory Coast (1983)	I	1808	22	59	19	194
	E	2067	77	12	11	222
Jamaica	I	1146	21	50	29	513
	E	747	83	13	4	339
Japan	I	136492	26	27	47	1137
	E	170132	2	98	0	1148
Jordan	I	2784	20	59	21	824
	E	752	50	50	0	222
Kenya	I	1502	10	53	37	73
	E	1083	59	14	27	52
Korea, South	I	30631	21	48	31	755
	E	29245	7	92	1	721
Kuwait	I	7697	16	83	1	4028
	E	10750	1	15	84	5625
Lebanon	I	287	28	65	7	82
	E	582	21	79	0	166
Liberia	I	363	24	49	27	288
	E	452	97	3	0	298
Libya (1982)	I	7178	20	79	1	2155
	E	13951	0	0	100	4189

Country I = Imports E = Exports		Total trade (million U.S. $)	Primary Comms. as a % of total trade	Manuf'd Goods as a % of total trade	Fuels as a % of total trade	Value of trade per capita (U.S. $)
Madagascar (1983)	I	387	19	57	24	41
	E	293	85	7	8	43
Malawi	I	268	13	70	17	47
	E	309	88	12	0	33
Malaysia (1982)	I	13987	15	70	15	963
	E	13917	43	28	29	958
Mali (1983)	I	344	17	67	16	46
	E	167	77	23	0	22
Malta	I	717	19	69	12	1887
	E	394	7	89	4	1037
Mauritania	I	246	33	59	8	106
	E	297	100	0	0	161
Mauritius	I	472	33	48	19	487
	E	373	69	31	0	385
Mexico	I	11302	32	65	3	146
	E	23603	11	15	74	306
Morocco	I	3861	25	48	27	169
	E	2095	60	36	4	92
Mozambique	I	486	10	82	8	65
	E	86	87	7	6	18
Netherlands	I	62136	20	55	25	4309
	E	65881	25	51	24	4569
Netherlands Antilles (1983)	I	4527	3	11	86	21831
	E	4409	0	1	99	19854
New Caledonia	I	311	24	53	23	2145
	E	207	17	83	0	1427
New Zealand	I	6010	12	70	18	1860
	E	5358	73	27	0	1659
Nicaragua	I	826	13	64	23	256
	E	385	91	8	1	124
Niger (1982)	I	442	21	58	15	79
	E	333	97	2	1	59
Nigeria (1983)	I	13440	19	79	2	151
	E	11317	4	1	95	127
Norway	I	13889	13	74	10	3355
	E	18892	10	37	53	4568
Pakistan	I	5873	20	52	28	63
	E	2614	34	64	2	35
Panama	I	1423	11	61	28	668
	E	256	78	10	12	120
Papua New Guinea*	I	968	20	61	19	305
	E	895	91	9	0	247
Paraguay*	I	563	14	57	19	146
	E	386	91	9	0	82
Peru	I	1870	20	78	2	97
	E	3131	32	42	26	163
Philippines	I	6099	12	61	27	153
	E	5293	13	69	18	96
Poland*	I	10633	19	55	26	269
	E	11687	13	69	18	304
Portugal	I	7797	21	52	27	767
	E	5184	20	74	6	510
Reunion	I	791	26	63	11	1451
	E	77	85	14	1	141
Romania* (1983)	I	9959	18	75	7	435
	E	13241	28	63	9	578
Rwanda	I	290	17	70	13	48
	E	83	100	0	0	14
Saudi Arabia (1983)	I	33696	14	85	1	3763
	E	36834	0	1	99	4505

Country I = Imports E = Exports		Total trade (million U.S. $)	Primary Comms. as a % of total trade	Manuf'd Goods as a % of total trade	Fuels as a % of total trade	Value of trade per capita (U.S. $)
Senegal	I	1039	28	42	30	164
	E	416	43	20	37	86
Sierra Leone	I	166	25	61	14	47
	E	148	41	59	0	42
Singapore	I	28712	12	57	31	11394
	E	24108	14	58	28	9567
Somali Republic	I	109	27	71	2	65
	E	45	100	0	0	39
South Africa*	I	14956	7	93	0	472
	E	9334	17	76	7	314
Spain	I	28812	21	39	40	744
	E	23544	19	72	9	608
Sri Lanka	I	1845	19	57	24	118
	E	1454	61	28	11	93
Sudan (1983)	I	1354	19	54	27	67
	E	624	93	2	5	31
Sweden	I	26408	11	66	23	3159
	E	29378	13	81	6	3518
Switzerland	I	29469	11	78	11	4577
	E	25863	5	95	0	4016
Syria	I	4116	17	45	38	414
	E	1853	15	10	75	187
Tanzania (1983)	I	1822	8	61	31	40
	E	366	87	13	0	18
Thailand	I	10398	10	66	24	203
	E	7413	63	37	0	147
Togo (1983)	I	284	28	64	8	103
	E	162	82	17	1	59
Trinidad and Tobago	I	2101	21	76	3	2245
	E	2194	2	14	84	2046
Tunisia	I	3128	21	58	21	452
	E	1797	12	34	54	269
Turkey	I	10822	8	49	43	224
	E	7086	48	48	4	147
Uganda (1980)	I	293	9	61	30	22
	E	345	97	2	1	26
U.S.S.R.*	I	80624	27	73	0	293
	E	91649	8	26	66	333
United Kingdom	I	105968	19	70	11	1775
	E	94508	10	68	22	1628
United States	I	341177	12	66	22	1376
	E	217888	24	71	5	921
Uruguay	I	776	11	57	32	265
	E	925	70	30	0	352
Venezuela*	I	6676	20	79	1	407
	E	14937	2	5	93	915
Yemen, North (1982)	I	1512	32	60	8	250
	E	39	24	76	0	6
Yemen, South (1981)	I	673	—	—	—	332
	E	430	—	—	—	212
Yugoslavia	I	11538	16	57	27	522
	E	9011	16	81	3	447
Zaïre	I	682	24	68	8	16
	E	1004	41	58	1	36
Zambia*	I	566	10	72	18	88
	E	648	1	98	1	100
Zimbabwe	I	959	5	78	17	120
	E	1008	62	36	2	136

Source: U.N. Monthly Bulletin of Statistics and U.N. Yearbook of International Trade Statistics

Primary Commodities refer to sections 0-4 of the Standard International Trade Classification (Revised)** and Manufactured Goods include sections 5-9. All the trade values used in the compilation of this table are at current prices. Unless otherwise stated imports are in terms of c.i.f. transaction values and exports f.o.b. transaction values. The latest figures available have been used and these are for 1984 unless otherwise stated.
* = Imports f.o.b.
** excluding Fuels, section 3

CLIMATIC STATISTICS

These four pages give temperature and precipitation statistics for over 80 stations, which are arranged by listing the continents and the places within each continent in alphabetical order. The elevation of each station, in metres above mean sea level, is stated beneath its name. The average monthly temperature, in degrees Celsius, and the average monthly precipitation, in millimetres, are given. To the right, the average yearly rainfall, the average yearly temperature, and the annual range of temperature (the difference between the warmest and the coldest months) are also stated.

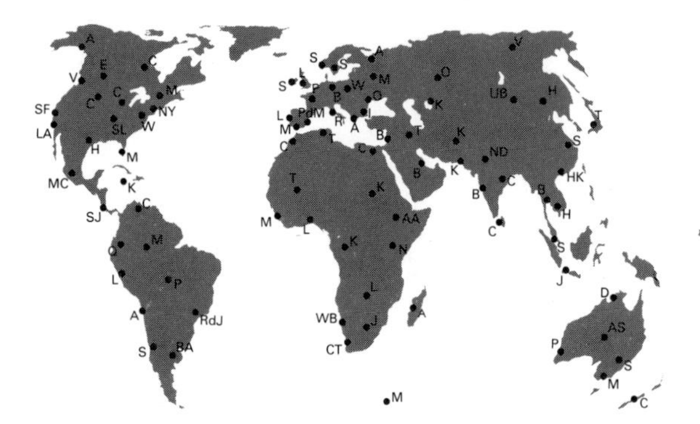

AFRICA		Jan.	Feb.	Mar.	Apr.	May	June	July	Aug.	Sept.	Oct.	Nov.	Dec.	Year	Annual Range
Addis Ababa, Ethiopia															
	Precipitation	201	206	239	102	28	<3	0	<3	3	25	135	213	1 151	
2 450 m	Temperature	19	20	20	20	19	18	18	19	21	22	21	20	20	4
Cairo, Egypt															
	Precipitation	5	5	5	3	3	<3	0	0	<3	<3	3	5	28	
116 m	Temperature	13	15	18	21	25	28	28	28	26	24	20	15	22	15
Cape Town, South Africa															
	Precipitation	15	8	18	48	79	84	89	66	43	31	18	10	508	
17 m	Temperature	21	21	20	17	14	13	12	13	14	16	18	19	17	9
Casablanca, Morocco															
	Precipitation	53	48	56	36	23	5	0	<3	8	38	66	71	404	
50 m	Temperature	13	13	14	16	18	20	22	23	22	19	16	13	18	10
Johannesburg, South Africa															
	Precipitation	114	109	89	38	25	8	8	8	23	56	107	125	709	
1 665 m	Temperature	20	20	18	16	13	10	11	13	16	18	19	20	16	10
Khartoum, Sudan															
	Precipitation	<3	<3	<3	<3	3	8	53	71	18	5	<3	0	158	
390 m	Temperature	24	25	28	31	33	34	32	31	32	32	28	25	29	9
Kinshasa, Zaïre															
	Precipitation	135	145	196	196	158	8	3	3	31	119	221	142	1 354	
325 m	Temperature	26	26	27	27	26	24	23	24	25	26	26	26	25	4
Lagos, Nigeria															
	Precipitation	28	46	102	150	269	460	279	64	140	206	69	25	1 836	
3 m	Temperature	27	28	29	28	28	26	26	25	26	26	28	28	27	4
Lusaka, Zambia															
	Precipitation	231	191	142	18	3	<3	<3	0	<3	10	91	150	836	
1 277 m	Temperature	21	22	21	21	19	16	16	18	22	24	23	22	21	8
Monrovia, Liberia															
	Precipitation	31	56	97	216	516	973	996	373	744	772	236	130	5 138	
23 m	Temperature	26	26	27	27	26	25	24	25	25	25	26	26	26	3
Nairobi, Kenya															
	Precipitation	38	64	125	211	158	46	15	23	31	53	109	86	958	
1 820 m	Temperature	19	19	19	19	18	16	16	16	18	19	18	18	18	3
Antananarivo, Madagascar															
	Precipitation	300	279	178	53	18	8	8	10	18	61	135	287	1 356	
1 372 m	Temperature	21	21	21	19	18	15	14	15	17	19	21	21	19	7
Timbuktu, Mali															
	Precipitation	<3	<3	3	<3	5	23	79	81	38	3	<3	<3	231	
301 m	Temperature	22	24	28	32	34	35	32	30	32	31	28	23	29	13
Tunis, Tunisia															
	Precipitation	64	51	41	36	18	8	3	8	33	51	48	61	419	
66 m	Temperature	10	11	13	16	19	23	26	27	25	20	16	11	18	17
Walvis Bay, South Africa															
	Precipitation	<3	5	8	3	3	<3	<3	3	<3	<3	<3	<3	23	
7 m	Temperature	19	19	19	18	17	16	15	14	14	15	17	18	18	5

AMERICA, NORTH															
Anchorage, Alaska, U.S.A.															
	Precipitation	20	18	15	10	13	18	41	66	66	56	25	23	371	
40 m	Temperature	−11	−8	−5	2	7	12	14	13	9	2	−5	−11	2	25
Cheyenne, Wyo., U.S.A.															
	Precipitation	10	15	25	48	61	41	53	41	31	25	13	13	376	
1 871 m	Temperature	−4	−3	1	5	10	16	19	19	14	7	1	−2	7	23
Chicago, Ill., U.S.A.															
	Precipitation	51	51	66	71	86	89	84	81	79	66	61	51	836	
251 m	Temperature	−4	−3	2	9	14	20	23	22	19	12	5	−1	10	27
Churchill, Man., Canada															
	Precipitation	15	13	18	23	32	44	46	58	51	43	39	21	402	
13 m	Temperature	−28	−26	−20	−10	−2	6	12	11	5	−2	−12	−22	−7	40

	Jan.	Feb.	Mar.	Apr.	May	June	July	Aug.	Sept.	Oct.	Nov.	Dec.	Year	Annual range
Edmonton, Alta., Canada														
Precipitation	25	19	19	22	43	77	89	78	39	17	16	25	466	
676 m Temperature	−15	−10	−5	4	11	15	17	16	11	6	−4	−10	3	32
Honolulu, Hawaii, U.S.A.														
Precipitation	104	66	79	48	25	18	23	28	36	48	64	104	643	
12 m Temperature	23	18	19	20	22	24	25	26	26	24	22	19	22	8
Houston, Tex., U.S.A.														
Precipitation	89	76	84	91	119	117	99	99	104	94	89	109	1 171	
12 m Temperature	12	13	17	21	24	27	28	29	26	22	16	12	21	17
Kingston, Jamaica														
Precipitation	23	15	23	31	102	89	38	91	99	180	74	36	800	
34 m Temperature	25	25	25	26	26	28	28	28	27	27	26	26	26	3
Los Angeles, Calif., U.S.A.														
Precipitation	79	76	71	25	10	3	< 3	< 3	5	15	31	66	381	
95 m Temperature	13	14	14	16	17	19	21	22	21	18	16	14	17	9
Mexico City, Mexico														
Precipitation	13	5	10	20	53	119	170	152	130	51	18	8	747	
2 309 m Temperature	12	13	16	18	19	19	17	18	18	16	14	13	16	7
Miami, Fla., U.S.A.														
Precipitation	71	53	64	81	173	178	155	160	203	234	71	51	1 516	
8 m Temperature	20	20	22	23	25	27	28	28	27	25	22	21	24	8
Montréal, Que., Canada														
Precipitation	72	65	74	74	66	82	90	92	88	76	81	87	946	
57 m Temperature	−10	−9	−3	−6	13	18	21	20	15	9	2	−7	6	31
New York, N.Y., U.S.A.														
Precipitation	94	97	91	81	81	84	107	109	86	89	76	91	1 092	
96 m Temperature	−1	−1	3	10	16	20	23	23	21	15	7	2	8	24
St. Louis, Mo., U.S.A.														
Precipitation	58	64	89	97	114	114	89	86	81	74	71	64	1 001	
173 m Temperature	0	1	7	13	19	24	26	26	22	15	8	2	14	26
San Francisco, Calif., U.S.A.														
Precipitation	119	97	79	38	18	3	< 3	< 3	8	25	64	112	561	
16 m Temperature	10	12	13	13	14	15	15	15	17	16	14	11	14	7
San José, Costa Rica														
Precipitation	15	5	20	46	229	241	211	241	305	300	145	41	1 798	
1 146 m Temperature	19	19	21	21	22	21	21	21	21	20	20	19	20	2
Vancouver, B.C., Canada														
Precipitation	154	115	101	60	52	45	32	41	67	114	150	182	1113	
14 m Temperature	3	5	6	9	12	15	17	17	14	10	6	4	10	14
Washington, D.C., U.S.A.														
Precipitation	86	76	91	84	94	99	112	109	94	74	66	79	1 064	
22 m Temperature	1	2	7	12	18	23	25	24	20	14	8	3	13	24

AMERICA, SOUTH

	Jan.	Feb.	Mar.	Apr.	May	June	July	Aug.	Sept.	Oct.	Nov.	Dec.	Year	Annual range
Antofagasta, Chile														
Precipitation	0	0	0	< 3	< 3	3	5	3	< 3	3	< 3	0	13	
94 m Temperature	21	21	20	18	16	15	14	14	15	16	18	19	17	7
Buenos Aires, Argentina														
Precipitation	79	71	109	89	76	61	56	61	79	86	84	99	950	
27 m Temperature	23	23	21	17	13	9	10	11	13	15	19	22	16	14
Caracas, Venezuela														
Precipitation	23	10	15	33	79	102	109	109	107	109	94	46	836	
1 042 m Temperature	19	19	20	21	22	21	21	21	21	21	20	20	21	3
Lima, Peru														
Precipitation	3	< 3	< 3	< 3	5	5	8	8	8	3	3	< 3	41	
120 m Temperature	23	24	24	22	19	17	17	16	17	18	19	21	20	8
Manaus, Brazil														
Precipitation	249	231	262	221	170	84	58	38	46	107	142	203	1 811	
44 m Temperature	28	28	28	27	28	28	28	28	29	29	29	28	28	2
Paraná, Brazil														
Precipitation	287	236	239	102	13	< 3	3	5	28	127	231	310	1 582	
260 m Temperature	23	23	23	23	23	21	21	22	24	24	24	23	23	3
Quito, Ecuador														
Precipitation	99	112	142	175	137	43	20	31	69	112	97	79	1 115	
2 879 m Temperature	15	15	15	15	15	14	14	15	15	15	15	15	15	1
Rio de Janeiro, Brazil														
Precipitation	125	122	130	107	79	53	41	43	66	79	104	137	1 082	
61 m Temperature	26	26	25	24	22	21	21	21	21	22	23	25	23	5
Santiago, Chile														
Precipitation	3	3	5	13	64	84	76	56	31	15	8	5	358	
520 m Temperature	21	20	18	15	12	9	9	10	12	15	17	19	15	12

ASIA

		Jan.	Feb.	Mar.	Apr.	May	June	July	Aug.	Sept.	Oct.	Nov.	Dec.	Year	Annual range
Bahrain															
5 m	Precipitation	8	18	13	8	<3	0	0	0	0	0	18	18	81	
	Temperature	17	18	21	25	29	32	33	34	31	28	24	19	26	16
Bangkok, Thailand															
2 m	Precipitation	8	20	36	58	198	160	160	175	305	206	66	5	1 397	
	Temperature	26	28	29	30	29	29	28	28	28	28	26	25	28	5
Beirut, Lebanon															
34 m	Precipitation	191	158	94	53	18	3	<3	<3	5	51	132	185	892	
	Temperature	14	14	16	18	22	24	27	28	26	24	19	16	21	14
Bombay, India															
11 m	Precipitation	3	3	3	<3	18	485	617	340	264	64	13	3	1 809	
	Temperature	24	24	26	28	30	29	27	27	27	28	27	26	27	6
Calcutta, India															
6 m	Precipitation	10	31	36	43	140	297	325	328	252	114	20	5	1 600	
	Temperature	20	22	27	30	30	30	29	29	29	28	23	19	26	11
Colombo, Sri Lanka															
7 m	Precipitation	89	69	147	231	371	224	135	109	160	348	315	147	2 365	
	Temperature	26	26	27	28	28	27	27	27	27	27	26	26	27	2
Harbin, China															
160 m	Precipitation	5	5	10	23	43	94	112	104	46	33	8	5	488	
	Temperature	−18	−15	−5	6	13	19	22	21	14	4	−6	−16	3	40
Ho Chi Minh City, Vietnam															
9 m	Precipitation	15	3	13	43	221	330	315	269	335	269	114	56	1 984	
	Temperature	26	27	29	30	29	28	28	28	27	27	27	26	28	4
Jakarta, Indonesia															
8 m	Precipitation	300	300	211	147	114	97	64	43	66	112	142	203	1 798	
	Temperature	26	26	27	27	27	27	27	27	27	27	27	26	27	1
Hong Kong															
33 m	Precipitation	33	46	74	137	292	394	381	361	257	114	43	31	2 162	
	Temperature	16	15	18	22	26	28	28	28	27	25	21	18	23	13
Kabul, Afghanistan															
1 815 m	Precipitation	31	36	94	102	20	5	3	3	<3	15	20	10	338	
	Temperature	−3	−1	6	13	18	22	25	24	20	14	7	3	12	28
Karachi, Pakistan															
4 m	Precipitation	13	10	8	3	3	18	81	41	13	<3	3	5	196	
	Temperature	19	20	24	28	30	31	30	29	28	28	24	20	26	12
New Delhi, India															
218 m	Precipitation	23	18	13	8	13	74	180	172	117	10	3	10	640	
	Temperature	14	17	23	28	33	34	31	30	29	26	20	15	25	20
Shanghai, China															
7 m	Precipitation	48	58	84	94	94	180	147	142	130	71	51	36	1 135	
	Temperature	4	5	9	14	20	24	28	28	23	19	12	7	16	24
Singapore															
10 m	Precipitation	252	173	193	188	173	173	170	196	178	208	254	257	2 413	
	Temperature	26	27	28	28	28	28	28	27	27	27	27	27	27	2
Tehran, Iran															
1 220 m	Precipitation	46	38	46	36	13	3	3	3	3	8	20	31	246	
	Temperature	2	5	9	16	21	26	30	29	25	18	12	6	17	28
Tokyo, Japan															
6 m	Precipitation	48	74	107	135	147	165	142	152	234	208	97	56	1 565	
	Temperature	3	4	7	13	17	21	25	26	23	17	11	6	14	23
Ulan Bator, Mongolia															
1 325 m	Precipitation	<3	<3	3	5	10	28	76	51	23	5	5	3	208	
	Temperature	−26	−21	−13	−1	6	14	16	14	8	−1	−13	−22	−3	42

AUSTRALIA, NEW ZEALAND and ANTARCTICA

		Jan.	Feb.	Mar.	Apr.	May	June	July	Aug.	Sept.	Oct.	Nov.	Dec.	Year	Annual range
Alice Springs, Australia															
579 m	Precipitation	43	33	28	10	15	13	8	8	8	18	31	38	252	
	Temperature	29	28	25	20	15	12	12	14	18	23	26	28	21	17
Christchurch, New Zealand															
10 m	Precipitation	56	43	48	48	66	66	69	48	46	43	48	56	638	
	Temperature	16	16	14	12	9	6	6	7	9	12	14	16	11	10
Darwin, Australia															
30 m	Precipitation	386	312	254	97	15	3	<3	3	13	51	119	239	1 491	
	Temperature	29	29	29	29	28	26	25	26	28	29	30	29	28	5
Mawson, Antarctica															
14 m	Precipitation	11	30	20	10	44	180	4	40	3	20	0	0	362	
	Temperature	0	−5	−10	−14	−15	−16	−18	−18	−19	−13	−5	−1	−11	18

		Jan.	Feb.	Mar.	Apr.	May	June	July	Aug.	Sept.	Oct.	Nov.	Dec.	Year	Annual Range
Melbourne, Australia															
	Precipitation	48	46	56	58	53	53	48	48	58	66	58	58	653	
35 m	Temperature	20	20	18	15	13	10	9	11	13	14	16	18	15	11
Perth, Australia															
	Precipitation	8	10	20	43	130	180	170	149	86	56	20	13	881	
60 m	Temperature	23	23	22	19	16	14	13	13	15	16	19	22	18	10
Sydney, Australia															
	Precipitation	89	102	127	135	127	117	117	76	73	71	73	73	1 181	
42 m	Temperature	22	22	21	18	15	13	12	13	15	18	19	21	17	10

EUROPE and U.S.S.R.

		Jan.	Feb.	Mar.	Apr.	May	June	July	Aug.	Sept.	Oct.	Nov.	Dec.	Year	Annual Range
Archangel, U.S.S.R.															
	Precipitation	31	19	25	29	42	52	62	56	63	63	47	41	530	
13 m	Temperature	−16	−14	−9	0	7	12	15	14	8	2	−4	−11	0	31
Athens, Greece															
	Precipitation	62	37	37	23	23	14	6	7	15	51	56	71	402	
107 m	Temperature	10	10	12	16	20	25	28	28	24	20	15	11	18	18
Berlin, Germany															
	Precipitation	46	40	33	42	49	65	73	69	48	49	46	43	603	
55 m	Temperature	−1	0	4	9	14	17	19	18	15	9	5	1	9	20
Istanbul, Turkey															
	Precipitation	109	92	72	46	38	34	34	30	58	81	103	119	816	
114 m	Temperature	5	6	7	11	16	20	23	23	20	16	12	8	14	18
Kazalinsk, U.S.S.R.															
	Precipitation	10	10	13	13	15	5	5	8	8	10	13	15	125	
63 m	Temperature	−12	−11	−3	6	18	23	25	23	16	8	−1	−7	7	37
Lisbon, Portugal															
	Precipitation	111	76	109	54	44	16	3	4	33	62	93	103	708	
77 m	Temperature	11	12	14	16	17	20	22	23	21	18	14	12	17	12
London, U.K.															
	Precipitation	54	40	37	37	46	45	57	59	49	57	64	48	593	
5 m	Temperature	4	5	7	9	12	16	18	17	15	11	8	5	11	14
Málaga, Spain															
	Precipitation	61	51	62	46	26	5	1	3	29	64	64	62	474	
33 m	Temperature	12	13	15	17	19	29	25	26	23	20	16	13	18	17
Moscow, U.S.S.R.															
	Precipitation	39	38	36	37	53	58	88	71	58	45	47	54	624	
156 m	Temperature	−13	−10	−4	6	13	16	18	17	12	6	−1	−7	4	31
Odessa, U.S.S.R.															
	Precipitation	57	62	30	21	34	34	42	37	37	13	35	71	473	
64 m	Temperature	−3	−1	2	9	15	20	22	22	18	12	9	1	10	25
Omsk, U.S.S.R.															
	Precipitation	15	8	8	13	31	51	51	51	28	25	18	20	318	
85 m	Temperature	−22	−19	−12	−1	10	16	18	16	10	1	−11	−18	−1	40
Palma de Mallorca, Spain															
	Precipitation	39	34	51	32	29	17	3	25	55	77	47	40	449	
10 m	Temperature	10	11	12	15	17	21	24	25	23	18	14	11	17	15
Paris, France															
	Precipitation	56	46	35	42	57	54	59	64	55	50	51	50	619	
75 m	Temperature	3	4	8	11	15	18	20	19	17	12	7	4	12	17
Rome, Italy															
	Precipitation	71	62	57	51	46	37	15	21	63	99	129	93	744	
17 m	Temperature	8	9	11	14	18	22	25	25	22	17	13	10	16	17
Shannon, Irish Republic															
	Precipitation	94	67	56	53	61	57	77	79	86	86	96	117	929	
2 m	Temperature	5	5	7	9	12	14	16	16	14	11	8	6	10	11
Stavanger, Norway															
	Precipitation	93	56	45	70	49	84	93	118	142	129	125	126	1 130	
85 m	Temperature	1	1	3	6	10	13	15	15	13	9	6	3	8	14
Stockholm, Sweden															
	Precipitation	43	30	25	31	34	45	61	76	60	48	53	48	554	
44 m	Temperature	−3	−3	−1	5	10	15	18	17	12	7	3	0	7	21
Verkhoyansk, U.S.S.R.															
	Precipitation	5	5	3	5	8	23	28	25	13	8	8	5	134	
100 m	Temperature	−50	−45	−32	−15	0	12	14	9	2	−15	−38	−48	−17	64
Warsaw, Poland															
	Precipitation	27	32	27	37	46	69	96	65	43	38	31	44	555	
110 m	Temperature	−3	−3	2	7	14	17	19	18	14	9	3	0	8	22

INDEX

The number printed in bold type against each index entry indicates the map page where the feature will be found. The geographical coordinates which follow the name are sometimes only approximate but are close enough for the place name to be located.

An open square □ signifies that the name refers to an administrative subdivision of a country while a solid square ■ follows the name of a country.

The alphabetical order of names composed of two or more words is governed primarily by the first word and then by the second. This rule applies even if the second word is a description or its abbreviation, R.,L.,I. for example. Names composed of a proper name (Gibraltar) and a description (Strait of) are positioned alphabetically by the proper name. If the same place name occurs twice or more times in the index and all are in the same country, each is followed by the name of the administrative subdivision in which it is located. The names are placed in the alphabetical order of the subdivisions. If the same place name occurs twice or more in the index and the places are in different countries they will be followed by their country names, the latter governing the alphabetical order. In a mixture of these situations the primary order is fixed by the alphabetical sequence of the countries and the secondary order by that of the country subdivisions.

A. R. – Autonomous Region
A. S. S. R. – Autonomous Soviet Socialist Republic
Afr. – Africa
Ala. – Alabama
Alas. – Alaska
Alta. – Alberta
Amer. – America
Arch. – Archipelago
Ariz. – Arizona
Ark. – Arkansas
Atl. Oc. – Atlantic Ocean
Austral. – Australia
B. – Bay, Bight
B.C. – British Columbia
Belg. – Belgium
Berks. – Berkshire
Br. – British
Bucks. – Buckinghamshire
C. – Cabo, Cape
C. Prov. – Cape Province
Calif. – California
Cambs. – Cambridgeshire
Can. – Canada
Cent. – Central
Chan. – Channel
Ches. – Cheshire
Co. – Country
Colo. – Colorado
Conn. – Connecticut
Cord. – Cordillera
Corn. – Cornwall
Cumb. – Cumbria
D.C. – District of Columbia
Del. – Delaware
Dep. – Dependency
Derby. – Derbyshire
Des. – Desert
Dist. – District
Dumf. & Gall. – Dumfries and Galloway
E. – East
Eng. – England
Fed. – Federal, Federation
Fla. – Florida
Fr. – France, French
Fs. – Falls
Ft. – Fort
G. – Golf, Golfo, Gulf
Ga. – Georgia
Germ. – Germany
Glam. – Glamorgan
Glos. – Gloucestershire
Gr. – Grande, Great, Greater, Group
Hants. – Hampshire
Harb. – Harbour
Hd. – Head
Here. & Worcs. – Hereford and Worcester
Herts. – Hertfordshire
I.o.M. – Isle of Man
I. of W. – Isle of Wight
I.(s). – Île, Island, Isle
Id. – Idaho
Ill. – Illinois
Ind. – Indiana
Ind. Oc. – Indian Ocean
J. – Jabal, Jabel, Jazira
Junc. – Junction
Kans. – Kansas
Ky. – Kentucky
L. – Lac, Lago, Lake, Loch, Lough
La. – Lousiana
Lancs. – Lancashire
Leics. – Leicestershire
Lim. – Limerick
Lincs. – Lincolnshire
Lr. – Lower
Malay. – Malaysia
Man. – Manitoba
Manch. – Manchester
Mass. – Massachusetts
Md. – Maryland
Me. – Maine
Mich. – Michigan
Mid. – Middle
Minn. – Minnesota
Miss. – Mississippi
Mo. – Missouri
Mont. – Montana
Mt.(e). – Mont, Monte, Monti, Montaña, Mountain
N. – North, Northern, Nouveau
N.B. – New Brunswick
N.C. – North Carolina
N.D. – North Dakota
N.H. – New Hampshire
N.I. – Northern Ireland
N.J. – New Jersey
N. Mex. – New Mexico
N.S. – Nova Scotia
N.S.W. – New South Wales
N.T. – Northern Territory
N.W.T. – North West Territory
N.Y. – New York
N.Z. – New Zealand
Nat. – National
Nat Park. – National Park
Nebr. – Nebraska
Neth. – Netherlands
Nev. – Nevada
Newf. – Newfoundland
Nic. – Nicaragua
Norf. – Norfolk
Northants. – Northamptonshire
Northumb. – Northumberland
Notts. – Nottinghamshire
Okla. – Oklahoma
Ont. – Ontario
Or. – Orientale
Oreg. – Oregon
Oxon. – Oxfordshire
Oz. – Ozero
P. – Pass
Pa. – Pennsylvania
Pac. Oc. – Pacific Ocean
Pak. – Pakistan
Pass. – Passage
Pen. – Peninsula
Pk. – Peak
Plat. – Plateau
P-ov. – Poluostrov
Port. – Portugal, Portuguese
Prov. – Province, Provincial
Pt. – Point
Pta. – Punta
Pte. – Pointe
Qué. – Québec
Queens. – Queensland
R. – Rio, River
R.I. – Rhode Island
R.S.F.S.R. – Russian Soviet Federative Socialist Republic
Ra.(s). – Range(s)
Raj. – Rajasthan
Reg. – Region
Rep. – Republic
Res. – Reserve, Reservoir
Rhld. – Pfz. – Rheinland – Pfalz
Rosc. – Roscommon
S. – San, South
S. Afr. – South Africa
S. Austral. – South Australia
S.C.. – South Carolina
S.D. – South Dakota
S.-Holst. – Schleswig-Holstein
S.S.R. – Soviet Socialist Republic
S.-U. – Sinkiang-Uighur
Sa. – Serra, Sierra
Sask. – Saskatchewan
Scot. – Scotland
Sd. – Sound
Som. – Somerset
St. – Saint
Sta. – Santa
Staffs. – Staffordshire
Sto. – Santo
Str. – Strait
Suff. – Suffolk
Switz. – Switzerland
Tas. – Tasmania
Tenn. – Tennessee
Terr. – Territory
Tex. – Texas
Tipp. – Tipperary
Trans. – Transvaal
U.K. – United Kingdom
U.S.A. – United States of America
U.S.S.R. – Union of Soviet Socialist Republics
Ukr. – Ukraine
Va. – Virginia
Vic. – Victoria
Viet. – Vietnam
Vol. – Volcano
Vt. – Vermont
W. – Wadi, West
W.A. – Western Australia
W. Isles – Western Isles
War. – Warwickshire
Wash. – Washington
Wilts. – Wiltshire
Wis. – Wisconsin
Wyo. – Wyoming
Yorks. – Yorkshire
Yug. – Yugoslavia
Zimb.– Zimbabwe

A

As Sohar Belomorsk

As Sohar 71 24 20N 56 40 E
Asaba 90 6 12N 6 38 E
Asahikawa 76 43 45N 142 30 E
Asansol 70 23 40N 87 1 E
Asbestos Mts. 94 29 0S 23 0 E
Ascension, I. 22 8 0S 14 15W
Áscoli Piceno 58 42 51N 13 34 E
Aseb 89 13 0N 42 40 E
Ash 40 51 14N 0 43W
Ashaira 71 21 40N 40 40 E
Ashan 73 41 3N 122 58 E
Ashbourne 44 53 2N 1 44W
Ashburton, N.Z. 83 43 53S 171 48 E
Ashburton, U.K. 43 50 31N 3 45W
Ashburton, R. 78 21 40S 114 56 E
Ashby-de-la-Zouch 45 52 45N 1 29W
Asheville 101 35 39N 82 30W
Ashford 40 51 8N 0 53 E
Ashington 44 55 12N 1 35W
Ashkhabad 71 38 0N 57 50 E
Ashland 101 38 25N 82 40W
Ashmûn 91 30 18N 30 55 E
Ashq'elon 89 31 42N 34 55 E
Ashton-in-Makerfield 45 53 29N 2 39W
Ashton-u.-Lyne 45 53 30N 2 8 E
Asia 66 45 0N 75 0 E
Asinara, G. of 58 41 0N 8 30 E
Asinara I. 58 41 5N 8 15 E
Asir □ 89 18 40N 42 30 E
Askeaton 51 52 37N 8 58W
Asmera 89 15 19N 38 55 E
Aspiring, Mt. 83 44 23S 168 46 E
Aspull 45 53 33N 2 36W
Assam □ 70 25 45N 92 30 E
Assen 52 53 0N 6 35 E
Assisi 58 43 4N 12 36 E
Asti 58 44 54N 8 11 E
Astorga 60 42 29N 6 8W
Astoria 105 46 16N 123 50W
Astrakhan 62 46 25N 48 5 E
Asturias 60 43 15N 6 0W
Asunción 108 25 21S 57 30W
Asunción, La 107 11 2N 63 53W
Aswân 91 24 4N 32 57 E
Aswan High Dam 91 24 5N 32 54 E
Asyût 91 27 11N 31 4 E
Asyûti, Wadi 91 27 18N 31 20 E
Atacama Desert 108 24 0S 69 20W
Atami 76 35 0N 139 55 E
Atbara 89 17 42N 33 59 E
'Atbara, R. 89 17 40N 33 56 E
Ath 52 50 38N 3 47 E
Athabasca 98 54 45N 113 20W
Athabasca, L. 98 59 15N 109 15W
Athabasca, R. 98 58 40N 110 50W
Athenry 51 53 18N 8 45W
Athens, Greece 89 37 58N 23 46 E
Athens, U.S.A. 101 33 56N 83 24W
Atherstone 41 52 35N 1 32W
Atherton, Austral. 79 17 17S 145 30 E
Atherton, U.K. 45 53 32N 2 30W
Athlone 51 53 26N 7 57W
Atholl, Forest of 48 56 51N 3 50W
Athy 51 53 0N 7 0W
Atlanta 101 33 50N 84 24W
Atlantic City 104 39 25N 74 25W
Atlantic Ocean 22 0 0 20 0W
Atsoum, mts. 90 7 0N 12 30 E
Attleborough 40 52 32N 1 1 E
Aube, R. 56 48 34N 3 17 E
Auburn 104 42 57N 76 39W
Auchi 90 7 6N 6 13 E
Auchterarder 47 56 18N 3 43W
Auchterderran 47 56 8N 3 16W
Auchtermuchty 47 56 18N 3 15W
Auckland 83 36 52S 174 46 E
Auckland Is. 23 51 0S 166 0 E
Aude, R. 56 44 13N 3 15 E
Audenshaw 45 53 29N 2 06W
Augathella 79 25 48S 146 35 E
Aughnacloy 50 54 25N 7 0W
Aughrim 51 52 52N 6 20W
Augsburg 54 48 22N 10 54 E
Augusta, Italy 58 37 14N 15 12 E
Augusta, Ga., U.S.A. 101 33 29N 81 59W
Augusta, Me., U.S.A. 101 44 20N 69 46W
Augustus, Mt. 78 24 20S 116 50 E
Aunis 56 46 0N 0 50W
Auob, R. 94 25 0S 18 50 E
Aurangabad 70 19 50N 75 23 E
Aurich 52 53 28N 7 30 E
Aurora 101 41 42N 88 12W
Austerlitz 54 49 10N 16 52 E
Austin 100 30 20N 97 45W
Austral Downs 79 20 30S 137 45 E
Australia ■ 78 23 0S 135 0 E
Australian Alps 79 36 30S 148 8 E
Australian Cap. Terr. □ 79 35 15S 149 8 E
Austria ■ 54 47 0N 14 0 E
Auvergne 56 45 20N 3 0 E
Auxerre 56 47 48N 3 32 E
Aveiro 60 40 37N 8 38W
Avellino 58 40 54N 14 46 E
Aversa 58 40 58N 14 11 E
Aveyron, R. 56 44 5N 1 16 E
Aviemore 48 57 11N 3 50W
Avignon 56 43 57N 4 50 E
Ávila 60 40 39N 4 43W
Avilés 60 43 35N 5 57W
Avon □ 42 51 30N 2 40W
Avon, R., Avon, U.K. 42 51 30N 2 43W
Avon, R., Warwick, U.K. 42 52 0N 2 9W
Avonmouth 43 51 30N 2 42W
Avranches 56 48 40N 1 20W
Awe, L. 46 56 15N 5 15W
Ax-les-Thermes 56 42 44N 1 50 E
Axbridge 42 51 17N 2 50W
Axe Edge 44 53 14N 2 2W
Axel Heiberg I. 98 80 0N 90 0W
Axminster 43 50 47N 3 1W
Ayabe 76 35 20N 135 20 E
Ayaguz 73 48 10N 80 0 E
Ayamonte 60 37 12N 7 24W
Ayers Rock 78 25 23S 131 5 E
Aylesbury 40 51 48N 0 49W
Aylsham 40 52 48N 1 16 E
Ayr, Austral. 79 19 35S 147 25 E
Ayr, U.K. 46 55 28N 4 37W
Ayr, R. 46 55 29N 4 40W
Ayre, Pt. of 42 54 27N 4 21W
Ayutla 106 16 58N 99 17W
Azare 90 11 55N 10 10 E
Azerbaijan S.S.R. □ 71 40 20N 48 0 E
Azores, Is. 22 38 44N 29 0W
Azov, Sea of 62 46 0N 36 30 E
Azuaga 60 38 16N 5 39W

B

Baarle Nassau 52 51 27N 4 56 E
Baarn 52 52 12N 5 17 E
Bāb el Māndeb 89 12 35N 43 25 E
Babine L. 105 54 48N 126 0W
Babura 90 12 51N 8 59 E
Babuyan Chan. 72 18 40N 121 30 E
Babuyan Is. 72 19 10N 121 40 E
Bacău 55 46 35N 26 55 E
Bacolod 72 10 40N 122 57 E
Bacup 45 53 42N 2 12W
Bad Godesberg 52 50 41N 7 4 E
Bad Honnef 52 50 39N 7 13 E
Bad Ischl 54 47 44N 13 38 E
Bad Kreuznach 52 49 47N 7 47 E
Bad Lands 100 43 40N 102 10W
Bad Zwischenahn 52 53 15N 8 0 E
Badagri 90 6 25N 2 55 E
Badajoz 60 38 50N 6 59W
Badalona 60 41 26N 2 15 E
Baden 54 48 1N 16 13 E
Baden-Baden 54 48 45N 8 15 E
Baden-Württemberg □ 54 48 40N 9 0 E
Badgastein 54 47 7N 13 9 E
Baffin Bay 99 72 0N 64 0W
Baffin I. 99 68 0N 75 0W
Bafia 90 4 40N 11 10 E
Bafoulabé 88 13 50N 10 55W
Bafut 90 6 6N 10 2 E
Bagamoyo 93 6 28S 38 55 E
Bagan Siapiapi 72 2 12N 100 50 E
Baghdād 71 33 20N 44 30 E
Bagshot 41 51 22N 0 41W
Baguio 72 16 26N 120 34 E
Bahamas, Is. 107 24 40N 74 0W
Bahamas ■ 107 24 0N 74 0W
Bahawalpur □ 71 29 5N 71 3 E
Bahia = Salvador 108 13 0S 38 30W
Bahia Blanca 108 38 35S 62 13W
Bahr el 'Arab, R. 89 10 0N 26 0 E
Bahr el Jebel 89 7 30N 30 30 E
Bahr Yûsef 91 28 25N 30 35 E
Bahra el Burullus 91 31 28N 30 48 E
Bahra el Manzala 91 31 28N 32 01 E
Bahrain ■ 71 26 0N 50 35 E
Bai Bung Pt. 72 8 35N 104 42 E
Baie Comeau 99 49 12N 68 10W
Baikal, L. = Baykal, Oz. 62 53 0N 108 0 E
Baildon 45 53 52N 1 46W
Baile Atha Cliath = Dublin 51 53 20N 6 18W
Baillieborough 50 53 55N 7 0W
Baillieston 47 55 51N 4 08W
Baird Inlet 98 64 49N 164 18W
Bairnsdale 79 37 48S 147 36 E
Baja 55 46 12N 18 59 E
Baja, Pte. 106 29 50N 116 0W
Baker 100 44 50N 117 55W
Baker Is. 22 0 10N 176 35 E
Baker Mt. 105 48 50N 121 49W
Bakersfield 105 35 25N 119 0W
Bakewell 44 53 13N 1 40W
Bakony Forest 55 47 10N 17 30 E
Baku 71 40 25N 49 45 E
Bala 42 52 54N 3 36W
Balaguer 60 41 50N 0 50 E
Balasore 70 21 35N 87 3 E
Balaton 55 46 50N 17 40 E
Balbriggan 51 53 35N 6 10W
Balclutha 83 46 15S 169 45 E
Baldoyle 51 53 24N 6 10W
Baldy Peak 100 33 50N 109 30W
Balearic Is. 60 39 30N 3 0 E
Balerno 47 55 53N 3 20W
Bali 90 5 54N 10 0 E
Bali, I. 72 8 20S 115 0 E
Balikesir 71 39 35N 27 58 E
Balikpapan 72 1 10S 116 55 E
Balkan Pen. 24 42 0N 22 0 E
Balkhash 62 46 50N 74 50 E
Balkhash, L. 62 46 0N 74 50 E
Ballaghaderreen 50 53 55N 8 35W
Ballantrae 46 55 6N 5 0W
Ballarat 79 37 33S 143 50 E
Ballater 48 57 2N 3 2W
Balleny Is. 112 66 30S 163 0 E
Ballina 50 54 7N 9 10W
Ballinamore 50 54 3N 7 48W
Ballinasloe 51 53 20N 8 12W
Ballingeary 51 51 51N 9 13W
Ballingry 47 56 09N 3 20W
Ballinrobe 50 53 36N 9 13W
Balloch 46 56 0N 4 35W
Ballybay 50 54 8N 6 52W
Ballybofey 50 54 48N 7 47W
Ballybunion 51 52 30N 9 40W
Ballycastle, Ireland 50 54 17N 9 24W
Ballycastle, U.K. 50 55 12N 6 15W
Ballyclare 50 54 46N 6 0W
Ballycotton 51 51 50N 8 0W
Ballygawley 50 54 27N 7 2W
Ballyhaunis 50 53 47N 8 47W
Ballyheige I. 51 52 22N 9 51W
Ballylongford 51 52 34N 9 30W
Ballymahon 51 53 35N 7 45W
Ballymena 50 54 53N 6 18W
Ballymena □ 50 54 53N 6 18W
Ballymoney 50 55 5N 6 30W
Ballymoney □ 50 55 5N 6 23W
Ballymote 50 54 5N 8 30W
Ballynahinch 50 54 24N 5 55W
Ballyragget 51 52 47N 7 20W
Ballysadare 50 54 12N 8 30W
Ballyshannon 50 54 30N 8 10W
Balmoral 48 57 3N 3 13W
Balranald 79 34 38S 143 33 E
Balsas, R. 106 18 30N 101 20W
Baltic Sea 61 56 0N 20 0 E
Baltim 91 31 35N 31 10 E
Baltimore, Ireland 51 51 29N 9 22W
Baltimore, U.S.A. 104 39 18N 76 37W
Baltinglass 51 52 57N 6 42W
Bam 71 29 7N 58 14 E
Bama 90 11 33N 13 33 E
Bamako 88 12 34N 7 55W
Bamberg 54 49 54N 10 53 E
Bamburgh 44 55 37N 1 43W
Bampton 43 50 59N 3 29W
Banbridge 50 54 21N 6 17W
Banbridge □ 50 54 21N 6 16W
Banchory 48 57 3N 2 30W
Banda Aceh 72 5 35N 95 20 E
Banda Is. 72 4 37S 129 50 E
Banda Sea 72 6 0S 130 0 E
Bandar-e Anzalī 71 37 30N 49 30 E
Bandar-e Torkeman 71 37 0N 54 10 E
Bandirma 71 40 20N 28 0 E
Bandon 51 51 44N 8 45W
Bandon, R. 51 51 40N 8 11W
Bandung 72 6 36S 107 48 E
Banff, Can. 98 51 10N 115 34W
Banff, U.K. 48 57 40N 2 32W
Bangalore 70 12 59N 77 40 E
Bangassou 89 4 55N 23 55 E
Banggai 72 1 40S 123 30 E
Banggai Arch. 72 2 0S 123 15 E
Bangka Str. 72 3 30S 105 30 E
Bangkok 72 13 45N 100 35 E
Bangladesh ■ 70 24 0N 90 0 E
Bangor, Ireland 50 54 09N 9 44W
Bangor, N.I., U.K. 50 54 40N 5 40W
Bangor, Wales, U.K. 42 53 13N 4 9W
Bangor, U.S.A. 101 44 48N 68 42W
Bangui 88 4 23N 18 35 E
Bangweulu, L. 92 11 0S 30 0 E
Bani 107 18 16N 70 22W
Banja Luka 58 44 49N 17 26 E
Banjul 88 13 28N 16 40W
Banks I. 105 53 20N 130 0W
Banks Peninsula 83 43 45S 173 15 E
Bann R. 50 54 30N 6 31W
Bannockburn 47 56 5N 3 55W
Banstead 41 51 19N 0 10W
Bantry 51 51 40N 9 28W
Bantry, B. 51 51 35N 9 50W
Baqûbah 71 33 45N 44 50 E
Bar Harbor 101 44 15N 68 20W
Bar-le-Duc 56 48 47N 5 10 E
Barahona 107 18 13N 71 7W
Barbados ■ 107 13 0N 59 30W
Barberton 94 25 42S 31 2 E
Barbuda I. 107 17 30N 61 40W
Barcaldine 79 23 33S 145 13 E
Barcelona, Spain 60 41 21N 2 10 E
Barcelona, Venez. 107 10 10N 64 40W
Bardiyah 91 31 45N 25 0 E
Bardsey 45 53 53N 1 26W
Bardsey, I. 42 52 46N 4 47W
Bareilly 70 28 22N 79 27 E
Barents Sea 62 73 0N 39 0 E
Bargoed 43 51 42N 3 22W
Bari 58 41 6N 16 52 E
Bâris 91 24 42N 30 31 E
Barisal 70 22 30N 90 20 E
Barito, R. 72 2 50S 114 50 E
Barking 41 51 31N 0 10 E
Barkley Sound 105 48 50N 125 10W
Barkly East 94 30 58S 27 33 E
Barkly Tableland 78 19 50S 138 40 E
Barkly West 94 28 5S 24 31 E
Barkol 73 43 37N 93 2 E
Bârlad 55 46 15N 27 38 E
Barlee, L. 78 29 15S 119 30 E
Barlee Ra. 78 23 30S 116 0 E
Barletta 58 41 20N 16 17 E
Barmer 70 25 45N 71 20 E
Barmouth 42 52 44N 4 3W
Barnard Castle 44 54 33N 1 55W
Barnaul 62 53 20N 83 40 E
Barnet 41 51 37N 0 15W
Barnoldswick 45 53 55N 2 11W
Barnsley 45 53 33N 1 29W
Barnstaple 42 51 5N 4 3W
Barquisimeto 107 9 58N 69 13W
Barra Hd. 49 56 47N 7 40W
Barra, I. 49 57 0N 7 30W
Barrancos 60 38 10N 6 58W
Barranquilla 107 11 0N 74 50W
Barre 101 42 26N 72 6W
Barreiro 60 38 40N 9 6W
Barrhead 47 55 48N 4 23W
Barrow, C. 98 71 10N 156 20W
Barrow Creek T.O. 78 21 30S 133 55 E
Barrow-in-Furness 44 54 8N 3 15W
Barrow, Pt. 98 71 22N 156 30W
Barrow Ra. 78 26 0S 127 40 E
Barrowford 45 53 51N 2 14W
Barry 43 51 23N 3 19W
Bartica 107 6 25N 58 40W
Bartle Frere, Mt. 79 17 27S 145 50 E
Barton-upon-Humber 44 53 41N 0 27W
Barvas 48 58 21N 6 31W
Barwell 41 52 35N 1 22W
Bashi Channel 73 21 15N 122 0 E
Basilan, I. 72 6 35N 122 0 E
Basildon 41 51 34N 0 29 E
Basilicata □ 58 40 30N 16 0 E
Basle 54 47 35N 7 35 E
Basoka 92 1 16N 23 40 E
Basque Provinces 60 42 50N 2 45W
Basra 71 30 30N 47 50 E
Bass Strait 79 39 15S 146 30 E
Bassano 58 50 48N 112 20W
Bassein 70 16 30N 94 30 E
Bastia 56 42 40N 9 30 E
Batabanó 107 22 40N 82 20W
Batan I. 72 20 58N 122 5 E
Batanes Is. 72 20 30N 122 0 E
Bataszék 55 46 10N 18 44 E
Batchelor 78 13 4S 131 1 E
Bath 101 43 50N 69 49W
Bathgate 47 55 54N 3 38W
Bathurst 79 33 25S 149 31 E
Bathurst, C. 98 70 34N 128 0W
Bathurst I. 98 76 0N 100 30W
Bathurst Inlet 98 66 50N 108 1W
Batley 45 53 43N 1 38W
Baton Rouge 101 30 30N 91 5W
Battambang 72 13 7N 103 12 E
Battle 40 50 55N 0 30 E
Battle Creek 104 42 20N 85 6W
Battle Harbour 99 52 16N 55 35W
Battleford 98 52 45N 108 15W
Batu Is. 72 0 30S 98 25 E
Batu Pahat 72 1 50N 102 56 E
Batumi 71 41 30N 41 30 E
Bauchi 90 10 22N 9 48 E
Bautzen 54 51 11N 14 25 E
Bavaria 54 49 7N 11 30 E
Bawean 72 5 46S 112 35 E
Bawtry 44 53 25N 1 1W
Bay City 104 43 35N 83 51W
Bay View 83 39 25S 176 50 E
Bayan Kara Shan 73 34 0N 98 0 E
Bayeux 56 49 17N 0 42W
Bayonne 56 43 30N 1 28W
Bayreuth 54 49 56N 11 35 E
Bayrut = Beirut 89 33 53N 35 31 E
Baza 60 37 30N 2 47W
Beachy Head 40 50 44N 0 16 E
Beaconsfield, Austral. 79 41 11S 146 48 E
Beaconsfield, U.K. 41 51 36N 0 39W
Bealey 83 43 2S 171 36 E
Beaminster 43 50 48N 2 44W
Bear I. 51 51 38N 9 50W
Bear L. 100 42 0N 111 20W
Beardmore Glacier 112 84 30S 170 0 E
Béarn 56 43 28N 0 36W
Bearsden 47 55 55N 4 21W
Bearsted 41 51 15N 0 35 E
Beauce, Plaines de 56 48 10N 1 45 E
Beaufort Sea 98 72 0N 140 0W
Beaufort-West 94 32 18S 22 36 E
Beaulieu 40 50 49N 1 27W
Beauly 48 57 29N 4 27W
Beauly Firth 48 57 30N 4 20W
Beauly, R. 48 57 26N 4 28W
Beaumaris 42 53 16N 4 7W
Beaumont 101 30 5N 94 8W
Beaune 56 47 2N 4 50 E
Beauvais 56 49 25N 2 8 E
Beawar 70 26 3N 74 18 E
Bebington 45 53 23N 3 1W
Beccles 40 52 27N 1 33 E
Béchar 88 31 38N 2 18W
Bechuanaland, reg. 94 26 30S 22 30 E
Beddau 43 51 33N 3 23W
Bedford 40 52 8N 0 29W
Bedford □ 40 52 4N 0 28W
Bedford Level 40 52 25N 0 5 E
Bedlington 44 55 8N 1 35W
Bedourie 79 24 30S 139 30 E
Bedwas 43 51 36N 3 10W
Bedworth 41 52 28N 1 29W
Beersheba 89 31 15N 34 48 E
Beeston, Ches. 45 53 07N 2 41W
Bega 79 36 41S 149 51 E
Behbehan 71 30 30N 50 15 E
Beighton 45 53 21N 1 21W
Beira 92 19 50S 34 52 E
Beira-Alta 60 40 35N 7 35W
Beira-Baixa 60 40 2N 7 30W
Beira-Litoral 60 40 5N 8 30W
Beirut 89 33 53N 35 31 E
Beit Shean 89 32 30N 35 30 E
Beith 47 55 45N 4 38W
Beja □ 60 37 55N 7 55W
Békéscsaba 55 46 40N 21 10 E
Bekily 92 24 13S 45 19 E
Bela Crkva 55 44 55N 21 27 E
Belaya Tserkov 55 49 45N 30 10 E
Belbroughton 41 52 23N 2 5W
Belcoo 50 54 18N 7 52W
Belem 108 1 20S 48 30W
Belfast 50 54 35N 5 56W
Belfast □ 50 54 35N 5 56W
Belfast, L. 50 54 40N 5 50W
Belford 46 55 36N 1 50W
Belfort 56 47 38N 6 50 E
Belgium ■ 52 51 30N 5 0 E
Belgorod Dnestrovskiy 55 46 11N 30 23 E
Belgrade 55 44 50N 20 37 E
Belize ■ 106 17 0N 88 30W
Belize City 106 17 25N 88 0W
Bell I. 99 50 46N 55 35W
Bella Coola 98 52 25N 126 40W
Bellananagh 50 53 55N 7 25W
Bellary 70 15 10N 76 56 E
Belle-Ile 56 47 20N 3 10W
Belle Isle 99 51 57N 55 25W
Belle Isle, Str. of 99 51 30N 56 30W
Belleville 104 44 10N 77 23W
Bellingham, U.K. 44 55 09N 2 16W
Bellingham, U.S.A. 105 48 45N 122 27W
Bellingshausen Sea 112 66 0S 80 0W
Bellinzona 54 46 11N 9 1 E
Belluno 58 46 8N 12 6 E
Belmont 45 53 38N 2 30W
Belmullet 50 54 13N 9 58W
Belo Horizonte 108 19 55S 43 56W
Belomorsk 61 64 35N 34 30 E

Beloye L **Brunswick**

Name	Page	Lat.	Long.
Beloye L.	**61**	60 10N	37 35E
Belper	**44**	53 2N	1 29W
Belshill	**47**	55 49N	4 01W
Belton	**100**	31 4N	97 30W
Beltsy	**55**	47 48N	28 0E
Belturbet	**50**	54 6N	7 28W
Bemidji	**101**	47 30N	94 50W
Ben Alder	**48**	56 50N	4 30W
Ben Cruachan	**48**	56 26N	5 8W
Ben Dearg	**48**	57 47N	4 58W
Ben Hope	**48**	58 24N	4 36W
Ben Klibreck	**48**	58 14N	4 25W
Ben Lawers	**48**	56 33N	4 13W
Ben Macdhui	**48**	57 4N	3 40W
Ben More, Mull, U.K.	**48**	56 26N	6 2W
Ben More, Perth, U.K.	**48**	56 23N	4 31W
Ben More Assynt	**48**	58 7N	4 51W
Ben Nevis	**48**	56 48N	5 0W
Ben Vorlich	**46**	56 22N	4 15W
Bena	**90**	11 20N	5 50E
Benalla	**79**	36 30S	146 0E
Bencubbin	**78**	30 48S	117 52E
Bend	**105**	44 2N	121 15W
Bendery	**55**	46 50N	29 50E
Bendigo	**79**	36 40S	144 15E
Benevento	**58**	41 7N	14 45E
Bengal, Bay of	**70**	18 0N	90 0E
Benghazi	**89**	32 11N	20 3E
Benguela	**92**	12 37S	13 25E
Benha	**91**	30 26N	31 8E
Beni	**93**	0 30N	29 27E
Benî Mazâr	**91**	28 32N	30 44E
Benî Suêf	**91**	29 5N	31 6E
Benin ■	**90**	10 0N	2 0E
Benin City	**90**	6 20N	5 31E
Benoni	**94**	26 11S	28 18E
Bentley	**44**	53 33N	1 9W
Benue □	**90**	7 30N	7 30E
Benue, R.	**90**	7 50N	6 30E
Benwee Hd.	**50**	54 20N	9 50W
Beograd = Belgrade	**55**	44 50N	20 37E
Beppu	**76**	33 15N	131 30E
Beragh	**50**	54 34N	7 10W
Berbera	**89**	10 30N	45 2E
Berdichev	**55**	49 57N	28 30E
Beregovo	**55**	48 15N	22 45E
Bérgamo	**58**	45 42N	9 40E
Bergen	**61**	60 23N	5 20E
Bergen-op-Zoom	**52**	51 30N	4 18E
Bergerac	**56**	44 51N	0 30E
Bergisch-Gladbach	**52**	50 59N	7 9E
Berhampore	**70**	24 2N	88 27E
Berhampur	**70**	19 15N	84 54E
Bering Sea	**62**	58 0N	167 0E
Bering Str.	**96**	66 0N	170 0W
Berkel, R.	**52**	52 8N	6 12E
Berkhamsted	**41**	51 45N	0 33W
Berkner I.	**112**	79 30S	50 0W
Berlin	**54**	52 32N	13 24E
Bermuda, I.	**96**	32 45N	65 0W
Bern	**54**	46 57N	7 28E
Bernay	**56**	49 5N	0 35E
Bernburg	**54**	51 40N	11 42E
Berneray I.	**49**	56 47N	7 40W
Bernina Pass	**54**	46 22N	9 54E
Berry	**56**	47 0N	2 0E
Berwick-upon-Tweed	**46**	55 47N	2 0W
Berwyn Mts.	**42**	52 54N	3 26W
Besalampy	**92**	16 43S	44 29E
Besançon	**56**	47 9N	6 0E
Bessarabiya	**55**	46 20N	29 0E
Bessemer	**101**	46 27N	90 0W
Bethal	**94**	26 27S	29 28E
Bethesda	**42**	53 11N	4 3W
Bethlehem, Jordan	**89**	31 43N	35 12E
Bethlehem, S. Afr.	**94**	28 14S	28 18E
Bethlehem, U.S.A.	**104**	40 39N	75 24W
Bethulie	**94**	30 30S	25 59E
Béthune	**56**	50 30N	2 38E
Betroka	**92**	23 16S	46 0E
Bettiah	**70**	26 48N	84 33E
Bettyhill	**48**	58 31N	4 12W
Betws-y-Coed	**42**	53 4N	3 49W
Beverley, Austral.	**78**	32 9S	116 56E
Beverley, U.K.	**44**	53 52N	0 26W
Beverwijk	**52**	52 28N	4 38E
Bewdley	**41**	52 23N	2 19W
Bexhill	**40**	50 51N	0 29E
Bexley	**41**	51 26N	0 10E
Beyneu	**71**	45 10N	55 3E
Béziers	**56**	43 20N	3 12E
Bhagalpur	**70**	25 10N	87 0E
Bhamo	**70**	24 15N	97 15E
Bharatpur	**70**	27 15N	77 30E
Bhaunagar	**70**	21 45N	72 10E
Bhilwara	**70**	25 25N	74 38E
Bhopal	**70**	23 20N	77 53E
Bhubaneswar	**70**	20 15N	85 50E
Bhutan ■	**70**	27 25N	89 50E
Biała Podlaska	**55**	52 4N	23 6E
Białystok	**55**	53 10N	23 10E
Biarritz	**56**	43 29N	1 33W
Biba	**91**	28 55N	31 0E
Bibaī	**76**	43 19N	141 52E
Bida	**90**	9 3N	5 58E
Biddulph	**44**	53 8N	2 11W
Bideford	**43**	51 1N	4 13W
Bideford Bay	**42**	51 5N	4 20W
Bié	**92**	12 22S	16 55E
Bié Plateau	**92**	12 0S	16 0E
Biel	**54**	47 8N	7 14E
Bielefeld	**54**	52 2N	8 31E
Biella	**58**	45 33N	8 3E
Bielsko-Biała	**55**	49 50N	19 8E
Bien Hoa	**72**	10 57N	106 49E
Big Belt Mts.	**100**	46 50N	111 30W
Big Delta	**98**	64 15N	145 0W
Big Sioux, R.	**101**	42 30N	96 25W
Big Spring	**100**	32 10N	101 25W
Biggar	**47**	55 37N	3 31W
Biggleswade	**40**	52 6N	0 16W
Bighorn Mts.	**100**	44 30N	107 30W
Bihar □	**70**	25 0N	86 0E
Bijapur	**70**	16 50N	75 55E
Bikaner	**70**	28 2N	73 18E
Bilbao	**60**	43 16N	2 56W
Billericay	**41**	51 38N	0 25E
Billingham	**44**	54 36N	1 18W
Billings	**100**	45 43N	108 29W
Billiton Is.	**72**	3 10S	107 50E
Biloela	**79**	24 24S	150 31E
Biloxi	**101**	30 30N	89 0W
Bilston	**41**	52 34N	2 5W
Binche	**52**	50 26N	4 10E
Bingen	**52**	49 57N	7 53E
Bingerville	**88**	5 18N	3 49W
Bingley	**45**	53 51N	1 50W
Bir Hirmas	**91**	28 57N	36 25E
Birdsville	**79**	25 51S	139 20E
Birdum	**78**	15 39S	133 13E
Birkdale	**45**	53 38N	3 2W
Birkenhead	**45**	53 24N	3 1W
Birkenshaw	**45**	53 45N	1 41W
Birket Qârûn	**91**	29 30N	30 40E
Birmingham, U.K.	**41**	52 30N	1 55W
Birmingham, U.S.A.	**101**	33 31N	86 50W
Birni Nkonni	**90**	13 55N	5 15E
Birnin Kebbi	**90**	12 32N	4 12E
Birnin Kudu	**90**	11 30N	9 29E
Birobidzhan	**73**	48 50N	132 50E
Birr	**51**	53 7N	7 55W
Birtley	**44**	54 53N	1 34W
Biscay, B. of	**56**	45 0N	2 0W
Bishop Auckland	**44**	54 40N	1 40W
Bishopbriggs	**47**	55 54N	4 14W
Bishop's Castle	**45**	52 29N	3 0W
Bishop's Stortford	**40**	51 52N	0 11E
Bishop's Waltham	**40**	50 57N	1 13W
Bishopton	**47**	55 54N	4 30W
Biskra	**88**	34 50N	5 44E
Bismarck	**100**	46 49N	100 49W
Bismarck Arch.	**79**	2 30S	150 0E
Bismarck Ra.	**79**	5 35S	145 0E
Bismarck Sea	**79**	4 10S	146 50E
Bissau	**88**	11 45N	15 45W
Bistriţa	**55**	47 9N	24 35E
Bitlis	**71**	38 20N	42 3E
Bitter Lakes	**91**	30 15N	32 40E
Bitterroot Range	**100**	46 0N	114 20W
Biu	**90**	10 40N	12 3E
Biwa-Ko	**76**	35 15N	135 45E
Biysk	**62**	52 40N	85 0E
Bjelovar	**58**	45 56N	16 49E
Blaby	**41**	52 34N	1 10W
Black Hills	**100**	44 0N	103 50W
Black Isle, dist.	**48**	57 35N	4 10W
Black Mts.	**42**	51 52N	3 5W
Black Range, Mts.	**100**	33 30N	107 55W
Black Sea	**71**	43 30N	35 0E
Blackall	**79**	24 25S	145 45E
Blackball	**83**	42 22S	171 26E
Blackburn, Lancs., U.K.	**45**	53 44N	2 30W
Blackburn, Lothian, U.K.	**47**	55 52N	3 38W
Blackdown Hills	**43**	50 57N	3 15W
Blackfoot	**100**	43 13N	112 12W
Blackford	**47**	56 15N	3 48W
Blackpool	**45**	53 48N	3 3W
Blackrock	**51**	53 18N	6 11W
Blackrod	**45**	53 35N	2 35W
Blacksod B.	**50**	54 6N	10 0W
Blackwater	**51**	52 26N	6 20W
Blackwater, R., Meath, Ireland	**50**	53 46N	7 0W
Blackwater, R., Munster, Ireland	**51**	51 55N	7 50W
Blackwater, R., Essex, U.K.	**40**	51 44N	0 53E
Blackwater, R., Ulster, U.K.	**50**	54 31N	6 35W
Blackwell	**100**	36 55N	97 20W
Blackwood	**43**	51 40N	3 13W
Blaenau Ffestiniog	**42**	53 0N	3 57W
Blaenavon	**43**	51 46N	3 5W
Blaengawr	**43**	51 37N	3 35W
Blagoveshchensk	**62**	55 1N	55 59E
Blaina	**43**	51 46N	3 10W
Blair Athol	**79**	22 42S	147 31E
Blair Atholl	**48**	56 46N	3 50W
Blairgowrie	**48**	56 36N	3 20W
Blakeney	**43**	51 45N	2 29W
Blanc, Mt.	**54**	45 48N	6 50E
Blanca Peak	**100**	37 35N	105 29W
Blanchland	**44**	54 50N	2 03W
Blanco, C.	**100**	42 50N	124 40W
Blandford Forum	**43**	50 25N	2 10W
Blankenberge	**52**	51 20N	3 9E
Blantyre	**92**	15 45S	35 0E
Blarney	**51**	51 57N	8 35W
Blaydon	**44**	54 56N	1 47W
Bleiburg	**61**	46 35N	14 49E
Blenheim	**83**	41 38S	174 5E
Blessington	**51**	53 10N	6 32W
Blewbury	**40**	51 33N	1 14W
Blida	**88**	36 30N	2 49E
Blitar	**72**	8 5S	112 11E
Bloemfontein	**94**	29 6S	26 14E
Bloemhof	**94**	27 38S	25 32E
Blois	**56**	47 35N	1 20E
Bloody Foreland	**50**	55 10N	8 18W
Bloubergstrand	**94**	33 48S	18 28E
Blue Mts., Austral.	**79**	33 40S	150 0E
Blue Mts., U.S.A.	**100**	45 15N	119 0W
Blue Nile, R.	**89**	12 30N	34 30E
Blue Ridge, Mts.	**101**	36 30N	80 15W
Blue Stack Mts.	**50**	54 46N	8 5W
Bluefield	**101**	37 18N	81 14W
Bluefields	**107**	12 0N	83 50W
Bluff	**83**	46 37S	168 20E
Blyth, Northumberland, U.K.	**44**	55 8N	1 32W
Blyth, Notts., U.K.	**43**	53 22N	1 2W
Bobo-Dioulasso	**88**	11 8N	4 13W
Bocholt	**52**	51 50N	6 35E
Bochum	**52**	51 28N	7 12E
Boddam	**48**	57 28N	1 46W
Boden	**61**	65 50N	21 42E
Bodmin	**43**	50 28N	4 44W
Bodmin Moor	**43**	50 33N	4 36W
Bodø	**61**	67 17N	14 24E
Boggeragh Mts.	**51**	52 2N	8 55W
Bognor Regis	**40**	50 47N	0 40W
Bogong, Mt.	**79**	36 47S	147 17E
Bogor	**72**	6 36S	106 48E
Bogotá	**107**	4 34N	74 0W
Bohemia	**54**	50 0N	14 0E
Bohemian Forest	**54**	49 20N	13 0E
Boholl, I.	**72**	9 50N	124 10E
Boise	**100**	43 43N	116 9W
Boju	**90**	7 22N	7 55E
Bokkos	**90**	9 17N	9 1E
Bokpyin	**70**	11 18N	98 42E
Bolbec	**56**	49 30N	0 30E
Boldon	**44**	54 57N	1 28W
Bolivia ■	**108**	17 6S	64 0W
Bolivian Plateau	**108**	20 0S	67 30W
Bollington	**45**	53 18N	2 05W
Bologna	**58**	44 30N	11 20E
Bologoye	**61**	57 55N	34 0E
Bolsena	**58**	42 40N	11 58E
Bolshevik I.	**62**	78 30N	102 0E
Bolsover	**45**	53 14N	1 18W
Bolton	**45**	53 35N	2 26W
Bolton-on-Dearne	**45**	53 31N	1 19W
Bolus Hd.	**51**	51 48N	10 20W
Bolzano	**58**	46 30N	11 20E
Boma	**92**	5 50S	13 4E
Bombala	**79**	36 56S	149 15E
Bombay	**70**	18 55N	72 50E
Bona Mt.	**98**	61 20N	140 0W
Bonaire, I.	**107**	12 10N	68 15W
Bonarbridge	**48**	57 53N	4 20W
Bonavista	**99**	48 40N	53 5W
Bo'ness	**47**	56 0N	3 38W
Bonhill	**47**	55 59N	4 34W
Bonifacio, Str. of	**58**	41 12N	9 15E
Bonin Is.	**23**	27 0N	142 0E
Bonn	**52**	50 43N	7 6E
Bonnie Rock	**78**	30 29S	118 22E
Bonnybridge	**47**	56 00N	3 53W
Bonnyrigg	**47**	55 52N	3 8W
Boom	**52**	51 6N	4 20E
Boothia, Gulf of	**98**	71 0N	91 0W
Boothia Pen.	**98**	71 0N	94 0W
Bootle	**45**	53 28N	3 1W
Borås	**61**	57 43N	12 56E
Bordeaux	**56**	44 50N	0 36W
Borders □	**46**	55 45N	2 50W
Borehamwood	**41**	51 40N	0 15W
Borger	**100**	35 40N	101 20W
Borisovka	**71**	43 15N	68 10E
Borneo, I.	**72**	1 0N	115 0E
Bornholm, I.	**61**	55 10N	15 0E
Borno □	**90**	12 30N	12 30E
Borough Green	**41**	51 17N	0 18E
Borovichi	**61**	58 25N	33 55E
Borrisokane	**51**	53 0N	8 8W
Borroloola	**79**	16 4S	136 17E
Borrowdale	**44**	54 31N	3 10W
Borzya	**73**	50 24N	116 31E
Bosa	**58**	40 17N	8 32E
Boshof	**94**	28 31S	25 13E
Bosna, R.	**58**	45 4N	18 29E
Bosnia	**58**	44 0N	18 0E
Boston, U.K.	**44**	52 59N	0 2W
Boston, U.S.A.	**104**	42 20N	71 0W
Bothnia, G. of	**61**	63 0N	21 0E
Bothwell	**47**	55 48N	4 04W
Botoşani	**55**	47 42N	26 41E
Botswana ■	**92**	22 0S	24 0E
Bottesford	**44**	52 57N	0 48W
Bottrop	**52**	51 34N	6 59E
Bouaké	**88**	7 40N	5 2W
Bougainville I.	**79**	6 0S	155 0E
Boulder, Austral.	**78**	30 46S	121 30E
Boulder, U.S.A.	**100**	40 3N	105 10W
Boulia	**79**	22 52S	139 51E
Boulogne-sur-Mer	**56**	50 42N	1 36E
Bountiful	**100**	40 57N	111 58W
Bourbonnais	**56**	46 28N	3 0E
Bourg-en-Bresse	**56**	46 13N	5 12E
Bourges	**56**	47 9N	2 25E
Bourke	**79**	30 8S	145 55E
Bourne	**45**	52 46N	0 22W
Bourton-on-the-Water	**42**	51 53N	1 45W
Boussu	**52**	50 26N	3 48E
Bovey Tracey	**43**	50 36N	3 40W
Bovingdon	**41**	51 43N	0 32W
Bowen	**79**	20 0S	148 16E
Bowes	**44**	54 31N	1 59W
Bowland, Forest of	**44**	54 0N	2 30W
Bowling Green	**101**	37 0N	86 25W
Bowmore	**46**	55 45N	6 18W
Bowness	**44**	54 57N	3 13W
Boyle	**50**	53 58N	8 19W
Boyne, R.	**50**	53 40N	6 34W
Boyoma Falls	**92**	0 12N	25 25E
Brabant □	**52**	50 46N	4 30E
Brač	**58**	43 20N	16 40E
Bracciano, L.	**58**	42 8N	12 11E
Bräcke	**61**	62 45N	15 26E
Brackley	**40**	52 3N	1 9W
Bracknell	**41**	51 24N	0 45W
Brad	**55**	46 10N	22 50E
Bradford	**45**	53 47N	1 45W
Bradford-on-Avon	**42**	51 20N	2 15W
Bradworthy	**43**	50 54N	4 22W
Braemar	**48**	57 2N	3 20W
Braemar, dist.	**48**	57 2N	3 20W
Braga	**60**	41 35N	8 25W
Brahmaputra, R.	**70**	26 30N	93 30E
Braich-y-pwll	**42**	52 47N	4 46W
Brăila	**55**	45 19N	27 59E
Brainerd	**101**	46 20N	94 10W
Braintree	**40**	51 53N	0 34E
Bramhall	**45**	53 22N	2 10W
Bramhope	**45**	53 53N	1 37W
Bramley	**45**	53 25N	1 16W
Brampton	**44**	54 56N	2 43W
Branco, C.	**108**	7 9S	34 47W
Brandenburg	**54**	52 24N	12 33E
Brandon, Can.	**98**	49 50N	99 57W
Brandon, Durham, U.K.	**44**	54 46N	1 37W
Brandon, Suffolk, U.K.	**40**	52 27N	0 37E
Brandon, Mt.	**51**	52 15N	10 15W
Brandvlei	**94**	30 25S	20 30E
Brantford	**104**	43 10N	80 15W
Brasilia	**108**	15 47S	47 55E
Braşov	**55**	45 38N	25 35E
Brass, R.	**90**	4 15N	6 13E
Brasschaat	**52**	51 19N	4 27E
Bratislava	**54**	48 10N	17 7E
Braunstone	**41**	52 36N	1 10W
Bray	**51**	53 12N	6 6W
Brazil ■	**108**	5 0N	20 0W
Brazilian Highlands	**108**	18 0S	46 30W
Brazzaville	**92**	4 9S	15 12E
Breadalbane	**48**	56 30N	4 15W
Bream	**43**	51 45N	2 34W
Brechin	**48**	56 44N	2 40W
Breckland	**40**	52 30N	0 40E
Brecon	**42**	51 57N	3 23W
Brecon Beacons	**42**	51 53N	3 27W
Breda	**52**	51 35N	4 45E
Bredasdorp	**94**	34 33S	20 2E
Bredbury	**45**	53 24N	2 08W
Bremen	**54**	53 4N	8 47E
Bremerhaven	**54**	53 34N	8 35E
Bremerton	**105**	47 30N	122 38W
Brendon Hills	**42**	51 6N	3 25W
Brenner Pass	**54**	47 0N	11 30E
Brent	**41**	51 33N	0 18W
Brentford	**41**	51 30N	0 19W
Breslau = Wrocław	**54**	51 5N	17 5E
Bressanone	**58**	46 43N	11 40E
Bressay	**49**	60 10N	1 6W
Brest, France	**56**	48 24N	4 31W
Brest, U.S.S.R.	**55**	52 10N	23 40E
Brett, C.	**83**	35 10S	174 20E
Brewood	**41**	52 41N	2 10W
Bricket Wood	**41**	51 42N	0 21W
Bridge of Allan	**47**	56 9N	3 57W
Bridge of Earn	**47**	56 20N	3 25W
Bridge of Orchy	**48**	56 29N	4 48W
Bridge of Weir	**47**	55 51N	4 35W
Bridgend	**43**	51 30N	3 35W
Bridgeport	**104**	41 12N	73 12W
Bridgetown, Austral.	**78**	33 58S	116 7E
Bridgetown, Barbados	**107**	13 0N	59 30W
Bridgetown, Can.	**99**	44 55N	65 18W
Bridgetown, Ireland	**51**	52 13N	6 33W
Bridgwater	**42**	51 7N	3 0W
Bridgwater B.	**42**	51 15N	3 15W
Bridlington	**44**	54 6N	0 11W
Bridport	**43**	50 43N	2 45W
Brierfield	**45**	53 49N	2 15W
Brierley Hill	**41**	52 29N	2 7W
Brig	**54**	46 18N	7 59E
Brigg	**44**	53 33N	0 30W
Brigham City	**100**	41 30N	112 1W
Brighouse	**45**	53 42N	1 47W
Brightlingsea	**40**	51 49N	1 1E
Brighton	**40**	50 50N	0 9W
Brindisi	**58**	40 39N	17 55E
Brisbane	**79**	27 25S	153 2E
Bristol	**43**	51 26N	2 35W
Bristol B.	**96**	58 0N	160 0W
Bristol Channel	**42**	51 18N	4 30W
British Columbia □	**98**	55 0N	125 15W
British Isles	**24**	55 0N	4 0W
Britstown	**94**	30 37S	23 30E
Brittany, reg.	**56**	48 0N	3 0W
Brittas	**51**	53 14N	6 29W
Brive-la-Gaillarde	**56**	45 10N	1 32E
Brixham	**43**	50 24N	3 31W
Brno	**54**	49 10N	16 35E
Broad Law, Mt.	**46**	55 30N	3 22W
Broad Sd.	**79**	22 0S	149 45E
Broadford	**48**	57 14N	5 55W
Broads, The	**40**	52 45N	1 30E
Broadsound Ra.	**79**	22 50S	149 30E
Broadstairs	**40**	51 21N	1 28E
Broadstone	**43**	50 45N	1 59W
Broadway	**42**	52 2N	1 51W
Brockenhurst	**40**	50 49N	1 34W
Brockton	**104**	42 8N	71 2W
Brod	**58**	41 35N	21 17E
Brodick	**46**	55 34N	5 9W
Brody	**55**	50 5N	25 10E
Broken Hill	**79**	31 58S	141 29E
Bromley	**41**	51 20N	0 5E
Bromsgrove	**42**	52 20N	2 3W
Bromyard	**42**	52 12N	2 30W
Brookings	**101**	44 20N	96 45W
Brookmans Park	**41**	51 43N	0 11W
Brooks Ra.	**98**	68 40N	147 0W
Brookton	**78**	32 22S	116 57E
Broom, L.	**48**	57 55N	5 15W
Broome	**78**	18 0S	122 15E
Brosna, R.	**51**	53 8N	8 0W
Brough, Cumbria, U.K.	**43**	54 32N	2 19W
Brough, Humberside, U.K.	**44**	53 44N	0 35W
Broughton	**45**	53 33N	0 36W
Brown, Mt.	**79**	32 30S	138 0E
Brown Willy	**43**	50 35N	4 34W
Brownhills	**41**	52 38N	1 57W
Brownsville	**101**	35 35N	89 15W
Brownwood	**100**	31 45N	99 0W
Broxbourne	**41**	51 44N	0 00
Broxburn	**47**	55 56N	3 23W
Bruay-en-Artois	**56**	50 29N	2 33E
Bruce, Mt.	**78**	22 37S	118 8E
Bruck a.d. Mur	**54**	47 24N	15 16E
Brue, R.	**42**	51 10N	2 59W
Bruges = Brugge	**52**	51 13N	3 13E
Brugge	**52**	51 13N	3 13E
Brühl	**52**	50 49N	6 51E
Brunei ■	**72**	4 50N	115 0E
Brunner	**83**	42 27S	171 20E
Brunswick, Ger.	**54**	52 17N	10 28E
Brunswick, U.S.A.	**101**	31 10N	81 30W

Brussels Chatham Is

Brussels 52 50 51N 4 21E
Bruton 42 51 6N 2 28W
Bruxelles = Brussels 52 50 51N 4 21E
Bryan 101 30 40N 96 27W
Brynamman 42 51 49N 3 52W
Brynmawr 43 51 48N 3 11W
Buabuq 91 31 29N 25 29E
Bucaramanga 107 7 0N 73 0W
Buchan 48 57 32N 2 8W
Buchanan 88 5 57N 10 2W
Bucharest 55 44 27N 26 10E
Buckfastleigh 43 50 28N 3 47W
Buckhaven 47 56 10N 3 02W
Buckie 48 57 40N 2 58W
Buckingham 40 52 0N 0 59W
Buckingham □ 40 51 50N 0 55W
București = Bucharest 55 44 27N 26 10E
Budapest 55 47 29N 19 5E
Bude 43 50 49N 4 33W
Bude Bay 43 50 50N 4 40W
Budleigh Salterton 43 50 37N 3 19W
Buea 90 4 10N 9 9E
Buenaventura 107 3 53N 77 4W
Buenos Aires 108 34 30S 58 20W
Buffalo, N.Y., U.S.A. 104 42 55N 78 50W
Buffalo, Wyo., U.S.A. 100 44 25N 106 50W
Bug, R., Poland 55 52 31N 21 5E
Bug, R., U.S.S.R. 55 48 0N 31 0E
Buga 107 4 0N 77 0W
Buguma 90 4 42N 6 55E
Bugun Shara 73 49 0N 104 0E
Builth Wells 42 52 10N 3 26W
Bujumbura 93 3 16S 29 18E
Bukama 92 9 10S 25 50E
Bukavu 92 2 20S 28 52E
Bukhara 71 39 48N 64 25E
Bukoba 93 1 20S 31 49E
Bukombe 93 3 31S 32 4E
Bukuru 90 9 42N 8 48E
Bulawayo 92 20 7S 28 32E
Bulgaria ■ 55 42 35N 25 30E
Bulkington 41 52 29N 1 25W
Bullfinch 78 30 58S 119 3E
Bulls 83 40 10S 175 24E
Bully-les-Mines 56 50 27N 2 44E
Bultfontein 94 28 18S 26 10E
Buna 79 8 42S 148 27E
Bunbeg 50 55 4N 8 18W
Bunbury 78 33 20S 115 35E
Bunclody 51 52 40N 6 40W
Buncrana 50 55 8N 7 28W
Bundaberg 79 24 54S 152 22E
Bundoran 50 54 24N 8 17W
Bungay 40 52 27N 1 26E
Bungo Channel 76 33 0N 132 15E
Bunia 93 1 35N 30 20E
Bununu Doss 90 10 6N 9 25E
Bunza 90 12 8N 4 0E
Bûr Fuad 91 31 15N 32 20E
Bûr Said = Port Said 91 31 16N 32 18E
Bura 93 1 4S 39 58E
Burbage 41 52 31N 1 20W
Burbank 100 34 9N 118 23W
Bure, R. 40 52 38N 1 45E
Bureya 73 50 35N 132 0E
Bureya, R. 73 51 30N 133 0E
Burford 40 51 48N 1 38W
Burg el Arab 91 30 54N 29 32E
Burgenland □ 54 47 20N 16 20E
Burgersdorp 94 31 0S 26 20E
Burgess Hill 40 50 57N 0 7W
Burghead 48 57 42N 3 30W
Burgos 60 42 21N 3 41W
Burgsteinfurt 52 52 9N 7 23E
Burgundy, reg. 56 47 0N 4 30E
Burketown 79 17 45S 139 33E
Burkina Faso ■ 88 12 30N 2 0W
Burlington 104 42 41N 88 18W
Burlyu-Tyube 62 46 30N 79 10E
Burma ■ 70 21 0N 96 30E
Burnham, Essex, U.K. 41 51 37N 0 50E
Burnham, Somerset, U.K. 42 51 14N 3 0W
Burnham Market 40 52 57N 0 43E
Burnie 79 41 4S 145 56E
Burnley 45 53 47N 2 15W
Burns Lake 105 54 20N 125 45W
Burnsall 44 54 03N 1 57W
Burntisland 46 56 4N 3 14W
Burra 79 33 40S 138 55E
Bursa 71 40 15N 29 5E
Burscough 45 53 36N 2 52W
Burton 45 53 15N 3 02W
Burton Latimer 40 52 23N 0 41W
Burton-upon-Trent 45 52 48N 1 39W
Buru, I. 72 3 30S 126 30E
Burundi ■ 93 3 15S 30 0E
Burutu 90 5 20N 5 29E
Bury 45 53 36N 2 19W
Bury St. Edmunds 40 52 15N 0 42E
Bushehr 71 28 55N 50 55E
Bushey 41 51 38N 0 20W
Bushmills 50 55 14N 6 32W
Busselton 78 33 42S 115 15E
Bussum 52 52 16N 5 10E
Buta 92 2 50N 24 53E
Bute, I. 46 55 48N 5 2W
Butte 100 46 0N 112 31W
Butterworth 72 5 24N 100 23E
Buttevant 51 52 14N 8 40E
Butung, I. 72 5 0S 122 45E
Buxton 44 53 16N 1 54W
Buzău, Pasul 55 45 35N 26 12E
Bydgoszcz 55 53 10N 18 0E
Byfleet 41 51 20N 0 28W
Bylot I. 99 73 13N 78 34W
Byrd Ld. 112 79 30S 125 0W
Byron, C. 79 28 38S 153 40E
Bytom 55 50 25N 19 0E

C

Cabinda 92 5 40S 12 11E
Cabonga Réservoir 99 47 20N 76 40W
Cabot Strait 99 47 15N 59 40W
Cabrera, I. 60 39 6N 2 59E
Cacak 55 43 54N 20 20E
Cáceres 60 39 26N 6 23W
Cader Idris 42 52 43N 3 56W
Cadillac 104 44 16N 85 25W
Cadishead 45 53 25N 2 25W
Cádiz 60 36 30N 6 20W
Cadiz, G. of 60 36 40N 7 0W
Caen 56 49 10N 0 22W
Caerleon 43 51 37N 2 57W
Caernarfon 42 53 8N 4 17W
Caernarfon B. 42 53 4N 4 40W
Caersws 42 52 32N 3 27W
Cagayan, I. 72 9 37N 121 12E
Cágliari 58 39 15N 9 6E
Cagliari, G. of 58 39 8N 9 10E
Caguas 107 18 14N 66 4W
Caha Mts. 51 51 45N 9 40W
Caher 51 52 23N 7 56W
Caher I. 51 53 44N 10 1W
Cahersiveen 51 51 57N 10 13W
Cahore Pt. 51 52 34N 6 11W
Cahors 56 44 27N 1 27E
Caicos Is. 107 21 40N 71 40W
Cairngorm Mts. 48 57 6N 3 42W
Cairnryan 46 54 59N 5 0W
Cairns 79 16 57S 145 45E
Cairo 91 30 1N 31 14E
Caister-on-Sea 40 52 38N 1 43E
Caistor 44 53 29N 0 20W
Caithness 48 58 25N 3 25W
Calabar 90 4 57N 8 20E
Calabozo 107 9 0N 67 20W
Calábria □ 58 39 24N 16 30E
Calais 56 50 57N 1 56E
Calamian Group 72 11 50N 119 55E
Calamocha 60 40 50N 1 17W
Calapan 72 13 25N 121 7E
Calatayud 60 41 20N 1 40W
Calcutta 70 22 36N 88 24E
Calder, R. 45 53 44N 1 21W
Calderbrook 45 53 39N 2 08W
Caldercruix 47 55 53N 3 53W
Caldicot 43 51 36N 2 45W
Caldwell 100 43 45N 116 42W
Caldy 45 53 21N 3 10W
Caldy I. 42 51 38N 4 42W
Caledon 94 34 14S 19 26E
Caledon, R. 94 30 0S 26 46E
Calgary 98 51 0N 114 10W
Cali 107 3 25N 76 35W
California □ 100 37 25N 120 0W
California, G. of 106 27 0N 111 0W
Callan 51 52 33N 7 25W
Callander 46 56 15N 4 14W
Callao 108 12 0S 77 0W
Callington 43 50 30N 4 18W
Calne 42 51 26N 2 0W
Caltagirone 58 37 13N 14 30E
Caltanissetta 58 37 30N 14 3E
Calvinia 94 31 28S 19 45E
Cam, R. 40 52 21N 0 16E
Camagüey 107 21 20N 78 0W
Camargue 56 43 34N 4 34E
Cambay, G. of 70 20 45N 72 30E
Camberley 41 51 20N 0 44W
Camberwell 41 51 28N 0 04W
Cambodia ■ 72 12 15N 105 0E
Camborne 43 50 13N 5 18W
Cambrai 56 50 11N 3 14E
Cambrian Mts. 42 52 25N 3 52W
Cambridge, N.Z. 83 37 54S 175 29E
Cambridge, U.K. 40 52 13N 0 8E
Cambridge, U.S.A. 104 42 20N 71 8W
Cambridge Gulf 78 14 45S 128 0E
Cambridgeshire □ 40 52 12N 0 7E
Cambusbarron 47 56 06N 3 58W
Cambuslang 47 55 49N 4 11W
Camden, U.K. 41 51 33N 0 10W
Camden, U.S.A. 104 39 57N 75 1W
Camelford 43 50 37N 4 41W
Cameroon ■ 90 3 30N 12 30E
Caminha 60 41 50N 8 50W
Camooweal 79 19 56S 138 7E
Campania □ 58 40 50N 14 45E
Campbell I. 23 52 30S 169 0E
Campbellton 99 47 57N 66 43W
Campbeltown 46 55 25N 5 36W
Campeche 106 19 50N 90 32W
Campeche, G. of 106 19 30N 93 0W
Campinas 108 22 50S 47 0W
Campo Grande 108 20 25S 54 40W
Campobasso 58 41 34N 14 40E
Campsie Fells 47 56 2N 4 20W
Camrose 98 53 0N 112 50W
Canada ■ 96 60 0N 100 0W
Canadian, R. 101 35 27N 95 3W
Canary Is. 88 29 30N 17 0W
Canaveral, C. 101 28 28N 80 31W
Canberra 79 35 15S 149 8E
Canfranc 60 42 42N 0 31W
Canik Mts. 71 40 30N 38 0E
Canna I. 48 57 3N 6 33W
Cannes 56 43 32N 7 0E
Canning Basin 78 19 50S 124 0E
Cannock 45 52 42N 2 2W
Cannock Chase, hills 45 52 43N 2 0W
Canon City 100 39 30N 105 20W
Canora 98 51 40N 102 30W
Canso 99 45 20N 61 0W
Cantabrian Mts. 60 43 0N 5 10W
Canterbury 40 51 17N 1 5E
Canterbury □ 83 43 45S 171 19E
Canterbury Bight 83 44 16S 171 55E
Canterbury Plains 83 43 55S 171 22E
Canton 104 40 32N 90 0W
* Canton I. 22 2 30S 172 0W

Renamed Abariringa

Canvey 41 51 32N 0 35E
Cap Haïtien 107 19 40N 72 20W
Cape Breton I. 99 46 0N 60 30W
Cape Coast 88 5 5N 1 15W
Cape Dorset 99 64 14N 76 32W
Cape Province □ 94 32 0S 23 0E
Cape Town 94 33 55S 18 22E
Cape Verde Is. 22 17 10N 25 20W
Cape York Peninsula 79 33 34S 115 33E
Cappoquin 51 52 9N 7 46W
Capri, I. 58 40 34N 14 15E
Caracal 55 44 8N 24 22E
Caracas 107 10 30N 66 55W
Caransebeş 55 45 28N 22 18E
Carbonara, C. 58 39 8N 9 30E
Carcassonne 56 43 13N 2 20E
Cardamom Hills 70 9 30N 77 15E
Cárdenas 107 23 0N 81 30W
Cardenden 47 56 08N 3 15W
Cardiff 43 51 28N 3 11W
Cardigan 42 52 6N 4 41W
Cardigan B. 42 52 30N 4 30W
Cardona 60 41 56N 1 40E
Cardross 47 55 58N 4 38W
Carei 55 47 40N 22 29E
Cargelligo, L. 79 33 17S 146 24F
Caribbean Sea 107 15 0N 75 0W
Cariboo Mts. 105 53 0N 121 0W
Caribou 101 46 55N 68 0W
Caribou Mts. 98 59 12N 115 40W
Carinthia □ 54 46 52N 13 30E
Caripito 107 10 8N 63 6W
Carletonville 94 26 23S 27 22E
Carlingford 50 54 3N 6 10W
Carlingford, L. 50 54 0N 6 5W
Carlisle 44 54 54N 2 55W
Carlow 51 52 50N 6 58W
Carlow □ 51 52 43N 6 50W
Carloway 48 58 17N 6 48W
Carlsbad 100 32 20N 104 7W
Carlton 45 53 35N 1 27W
Carluke 47 55 44N 3 50W
Carmarthen 42 51 52N 4 20W
Carmarthen B. 42 51 40N 4 30W
Carmel Hd. 42 53 24N 4 34W
Carmel Mt. 89 32 45N 35 3E
Carmona 60 37 28N 5 42W
Carmunnock 47 55 47N 4 16W
Carmyle 47 55 50N 4 11W
Carn Eige 48 57 17N 5 9W
Carnarvon, Austral. 78 24 51S 113 42E
Carnarvon, S. Afr. 94 30 56S 22 8E
Carndonagh 50 55 15N 7 16W
Carnegie, L. 78 26 5S 122 30E
Carnforth 44 54 8N 2 47W
Carnic Alps 54 46 34N 12 50E
Carnoustie 48 56 30N 2 41W
Carnsore Pt. 51 52 10N 6 20W
Carnwath 47 55 42N 3 38W
Caroline Is. 23 8 0N 150 0E
Carpathians, Mts. 55 46 20N 26 0E
Carpentaria, G. of 79 14 0S 139 0E
Carra, L. 51 53 41N 9 12W
Carrara 58 44 5N 10 7E
Carrbridge 48 57 17N 3 50W
Carrick 46 55 12N 4 38W
Carrick-on-Shannon 50 53 57N 8 7W
Carrick-on-Suir 51 52 22N 7 30W
Carrickfergus 50 54 43N 5 50W
Carrickfergus □ 50 54 43N 5 49W
Carrickmacross 50 54 0N 6 43W
Carron, R. 47 56 02N 3 44W
Carshalton 41 51 21N 0 09W
Carson City 100 39 12N 119 46W
Carson Sink 105 39 50N 118 40W
Carstairs 47 55 42N 3 41W
Cartagena, Colomb. 107 10 25N 75 33W
Cartagena, Spain 60 37 38N 0 59W
Cartago 107 4 45N 75 55W
Carterton 83 41 2S 175 31E
Carthage 101 37 10N 94 20W
Carúpano 107 10 45N 63 15W
Casablanca 88 33 36N 7 36W
Casale Monferrato 58 45 8N 8 28E
Cascade Ra. 100 45 0N 121 30W
Cascais 60 38 41N 9 25W
Caserta 58 41 5N 14 20E
Cashel 51 52 31N 7 53W
Casino 79 28 52S 153 3E
Casiquiare, R. 107 2 45N 66 20W
Caspe 60 41 14N 0 1W
Casper 100 42 52N 106 27W
Caspian Sea 62 43 0N 50 0E
Cassiar Mts. 98 59 30N 130 30W
Castellammare, G. 58 38 5N 12 55E
Castellón □ 60 40 15N 0 5W
Castelnaudary 56 43 20N 1 58E
Castelo Branco 60 39 50N 7 31W
Castelvetrano 58 37 40N 12 46E
Castle Bromwich 41 52 30N 1 47W
Castle Cary 42 51 5N 2 32W
Castle Douglas 46 54 57N 3 57W
Castlebar 50 53 52N 9 17W
Castlebay 49 56 57N 7 30W
Castleblaney 50 54 7N 6 44W
Castlecomer 51 52 49N 7 13W
Castlederg 50 54 43N 7 35W
Castleford 45 53 43N 1 21W
Castleisland 51 52 14N 9 28W
Castlemaine 79 37 2S 144 12E
Castlepollard 51 53 40N 7 20W
Castlerea 50 53 46N 8 29W
Castlereagh 50 53 47N 8 30W
Castlereagh □ 50 54 33N 5 33W
Castlerock 50 55 09N 6 47W
Castleton, Gwent, U.K. 42 51 33N 3 05W
Castleton, Manch., U.K. 45 53 35N 2 11W
Castletown 48 58 35N 3 22W
Castletown Bearhaven 51 51 40N 9 54W
Castres 56 43 37N 2 13E
Catalonia □ 60 41 40N 1 15E
Catanduanas, Is. 72 13 50N 124 20E
Catánia 58 37 31N 15 4E

Caterham 41 51 16N 0 4W
Cathcart, S. Afr. 94 32 18S 27 10E
Cathcart, U.K. 47 55 48N 4 15W
Catoche, C. 106 21 40N 87 0W
Catorce 106 23 50N 100 55W
Catskill Mts. 104 42 15N 74 15W
Catterick 44 54 23N 1 38W
Caucasus Ra. 71 43 0N 44 0E
Caura, R. 107 6 20N 64 30W
Cavan 50 54 0N 7 22W
Cavan □ 50 53 58N 7 10W
Cayenne 108 5 0N 52 18W
Cayes, Les 107 18 15N 73 46W
Ceanannus Mor 50 53 42N 6 53W
Ceará = Fortaleza 108 3 35S 38 35W
Cebú 72 10 18N 123 54E
Cedar City 100 37 41N 113 3W
Cedar Falls 101 42 39N 92 29W
Cedar L. 98 53 20N 100 10W
Cedar Rapids 101 42 0N 91 38W
Ceduna 78 32 7S 133 46E
Cefalù 58 38 3N 14 1E
Cegléd 55 47 11N 19 47E
Ceiba, La 106 15 40N 86 50W
Celaya 106 20 31N 100 37W
Celbridge 51 53 20N 6 33W
Celebes Sea 72 3 0N 123 0E
Celje 58 46 16N 15 18E
Celle 54 52 37N 10 4E
Cemmaes Road 42 52 39N 3 41W
Central □, Scotland 46 56 0N 4 30W
Central African Republic ■ 89 7 0N 20 0E
Central America 96 10 0N 83 0W
Central Auckland □ 83 37 30S 175 30E
Central Makran Range 71 26 30N 64 15E
Central Russian Uplands 24 54 0N 36 0E
Central Siberian Plateau 62 65 0N 105 0E
Centralia 100 46 46N 122 59W
Ceram Sea 72 2 30S 128 30E
Ceres, S. Afr. 94 33 21S 19 18E
Ceres, U.K. 47 56 18N 2 57W
Cerignola 58 41 17N 15 53E
Cerne Abbas 43 50 49N 2 29W
Cervera 60 41 40N 1 16E
Cesena 58 44 9N 12 14E
České Budějovice 54 48 55N 14 25E
Český Těšín 55 49 45N 18 39E
Cessnock 79 32 50S 151 21E
Ceuta 88 35 52N 5 18W
Cévennes, mts. 56 44 10N 3 50E
Chad ■ 88 12 30N 17 15E
Chad, L. 88 13 30N 14 30E
Chadderton 45 53 27N 2 08W
Chadron 100 42 50N 103 0W
Chadwell St. Mary 41 51 28N 0 22E
Chafe 90 11 59N 6 50E
Chagos Arch. 23 6 0S 72 0E
Chake Chake 93 5 15S 39 45E
Chalatun 73 48 0N 122 40E
Chalfont St. Peter 41 51 36N 0 33W
Chalfont St. Giles 41 51 37N 0 34W
Chalon-sur-Saône 56 46 48N 4 50E
Châlons-sur-Marne 56 48 58N 4 20E
Chama, R. 100 36 57N 106 37W
Chambéry 56 45 34N 5 55E
Champagne 56 49 0N 4 40E
Chandernagore 70 22 52N 88 24E
Changchih 73 36 11N 113 6E
Changchow, Fukien, China 73 24 32N 117 44E
Changchow, Shantung, China 73 36 55N 118 3E
Changchun 73 43 58N 125 19E
Changhua 73 30 10N 119 15E
Changkiakow 73 40 52N 114 45E
Changkiang (Shihlu) 73 19 25N 108 57E
Changpai Shan 73 42 25N 129 0E
Changsha 73 28 5N 113 1E
Changteh 73 29 12N 111 43E
Channel Is. 43 49 30N 2 40W
Chanthaburi 72 12 38N 102 12E
Chantilly 56 49 12N 2 29E
Chaochow 73 23 45N 116 32E
Chaotung 73 27 19N 103 42E
Chapel-en-le-Frith 45 53 19N 1 54W
Chapelhall 47 55 50N 3 56W
Chapeltown 45 53 27N 1 28W
Charambirá, Punta 107 4 16N 77 32W
Charcas 106 23 10N 101 20W
Charchan 73 38 4N 85 16E
Charchan, R. 73 39 0N 86 0E
Charcot I. 112 70 0S 75 0W
Chard 43 50 52N 2 59W
Chardzhou 71 39 6N 63 34E
Chârib, G. 91 28 6N 32 54E
Charing 40 51 12N 0 49E
Charleroi 52 50 24N 4 27E
Charles, C. 101 37 10N 75 52W
Charleston 101 32 47N 79 56W
Charlestown 50 53 58N 8 48W
Charlestown of Aberlour 48 57 27N 3 13W
Charleville, Austral. 79 26 24S 146 15E
Charleville, Ireland 51 52 21N 8 40W
Charleville-Mézières 56 49 44N 4 40E
Charlotte 101 35 16N 80 46W
Charlotte Waters 78 25 56S 134 54E
Charlottenburg 54 52 31N 13 15E
Charlottesville 101 38 1N 78 30W
Charlottetown 99 46 14N 63 8W
Charnwood Forest 45 52 43N 1 18W
Charters Towers 79 20 5S 146 13E
Chartres 56 48 29N 1 30E
Chasetown 41 52 40N 1 55W
Châteaubriant 56 47 43N 1 23W
Châteauroux 56 46 50N 1 40E
Châtellerault 56 46 50N 0 30E
Chatham, N.B., Can. 99 47 2N 65 28W
Chatham, Ont., Can. 104 42 24N 82 11W
Chatham, U.K. 41 51 22N 0 32E
Chatham Is. 22 44 0S 176 40W

Chattanooga **Cullen**

Place	Page	Lat.	Long.
Chattanooga	101	35 2N	85 17W
Chatteris	40	52 27N	0 3E
Chaumont	56	48 7N	5 8E
Chaux de Fonds, La	54	47 7N	6 50E
Cheadle	45	53 23N	2 14W
Cheadle Hulme	45	53 22N	2 12W
Cheb	54	50 9N	12 20E
Cheboygan	101	45 38N	84 29W
Cheddar	42	51 16N	2 47W
Chehalis	100	46 44N	122 59W
Cheju	73	33 28N	126 30E
Cheju Do	73	33 29N	126 34E
Chekiang □	73	29 30N	120 0E
Chełm	55	51 8N	23 30E
Chelmer, R.	41	51 45N	0 42E
Chelmsford	41	51 44N	0 29E
Chełmża	55	53 10N	18 39E
Cheltenham	42	51 55N	2 5W
Chelyabinsk	62	55 10N	61 24E
Chengchow	73	34 47N	113 46E
Chengteh	73	41 0N	117 55E
Chengtu	73	30 45N	104 0E
Chenyüan	73	27 0N	108 20E
Chepstow	43	51 38N	2 40W
Chér, R.	56	47 10N	2 10E
Cherbourg	56	49 39N	1 40W
Cheremkhovo	62	53 32N	102 40E
Cherepovets	61	59 5N	37 55E
Chernovtsy	55	48 0N	26 0E
Cherskiy Ra.	62	65 0N	143 0E
Chertsey	41	51 23N	0 30W
Chesapeake Bay	101	38 0N	76 12W
Chesham	41	51 42N	0 36W
Cheshire □	44	53 14N	2 30W
Cheshunt	41	51 42N	0 1W
Chesil Beach	43	50 37N	2 33W
Cheslyn Hay	41	52 41N	2 01W
Chester, U.K.	44	53 12N	2 53W
Chester, U.S.A.	101	39 54N	75 20W
Chester-le-Street	44	54 53N	1 34W
Chesterfield	44	53 14N	1 26W
Chesterfield Inlet	98	63 30N	90 45W
Chesterfield, Is.	79	19 52S	158 15E
Cheviot Hills	46	55 20N	2 30W
Chew Bahir	89	4 40N	30 50E
Cheyenne	100	41 9N	104 49W
Chiai	73	23 29N	120 25E
Chiang Mai	72	18 47N	98 59E
Chiapa, R.	106	16 42N	93 0W
Chiávari	58	44 20N	9 20E
Chiba	76	35 30N	140 7E
Chibougamau	99	49 56N	74 24W
Chibuk	90	10 52N	12 50E
Chicago	104	41 53N	87 40W
Chichagof I.	98	58 0N	136 0W
Chichester	40	50 50N	0 47W
Chichibu	76	36 5N	139 10E
Chickasha	100	35 0N	98 0W
Chiclayo	108	6 42S	79 50W
Chico	100	39 45N	121 54W
Chicoutimi	99	48 28N	71 5W
Chidley C.	99	60 23N	64 26W
Chigwell	41	51 37N	0 05E
Chihuahua	106	28 40N	106 3W
Chilapa	106	17 40N	99 20W
Childress	100	34 30N	100 15W
Chile ■	108	35 0S	71 15W
Chillagoe	79	17 14S	144 33E
Chillicothe	101	40 55N	89 32W
Chiloé I.	108	42 50S	73 45W
Chilpancingo	106	17 30N	99 40W
Chiltern Hills	40	51 44N	0 42W
Chilumba	93	10 28S	34 12E
Chilung	73	25 3N	121 45E
Chimborazo	108	1 20S	78 55W
Chimkent	71	42 18N	69 36E
China ■	73	30 0N	110 0E
Chinchow	73	41 10N	121 2E
Chinde	92	18 45S	36 30E
Chindwin, R.	70	21 26N	95 15E
Chinju	73	35 12N	128 2E
Chinkiang	73	32 2N	119 29E
Chins Division □	70	22 30N	93 30E
Chinteche	93	11 50S	34 5E
Chinwangtao	73	40 0N	119 31E
Chióggia	58	45 13N	12 15E
Chipata	92	13 38S	32 28E
Chippenham	42	51 27N	2 7W
Chipping Campden	42	52 4N	1 48W
Chipping Norton	40	51 56N	1 32W
Chipping Ongar	41	51 42N	0 11E
Chipping Sodbury	43	51 31N	2 23W
Chita	62	52 0N	113 25E
Chittagong	70	22 19N	91 55E
Chiusi	58	43 1N	11 58E
Chivasso	58	45 10N	7 52E
Chobham	41	51 20N	0 36W
Choctawhatchee B.	101	30 13N	86 30W
Choiseul I.	79	7 0S	156 40E
Chojnice	55	53 42N	17 40E
Cholet	56	47 4N	0 52W
Chomutov	54	50 28N	13 23E
Chŏnju	73	35 50N	127 4E
Chonos Arch.	108	45 0S	75 0W
Chorley	45	53 39N	2 39W
Chorleywood	41	51 39N	0 29W
Chortkov	55	49 2N	25 46E
Chorzów	55	50 18N	19 0E
Choshi	76	35 45N	140 45E
Choszczno	54	53 7N	15 25E
Chott Djerid, L.	88	33 30N	8 30E
Choybalsan	73	48 4N	114 30E
Christchurch	83	43 33S	172 47E
Christiana	94	27 52S	25 8E
Christmas I., Ind. Oc.	72	10 0S	105 40E
* Christmas I., Pac. Oc.	22	1 58N	157 27W
Chu	73	43 36N	73 42E
Chuanchow	73	24 57N	118 31E
Chubut, R.	108	43 0S	70 0W
Chuchou	73	27 50N	113 10E
Chudskoye, L.	61	58 13N	27 30E
Chühsien	73	35 35N	118 49E
Chuka	93	0 23S	37 38E
Chukot Ra.	62	68 0N	175 0E
Chulmleigh	43	50 55N	3 52W
Chumatien	73	33 0N	114 4E
Chumphon	72	10 35N	99 14E
Chunchŏn	73	37 58N	127 44E
Chungking	73	29 30N	106 30E
Chunya	93	8 30S	33 27E
Chur	54	46 52N	9 32E
Church Stretton	45	52 32N	2 49W
Churchill Falls	99	53 36N	64 19W
Churchill Pk.	98	58 10N	125 10W
Churchill, R.	98	53 19N	60 10W
Cienfuegos	107	22 10N	80 30W
Cieza	60	38 17N	1 23W
Cimone, Mte.	58	44 10N	10 40E
Cîmpina	55	45 10N	25 45E
Cincinnati	104	39 10N	84 26W
Cinderford	43	51 49N	2 30W
Cinto, Mt.	56	42 24N	8 54E
Cirebon	72	6 45S	108 32E
Cirencester	42	51 43N	1 59W
Ciudad del Carmen	106	18 20N	97 50W
Ciudad Guayana	107	8 22N	62 40W
Ciudad Juárez	106	31 40N	106 28W
Ciudad Madero	106	22 19N	97 50W
Ciudad Mante	106	22 50N	99 0W
Ciudad Obregón	106	27 28N	109 59W
Ciudad Real	60	38 59N	3 55W
Ciudad Rodrigo	60	40 35N	6 32W
Ciudad Victoria	106	23 41N	99 9W
Civitanova Marche	58	43 18N	13 41E
Civitavécchia	58	42 6N	11 46E
Clackmannan	47	56 10N	3 50W
Clacton-on-Sea	40	51 47N	1 10E
Clanwilliam	94	32 11S	18 52E
Claonaig	46	55 45N	5 24W
Clare □	51	52 20N	7 38W
Clare I.	50	53 48N	10 0W
Clare, R.	51	53 20N	9 0W
Claremorris	50	53 45N	9 0W
Clarence	83	42 9S	173 56E
Clarksburg	104	39 18N	80 21W
Clarkston	47	55 47N	4 17W
Clarksville	101	36 32N	87 20W
Claunie, L.	48	57 8N	5 6W
Clay Cross	44	53 11N	1 26W
Clayton-le-Moors	45	53 46N	2 23W
Clear, C.	51	51 26N	9 30W
Clear Lake Res.	105	41 55N	121 10W
Clearwater	101	27 58N	82 45W
Cleator Moor	44	54 30N	3 32W
Cleckheaton	45	53 43N	1 43W
Clee Hills	45	52 26N	2 35W
Cleethorpes	44	53 33N	0 2W
Cleeve Cloud	42	51 56N	2 0W
Clermont	79	22 49S	147 39E
Clermont-Ferrand	56	45 46N	3 4E
Clevedon	43	51 26N	2 52W
Cleveland, U.K.	44	54 29N	1 0W
Cleveland, U.S.A.	104	41 28N	81 43W
Cleveland Hills	44	54 25N	1 11W
Cleveleys	45	53 52N	3 03W
Clew Bay	50	53 54N	9 50W
Clifden, Ireland	51	53 30N	10 2W
Clifden, N.Z.	83	46 1S	167 42E
Clifton	45	51 28N	2 38W
Clinch Mts.	101	36 30N	83 0W
Clinton, N.Z.	83	46 12S	169 23E
Clinton, Iowa, U.S.A.	101	41 50N	90 12W
Clinton, Okla., U.S.A.	100	35 30N	99 0W
Clisham, Mt.	48	57 57N	6 49W
Clitheroe	45	53 52N	2 23W
Cloghan	51	53 13N	7 53W
Clogher Hd.	50	53 48N	6 15W
Cloghran	51	53 26N	6 14W
Clonakilty	51	51 37N	8 53W
Cloncurry	79	20 40S	140 28E
Clondalkin	51	53 20N	6 25W
Clonee	51	53 25N	6 28W
Clones	50	54 10N	7 13W
Clonmel	51	52 22N	7 42W
Clonroche	51	52 27N	6 42W
Clontarf	51	53 22N	6 10W
Cloppenburg	52	52 50N	8 3E
Clovelly	43	51 0N	4 25W
Clovis	100	36 54N	119 45W
Cloyne	51	51 52N	8 7W
Cluj	55	46 47N	23 38E
Clutha, R.	83	46 20S	169 49E
Clwyd □	42	53 5N	3 20W
Clwyd, R.	42	53 12N	3 30W
Clwydian Range	42	53 10N	3 15W
Clydach	43	51 42N	3 54W
Clyde	83	45 12S	169 20E
Clyde, Firth of	46	55 20N	5 0W
Clyde, R.	46	55 46N	4 58W
Clydebank	47	55 54N	4 25W
Coalisland	50	54 33N	6 42W
Coalville	45	52 43N	1 21W
Coast Mts.	98	52 0N	126 0W
Coast Range	100	40 0N	124 0W
Coastal Plains Basin	78	30 10S	115 30E
Coatbridge	47	55 52N	4 2W
Coats I.	99	62 30N	83 0W
Coats Land	112	77 0S	25 0W
Coatzacoalcos	106	18 7N	94 35W
Cobalt	99	47 25N	79 42W
Cobar	79	31 27S	145 48E
Cobh	51	51 50N	8 18W
Cobham	41	51 19N	0 25W
Coburg	54	50 15N	10 58E
Cochabamba	108	17 15S	66 20W
Cochin	70	9 55N	76 22E
Cochrane	99	49 0N	81 0W
Cockenzie	47	55 58N	2 59W
Cockermouth	44	54 40N	3 22W
Cocos, Is.	72	10 12S	96 50E
Cocos (Keeling) Is.	72	12 12S	96 54E
Codsall	41	52 38N	2 12W
Coen	79	13 52S	143 12E
Coeur d'Alene	100	47 45N	116 51W
Coffs Harbour	79	30 16S	153 5E
Coggeshall	40	51 53N	0 41E
Coiba I.	107	7 30N	81 40W
Coimbatore	70	11 2N	76 59E
Coimbra	60	40 15N	8 27W
Colac	79	38 21S	143 35E
Colchester	40	51 54N	0 55E
Coldstream	46	55 39N	2 14W
Coleford	42	51 46N	2 38W
Coleraine	50	55 8N	6 40E
Coleraine □	50	55 8N	6 40E
Colesberg	94	30 45S	25 5E
Coleshill	41	52 30N	1 42W
Colima	106	19 10N	103 40W
Colinton	47	55 54N	3 15W
Colintraive	46	55 55N	5 09W
Coll, I.	48	56 40N	6 35W
Collie	78	33 22S	116 8E
Collier B.	78	16 10S	124 15E
Collingwood	83	40 25S	172 40E
Collinsville	79	20 30S	147 56E
Collon	50	53 46N	6 29W
Collooney	50	54 11N	8 28W
Colmar	56	48 5N	7 20E
Colne	45	53 51N	2 11W
Colne, R.	41	51 36N	0 30W
Cologne	52	50 56N	9 58E
Colombia ■	107	3 45N	73 0W
Colombo	70	6 56N	79 58E
Colón	107	9 20N	80 0W
Colonsay, I.	46	56 4N	6 12W
Colorado □	100	37 40N	106 0W
Colorado Desert	100	34 20N	116 0W
Colorado Plateau	100	36 40N	110 30W
Colorado, R.	108	37 30S	69 0W
Colorado Springs	100	38 55N	104 50W
Coltishall	40	52 44N	1 21E
Columbia, Mo., U.S.A.	101	38 58N	92 20W
Columbia, S.C., U.S.A.	101	34 0N	81 0W
Columbia, Tenn., U.S.A.	101	35 40N	87 0W
Columbia Plateau	100	47 30N	118 30W
Columbia, R.	100	46 15N	124 5W
Columbretes, Is.	60	39 50N	0 50E
Columbus, Ga., U.S.A.	101	32 30N	84 58W
Columbus, Miss., U.S.A.	101	33 30N	88 26W
Colwyn Bay	42	53 17N	3 44W
Comácchio	58	44 41N	12 10E
Comayagua	106	14 25N	87 37W
Combe Martin	42	51 12N	4 2W
Comber	50	54 33N	5 45W
Comeragh Mts.	51	52 17N	7 35W
Communism Pk.	71	38 40N	72 20E
Como	58	45 48N	9 5E
Como, L.	58	46 5N	9 17E
Comodoro Rivadavia	108	45 50S	67 40W
Comorin, C.	70	8 3N	77 40E
Comoro Is.	23	12 10S	44 15E
Compiègne	56	49 24N	2 50E
Comrie	46	56 22N	4 0W
Con Son, Is.	72	8 41N	106 37E
Conakry	88	9 29N	13 49W
Concepción	108	36 50S	73 0W
Concepción del Oro	106	24 40N	101 30W
Conception, Pt.	105	34 27N	120 28W
Conchos, R.	106	29 20N	105 0W
Concord	104	43 12N	71 30W
Concordia	100	39 35N	97 40W
Condobolin	79	33 4S	147 6E
Congleton	44	53 10N	2 12W
Congo ■	92	1 0S	16 0E
Congo, R. = Zaïre, R.	92	1 30N	28 0E
Congresbury	42	51 20N	2 49W
Coniston	44	54 22N	3 6W
Coniston Water	44	54 20N	3 5W
Conn, L.	50	54 3N	9 15W
Connacht	50	53 23N	8 40W
Connah's Quay	45	53 13N	3 6W
Connecticut □	101	41 40N	72 40W
Connecticut, R.	101	41 17N	72 21W
Connemara	51	53 29N	9 45W
Connersville	101	39 40N	85 10W
Consett	44	54 52N	1 50W
Constanţa	55	44 14N	28 38E
Constantia	94	34 2S	18 26E
Constantine	88	36 25N	6 42E
Conwy	42	53 17N	3 50W
Conwy, R.	42	53 18N	3 50W
Cook Is.	22	20 0S	160 0W
Cook, Mt.	83	43 36S	170 9E
Cook Strait	83	41 15S	174 29E
Cookhouse	94	32 44S	25 47E
Cookstown	50	54 40N	6 43W
Cookstown □	50	54 40N	6 43W
Cooktown	79	15 30S	145 16E
Coolgardie	78	30 55S	121 8E
Cooma	79	36 12S	149 8E
Coonamble	79	30 56S	148 27E
Coopers Creek, R.	79	28 0S	141 0E
Cootamundra	79	34 36S	148 1E
Cootehill	50	54 5N	7 5W
Copenhagen	61	55 41N	12 34E
Copley	79	30 24S	138 26E
Coppermine	98	67 50N	115 5W
Coppull	45	53 37N	2 39W
Coquet, R.	46	55 18N	1 45W
Corabia	55	43 48N	24 30E
Coral Gables	101	25 46N	80 16W
Coral Sea	79	15 0S	150 0E
Corbeil-Essonnes	56	48 36N	2 26E
Corbridge	44	54 58N	2 0W
Corby	45	52 49N	0 31W
Córdoba, Argent.	108	31 20S	64 10W
Córdoba, Spain	60	37 50N	4 50W
Cordova	98	60 36N	145 45W
Corfe Castle	43	50 37N	2 3W
Corigliano	58	39 36N	16 32E
Cork	51	51 54N	8 30W
Cork □	51	51 50N	8 50W
Cork Harbour	51	51 46N	8 16W
Corner Brook	99	48 57N	57 58W
Cornforth	44	54 42N	1 28W
Cornwall □	43	50 26N	4 40W
Coromandel	83	36 45S	175 31E
Coromandel Coast	70	12 30N	81 0E
Coronation Gulf	98	68 25N	112 0W
Corozal, Belize	106	18 30N	88 30W
Corozal, Colomb.	107	9 19N	75 18W
Corpus Christi	100	27 50N	97 28W
Corran	48	56 44N	5 14W
Corrib, L.	51	53 25N	9 10W
Corrientes, C.	106	20 25N	105 42W
Corringham	41	51 30N	0 26E
Corse, C.	56	43 1N	9 25E
Corsham	42	51 25N	2 11W
Corsica, I.	56	42 0N	9 0E
Corsicana	101	32 5N	96 30W
Corté	56	42 19N	9 11E
Corumbá	108	19 0S	57 30W
Coruña, La	60	43 20N	8 25W
Corvallis	100	44 36N	123 15W
Corwen	42	52 59N	3 23W
Coseley	41	52 33N	2 6W
Cosenza	58	39 17N	16 14E
Costa Rica ■	107	10 0N	84 0W
Côte d'Or □	56	47 30N	4 50E
Coteau du Missouri, Plat. du	100	47 0N	101 0W
Cotonou	90	6 20N	2 25E
Cotopaxi, Vol.	108	0 30S	78 30W
Cotswold Hills	42	51 42N	2 10W
Cottbus	54	51 44N	14 20E
Council Bluffs	101	41 20N	95 50W
Coundon	44	54 40N	1 39W
Coupar Angus	48	56 33N	3 17W
Courtenay	98	49 45N	125 0W
Courtmacsherry	51	51 38N	8 43W
Courtown	51	52 39N	6 14W
Coventry	41	52 25N	1 31W
Covington	101	39 5N	84 30W
Cowal.	46	56 5N	5 8W
Cowan, L.	78	31 45S	121 45E
Cowbridge	43	51 28N	3 28W
Cowdenbeath	47	56 7N	3 20W
Cowie	47	56 04N	3 52W
Cowra	79	33 49S	148 42E
Coxhoe	44	54 43N	1 30W
Cradock	94	32 8S	25 36E
Craig-y-nos	42	51 49N	3 41W
Craigavon □	50	54 30N	6 25W
Craigellachie	48	57 29N	3 9W
Craignure	48	56 28N	5 43W
Crail	46	56 16N	2 38W
Craiova	55	44 21N	23 48E
Cramlington	44	55 5N	1 36W
Cranbrook	40	51 6N	0 33E
Cranleigh	40	51 8N	0 29W
Crathie	48	57 3N	3 12W
Craven Arms	45	52 27N	2 49W
Crawcrook	44	54 57N	1 50W
Crawford	46	55 28N	3 40W
Crawley	40	51 7N	0 10W
Crediton	43	50 47N	3 39W
Creil	56	49 15N	2 34E
Cremona	58	45 8N	10 2E
Cres	58	44 58N	14 25E
Crete, I.	89	35 15N	25 0E
Creus, C.	60	42 20N	3 19E
Creuse, R.	56	47 0N	0 34E
Creusot, Le	56	46 48N	4 26E
Crewe	44	53 6N	2 28W
Crewkerne	43	50 53N	2 48W
Crianlarich	46	56 24N	4 37W
Criccieth	42	52 55N	4 15W
Crickhowell	42	51 52N	3 8W
Cricklade	42	51 38N	1 50W
Crieff	46	56 22N	3 50W
Criffell	44	54 56N	3 38W
Crigglestone	45	53 38N	1 32W
Crimea	62	45 0N	34 0E
Crinan	46	56 5N	5 34W
Croatia	58	45 20N	16 0E
Crofton	45	53 39N	1 25W
Cromarty	48	57 40N	4 2W
Cromarty Firth	48	57 40N	4 15W
Cromer	40	52 56N	1 18E
Cromwell	83	45 3S	169 14E
Crook	43	54 43N	1 45W
Crookhaven	51	51 28N	9 43W
Croom	51	52 32N	8 43W
Crosby	45	53 30N	3 2W
Cross Fell	44	54 44N	2 29W
Cross Hands	42	51 48N	4 04W
Crosse, La	101	43 48N	91 13W
Crosshaven	51	51 48N	8 19W
Crosskeys	43	51 37N	3 07W
Crossmaglen	50	54 5N	6 37W
Crossmolina	50	54 6N	9 21W
Crotone	58	39 5N	17 6E
Crouch, R.	40	51 37N	0 53E
Crowborough	40	51 3N	0 9E
Crowsnest Pass	98	49 40N	114 40W
Crowthorne	41	51 22N	0 45W
Croyde	42	51 7N	4 13W
Croydon, Austral.	79	18 13S	142 14E
Croydon, U.K.	41	51 18N	0 5W
Crozet, I.	23	46 27S	52 0E
Cuando, R.	92	14 0S	19 30E
Cuanza R.	92	9 2S	13 30E
Cuba ■	107	22 0N	79 0W
Cubango, R.	92	16 15S	17 45E
Cuckfield	40	51 0N	0 8W
Cúcuta	107	7 54N	72 31W
Cudworth	45	53 35N	1 25W
Cue	78	27 25S	117 54E
Cuenca, Ecuador	108	2 50S	79 9W
Cuenca, Spain	60	40 5N	2 10W
Cuenca, Serrania de	60	39 55N	1 50W
Cuernavaca	106	18 50N	99 20W
Cuevas del Almanzora	60	37 18N	1 58W
Cuffley	41	51 43N	0 09W
Cuiabá, R.	108	16 50S	56 30W
Cuidad Bolivar	107	8 21N	70 34W
Cuillin Hills	48	57 14N	6 15W
Cuillin Sd.	48	57 4N	6 20W
Cŭlaraşi	55	44 14N	27 23E
Culcheth	45	53 28N	2 30W
Culiacán	106	24 50N	107 40W
Cullen	48	57 45N	2 50W

** Renamed Kiritimati*

Cullera | Eday, I

Cullera 60 39 9N 0 17W
Cullercoats 44 55 02N 1 26W
Cullompton 43 50 52N 3 23W
Culross 47 56 4N 3 38W
Culverden 83 42 47S 172 49E
Cumaná 107 10 30N 64 5W
Cumberland 104 39 40N 78 43W
Cumberland Pen. 99 67 0N 64 0W
Cumberland Plat. 101 36 0N 84 30W
Cumberland Sd. 99 65 30N 66 0W
Cumbria □ 44 54 35N 2 55W
Cumbrian Mts. 44 54 30N 3 0W
Cunene, R. 92 17 0S 15 0E
Cúneo 58 44 23N 7 31E
Cunnamulla 79 28 2S 145 38E
Cunninghame 46 55 38N 4 35W
Cupar 47 56 20N 3 0W
Curaçao, I. 107 12 10N 69 0W
Curitiba 108 25 20S 49 10W
Currie 47 55 53N 3 17W
Cushendall 50 55 5N 6 3W
Cuttack 70 20 25N 85 57E
Cuxhaven 54 53 51N 8 41E
Cuyuni, R. 107 7 0N 59 30W
Cuzco 108 13 32S 72 0W
Cwm 43 51 44N 3 11W
Cymmer 43 51 37N 3 38W
Cyprus ■ 71 35 0N 33 0E
Czechoslovakia ■ 55 49 0N 17 0E
Czeremcha 55 52 32N 23 20E
Częstochowa 55 50 49N 19 7E

D

Da Nang 72 16 4N 108 13E
Dab'a, Râs el 91 31 3N 28 31E
Dabai 90 11 25N 5 15E
Dąbie 54 53 27N 14 45E
Dacca 70 23 43N 90 26E
Dadiya 90 9 35N 11 24E
Dagenham 41 51 33N 0 10E
Dagupan 72 16 3N 120 20E
Dahlak Arch 71 15 40N 40 5E
Daingean 51 53 18N 7 15W
Dairût 91 27 34N 30 43E
Dajarra 79 21 42S 139 30E
Dakar 88 14 34N 17 29W
Dakhla Oasis 91 25 30N 28 50E
Dakingari 90 11 37N 4 1E
Dal, R. 61 60 12N 16 40E
Dalbandin 71 29 0N 64 23E
Dalbeattie 46 54 55N 3 50W
Dalby 79 27 10S 151 17E
Dalga 91 27 39N 30 41E
Dalkeith 47 55 54N 3 5W
Dalkey 51 53 16N 6 7W
Dallas, Oregon, U.S.A. 100 45 0N 123 15W
Dallas, Texas, U.S.A. 101 32 50N 96 50W
Dalmally 46 56 25N 5 0W
Dalmatia, dist. 58 43 20N 17 0E
Dalneretchensk 73 45 50N 133 40E
Dalry 47 55 44N 4 42W
Dalton 45 54 09N 3 10W
Dalwhinnie 48 56 56N 4 14W
Daly, R. 78 13 21S 130 18E
Daly Waters 78 16 15S 133 24E
Dam 71 20 30N 44 35E
Daman 70 20 25N 72 57E
Damanhûr 91 31 0N 30 30E
Damaraland 92 21 0S 17 0E
Damascus 71 33 30N 36 18E
Dâmbovița, R. 55 44 40N 26 0E
Damietta 91 31 24N 31 48E
Damme 52 52 32N 8 12E
Dampier 78 20 41S 116 42E
Dampier Downs 78 18 24S 123 5E
Dan Gulbi 90 11 40N 6 15E
Danger Pt. 94 34 40S 19 17E
Dangora 90 11 30N 8 7E
Danja 90 11 29N 7 30E
Dankama 90 13 20N 7 44E
Dannemora 61 60 12N 17 51E
Dannevirke 83 40 12S 176 8E
Danube, R. 55 45 0N 28 20W
Danville, Ill., U.S.A. 101 40 10N 87 40W
Danville, Ky., U.S.A. 101 37 40N 84 45W
Dar-es-Salaam 93 6 50S 39 12E
Darbhanga 70 26 15N 86 8E
Dardanelles 71 40 0N 26 20E
Darent, R. 41 51 22N 0 12E
Darfield 45 53 32N 1 23W
Dargaville 83 35 57S 173 52E
Darien, G. of 96 9 0N 77 0W
Darjeeling 70 27 3N 88 18E
Darlaston 41 52 35N 2 1W
Darling, R. 79 34 4S 141 54E
Darling Ra. 78 32 30S 116 0E
Darlington 44 54 33N 1 33W
Darmstadt 54 49 51N 8 40E
Darnley, C. 112 68 0S 69 0E
Dart, R. 43 50 24N 3 36W
Dartford 41 51 26N 0 15E
Dartmoor 43 50 36N 4 0W
Dartmouth, Can. 99 44 40N 63 30W
Dartmouth, U.K. 43 50 21N 3 35W
Darton 45 53 36N 1 32W
Daru 79 9 3S 143 13E
Darvel 47 55 37N 4 20W
Darwen 45 53 42N 2 29W
Darwin 78 12 25S 130 51E
Dasht-e Lüt 71 31 30N 58 0E
Dasht-i-Margo 71 30 40N 62 30E
Dasht, R. 71 25 40N 62 20E
Datchet 41 51 28N 0 34W
Datteln 52 51 39N 7 23E
Daugavpils 61 55 53N 26 32E
Dauphin 98 51 9N 100 5W
Dauphiné 56 45 15N 5 25E
Daura 90 11 31N 11 24E
Davangere 70 14 25N 75 50E
Davao 72 7 0N 125 40E
Davao, G. of 72 6 30N 125 48E
Davenport 101 41 30N 90 40W
Davis Str. 96 65 0N 58 0W
Dawlish 43 50 34N 3 28W
Dawson 98 64 10N 139 30W
Dawson Creek 98 55 45N 120 15W
Dax 56 43 43N 1 3W
Dayton, Ohio, U.S.A. 104 39 45N 84 10W
Dayton, Wash., U.S.A. 100 46 20N 118 10W
Daytona Beach 101 29 14N 81 0W
De Aar 94 30 39S 24 0E
De Grey 78 20 12S 119 12E
Dead Sea 89 31 30N 35 30E
Deakin 78 30 46S 129 58E
Deal 40 51 13N 1 25E
Dean, Forest of 42 51 50N 2 35W
Death Valley 100 36 27N 116 52W
Debre Tabor 89 11 50N 38 26E
Debrecen 55 47 33N 21 42E
Decatur, Ala., U.S.A. 101 34 35N 87 0W
Decatur, Ga., U.S.A. 101 33 47N 84 17W
Decatur, Ill., U.S.A. 101 39 50N 89 0W
Decazeville 56 44 34N 2 15E
Deccan 70 14 0N 77 0E
Dee, R., Scot., U.K. 48 57 4N 2 7W
Dee, R., Wales, U.K. 44 53 15N 3 7W
Deggendorf 54 48 49N 12 59E
Dehiwala 70 6 50N 79 51E
Dej 55 47 10N 23 52E
Del Rio 100 29 15N 100 50W
Delagoa B. 92 25 50S 32 45E
Delareyville 94 26 41S 25 26E
Delaware □ 101 39 0N 75 40W
Delaware B. 101 38 50N 75 0W
Delft 52 52 1N 4 22E
Delfzijl 52 53 20N 6 55E
Delhi 70 28 38N 77 17E
Demanda, Sierra de la 60 42 15N 3 0W
Deming 100 32 10N 107 50W
Den Helder 52 52 57N 4 45E
Denain 56 50 20N 3 22E
Denbigh 42 53 12N 3 26W
Denby Dale 45 53 35N 1 40W
Dengi 90 9 25N 9 55E
Denham 78 25 56S 113 31E
Denholme 45 53 47N 1 54W
Deniliquin 79 35 30S 144 58E
Denmark 78 34 59S 117 18E
Denmark ■ 61 55 30N 9 0E
Denmark Str. 96 66 0N 30 0W
Denny 47 56 1N 3 55W
Denpasar 72 8 45S 115 5E
Denton, U.K. 45 53 26N 2 10W
Denton, U.S.A. 101 33 12N 97 10W
Denver 100 39 45N 105 0W
Der'a 89 32 36N 36 7E
Dera Ismail Khan 71 31 50N 70 50E
Derbent 71 42 5N 48 15E
Derby 45 52 55N 1 28W
Derby □ 45 52 55N 1 28W
Derg, L., Ireland 51 53 0N 8 20W
Derg, L., Donegal, Ireland 50 54 37N 7 53W
Derna 89 32 40N 22 35E
Derry = Londonderry 50 55 0N 7 19W
Derrynasaggart Mts. 51 51 58N 9 15W
Derryveagh Mts. 50 55 0N 8 40W
Dersingham 40 52 51N 0 30E
Derwent, R., N. Yorks., U.K. 44 53 45N 0 57W
Derwent, R., Tyne & Wear, U.K. 44 54 58N 1 40W
Derwentwater, L. 44 54 34N 3 9W
Des Moines 101 41 35N 93 37W
Desborough 45 52 27N 0 50W
Deschutes, R. 105 45 30N 121 0W
Desford 41 52 38N 1 19W
Desna, R. 55 52 0N 33 15E
Dessau 54 51 49N 12 15E
Detroit 104 42 13N 83 22W
Deurne 52 51 27N 5 49E
Deventer 52 52 15N 6 10E
Deveron, R. 48 57 40N 2 31W
Devil's Bridge 42 52 23N 3 50W
Devils Lake 100 48 5N 98 50W
Devizes 42 51 21N 2 0W
Devon I. 98 75 47N 88 0W
Devon, R. 47 56 07N 3 51W
Devonport, Austral. 79 41 10S 146 22E
Devonport, N.Z. 83 36 49S 174 49E
Devonshire □ 43 50 50N 3 40W
Dewsbury 45 53 42N 1 38W
Dezfūl 71 32 20N 48 30E
Dhaulagiri Mt. 70 28 45N 83 45E
Diapaga 90 12 5N 1 46E
Dickinson 100 46 50N 102 40W
Didcot 40 51 36N 1 14W
*Diégo Suarez 92 12 25S 49 20E
Dieppe 56 49 54N 1 4E
Dieren 52 52 3N 6 6E
Differdange 52 49 81N 5 54E
Digby 99 44 38N 65 50W
Digges Is. 99 62 40N 77 50W
Dijon 56 47 20N 5 0E
Dikwa 90 12 4N 13 30E
Dila 89 6 14N 38 22E
Dili 72 8 39S 125 34E
Dillingen 52 49 22N 6 42E
Dinan 56 48 28N 2 2W
Dinaric Alps 58 44 0N 17 30E
Dinas Mawddwy 42 52 44N 3 41W
Dingle 51 52 9N 10 17W
Dingle B. 51 52 3N 10 20W
Dingwall 48 57 36N 4 26W
Dire Dawa 89 9 35N 41 45E
Dirranbandi 79 28 33S 148 17E
Disappointment, C. 100 46 20N 124 0W
Disappointment L. 78 23 20S 122 40E
Dishna 91 26 9N 32 32E
Disko 96 69 45N 53 30W
Diss 40 52 23N 1 6E
Disûq 91 31 8N 30 35E
Diu, I. 70 20 45N 70 58E
Diyarbakir 71 37 55N 40 18E
Djelfa 88 34 40N 3 15E
Djibouti 89 11 30N 43 5E
Djibouti ■ 89 11 30N 42 15E
Djougou 90 9 40N 1 45E
Dnepropetrovsk 62 48 30N 35 0E
Dnestr, R. 55 48 30N 26 30E
Dniepr, R. 62 52 29N 35 10E
Dobbyn 79 19 44S 139 59E
Dobrogea 55 44 30N 28 15E
Docking 40 52 55N 0 39E
Dodge City 100 37 42N 100 0W
Dodoma 93 6 8S 35 45E
Dodworth 45 53 32N 1 32W
Doetinchem 52 51 59N 6 18E
Dògondoutchi 90 13 38N 4 2E
Doha 71 25 15N 51 32E
Dole 56 47 7N 5 31E
Dolgellau 42 52 44N 3 53W
Dollar 47 56 9N 3 41W
Dolomites 58 46 30N 11 40E
Doma 90 8 25N 8 18E
Dominica ■ 107 15 20N 61 20W
Dominican Rep. ■ 107 19 0N 70 30W
Domodóssola 58 46 6N 8 19E
Don, R., Eng., U.K. 44 53 41N 0 51W
Don, R., Scot., U.K. 48 57 14N 2 5W
Don, R., U.S.S.R. 62 49 35N 41 40E
Donaghadee 50 54 38N 5 32W
Doncaster 44 53 31N 1 9W
Dondra Head 70 5 55N 80 40E
Donegal 50 54 39N 8 8W
Donegal □ 50 54 53N 8 0W
Donegal B. 50 54 30N 8 35W
Donetsk 62 48 0N 37 45E
Dongara 78 29 14S 114 57E
Donington 45 52 54N 0 12W
Donnelly's Crossing 83 35 42S 173 38E
Dooagh 50 53 59N 10 7W
Doon, R. 46 55 26N 4 41W
Dora Báltea, R. 58 45 11N 8 5E
Dorchester, Dorset, U.K. 43 50 42N 2 28W
Dorchester, Oxon., U.K. 40 51 38N 1 10W
Dordogne, R. 56 45 2N 0 36W
Dordon 41 52 36N 1 39W
Dordrecht 52 51 48N 4 39E
Dorking 40 51 14N 0 20W
Dornoch 48 57 52N 4 0W
Dornoch Firth 48 57 52N 4 0W
Dorridge 41 52 22N 1 45W
Dorset □ 43 50 48N 2 25W
Dortmund 52 51 32N 7 28E
Dothan 101 31 10N 85 25W
Douai 56 50 21N 3 4E
Douala 90 4 0N 9 45E
Doubtful B. 78 34 15S 119 28E
Douglas, S. Afr. 94 29 4S 23 46E
Douglas, U.K. 42 54 9N 4 29W
Douglas, U.S.A. 100 31 21N 109 30W
Doune 46 56 12N 4 3W
Dounreay 48 58 40N 3 28W
Douro Litoral □ 60 41 10N 8 20W
Douro, R. 60 41 1N 8 16W
Dove, R. 44 52 51N 1 36W
Dover, U.K. 40 51 7N 1 19E
Dover, U.S.A. 104 39 10N 75 31W
Dover, Str. of 40 51 0N 1 30E
Dovey, R. 42 52 32N 4 0W
Dovrefjell 61 62 15N 9 33E
Downham Market 40 52 36N 0 22E
Downpatrick 50 54 20N 5 43W
Drake Passage 112 58 0S 68 0W
Drakensberg 94 31 0S 25 0E
Drammen 61 59 42N 10 12E
Draperstown 50 54 48N 6 47W
Drava, R. 55 45 33N 18 55E
Drenthe □ 52 52 52N 6 40E
Dresden 54 51 2N 13 45E
Dreux 56 48 44N 1 23E
†Driffield 44 54 0N 0 25W
Drighlington 45 53 45N 1 40W
Drina, R. 55 44 53N 19 21E
Drogheda 50 53 45N 6 20W
Droichead Nua 51 53 11N 6 50W
Droitwich 42 52 16N 2 10W
Dromore, Down, U.K. 50 54 24N 6 10W
Dromore, Tyrone, U.K. 50 54 31N 7 28W
Dronfield 45 53 18N 1 29W
Dronne, R. 56 45 2N 0 12W
Drumcondra 51 53 50N 6 40W
Drumheller 98 51 25N 112 40W
Drummond Ra. 79 23 45S 147 10E
Drummondville 104 45 55N 72 25W
Drummore 46 54 41N 4 53W
Drumshanbo 50 54 2N 8 4W
Drybrook 43 51 51N 2 30W
Drygalski I. 112 66 0S 92 0E
Dschang 90 5 32N 10 3E
Dubawnt, R. 98 64 33N 100 6W
Dubbo 79 32 11S 148 35E
Dublin 51 53 20N 6 18W
Dublin □ 51 53 24N 6 20W
Dublin, B. 51 53 24N 6 20W
Dubrovnik 58 42 39N 18 6E
Duchess 79 21 20S 139 50E
Ducie I. 22 24 47S 124 40W
Duddon, R. 44 54 12N 3 15W
Dudley 41 52 30N 2 5W
Duero, R. 60 41 37N 4 25W
Duffel 52 51 6N 4 30E
Dugi Otok 58 44 0N 15 0E
Duisburg 52 51 27N 6 42E
Duiwelskloof 94 23 42S 30 10E
Dukana 93 3 59N 37 20E
Dukinfield 45 53 29N 2 05W
Duku 90 11 11N 4 55E
Dülmen 52 51 49N 7 18E
Duluth 101 46 48N 92 10W
Dulverton 43 51 2N 3 33W
Dumbarton 47 55 58N 4 35W
Dumfries 46 55 4N 3 37W
Dumfries & Galloway □ 46 55 0N 4 0W
Dun Laoghaire 51 53 17N 6 9W
Dunback 83 45 23S 170 36E
Dunbar 46 56 0N 2 32W
Dunbeath 48 58 15N 3 25W
Dunblane 47 56 10N 3 58W
Dunboyne 51 53 25N 6 30W
Duncan 100 34 25N 98 0W
Duncansby 48 58 37N 3 3W
Dundalk 50 54 1N 6 45W
Dundalk B. 50 53 55N 6 15W
Dundas 99 43 17N 79 59W
Dundas, L. 78 32 35S 121 50E
Dundee, S. Afr. 94 28 11S 30 15E
Dundee, U.K. 47 56 29N 3 0W
Dundrum, Ireland 51 53 17N 6 15W
Dundrum, U.K. 50 54 17N 5 50W
Dunedin 83 45 50S 170 33E
Dunfermline 47 56 5N 3 28W
Dungannon 50 54 30N 6 47W
Dungannon □ 50 54 30N 6 55W
Dungarvan 51 52 6N 7 40W
Dungbure Shan 73 35 0N 90 0E
Dungeness 40 50 54N 0 59E
Dungiven 50 54 55N 6 56W
Dunglow 50 54 57N 8 20W
Dunipace 47 56 01N 3 55W
Dunkeld 48 56 34N 3 36W
Dunkery Beacon 42 51 15N 3 37W
Dunleary = Dun Laoghaire 51 53 17N 6 8W
Dunleer 50 53 50N 6 23W
Dunlop 47 55 43N 4 32W
Dunmanus B. 51 51 31N 9 50W
Dunmanway 51 51 43N 9 8W
Dunnet Hd. 48 58 38N 3 22W
Dunoon 47 55 57N 4 56W
Duns 46 55 47N 2 20W
Dunshaughlin 51 53 31N 6 32W
Dunstable 40 51 53N 0 31W
Dunstan Mts. 83 44 53S 169 35E
Dunster 42 51 11N 3 28W
Dunvegan 48 57 26N 6 35W
Durack Ra. 78 16 50S 127 40E
Durance, R. 56 43 55N 4 45E
Durango, Mexico 106 24 3N 104 39W
Durango, Spain 60 43 13N 2 40W
Durango, U.S.A. 100 37 10N 107 50W
Durban 94 29 49S 31 1E
Düren 52 50 48N 6 30E
Durham, U.K. 44 54 47N 1 34W
Durham, U.S.A. 101 36 0N 78 55W
Durham □ 43 54 42N 1 45W
Durrow, Ireland 51 52 51N 7 23W
Durrow, Offaly, Ireland 51 53 20N 7 31W
Dursey I. 51 51 36N 10 12W
Dursley 43 51 41N 2 21W
D'Urville Island 83 40 50S 173 55E
Dushanbe 71 38 33N 68 48E
Düsseldorf 52 51 15N 6 46E
Dvina, N., R. 62 61 40N 45 30E
Dyce 48 57 12N 2 11W
Dyfed □ 42 52 0N 4 30W
Dymchurch 40 51 2N 1 0E
Dysart 47 56 8N 3 8W
Dzamin Und 73 44 0N 111 0E
Dzhalal-Abad 71 40 56N 73 0E
Dzhambul 71 42 54N 71 22E
Dzhungarskiye Vorota 73 45 0N 82 0E
Dzungaria 73 44 10N 88 0E
Dzuunmod 73 47 45N 106 58E

E

Eagle Pass 100 28 45N 100 35W
Eaglescliffe 44 54 32N 1 21W
Eagleshan 47 55 44N 4 18W
Ealing 41 51 30N 0 19W
Earby 45 53 55N 2 8W
Earl Shilton 41 52 35N 1 20W
Earn, L. 46 56 23N 4 14W
Earn, R. 46 56 20N 3 19W
Earnslaw, Mt. 83 44 32S 168 27E
Easington 44 54 50N 1 24W
Easington Lane 44 54 47N 1 21W
Easingwold 44 54 8N 1 11W
East Beskids, mts. 55 49 30N 18 45E
East C. 83 37 42S 178 35E
East China Sea 73 30 5N 126 0E
East Dereham 40 52 40N 0 57E
East Grinstead 40 51 8N 0 1W
East Horsley 41 51 15N 0 26W
East Keswick 45 53 54N 1 26W
East Linton 46 56 0N 2 40W
East London 94 33 0S 27 55E
East Point 101 33 40N 84 28W
East Retford 44 53 19N 0 55W
East Schelde, R. 52 51 38N 3 40E
E. Siberian Sea 62 73 0N 160 0E
East Sussex □ 40 51 0N 0 20E
East Wemyss 47 56 8N 3 5W
East Wittering 40 50 46N 0 53W
Eastbourne, N.Z. 83 41 19S 174 55E
Eastbourne, U.K. 40 50 46N 0 18E
Eastern Ghats 70 15 0N 80 0E
Eastmain, R. 99 52 27N 72 26W
Eau Claire 101 44 46N 91 30W
Ebbw Vale 43 51 47N 3 12W
Eboli 58 40 39N 15 2E
Ebro, R. 60 41 49N 1 5W
Ecclefechan 46 55 3N 3 18W
Eccles 45 53 29N 2 20W
Ecclesfield 45 53 26N 1 28W
Eccleshall 45 52 52N 2 14W
Eccleston 45 53 29N 2 43W
Echo Bay 98 66 10N 117 40W
Echuca 79 36 3S 144 46E
Ecija 60 37 30N 5 10W
Eckington 45 53 19N 1 21W
Ecuador ■ 108 2 0S 78 0W
Ed Dâmer 89 17 27N 34 0E
Eday, I. 49 59 11N 2 47W

* Renamed Antsiranana

† Renamed Great Driffield

Eddrachillis B
Fremont

Eddrachillis B. 48 58 16N 5 10W
Ede, Neth. 52 52 4N 5 40E
Ede, Nigeria 90 7 45N 4 29E
Eden, R. 44 54 57N 3 2W
Edenbridge 40 51 12N 0 4E
Edenburg 94 29 43S 25 58E
Edenderry 51 53 21N 7 3W
Edgerston Tofts 46 55 24N 2 29W
Edgworth 45 53 38N 2 23W
Edievale 83 45 49S 169 22E
Edinburgh 47 55 57N 3 12W
Edmonton 98 53 30N 113 30W
Edmundston 99 47 23N 68 20W
Edward, L. = Idi Amin Dada, L. 93 0 25S 29 40E
Edwards Plat. 100 30 30N 101 5W
Eekloo 52 51 11N 3 33E
Eersterus 94 25 44S 28 21E
Egadi Is. 58 37 55N 12 10E
Egersund, Norway 61 58 26N 6 1E
Egerton, Mt. 78 24 42S 117 44E
Egglescliffe 44 54 31N 1 21W
Egham 41 51 25N 0 33W
Egmont, C. 83 39 16S 173 45E
Egmont, Mt. 83 39 17S 174 5E
Egremont 44 54 28N 3 33W
Egume 90 7 30N 7 14E
Egypt ■ 91 28 0N 31 0E
Eha Amufu 90 6 30N 7 40E
Eifel 52 50 10N 6 45E
Eigg, I. 48 56 54N 6 10W
Eil 89 8 0N 49 50E
Eindhoven 52 51 26N 5 30E
Eitorf 52 50 46N 7 28E
Eket 90 4 38N 7 56W
Eketahuna 83 40 38S 175 43E
El 'Aiyat 91 29 36N 31 15E
El Alamein 91 30 48N 28 58E
El 'Arish 91 31 8N 33 50E
El Badâri 91 27 4N 31 25E
El Ballâs 91 26 2N 32 43E
El Balyana 91 26 10N 32 3E
El Bawiti 91 28 25N 28 45E
El Blanco 107 10 28N 74 52W
El Callao 107 7 25N 61 50W
El Centro 100 32 50N 115 40W
El Dab'a 91 31 0N 28 27E
El Djouf 88 20 0N 11 30E
El Dorado, Ark., U.S.A. 101 33 10N 92 40W
El Dorado, Kans., U.S.A. 101 37 55N 96 56W
El Escorial 60 40 35N 4 7W
El Faiyûm 91 29 19N 30 50E
El Fâsher 89 13 33N 25 26E
El Fashn 91 28 50N 30 54E
El Fuerte 106 26 30N 108 40W
El Gedida 91 25 40N 28 30E
El Geneina 89 13 27N 22 45E
El Gîza 91 30 0N 31 10E
El Hammam 91 30 52N 29 25E
El Iskandarîya = Alexandria 91 31 0N 30 0E
El Khârga 91 25 30N 30 33E
El Kuntilla 91 30 1N 34 45E
El Mahalla el Kubra 91 31 0N 31 0E
El Mahâriq 91 25 35N 30 35E
El Mansûra 91 31 0N 31 19E
El Manzala 91 31 10N 31 50E
El Matariya 91 31 15N 32 0E
El Minyâ 91 28 7N 30 33E
El Obeid 89 13 8N 30 10E
El Paso 100 31 50N 106 30W
El Qâhira = Cairo 91 30 1N 31 14E
El Qantara 91 30 51N 32 20E
El Qasr 91 25 44N 28 42E
El Quseima 91 30 40N 34 15E
El Qusiya 91 27 29N 30 44E
El Râshda 91 25 36N 28 57E
El Real 107 8 0N 77 40W
El Reis 91 27 50N 28 40E
El Reno 100 35 30N 98 0W
El Saff 91 29 34N 31 16E
El Thamad 91 29 40N 34 28E
El Tigre 107 8 55N 64 15W
El Wak 93 2 49N 40 56E
El Waqf 91 25 45N 32 15E
El Wâsta 91 29 19N 31 12E
Elat 91 29 30N 34 56E
Elâziğ 71 38 37N 39 22E
Elba. 58 42 48N 10 15E
Elbe, R. 54 53 50N 9 0E
Elbert, Mt. 100 39 12N 106 36W
Elbeuf 56 49 17N 1 2E
Elblag 55 54 10N 19 25E
Elbrus, Mt. 62 43 30N 42 30E
Elburz Mts. 71 36 0N 52 0E
Elche 60 38 15N 0 42W
Eldoret 93 0 30N 35 25E
Electra 100 34 0N 99 0W
Elephant I. 112 61 0S 55 0W
Eleuthera I. 107 25 0N 76 20W
Elgin, U.K. 48 57 39N 3 20W
Elgin, U.S.A. 101 42 0N 88 20W
Elgol 48 57 9N 6 6W
Elgon, Mt. 93 1 10N 34 30E
Elizabeth, Austral. 79 34 42S 138 41E
Elizabeth, U.S.A. 104 40 37N 74 12W
Elkhart 101 41 42N 85 55W
Elkins 101 38 53N 79 53W
Elko 100 40 50N 115 50W
Elland 45 53 41N 1 49W
Ellensburg 100 47 0N 120 30W
Ellesmere 45 52 55N 2 53W
Ellesmere I. 99 79 30N 80 0W
Ellesmere Port 45 53 17N 2 55W
Ellesworth Land 22 74 0S 85 0W
Ellon 48 57 21N 2 5W
Ellsworth Land 112 76 0S 89 0W
Elmenteita 93 0 32S 36 14E
Elmina 88 5 5N 1 21W
Elmira 104 42 8N 76 49W
Eltham 83 39 26S 174 19E
Eluru 70 16 48N 81 8E
Elvas 60 38 50N 7 17W
Ely, Cambs., U.K. 40 52 24N 0 16E
Ely, S. Glam., U.K. 43 51 28N 3 12W
Ely, U.S.A. 100 39 10N 114 50W
Elyria 101 41 22N 82 8W
Embu 93 0 32S 37 38E
Emden 52 53 22N 7 12E
Emerald 79 23 32S 148 10E
Emilia-Romagna □ 58 44 33N 10 40E
Emmen 52 52 48N 6 57E
Emmerich 52 51 50N 6 12E
Empangeni 94 28 50S 31 52E
Ems, R. 52 53 25N 7 0E
Emsdetten 52 52 11N 7 31E
Enard B. 48 58 5N 5 20W
Endeavour Str. 79 10 45S 142 0E
Enderby 41 52 35N 1 15W
Enderby Land 112 66 0S 53 0E
Endicott Mts. 98 68 0N 152 30W
Enfield 41 51 39N 0 4W
Enggano, I. 72 5 20S 102 40E
England □ 44 53 0N 2 0W
Englefield Green 41 51 26N 1 06W
English Channel 24 50 0N 2 0W
Enid 100 36 26N 97 52W
Enid, Mt. 78 21 43S 116 25E
Enna 58 37 34N 14 15E
Ennell, L. 51 53 29N 7 25W
Ennis 51 52 51N 8 59W
Enniscorthy 51 52 30N 6 35W
Enniskillen 50 54 20N 7 40W
Ennistimon 51 52 56N 9 18W
Enschede 52 52 13N 6 53E
Ensenada 106 31 50N 116 50W
Entebbe 93 0 4N 32 28E
Entre Rios □ 108 30 30S 58 30W
Entrecasteaux, Pt. d' 79 34 50S 115 56E
Enugu 90 6 30N 7 30E
Enugu Ezike 90 7 0N 7 29E
Epe 90 6 36N 3 59E
Epernay 56 49 3N 3 56E
Epinal 56 48 19N 6 27E
Epping 41 51 42N 0 8E
Eppynt, Mynydd 42 52 07N 3 30W
Epsom 41 51 19N 0 16W
Epworth 44 53 30N 0 50W
Equatorial Guinea ■ 88 2 0S 8 0E
Erebus, Mt. 112 77 35S 167 0E
Ereğli 71 41 15N 31 30E
Erfurt 54 50 58N 11 2E
Erg Chech, dist. 88 50 59N 11 0E
Eriboll, L. 48 58 28N 4 41W
Ericht, L. 48 56 50N 4 25W
Erie 104 42 10N 80 7W
Erie, L. 104 42 15N 81 0W
Eriskay, I. 49 57 4N 7 18W
Eritrea □ 89 14 0N 41 0E
Ermelo 94 26 31S 29 59E
Erne, Lough 50 54 26N 7 46W
Errigal, Mt. 50 55 2N 8 8W
Errol 47 56 24N 3 13W
Erz Gebirge, mts. 54 50 25N 13 0E
Erzincan 71 39 46N 39 30E
Erzurum 71 39 57N 41 15E
Esbjerg 61 55 29N 8 29E
Esch 52 51 37N 5 17E
Eschweiler 52 50 49N 6 14E
Escravos, R. 90 5 30N 5 10E
Escuinapa 106 22 50N 105 50W
Esfahân □ 71 33 0N 53 0E
Esh Winning 44 54 47N 1 42W
Esher 41 51 21N 0 22W
Eshowe 94 28 50S 31 30E
Esk, R., Dumfries, U.K. 48 54 58N 3 4W
Esk, R., N. Yorks., U.K. 44 54 27N 0 36W
Eskbank 47 55 52N 3 04W
Eskilstuna 61 59 22N 16 32E
Eskişehir 71 39 50N 30 35E
Esperance 78 33 45S 121 55E
Essen 52 51 28N 6 59E
Essequibo, R. 107 5 45N 58 50W
Essex □ 40 51 48N 0 30E
Estcourt 94 28 58S 29 53E
Eston 44 54 33N 1 6W
Estonian S.S.R. □ 61 48 30N 25 30E
Estrêla, Serra da 60 40 10N 7 45W
Estremadura 60 39 0N 9 0W
Étampes 56 48 26N 2 10E
Eteh 90 7 2N 7 28E
Ethiopia ■ 89 8 0N 40 0E
Ethiopian Highlands 89 10 0N 37 0E
Etive, L. 48 56 30N 5 12W
Etna, Mt. 58 37 45N 15 0E
Etoshapan 92 18 40S 16 30E
Ettrick 46 55 31N 2 55W
Euclid 101 41 32N 81 31W
Eugene 105 44 0N 123 8W
Eupen 52 50 37N 6 3E
Euphrates, R. 71 31 0N 47 25E
Eureka 105 40 50N 124 0W
Europa Pt. 60 36 2N 6 32W
Europe 24 20 0N 20 0E
Euskirchen 52 50 40N 6 45E
Euxton 45 53 41N 2 42W
Evanston 104 42 0N 87 40W
Evansville 101 38 0N 87 35W
Evanton 48 57 40N 4 20W
Everard Ras. 78 27 5S 132 28E
Everest, Mt. 70 28 5N 86 58E
Everett 105 48 0N 122 10W
Everglades 101 26 0N 80 30W
Évora 60 38 33N 7 57W
Évreux 56 49 0N 1 8E
Ewe, L. 48 57 49N 5 38W
Ewell 41 51 20N 0 15W
Exe, R. 43 50 38N 3 27W
Exeter 43 50 43N 3 31W
Exmoor 42 51 10N 3 59W
Exmouth, Austral. 78 22 6S 114 0E
Exmouth, U.K. 43 50 37N 3 26W
Exmouth G. 78 22 15S 114 15E
Expedition Range 79 24 30S 149 12E
Eyam 45 53 17N 1 40W
Eye, Camb., U.K. 40 52 36N 0 11W
Eye, Suff., U.K. 40 52 19N 1 09E
Eye Pen. 48 58 13N 6 10W
Eyemouth 46 55 53N 2 5W
Eyre 78 32 15S 126 18E
Eyre, L. 79 29 30S 137 26E
Eyre Mts. 83 45 25S 168 25E
Eyre Pen. 79 33 30S 137 17E

F

Fabriano 58 43 20N 12 52E
Faenza 58 44 17N 11 53E
Fagam 90 11 1N 10 1E
Fahūd 71 22 18N 56 28E
Failsworth 45 53 31N 2 08W
Fairbanks 98 64 59N 147 40W
Fairlie, N.Z. 83 44 5S 170 49E
Fairlie, U.K. 47 55 44N 4 52W
Fairmont 101 39 29N 80 10W
Faizabad, Afghan. 71 37 7N 70 33E
Faizabad, India 70 26 45N 82 10E
Fakenham 40 52 50N 0 51E
Falkirk 47 56 0N 3 47W
Falkland 47 56 15N 3 13W
Falkland Is. 108 51 30S 59 0W
Falkland Is. Dep. 112 57 0S 40 0W
Fall River 104 41 45N 71 5W
Falmouth 43 50 9N 5 5W
False B. 94 34 15S 18 40E
Falun 61 60 37N 15 37E
Fannich, L. 48 57 40N 5 0W
* Fanning I. 22 3 51N 159 22W
Fano 58 43 50N 13 0E
Farâfra Oasis 91 27 15N 28 20E
Farah 71 32 20N 62 7E
Faraid, Gebel 91 23 33N 35 19E
Farasan Is. 71 16 45N 41 55E
Fareham 40 50 52N 1 11W
Farewell, C., Greenl. 96 59 48N 43 55W
Farewell, C., N.Z. 83 40 29S 172 43E
Fargo 101 47 0N 97 0W
Faribault 101 44 15N 93 19W
Faringdon 40 51 39N 1 34W
Farmington 100 36 45N 108 28W
Farnborough 41 51 17N 0 46W
Farnham 40 51 13N 0 49W
Farnworth 45 53 33N 2 24W
Faro 60 37 2N 7 55W
Faroe Is. 22 62 0N 7 0W
Fasā 71 29 0N 53 32E
Fastov 55 50 7N 29 57E
Fatshan 73 23 0N 113 4E
Fauldhouse 47 55 50N 3 44W
Faversham 40 51 18N 0 54E
Fawley 40 50 49N 1 20W
Fayetteville 101 36 0N 94 5W
Fazeley 41 52 36N 1 42W
Featherston 83 41 6S 175 20E
Featherstone 45 53 42N 1 22W
Fehmarn 54 54 26N 11 10E
Fehmarn Belt 54 54 35N 11 20E
Feilding 83 40 13S 175 35E
Feldkirch 54 47 15N 9 37E
Felixstowe 40 51 58N 1 22E
Felling 44 54 57N 1 33W
Feltham 41 51 27N 0 25W
† Fénérive 92 17 22S 49 25E
Fengkieh (Kweichow) 73 31 0N 109 33E
Fens, The 40 52 45N 0 2E
Fenyang 73 37 19N 111 46E
Feolin Ferry 46 55 50N 6 05W
Ferbane 51 53 17N 7 50W
Fergana 71 40 23N 71 46E
Fergus, R. 51 52 45N 9 0W
Fermanagh □ 50 54 21N 7 40W
Fermoy 51 52 4N 8 18W
Fernando de Noronha, I. 108 4 0S 33 10W
Ferndale 43 51 40N 3 27W
Ferns 51 52 35N 6 30W
Ferozepore 71 30 55N 74 40E
Ferrara 58 44 50N 11 36E
Ferret, C. 56 44 38N 1 15W
Ferrol 60 43 29N 8 15W
Ferrybridge 45 53 42N 1 16W
Ferryhill 44 54 42N 1 32W
Fès 88 34 0N 5 0W
Fethard 51 52 29N 7 42W
Fetlar, I. 49 60 36N 0 52W
Fettercairn 48 56 50N 2 33W
Ffestiniog 42 52 58N 3 56W
Fianarantsoa 92 21 20S 46 45E
Fichtelgebirge 54 50 10N 12 0E
Fiditi 90 7 45N 3 53E
Fife □ 46 56 13N 3 2W
Fife Ness 46 56 17N 2 35W
Figueras 60 42 18N 2 58E
Fiji ■ 23 17 20S 179 0E
Filey 44 54 13N 0 18W
Findhorn 48 57 39N 3 36W
Findhorn, R. 48 57 38N 3 38W
Finglas 51 53 22N 6 18W
Finisterre, C. 60 42 50N 9 19W
Finland ■ 61 64 0N 27 0E
Finland, G. of 61 60 0N 26 0E
Finlay, R. 98 56 50N 125 10W
Fintona 50 54 30N 7 20W
Fionnphort 46 56 19N 6 23W
Fishguard 42 51 59N 4 59W
Fitzroy Crossing 78 18 9S 125 38E
Fitzroy, R. 78 17 25S 124 0E
Flagstaff 100 35 10N 111 40W
Flamborough Hd. 44 54 8N 0 4W
Flanders 56 51 10N 3 15E
Flandre Occidental □ 52 51 0N 3 0E
Flandre Orientale □ 52 51 0N 4 0E
Flattery, C. 100 48 21N 124 43W
Flèche, La 56 47 42N 0 4W
Fleet 41 51 16N 0 50W
Fleetwood 45 53 55N 3 00W
Flensburg 54 54 46N 9 28E
Flers 56 48 47N 0 33W
Flin Flon 98 54 46N 101 53W
Flinders B. 78 34 19S 115 9E
Flinders I. 79 40 0S 148 0E
Flinders, R. 79 17 36S 140 36E
Flinders Ranges 79 31 30S 138 30E
Flint 104 43 5N 83 40W
Flint □ 45 53 15N 3 12W
Florence, Italy 58 43 47N 11 15E
Florence, Ala., U.S.A. 101 34 50N 87 50W
Florence, S.C., U.S.A. 101 34 5N 79 50W
Flores, I. 72 8 35S 121 0E
Flores Sea 72 6 30S 124 0E
Florida □ 101 28 30N 82 0W
Florida B. 101 25 0N 81 20W
Florida, Strait of 107 25 0N 80 0W
Flushing = Vlissingen 52 51 26N 3 34E
Fochabers 48 57 37N 3 7W
Focşani 55 45 41N 27 15E
Fóggia 58 41 28N 15 31E
Foggo 90 11 21N 9 57E
Foligno 58 42 58N 12 40E
Folkestone 40 51 5N 1 11E
Fond du lac 101 43 46N 88 26W
Fonesca, G. of 106 13 10N 87 40W
Fontainebleau 56 48 24N 2 40E
Fontem 90 5 32N 9 52E
Foochow (Minhow) 73 26 2N 119 25E
Forbach 56 49 10N 6 52E
Forbes 79 33 22S 148 0E
Forcados 90 5 26N 5 26E
Fordingbridge 40 50 56N 1 48W
Forfar 48 56 40N 2 53W
Forli 58 44 14N 12 2E
Formartine, dist. 48 57 20N 2 15W
Formby 45 53 33N 3 03W
Formentera, I. 60 38 40N 1 30E
Formosa = Taiwan ■ 73 24 0N 121 0E
Forres 48 57 37N 3 38W
Forrest 78 30 51S 128 6E
Forsayth 79 18 33S 143 34E
Forst 54 51 43N 14 37E
Fort Albany 99 52 15N 81 35W
Fort Augustus 48 57 9N 4 40W
Fort Beaufort 94 32 46S 26 40E
Fort Bragg 105 39 28N 123 50W
Fort Chimo 99 58 6N 68 25W
Fort Collins 100 40 30N 105 4W
* Fort-Dauphin 92 25 2S 47 0E
Fort-de-France 107 14 36N 61 2W
Fort Dodge 101 42 29N 94 10W
Fort George 99 53 50N 79 0W
Fort Good-Hope 98 66 14N 128 40W
Fort Gouraud 88 22 40N 12 45W
Fort Lauderdale 101 26 10N 80 5W
Fort Liard 98 60 20N 123 30W
Fort McMurray 98 56 44N 111 23W
Fort McPherson 98 67 30N 134 55W
Fort Morgan 100 40 10N 103 50W
Fort Myers 101 26 30N 81 50W
Fort Norman 98 64 57N 125 30W
Fort Portal 93 0 40N 30 20E
Fort Providence 98 61 21N 117 40W
Fort Resolution 98 61 10N 113 40W
Fort Rupert 99 51 30N 78 40W
Fort Shevchenko 71 44 30N 50 10E
Fort Simpson 98 61 45N 121 23W
Fort Smith 101 35 25N 94 25W
Fort Trinquet 88 25 10N 11 25W
Fort Vermilion 96 58 24N 116 0W
† Fort Victoria 92 20 8S 30 55E
Fort Wayne 104 41 5N 85 10W
Fort William 48 56 48N 5 8W
Fort Worth 100 32 45N 97 25W
Fortaleza 108 3 35S 38 35W
Forth 47 55 45N 3 41W
Forth, Firth of 47 56 5N 2 55W
Fortrose 48 57 35N 4 10W
Fougères 56 48 21N 1 14W
Foulness I. 40 51 36N 0 55E
Foulridge 45 53 52N 2 10W
Foumban 90 5 45N 10 50E
Fouta Djalon 88 11 20N 12 10W
Foveaux Str. 83 46 42S 168 10E
Fowey 43 50 20N 4 39W
Foxe Basin 99 68 30N 77 0W
Foxe Channel 99 66 0N 80 0W
Foxford 50 54 0N 9 7W
Foxton 83 40 29S 175 18E
Foyers 48 57 15N 4 30W
Foyle, Lough 50 55 6N 7 8W
Foynes 51 52 30N 9 5W
Frampton on Severn 43 51 46N 2 22W
Framwellgate Moor 44 54 47N 1 34W
Francavilla Fontana 58 40 32N 17 35E
France ■ 56 47 0N 3 0E
Franceville 92 1 40S 13 32E
Franche Comté 56 46 30N 5 50E
François L. 105 54 0N 125 30W
Franeker 52 53 12N 5 33E
Frankford 51 53 10N 7 43W
Frankfort 94 27 17S 28 29E
Frankfurt □ 54 52 30N 14 0E
Frankfurt am Main 54 50 7N 8 40E
Franklin □ 98 71 0N 99 0W
Franklin Mts. 98 65 0N 125 0W
Franz 99 48 25N 84 30W
Franz Josef Land 112 81 0N 60 0E
Fraser I. 79 25 15S 153 10E
Fraser, R. 105 49 7N 123 11W
Fraserburg 94 31 55S 21 30E
Fraserburgh 48 57 41N 2 0W
Freckleton 45 53 45N 2 52W
Fredericksburg 101 38 16N 77 29W
Fredericton 99 45 57N 66 40W
Frederikshåb 96 62 0N 49 30W
Fredrikstad 61 59 13N 10 57E
Freeport 107 25 45N 88 30E
Freetown 88 8 30N 13 10W
Freiberg 54 50 55N 13 20E
Fremantle 78 32 1S 115 47E
Fremont 101 41 30N 96 30W

† *Renamed Fenoarivo*
* *Renamed Tabuaeran*
* *Renamed Taolanaro*
† *Renamed Nyanda*

French Guiana ■ Gujarat □

French Guiana ■ 108 4 0N 53 0W
Frenchpark 50 53 53N 8 25W
Fresnillo 106 23 10N 103 0W
Fresno 105 36 47N 119 50W
Freuchie 47 56 14N 3 8W
Fribourg 54 46 49N 7 9E
Friedrichshafen 54 47 39N 9 29E
Friendly, Is. = Tonga 22 19 50S 174 30W
Friesland □ 52 53 5N 5 50E
Frimley 41 51 18N 0 43W
Frinton-on-Sea 40 51 50N 1 16E
Frio, C., Brazil 108 22 50S 41 50W
Frio, C., Namibia 92 18 0S 12 0E
Frisian Is. 54 53 30N 6 0E
Friuli-Venezia Giulia □ 58 46 0N 13 0E
Frobisher B. 99 63 0N 67 0W
Frodsham 45 53 18N 2 44W
Frome 42 51 16N 2 17W
Frome, R. 43 50 44N 2 5W
Front Range 100 40 0N 105 40W
Frosinone 58 41 38N 13 20E
Froward C. 108 55 0S 71 0W
Frunze 71 42 54N 74 36E
Fuchow 73 27 50N 116 14E
Fuente Ovejuna 60 38 15N 5 25W
Fuerte, R. 106 26 0N 109 0W
Fuji-no-miya 76 35 20N 138 40E
Fuji-San 76 35 22N 138 44E
Fujisawa 76 35 22N 139 29E
Fukien □ 73 26 0N 117 30E
Fukui 76 36 0N 136 10E
Fukuoka 76 33 30N 130 30E
Fukushima 76 37 30N 140 15E
Fukuyama 76 34 35N 133 20E
Fulda 54 50 32N 9 41E
Fulda, R. 54 50 37N 9 40E
Fulwood 45 53 21N 1 33W
Fundy, B. of 99 45 0N 66 0W
Furneaux Group 79 40 10S 147 50E
Fürth 54 49 29N 11 0E
Fushun 73 41 50N 123 55E
Fusin 73 42 12N 121 33E
Fuwa 91 31 12N 30 33E
Fuyü 73 45 10N 124 50E
Fuyuan 73 48 9N 134 3E
Fylde 44 53 50N 2 58W
Fyn 61 55 20N 10 30E
Fyne, L. 46 56 0N 5 20W
Fyvie 48 57 26N 2 24W

G

Gaanda 90 10 10N 12 27E
Gabès 88 33 53N 10 2E
Gabon ■ 88 0 10S 10 0E
Gaborone 94 24 37S 25 57E
Gada 90 13 38N 5 36E
Gadsden 101 34 1N 86 0W
Gaeta 58 41 12N 13 35E
Gagnon 99 51 50N 68 5W
Gainesville 101 29 38N 82 20W
Gainsborough 44 53 23N 0 46W
Gairdner L. 79 31 30S 136 0E
Gairloch 48 57 42N 5 40W
Galapagos Is. 108 0 0 89 0W
Galashiels 46 55 37N 2 50W
Galaţi 55 45 27N 28 2E
Galatina 58 40 10N 18 10E
Galdhøpiggen 61 61 38N 8 18E
Galeana 106 24 50N 100 4W
Galesburg 101 40 57N 90 23W
Galicia, Poland 55 49 30N 23 0E
Galicia, Spain 60 42 43N 8 0W
Galilee, Sea of 89 32 53N 35 18E
Galle 70 6 5N 80 10E
Galley Hd. 51 51 32N 8 56W
Gallinas, Pta. 107 12 28N 71 40W
Gallipoli 58 40 8N 18 0E
Gällivare 61 67 9N 20 40E
Galloway 46 55 0N 4 25W
Galloway, Mull of 46 54 38N 4 50W
Gallup 100 35 30N 108 54W
Galty Mts. 51 52 22N 8 10W
Galtymore 51 52 22N 8 12W
Galula 93 8 40S 33 0E
Galveston 101 29 15N 94 48W
Galway 51 53 16N 9 4W
Galway □ 51 53 16N 9 3W
Galway B. 51 53 10N 9 20W
Gamawa 90 12 10N 10 31E
Gambia ■ 88 13 25N 16 0W
Gamtoos 94 33 52S 24 55E
Gander 99 48 58N 54 35W
Gandi 90 12 55N 5 49E
Ganga, R. 70 25 0N 88 0E
Ganganagar 70 29 56N 73 56E
Ganges = Ganga, R. 70 25 0N 88 0E
Gangtok 70 27 20N 88 37E
Gannett Pk. 100 43 15N 109 47W
Gara, L. 50 53 57N 8 26W
Garda, L. 58 45 40N 10 40E
Garden City 100 38 0N 100 45W
Gare, L. 47 56 1N 4 50W
Garelochhead 46 56 7N 4 50W
Garforth 45 53 48N 1 22W
Gargans, Mt. 58 45 37N 1 39E
Garies 94 30 32S 17 59E
Garissa 93 0 25S 39 40E
Garko 90 11 45N 8 53E
Garnock, R. 47 55 36N 4 42W
Garonne, R. 56 45 2N 0 36W
Garoua 90 9 19N 13 21E
Garry, L. 48 57 5N 4 52W
Garstang 45 53 53N 2 47W
Garston 45 53 21N 2 55W
Garvagh 50 55 0N 6 41W
Garvie Mts. 83 45 30S 168 50E
Gary 104 41 35N 87 20W
Gascony 56 43 45N 0 20E
Gascoyne, R. 78 24 52S 113 37E
Gashaka 90 7 20N 11 29E
Gashua 90 12 54N 11 0E
Gaspé Pen. 99 48 45N 65 40W
Gata, C. de 60 36 41N 2 13W
Gata, Sierra de 60 40 20N 6 20W
Gatehouse of Fleet 46 54 53N 4 10W
Gateshead 44 54 57N 1 37W
Gatley 45 53 25N 2 15W
Gävle 61 60 40N 17 9E
Gawler 79 34 30S 138 42E
Gaya 70 24 47N 85 4E
Gayndah 79 25 35S 151 39E
Gaza 89 31 30N 34 28E
Gaziantep 71 37 6N 37 23E
Gboko 90 7 17N 9 4E
Gbongan 90 7 28N 4 20E
Gdańsk 55 54 22N 18 40E
Gdańsk □ 55 54 10N 18 30E
Gdynia 55 54 35N 18 33E
Gebe, I. 72 0 5N 129 25E
Gedser 61 54 35N 11 55E
Geel 52 51 10N 4 59E
Geelong 79 38 10S 144 22E
Geelvink B. 72 3 0S 135 20E
Gela 58 37 6N 14 18E
Gelderland □ 52 52 5N 6 10E
Geleen 52 50 57N 5 49E
Gelligaer 43 51 40N 3 15W
Gelsenkirchen 52 51 30N 7 5E
Gembloux 52 50 34N 4 43E
Geneina, Gebel 91 29 2N 33 55E
Geneva 54 46 12N 6 9E
Geneva, L. 54 46 26N 6 30E
Genk 52 50 58N 5 32E
Genoa 58 44 24N 8 57E
Genoa, G. of 58 44 0N 9 0E
Gent 52 51 2N 3 37E
Geographe Chan. 78 24 30S 113 0E
George 94 33 58S 22 29E
George Town 72 5 25N 100 19E
George V Coast 112 67 0S 148 0E
Georgetown 107 6 50N 58 12W
Georgia, Str. of 105 49 25N 124 0W
Georgian B. 99 45 15N 81 0W
Georgian S.S.R. □ 71 41 0N 45 0E
Gera 54 50 53N 12 5E
Geraldton 78 28 48S 114 32E
Germany, East ■ 54 52 0N 12 0E
Germiston 94 26 11S 28 10E
Gerona 60 41 58N 2 46E
Gerrards Cross 41 51 35N 0 32W
Gevelsberg 52 51 21N 7 7E
Ghadames 88 30 11N 9 29E
Ghana ■ 88 6 0N 1 0W
Ghard Abû Muharik 91 26 50N 30 0E
Ghot Ogrein 91 31 10N 25 20E
Giant's Causeway 50 55 15N 6 30W
Giarre 58 37 44N 15 10E
Gibeon 94 25 7S 17 45E
Gibraltar 60 36 7N 5 22W
Gibraltar, Str. of 60 35 55N 5 40W
Gibson Des. 78 24 0S 126 0E
Gien 56 47 40N 2 36E
Giessen 54 50 34N 8 40E
Gifatin, Geziret 91 27 10N 33 50E
Gifu 76 35 30N 136 45E
Gigha, I. 46 55 42N 5 45W
Gijón 60 43 32N 5 42W
Gila, R. 100 32 43N 114 33W
Gilbedi 90 13 40N 5 45E
Gilbert Is. 23 1 0S 176 0E
Gilbert, R. 79 16 35S 141 15E
Gilf Kebir Plat. 91 23 0N 26 0E
Gilfach Goch 43 51 36N 3 29W
Gilford 50 54 23N 6 20W
Gilgandra 79 31 43S 148 39E
Gilgil 93 0 30S 36 20E
Gill, L. 50 54 15N 8 25W
Gillingham 41 51 23N 0 34E
Gilly 52 50 25N 4 30E
Ginowan 76 26 15N 127 47E
Gippsland 79 37 45S 147 15E
Girardot 107 4 18N 74 48W
Girga 91 26 17N 31 55E
Gironde, R. 56 45 27N 0 53W
Girvan 46 55 15N 4 50W
Gisborne 83 38 39S 178 5E
Gitega (Kitega) 93 3 26S 29 56E
Giurgiu 55 43 52N 25 57E
Gizhiga 62 62 0N 150 27E
Giżycko 55 54 2N 21 48E
Glace Bay 99 46 11N 59 58W
Glacier National Park 100 48 35N 113 40W
Glacier Peak 105 48 7N 121 7W
Glanaman 43 51 48N 3 56W
Glas Maol 48 56 52N 3 20W
Glasgow 47 55 52N 4 14W
Glastonbury 42 51 9N 2 42W
Glauchau 54 50 50N 12 33E
Gleadless Townend 45 53 20N 1 24W
Glen Affric 48 57 15N 5 0W
Glen Innes 79 29 40S 151 39E
Glen Parva 41 52 34N 1 07W
Glen Shiel 48 57 8N 5 20W
Glenarm 50 54 58N 5 58W
Glencairn 94 34 11S 18 26E
Glencoe 94 28 11S 30 11E
Glendale 105 34 7N 118 18W
Glendive 100 47 7N 104 40W
Glenfarg 47 56 16N 3 24W
Glenfield 41 52 38N 1 12W
Glenfinnan 48 56 52N 5 28W
Glenluce 46 54 53N 4 50W
Glenorchy 79 42 49S 147 18E
Glenridding 44 54 32N 2 57W
Glenties 50 54 48N 8 18W
Glin 51 52 34N 9 17W
Gliwice 55 50 22N 18 41E
Globe 100 33 25N 110 53W
Głogów 54 51 37N 16 5E
Glossop 45 53 27N 1 56W
Gloucester 43 51 52N 2 15W
Gloucestershire □ 42 51 44N 2 10W
Glyncorrwg 43 51 40N 3 39W
Gmünd 54 48 45N 15 0E
Gmunden 54 47 55N 13 48E
Gniezno 55 52 30N 17 35E
Gnowangerup 78 33 58S 117 59E
Goa 70 15 33N 73 59E
Goat Fell 46 55 37N 5 11W
Gobi, desert 73 44 0N 111 0E
Goch 52 51 40N 6 9E
Godalming 40 51 12N 0 37W
Godavari, R. 70 19 5N 79 0E
Godmanchester 40 52 19N 0 11W
Godstone 41 51 15N 0 3W
Godthåb 96 64 10N 51 46W
Godwin Austen (K2) 70 36 0N 77 0E
Goeree 52 51 50N 4 0E
Goiânia 108 16 35S 49 20W
Golborne 45 53 28N 2 36W
Golden Bay 83 40 40S 172 50E
Golden Gate 105 37 54N 122 30W
Golden Hinde, mt. 105 49 40N 125 44W
Golden Vale 51 52 33N 8 17W
Goldsworthy 78 20 21S 119 30E
Goldthorpe 45 53 32N 1 19W
Goleniów 54 53 35N 14 50E
Golspie 48 57 58N 3 58W
Gombe 90 10 19N 11 2E
Gomel 62 52 28N 31 0E
Gomersal 45 53 46N 1 49W
Gómez Palacio 106 25 40N 104 40W
Gonaives, Gulf of 107 19 29N 72 42W
Gongola □ 90 8 0N 12 0E
Good Hope, C. of 94 34 24S 18 30E
Goole 44 53 42N 0 52W
Goondiwindi 79 28 30S 150 21E
Goose Bay 99 53 15N 60 20W
Goose L. 105 42 0N 120 30W
Gorakhpur 70 26 47N 83 32E
Gordon Downs 78 18 48S 128 40E
Gore 83 46 5S 168 58E
Gorebridge 47 55 51N 3 2W
Gorey 51 52 41N 6 18W
Goring, Oxon, U.K. 40 51 31N 1 8W
Goring, W. Sussex, U.K. 40 50 49N 0 26W
Gorizia 58 45 56N 13 37E
Gorki 62 56 20N 44 0E
Gorodok 55 49 46N 23 32E
Gorseinon 43 51 40N 4 2W
Gort 51 53 4N 8 50W
Gorumna I. 51 53 15N 9 44W
Gosforth 44 55 00N 1 37W
Gospič 58 44 35N 15 23E
Gota Canal 61 58 35N 14 15E
Göteborg 61 57 43N 11 59E
Gotha 54 50 56N 10 42E
Gotland 61 57 30N 18 30E
Göttingen 54 51 31N 9 55E
Gottwaldov (Zlin) 55 49 14N 17 40E
Gouda 52 52 1N 4 42E
Gouin Rés. 99 48 35N 74 40W
Goulburn 79 34 44S 149 44E
Gourock 47 55 58N 4 49W
Govan 47 55 51N 4 19W
Gower, The 42 51 35N 4 10W
Gowrie, Carse of 47 57 30N 4 35W
Gozo, I. 58 36 0N 14 13E
Gračac 58 44 18N 15 57E
Grado 60 43 23N 6 4W
Graff-Reinet 94 32 15S 24 30E
Grafton 79 29 38S 152 58E
Graham Land 112 65 0S 64 0W
Grahamstown 94 33 19S 26 31E
Graiguenamanagh 51 52 32N 6 58W
Grampian □ 48 57 0N 3 0W
Grampian Mts. 48 56 50N 4 0W
Gran Chaco 108 25 0S 61 0W
Gran Paradiso 58 45 33N 7 17E
Gran Sasso d'Italia, Mt. 58 42 25N 13 30E
Granada 60 37 10N 3 35W
Granard 50 53 47N 7 30W
Grand Bahama I. 107 26 40N 78 30W
Grand Canyon National Park 100 36 15N 112 20W
Grand Cayman 107 19 20N 81 20W
Grand Cess 88 4 40N 8 12W
Grand Falls 99 48 56N 55 40W
Grand Forks 101 48 0N 97 3W
Grand Island 100 40 59N 98 25W
Grand Junction 100 39 0N 108 30W
Grand Rapids, Can. 98 53 12N 99 19W
Grand Rapids, U.S.A. 104 42 57N 85 40W
Grand Teton 100 43 54N 111 57W
Grande Prairie 98 55 10N 118 50W
Grange, La, Austral. 78 18 45S 121 43E
Grange, La, U.S.A. 101 33 4N 85 0W
Grange-over-Sands 44 54 12N 2 55W
Grangemouth 47 56 1N 3 43W
Grangeville 100 45 57N 116 4W
Granity 83 41 39S 171 51E
Granollers 60 41 39N 2 18E
Grantham 45 52 55N 0 39W
Grantown-on-Spey 48 57 19N 3 36W
Grants 100 35 14N 107 57W
Grants Pass 105 42 30N 123 22W
Grappenhall 45 53 22N 2 34W
Graskop 94 24 56S 30 49E
Grasmere 44 54 28N 3 2W
Grasse 56 43 38N 6 56E
Grassington 44 54 5N 2 0W
Gravesend 41 51 25N 0 22E
Grays Harbor 100 46 55N 124 8W
Grays-Thurrock 41 51 28N 0 23E
Graz 54 47 4N 15 27E
Greasbrough 45 53 27N 1 22W
Greasby 45 53 24N 3 04W
Great Abaco I. 107 26 15N 77 10W
Great Australian Bight 78 33 30S 130 0E
Great Baddow 41 51 43N 0 31E
Great Barrier I. 83 36 11S 175 25E
Great Barrier Reef 79 19 0S 149 0E
Great Basin 100 40 0N 116 30W
Great Bear L. 98 65 30N 120 0W
Great Belt 61 55 20N 11 0E
Great Bend 100 38 25N 98 55W
Great Bernera, I. 48 58 15N 6 50W
Great Britain 24 54 0N 2 15W
Great Bushman Land 94 29 20S 19 20E
Great Chesterford 40 52 4N 0 11E
Great Cumbrae I. 47 55 46N 4 57W
Great Divide 79 23 0S 146 0E
Great Dunmow 40 51 52N 0 22E
Great Falls 100 47 27N 111 12W
Great Fish R. 92 33 28S 27 5E
Great Harwood 45 53 47N 2 25W
Gt. Karas Mts. 94 27 10S 18 45E
Gt. Karoo, reg. 94 32 30S 23 0E
Great Kei, R. 94 32 15S 27 45E
Great Malvern 42 52 7N 2 19W
Great Missenden 41 51 42N 0 42W
Great Ouse, R. 40 52 20N 0 8E
Gt. St. Bernard P. 58 45 50N 7 10E
Great Salt Lake 100 41 0N 112 30W
Great Salt Lake Desert 100 40 20N 113 50W
Great Sandy Desert 78 21 0S 124 0E
Great Sankey 45 53 23N 2 37W
Great Shelford 40 52 9N 0 9E
Great Slave L. 98 61 23N 115 38W
Great Torrington 43 50 57N 4 9W
Gt. Victoria Des. 78 29 30S 126 30E
Great Wall 73 38 30N 109 30E
Gt. Winterberg, mt. 94 32 20S 26 20E
Great Winterhoek, mt. 94 33 07S 19 10E
Great Wyrley 41 52 40N 2 1W
Great Yarmouth 40 52 40N 1 45E
Greater Antilles 107 17 40N 74 0W
Greater London □ 40 51 31N 0 6W
Greater Manchester □ 44 53 30N 2 15W
Greater Sunda Is. 72 2 30S 110 0E
Greatham 44 54 38N 1 14W
Gredos, Sierra de 60 40 20N 5 0W
Greece ■ 23 40 0N 23 0E
Greeley 100 40 30N 104 40W
Green Bay 104 44 30N 88 0W
Green Island 83 45 55S 170 26E
Greenfield 45 53 32N 2 01W
Greenland 96 66 0N 45 0W
Greenland Sea 96 73 0N 10 0W
Greenlaw 46 55 42N 2 28W
Greenock 47 55 57N 4 46W
Greenore 50 54 2N 6 8W
Greensboro 101 36 7N 79 46W
Greenville, Miss., U.S.A. 101 33 25N 91 0W
Greenville, S.C., U.S.A. 101 34 54N 82 24W
Greenville, Tex., U.S.A. 101 33 5N 96 5W
Greenwich 41 51 30N 0 0
Greenwood 101 33 30N 90 4W
Gregory, L. 78 28 55S 139 0E
Gregory Ra. 79 19 30S 143 40E
Grenada ■ 107 12 10N 61 40W
Grenoble 56 45 12N 5 42E
Gretna 46 54 59N 3 4W
Grevenbroich 52 51 6N 6 32E
Grey, R. 83 42 27S 171 12E
Grey Range 79 27 0S 143 30E
Greyabbey 50 54 32N 5 35W
Greymouth 83 42 29S 171 13E
Greystones 51 53 9N 6 4W
Greytown, N.Z. 83 41 5S 175 29E
Greytown, S. Afr. 94 29 1S 30 36E
Griekwastad 94 28 49S 23 15E
Griffith 79 34 18S 146 2E
Griffithstown 43 51 41N 3 01W
Grimethorpe 45 53 34N 1 24W
Grimsby 44 53 35N 0 5W
Griqualand East 94 30 30S 29 0E
Griqualand West 94 28 40S 23 30E
Gris Nez, C. 56 50 52N 1 35E
Groblersdal 94 25 15S 29 25E
Grodno 55 53 42N 23 52E
Gronau 52 52 13N 7 2E
Groningen 52 53 15N 6 35E
Groningen □ 52 53 16N 6 40E
Groot-Brakrivier 94 34 2S 22 18E
Groot-Vloer 94 30 0S 20 40E
Groote Eylandt 79 14 0S 136 50E
Grootfontein 92 19 31S 18 6E
Gross Glockner 54 47 5N 12 40E
Grossenbrode 54 54 21N 11 4E
Grossenhain 54 51 17N 13 32E
Groznyy 71 43 20N 45 45E
Grudziądz 55 53 30N 18 47E
Gruinard B. 48 57 56N 5 35W
Grünau 94 27 45S 18 26E
Grutness 49 59 53N 1 17W
Guadalajara 106 20 40N 103 20W
Guadalcanal, I. 79 9 32S 160 12E
Guadalhorce, R. 60 36 50N 4 42W
Guadalquivir, R. 60 38 0N 4 0W
Guadalupe 60 39 27N 5 17W
Guadarrama, Sierra de 60 41 0N 4 0W
Guadeloupe, I. 107 16 20N 61 40W
Guadiana, R. 60 37 45N 7 35W
Guadix 60 37 18N 3 11W
Guam I. 23 13 27N 144 45E
Guamuchil 106 25 25N 108 3W
Guanajuato 106 21 0N 101 20W
Guanare 107 8 42N 69 12W
Guantánamo 107 20 10N 75 20W
Guaporé, R. 108 13 0S 63 0W
Guarda 60 40 32N 7 20W
Guardafui, C. 89 11 55N 51 10E
Guatemala 106 14 40N 90 30W
Guatemala ■ 106 15 40N 90 30W
Guaviare, R. 107 3 30N 71 0W
Guayaquil 108 2 15S 79 52W
Guaymas 106 27 50N 111 0W
Guecho 60 43 21N 2 59W
Guernsey I. 43 49 30N 2 35W
Guider 90 9 55N 13 59E
Guildford 41 51 14N 0 34W
Guinea ■ 88 10 20N 10 0W
Guinea Bissau ■ 88 12 0N 15 0W
Guinea, Gulf of 88 3 0N 2 30E
Guinea, Port. 88 12 0N 15 0W
Guiseley 45 53 52N 1 43W
Gujarat □ 70 23 20N 71 0E

Gujranwala **Hythe**

Gujranwala 71 32 10N 74 12E
Gulbargâ 70 17 20N 76 50E
Gulfport 101 30 28N 89 3W
Gulu 93 2 48N 32 17E
Gummersbach 52 51 2N 7 32E
Gummi 90 12 4N 5 9E
Gunnedah 79 30 59S 150 15E
Gunnislake 43 50 32N 4 12W
Guryev 62 47 5N 52 0E
Gusau 90 12 18N 6 31E
Güstrow 54 53 47N 12 12E
Guthrie 100 35 55N 97 30W
Guyana ■ 107 5 0N 59 0W
Guyenne 56 44 30N 0 40E
Gwadabawa 90 13 20N 5 15E
Gwalior 70 26 12N 78 10E
Gwanda 92 20 55S 29 0E
Gwaram 90 11 15N 9 51E
Gwarzo 90 12 20N 8 55E
Gwasero 90 9 30N 8 30E
Gweebarra B. 50 54 52N 8 21W
*Gwelo 92 19 28S 29 45E
Gwent □ 42 51 45N 2 55W
Gwynedd □ 42 53 0N 4 0W
Gympie 79 26 11S 152 38E
Gyoda 76 36 10N 139 30E
Györ 55 47 41N 17 40E
Gypsumville 98 51 45N 98 40W
Gyula 55 46 38N 21 17E

H

Ha Tinh 72 18 20N 105 54E
Haarlem 52 52 23N 4 39E
Hachinohe 76 40 30N 141 29E
Hackney 41 51 33N 0 2W
Hadarba, Ras 91 22 4N 36 51E
Haddenham 41 51 46N 0 56W
Haddington 46 55 57N 2 48W
Hadeija 90 12 30N 10 5E
Hadfield 45 53 28N 1 58W
Hadhramaut 71 15 30N 49 30E
Hadiya 71 25 30N 36 56E
Hadleigh 41 52 3N 0 58E
Haeju 73 38 3N 125 45E
Hagen 52 51 21N 7 29E
Hagi 76 34 30N 131 30E
Hague, C. de la 56 49 44N 1 56W
Hague, The 52 52 7N 4 17E
Haifa 89 32 46N 35 0E
Haikow 73 20 0N 110 20E
Hā'il 71 27 28N 42 2E
Hailar 73 49 12N 119 37E
Hailey 100 43 30N 114 15W
Hailsham 40 50 52N 0 17E
Hailun 73 47 24N 127 0E
Hailung 73 42 30N 125 40E
Hainan, I. 72 19 0N 110 0E
Hainaut □ 52 50 30N 4 0E
Haiphong 72 20 47N 106 35E
Haiti ■ 107 19 0N 72 30W
Hakodate 76 41 45N 140 44E
Halberstadt 54 51 53N 11 2E
Halcombe 83 40 8S 175 30E
Hale 45 53 24N 2 21W
Halesowen 41 52 27N 2 2W
Halesworth 40 52 21N 1 30E
Halewood 45 53 22N 2 49W
Halifax, Can. 99 44 38N 63 35W
Halifax, U.K. 45 53 43N 1 51W
Halifax B. 79 18 50S 147 0E
Halkirk 48 58 30N 3 30W
Halle, Belg. 52 50 44N 4 13E
Halle, Ger. 54 51 29N 12 0E
Halley Bay 112 75 31S 26 36W
Hall's Creek 78 18 16S 127 46E
Halmahera, I. 72 0 40N 128 0E
Halmstad 61 56 41N 12 52E
Halstead 40 51 59N 0 39E
Haltern 52 51 44N 7 10E
Haltwhistle 44 54 58N 2 27W
Hamada 76 34 50N 132 10E
Hamadān 71 34 52N 48 32E
Hamamatsu 76 34 45N 137 45E
Hamar 61 60 48N 11 7E
Hamâta, Gebel 91 24 17N 35 0E
Hambantota 70 6 10N 81 10E
Hambleton Hills 44 54 17N 1 12W
Hamburg 54 53 32N 9 59E
Hämeenlinna 61 61 0N 24 28E
Hameln 54 52 7N 9 24E
Hamersley Ra. 78 22 0S 117 45E
Hamhung 73 40 0N 127 30E
Hami 73 42 47N 93 32E
Hamilton, Austral. 79 37 45S 142 2E
Hamilton, Can. 104 43 15N 79 50W
Hamilton, N.Z. 83 37 47S 175 19E
Hamilton, U.K. 47 55 47N 4 2W
Hamm 52 51 40N 7 58E
Hammerfest 61 70 39N 23 41E
Hammersmith 41 51 30N 0 15W
Hammond 104 41 40N 87 30W
Hampden 83 45 18S 170 50E
Hamun-i-Mashkel 71 28 30N 63 0E
Hanamaki 76 39 23N 141 7E
Hanchung 73 33 10N 107 2E
Handeni 93 5 25S 38 2E
Hanford 100 36 25N 119 39W
Hangayn Nuruu 73 47 30N 100 0E
Hangchow 73 30 12N 120 1E
Hangchow Wan 73 30 30N 121 30E
Hanko 61 59 59N 22 57E
Hanku 73 39 16N 117 50E
Hanmer 83 42 32S 172 50E
Hannibal 101 39 42N 91 22W
Hanoi 72 21 5N 105 55E
Hanover, Ger. 54 52 23N 9 43E
Hanover, S. Afr. 94 31 4S 24 29E
Hantan 73 36 42N 114 30E
Haparanda 61 65 52N 24 8E
Harbin 73 45 46N 126 51E
Harburg 54 53 27N 9 58E
Hardanger Fjord, Norway 61 60 15N 6 0E
Hardanger Fjord, Norway 61 60 15N 6 0E
Hardap Dam 92 24 32S 17 50E
Harding 94 30 22S 29 55E
Harelbeke 52 50 52N 3 20E
Hari, R. 72 1 10S 101 50E
Haringey 41 51 35N 0 7W
Harlech 42 52 52N 4 7W
Harleston 40 52 25N 1 18E
Harlingen, Neth. 52 53 11N 5 25E
Harlingen, U.S.A. 100 26 20N 97 50W
Harlow 41 51 47N 0 9E
Harpenden 40 51 48N 0 20W
Harrat al Uwairidh 91 26 50N 38 0E
Harris 48 57 50N 6 55W
Harris, Sd. of 48 57 44N 7 6W
Harrisburg 104 40 18N 76 52W
Harrismith 94 28 15S 29 8E
Harrison, C. 99 54 55N 57 55W
Harrogate 44 53 59N 1 32W
Harrow 41 51 35N 0 15W
Hartford 104 41 47N 72 41W
Harthill, U.K. 47 55 52N 3 45W
Harthill, Ches., U.K. 45 53 05N 2 45W
Hartland 43 50 59N 4 29W
Hartland Pt. 43 51 2N 4 32W
Hartlepool 44 54 42N 1 11W
Harts R. 94 27 15S 25 12E
Hartshill 41 51 32N 1 31W
Harwich 40 51 56N 1 18E
Haryana □ 70 29 0N 76 10E
Harz 54 51 40N 10 40E
Hasa 71 26 0N 49 0E
Haslemere 40 51 5N 0 41W
Haslingden 45 53 43N 2 20W
Hasselt 52 50 56N 5 21E
Hastings, N.Z. 83 39 39S 176 52E
Hastings, U.K. 40 50 51N 0 36E
Hastings, U.S.A. 100 40 34N 98 22W
Hatfield 41 51 46N 0 11W
Hatherleigh 43 50 49N 4 4W
Hathersage 45 53 20N 1 39W
Hatteras, C. 101 35 10N 75 30W
Hattiesburg 101 31 20N 89 20W
Hatvan 55 47 40N 19 45E
Haugesund 61 59 23N 5 13E
Hauraki Gulf 83 36 35S 175 5E
Havana 107 23 8N 82 22W
Havant 40 50 51N 0 59W
Havasu, L. 100 34 18N 114 28W
Havel, R. 54 52 40N 12 15E
Havelock 83 41 17S 173 48E
Haverfordwest 42 51 48N 4 59W
Haverhill 40 52 6N 0 27E
Havering 41 51 33N 0 20E
Havre 100 48 40N 109 34W
Havre, Le 56 49 30N 0 5E
Hawaii I. 22 20 0N 155 0W
Hawaiian Is. 22 20 30N 156 0W
Hawarden 45 53 11N 3 2W
Hawea Lake 83 44 28S 169 19E
Hawera 83 39 35S 174 19E
Hawes 44 54 18N 2 12W
Haweswater 44 54 32N 2 48W
Hawick 46 55 25N 2 48W
Hawke B. 83 39 25S 177 20E
Hawke's Bay □ 83 39 45S 176 35E
Hawkhurst 40 51 2N 0 31E
Hawkshead 44 54 23N 3 0W
Hawkwell 41 51 35N 0 40E
Haworth 45 53 50N 1 57W
Hay 79 34 30S 144 51E
Hay-on-Wye 42 52 4N 3 9W
Hay River 98 60 51N 115 44W
Hayange 56 49 20N 6 2E
Haydock 45 53 27N 2 42W
Hayle 43 50 12N 5 25W
Hays 100 38 55N 99 25W
Hayward's Heath 40 51 0N 0 5W
Hazel Grove 45 53 23N 2 07W
Hazelton 98 55 20N 127 42W
Headcorn 40 51 10N 0 39E
Headford 51 53 28N 9 6W
Heald Green 45 53 23N 2 12W
Heanor 44 53 1N 1 20W
Heard I. 23 53 0S 74 0E
Hearst 99 49 40N 83 41W
Heathfield 40 50 58N 0 18E
Hebburn 44 54 59N 1 30W
Hebden Bridge 45 53 45N 2 0W
Hebrides, Inner Is. 48 57 20N 6 40W
Hebron 99 58 12N 62 38W
Heckmondwike 45 53 42N 1 40W
Hedgehope 83 46 12S 168 34E
Heemstede 52 52 22N 4 37E
Heerenveen 52 52 57N 5 55E
Heerlen 52 50 55N 6 0E
Heidelberg, Ger. 54 49 23N 8 41E
Heidelberg, C. Prov., S. Afr. 94 34 6S 20 59E
Heidelberg, Trans., S. Afr. 94 26 30S 28 23E
Heilbron 94 27 16S 27 59E
Heilbronn 54 49 8N 9 13E
Heilungkiang □ 73 47 30N 129 0E
Hejaz 71 26 0N 37 30E
Helena 100 46 40N 112 0W
Helensburgh 47 56 0N 4 44W
Helensville 83 36 41S 174 29E
Heligoland 54 54 10N 7 51E
Heliopolis 91 30 6N 31 17E
Hell-Ville 92 13 25S 48 16E
Hellin 60 38 31N 1 40W
Helmand, R. 71 34 0N 67 0E
Helmond 52 51 29N 5 41E
Helmsdale 48 58 7N 3 40W
Helmsley 44 54 15N 1 2W
Helsby 45 53 16N 2 47W
Helsingborg 61 56 3N 12 42E
Helsinki 61 60 15N 25 3E
Helvellyn 44 54 31N 3 1W
Helwân 91 29 50N 31 20E
Hemel Hempstead 41 51 45N 0 28W
Hemsworth 45 53 37N 1 21W
Henares, R. 60 40 55N 3 0W
Henfield 40 50 56N 0 17W
Hengelo 52 52 16N 6 48E
Hengoed 43 51 39N 3 14W
Hengyang 73 26 51N 112 30E
Henley 40 51 32N 0 53W
Hentiyn Nuruu 73 48 30N 108 30E
Henzada 70 17 38N 95 35E
Heptonstall 45 53 45N 2 01W
Herāt 71 34 20N 62 7E
Hercegnovi 58 42 30N 18 33E
Hercegovina 58 43 20N 18 0E
Hereford 42 52 4N 2 42W
Hereford and Worcester □ 42 52 10N 2 30W
Herentals 52 51 12N 4 51E
Herford 54 52 7N 8 40E
Herm I. 43 49 30N 2 28W
Herma Ness 49 60 50N 0 54W
Hermanus 94 34 27S 19 12E
Hermitage 83 43 44S 170 5E
Hermon, Mt. 89 33 20N 36 0E
Hermosillo 106 29 10N 111 0W
Herne 52 51 33N 7 12E
Herne Bay 40 51 22N 1 8E
Heron Bay 99 48 40N 86 25W
Herschel I. 98 69 35N 139 5W
Herstal 52 50 40N 5 38E
Hertford 41 51 47N 0 4W
Hertford □ 40 51 51N 0 5W
's-Hertogenbosch 52 51 42N 5 17E
Hervey B. 79 25 0S 152 52E
Herzliyya 89 32 10N 34 50E
Hesketh Bank 45 53 42N 2 51W
Hessen □ 54 50 40N 9 20E
Hessle 44 53 44N 0 28E
Heswall 45 53 19N 3 06W
Hetton-le-Hole 44 54 49N 1 26W
Hewett, C. 99 70 16N 67 45W
Hex River 94 33 30S 19 35E
Hexham 44 54 58N 2 7W
Heybridge 41 51 44N 0 42E
Heysham 44 54 5N 2 53W
Heywood 45 53 36N 2 13W
Hibbing 101 47 30N 93 0W
High Atlas, Mts. 88 32 30N 5 0W
High Blantyre 47 55 46N 4 06W
High Tatra 55 49 30N 20 00E
High Wycombe 41 51 37N 0 45W
Higham Ferrers 40 52 18N 0 36W
Highbridge 42 51 13N 2 59W
Higher Penwortham 45 53 45N 2 44W
Highland □ 48 57 30N 5 0W
Highworth 42 51 38N 1 42W
Hiiumaa 61 58 50N 22 45E
Hildesheim 54 52 9N 9 55E
Hillegom 52 52 18N 4 35E
Hillingdon 41 51 33N 0 29W
Hillston 79 33 30S 145 31E
Hilversum 52 52 14N 5 10E
Himachal Pradesh □ 70 31 30N 77 0E
Himalaya, mts. 70 29 0N 84 0E
Himeji 76 34 50N 134 40E
Hims 71 34 40N 36 45E
Hinckley 41 52 33N 1 21W
Hindhead 40 51 6N 0 42W
Hindley 45 53 32N 2 35W
Hindu Kush 71 36 0N 71 0E
Hindupur 70 13 49N 77 32E
Hines Creek 98 56 20N 118 40W
Hirara 76 24 48N 125 17E
Hirosaki 76 40 34N 140 28E
Hiroshima 76 34 30N 132 30E
Hirson 56 49 55N 4 4E
Hirwaun 43 51 43N 3 30W
Hispaniola, I. 107 19 0N 71 0W
Hitachi 76 36 36N 140 39E
Hitchin 40 51 57N 0 16W
Hjälmaren 61 59 18N 15 40E
Ho Chi Minh City 72 10 58N 106 40E
Hobart 79 42 50S 147 21E
Hobbs 100 32 40N 103 3W
Hoboken 52 51 11N 4 21E
Hoch'ih 73 24 43N 108 2E
Hochwan 73 30 0N 106 15E
Hockley 41 51 35N 0 39E
Hoddesdon 41 51 45N 0 1W
Hódmezövásárhely 55 46 28N 20 22E
Hodonín 54 48 50N 17 0E
Hoek van Holland 52 52 0N 4 7E
Hof 54 50 18N 11 55E
Hofei 73 31 52N 117 15E
Hofmeyr 94 31 39S 25 50E
Hoggar, Mts. 88 23 0N 6 30E
Hohenlimburg 52 51 21N 7 35E
Hokang 73 47 36N 130 28E
Hokitika 83 42 42S 171 0E
Hokkaidō □ 76 43 30N 143 0E
Holbeach 45 52 48N 0 1E
Holderness 44 53 45N 0 5W
Holdrege 100 40 25N 99 30W
Holguin 107 20 50N 76 20W
Hollywood 100 34 7N 118 25W
Holme 45 53 34N 1 50W
Holmfirth 45 53 34N 1 48W
Holsteinsborg 96 66 40N 53 30W
Holt 40 52 55N 1 4E
Holy I., Scotland, U.K. 46 55 31N 5 4W
Holy I., Wales, U.K. 42 53 17N 4 37W
Holyhead 42 53 18N 4 38W
Holyoke 104 42 14N 72 37W
Holywell 42 53 16N 3 14W
Holywood 50 54 38N 5 50W
Homburg 52 49 19N 7 21E
Home Hill 79 19 43S 147 25E
Honan □ 73 33 50N 113 15E
Honda 107 5 12N 74 45W
Honduras ■ 106 14 40N 86 30W
Honduras, G. of 106 16 50N 87 0W
Honey L. 105 40 13N 120 14W
Hong Kong ■ 73 22 11N 114 14E
Honiton 43 50 48N 3 11W
Honjo 76 39 23N 140 3E
Honkorâb, Ras 91 24 35N 35 10E
Honley 45 53 36N 1 48W
Honolulu 22 21 19N 157 52W
Honshū 76 36 0N 138 0E
Hood Mt. 100 45 30N 121 50W
Hoogeveen 52 52 44N 6 30E
Hook 41 51 17N 0 55W
Hook Hd. 51 52 8N 6 57W
Hoorn 52 52 38N 5 4E
Hopedale 99 55 28N 60 13W
Hopefield 94 33 3S 18 22E
Hopei □ 73 39 25N 116 45E
Hopetoun 78 33 57S 120 7E
Hopetown 94 29 34S 24 3E
Hoquiam 100 46 50N 123 55W
Horbury 45 53 40N 1 33W
Horley 40 51 10N 0 10W
Horn, C. 108 55 50S 67 30W
Horncastle 44 53 13N 0 8W
Hornsea 44 53 55N 0 10W
Horsforth 45 53 50N 1 39W
Horsham, Austral. 79 36 44S 142 13E
Horsham, U.K. 40 51 4N 0 20W
Horwich 45 53 37N 2 33W
Hospital 51 52 30N 8 28W
Hospitalet de Llobregat 60 41 21N 2 6E
Hot Springs, Ark., U.S.A. 101 34 30N 93 0W
Hot Springs, S.D., U.S.A. 100 43 25N 103 30W
Hotien (Khotan) 73 37 6N 79 59E
Houghton-le-Spring 44 54 51N 1 28W
Houhora 83 34 49S 173 9E
Houma 101 29 35N 90 50W
Hounslow 41 51 29N 0 20W
Hourn, L. 48 57 7N 5 35W
Houston 101 29 50N 95 20W
Hovd (Jargalant) 73 48 2N 91 37E
Hove 40 50 50N 0 10W
Howden 44 53 45N 0 52W
Howick 94 29 28S 30 14E
Howrah 70 22 37N 88 27E
Howth 51 53 23N 6 3W
Hoy I. 49 58 50N 3 15W
Høyanger 61 61 25N 6 50E
Hoylake 45 53 24N 3 11W
Hoyland Nether 45 53 30N 1 27W
Hrádec Králové 54 50 15N 15 50E
Hsiamen 73 24 30N 118 7E
Hsinchu 73 24 48N 120 58E
Hsuchang 73 34 1N 113 53E
Hualien 73 24 0N 121 30E
Huatabampo 106 26 50N 109 50W
Hubli-Dharwar 70 15 22N 75 15E
Hucknall 44 53 3N 1 12W
Huddersfield 45 53 38N 1 49W
Hudiksvall 61 61 43N 17 10E
Hudson Bay 99 60 0N 86 0W
Hudson, R. 101 40 42N 74 2W
Hudson Str. 99 62 0N 70 0W
Huê 72 16 30N 107 35E
Huelva 60 37 18N 6 57W
Huesca 60 42 8N 0 25W
Hughenden 79 20 52S 144 10E
Huhehot 73 40 52N 111 36E
Huixtla 106 15 9N 92 28W
Hulan 73 46 0N 126 44E
Huld 73 45 5N 105 30E
Hull, Can. 104 45 25N 75 44W
Hull, U.K. 44 53 45N 0 20W
Hull, R. 44 53 55N 0 23W
Humansdorp 94 34 2S 24 46E
Humber, R. 44 53 40N 0 10W
Humberside □ 44 53 50N 0 30W
Humboldt, R. 100 40 2N 118 31W
Humphreys Pk. 100 35 24N 111 38W
Hunan □ 73 27 30N 111 30E
Hunedoara 55 45 40N 22 50E
Hungary ■ 55 47 20N 19 20E
Hungerford 40 51 25N 1 30W
Hungkiang 73 27 0N 109 49E
Hungshui Ho, R. 73 23 24N 110 12E
Hungtse Hu 73 33 15N 118 45E
Hunsrück 52 49 30N 7 0E
Hunstanton 40 52 57N 0 30E
Hunterville 83 39 56S 175 35E
Huntingdon 40 52 20N 0 11W
Huntington, Ind., U.S.A. 101 40 52N 85 30W
Huntington, W. Va., U.S.A. 101 38 20N 82 30W
Huntly, N.Z. 83 37 34S 175 11E
Huntly, U.K. 48 57 27N 2 48W
Huonville 79 43 0S 147 5E
Hupei □ 73 31 5N 113 5E
Hurghada 91 27 15N 33 50E
Huron 100 44 30N 98 20W
Huron, L. 104 45 0N 83 0W
Hursley 40 51 1N 1 23W
Hurstpierpoint 40 50 56N 0 11W
Hutchinson 100 38 3N 97 59W
Huy 52 50 31N 5 15E
Huyton-with-Roby 45 53 24N 2 51W
Hvar, I. 58 43 11N 16 28E
Hwai Ho 73 32 20N 114 30E
Hwainan 73 32 44N 117 1E
Hwang-ho, R. 73 40 50N 107 30E
Hwangshih 73 30 27N 115 0E
Hyde 45 53 26N 2 6W
Hyderabad, India 70 17 10N 78 29E
Hyderabad, Pak. 71 25 23N 68 36E
Hyères 56 43 8N 6 9E
Hyeres, Is. 56 43 0N 6 28E
Hythe 40 51 4N 1 5E

* Renamed Gweru

Ialomita, R Kearsley

I

Ialomiţa, R. 55 44 45N 27 57E
Iaşi 55 47 10N 27 40E
Ibadan 90 7 22N 3 58E
Ibagué 107 4 27N 73 14W
Ibbenbüren 52 52 16N 7 41E
Iberian Peninsula 24 40 0N 5 0W
Ibiza 60 38 54N 1 26E
Ibiza, I. 60 39 0N 1 30E
Ibshawâi 91 29 21N 30 40E
Iceland, I. ■ 96 65 0N 19 0W
Ichang 73 30 48N 111 29E
Ichikawa 76 35 44N 139 55E
Ichinomiya 76 35 18N 136 48E
Ichinoseki 76 38 55N 141 8E
Ichun 73 47 42N 129 8E
Idah 90 6 10N 6 40E
Idaho □ 100 44 10N 114 0W
Idaho Falls 100 43 30N 112 10W
Idar-Oberstein 52 49 43N 7 19E
Idfû 91 25 0N 32 49E
• Idi Amin Dada, L. 93 0 25S 29 40E
Idutywa 94 32 8S 28 18E
Ife 90 7 30N 4 31E
Igbetti 90 8 44N 4 8E
Igbo-Ora 90 7 10N 3 15E
Igboho 90 8 40N 3 50E
Iglésias 58 39 19N 8 27E
Iguaçu Falls 108 25 41S 54 26W
Iguala 106 18 20N 99 40W
Igualada 60 41 37N 1 37E
Ihiala 90 5 40N 6 55E
Iisalmi 61 63 32N 27 10E
Ijebu-Igbo 90 6 56N 4 1E
Ijebu-Ode 90 6 47N 3 52E
IJmuiden 52 52 28N 4 35E
IJsselmeer 52 52 45N 5 20E
Ikare 90 7 18N 5 40E
Ikeja 90 6 28N 3 45E
Ikerre 90 7 25N 5 19E
Ikire 90 7 10N 4 15E
Ikot Ekpene 90 5 12N 7 40E
Ila 90 8 0N 4 51E
Ilaro Agege 90 6 53N 3 3E
Ilchester 43 51 0N 2 41W
Île de France □ 56 49 0N 2 20E
Ilebo 92 4 17S 20 47E
Ilero 90 8 0N 3 20E
Ilesha 90 7 37N 4 40E
Ilfracombe 42 51 13N 4 8W
Ilkeston 45 52 59N 1 19W
Ilkhuri Shan 73 51 30N 124 0E
Ilkley 45 53 56N 1 49W
Illinois □ 101 40 15N 89 30W
Ilmen L. 61 58 15N 31 10E
Ilminster 43 50 55N 2 56W
Ilobu 90 7 45N 4 25E
Iloilo 72 10 45N 122 33E
Ilora 90 7 45N 3 50E
Ilorin 90 8 30N 4 35E
Imatra 61 61 12N 28 48E
Imbaba 91 30 5N 31 12E
Immingham 44 53 37N 0 12W
Imo □ 90 5 15N 7 20E
Imola 58 44 20N 11 42E
Impéria 58 43 52N 8 0E
Imphal 70 24 48N 93 56E
In Salah 88 27 10N 2 32E
Ina 76 35 50N 138 0E
Inangahua Junc. 83 41 52S 171 59E
Inari 61 68 54N 27 5E
Inari, L. 61 69 0N 28 0E
Inca 60 39 43N 2 54E
Inch'ŏn 73 37 27N 126 40E
Indal 61 62 35N 17 5E
India ■ 70 20 0N 80 0E
Indian Harbour 99 54 27N 57 13W
Indian Ocean 23 5 0S 75 0E
Indiana □ 101 40 0N 86 0W
Indianapolis 101 39 42N 86 10W
Indigirka, R. 62 69 0N 147 0E
Indonesia ■ 72 5 0S 115 0E
Indore 70 22 42N 75 53E
Indre, R. 56 47 2N 1 8E
Indus, R. 71 28 40N 70 10E
Ingatestone 41 51 40N 0 23W
Inglefield Land 99 78 30N 70 0W
Inglewood 83 39 9S 174 14E
Ingolstadt 54 48 45N 11 26E
Inhambane 92 23 54S 35 30E
Ining (Kuldja) 73 43 57N 81 20E
Inishbofin I. 51 53 35N 10 12W
Inishcrone 50 54 13N 9 05W
Inishmore, I. 51 53 8N 9 45W
Inishowen, Pen. 50 55 14N 7 15W
Inishturk I. 50 53 42N 10 8W
Inistioge 51 52 30N 7 5W
Inn, R. 54 48 35N 13 28E
Innellan 47 55 54N 4 58W
Inner Mongolia □ 73 44 50N 117 40E
Innisfail 79 17 33S 146 5E
Innsbruck 54 47 16N 11 23E
Inoucdjouac 99 58 27N 78 6W
Inowrocław 55 52 50N 18 20E
Interlaken 54 46 41N 7 50E
Inuvik 98 68 16N 133 40W
Inveraray 46 56 13N 5 5W
Inverbervie 48 56 50N 2 17W
Invercargill 83 46 24S 168 24E
Inverell 79 29 45S 151 8E
Invergarry 48 57 5N 4 48W
Invergordon 48 57 41N 4 10W
Inverkeithing 47 56 2N 3 24W
Invermoriston 48 57 13N 4 38W
Inverness 48 57 29N 4 12W
Inverurie 48 57 15N 2 21W
Investigator Str. 79 35 30S 137 0E
Iona I. 46 56 20N 6 25W
Ionian Sea 58 37 30N 17 30E
Iowa □ 101 42 18N 93 30W
Ipin 73 28 48N 104 33E

• *Renamed Edward, L.*

Ipoh 72 4 35N 101 5E
Ipswich, Austral. 79 27 35S 152 46E
Ipswich, U.K. 40 52 4N 1 9E
Iquique 108 20 19S 70 5W
Iquitos 108 3 45S 73 10W
Iran ■ 71 33 0N 53 0E
Iran Ra. 72 2 20N 114 50E
Irapuato 106 20 40N 101 40W
Iraq ■ 71 33 0N 44 0E
Ireland ■ 51 53 0N 8 0W
Ireland's Eye 51 53 25N 6 4W
Irele 90 7 40N 5 40E
Irian Jaya □ 72 4 0S 137 0E
Iringa 93 7 48S 35 43E
Iriomote-Jima 76 24 19N 123 48E
Irish Sea 24 54 0N 5 0W
Irkutsk 62 52 10N 104 20E
Irlam 45 53 26N 2 27W
Iron Knob 79 32 46S 137 8E
Ironbridge 45 52 38N 2 29W
Irrawaddy, R. 70 15 50N 95 6E
Irthlingborough 40 52 20N 0 37W
Irtysh, R. 62 53 36N 75 30E
Irún 60 43 20N 1 52W
Irvine 47 55 37N 4 40W
Irvinestown 50 54 28N 7 38W
Isahaya 76 32 52N 130 2E
Isar, R. 54 48 40N 12 30E
Isbister 49 60 22N 0 54W
Ischia, I. 58 40 45N 13 51E
Isère, R. 56 45 15N 5 30E
Iserlohn 52 51 22N 7 40E
Iseyin 90 8 0N 3 36E
Ishigaki 76 24 20N 124 10E
Ishikari-Wan 76 43 20N 141 20E
Ishikawa 76 26 25N 127 48E
Ishinomaki 76 38 32N 141 20E
Isiolo 93 0 24N 37 33E
Isipingo 94 30 00S 30 57E
Isipingo Beach 94 30 00S 30 57E
Isiro 92 2 53N 27 58E
Iskenderun 71 36 32N 36 10E
Isla, R. 48 56 32N 3 20W
Islamabad 71 33 40N 73 0E
Islay, I. 46 55 46N 6 10W
Islington 41 51 32N 0 06W
Ismâ'ilîya 91 30 37N 32 18E
Isna 91 25 17N 32 30E
Isoka 93 10 4S 32 42E
Ispica 58 36 47N 14 53E
Israel ■ 89 32 0N 34 50E
Issoire 56 45 32N 3 15E
Issyk-Kul, L. 71 42 25N 77 15E
Istanbul 71 41 0N 29 0E
Istra 58 45 10N 14 0E
Ithaca 104 42 25N 76 30W
Itsa 91 29 15N 30 40E
Itu 90 5 10N 7 58E
Ivanhoe 79 32 56S 144 20E
Ivano-Frankovsk 55 49 0N 24 40E
Ivanovo 62 57 5N 41 0E
Ivory Coast ■ 88 7 30N 5 0W
Ivrea 58 45 30N 7 52E
Ivugivik 99 62 24N 77 55W
Ivybridge 43 50 24N 3 56W
Iwaki 76 37 3N 140 55E
Iwamisawa 76 43 12N 141 46E
Iwanai 76 42 58N 140 30E
Iwanuma 76 38 7N 140 58E
Iwate-San 76 39 51N 141 0E
Iwo 90 7 39N 4 9E
Izegem 52 50 55N 3 12E
• Izhevsk 62 56 51N 53 14E
Izmail 55 45 22N 28 46E
Izmir 71 38 25N 27 8E
Izumo 76 35 20N 132 55E

J

Jabalpur 70 23 9N 79 58E
Jaca 60 42 35N 0 33W
Jackson, Mich., U.S.A. 104 42 18N 84 25W
Jackson, Miss., U.S.A. 101 32 20N 90 10W
Jackson, Tenn., U.S.A. 101 35 40N 88 50W
Jacksons 83 42 46S 171 32E
Jacksonville, Fla., U.S.A. 101 30 15N 81 38W
Jacksonville, Ill., U.S.A. 101 39 42N 90 15W
Jacobabad 71 28 20N 68 29E
Jaén 60 37 44N 3 43W
Jaffna 70 9 45N 80 2E
Jagdalpur 70 19 3N 82 6E
Jagersfontein 94 29 44S 25 27E
Jahrom 71 28 30N 53 31E
Jaipur 70 27 0N 76 10E
Jakarta 72 6 9S 106 49E
Jalalabad 71 34 30N 70 29E
Jalapa 106 19 30N 96 50W
Jalgaon 70 21 0N 75 42E
Jalingo 90 8 55N 11 25E
Jalón, R. 60 41 20N 1 40W
Jamaari 90 11 44N 9 53E
Jamaica, I. ■ 107 18 10N 77 30W
Jambi 72 1 38S 103 30E
Jamestown, Austral. 79 33 10S 138 32E
Jamestown, N.D., U.S.A. 100 47 0N 98 45W
Jamestown, N.Y., U.S.A. 104 42 5N 79 18W
Jammu 70 32 43N 74 54E
Jammu & Kashmir □ 70 34 25N 77 0E
Jamnagar 70 22 30N 70 0E
Jamshedpur 70 22 44N 86 20E
Jan Mayen Is. 112 71 0N 11 0W
Jansenville 94 32 57S 24 39E
Japan ■ 76 36 0N 136 0E
Japan, Sea of 76 40 0N 135 0E
Japurá, R. 108 3 8S 64 46W
Jarosław 55 50 2N 22 42E
Jarrow 44 54 58N 1 28W
Jarvis I. 22 0 15S 159 55W

• *Renamed Ustinov*

Jasło 55 49 45N 21 30E
Játiva 60 39 0N 0 32W
Java, I. 72 7 0S 110 0E
Java Sea 72 4 35S 107 15E
Jebba 90 9 9N 4 48E
Jedburgh 46 55 28N 2 33W
Jędrzejów 55 50 35N 20 15E
Jefferson, Mt. 105 44 45N 121 50W
Jega 90 12 15N 4 23E
Jelenia Góra 54 50 50N 15 45E
Jelgava 61 56 41N 22 49E
Jemappes 52 50 27N 3 54E
Jemeppe 52 50 37N 5 30E
Jena 54 50 56N 11 33E
Jérez 60 36 41N 6 7W
Jersey City 104 40 41N 74 8W
Jersey, I. 43 49 13N 2 7W
Jerusalem 89 31 47N 35 10E
Jervis Bay 79 35 8S 150 43E
Jhansi 70 25 30N 78 36E
Jhelum 71 33 0N 73 45E
Jiddah 71 21 29N 39 16E
Jido 70 29 2N 94 58E
Jihlava 54 49 28N 15 35E
Jiloca, R. 60 41 0N 1 20W
Jinja 93 0 25N 33 12E
Jiu, R. 55 44 50N 23 20E
João Pessoa 108 7 10S 34 52W
Jodhpur 70 26 23N 73 2E
Joensuu 61 62 37N 29 49E
Johannesburg 94 26 10S 28 8E
John Day, R. 105 45 44N 120 39W
John o' Groats 48 58 39N 3 3W
Johnston Lakes 78 32 20S 120 45E
Johnstone 47 55 50N 4 31W
Johnstown, Ireland 51 52 46N 7 34W
Johnstown, U.S.A. 104 40 19N 78 53W
Johor Baharu 72 1 28N 103 46E
Joliet 104 41 30N 88 0W
Joliette 99 46 3N 73 24W
Jolo I. 72 6 0N 121 0E
Jones Sound 99 76 0N 89 0W
Jönköping 61 57 45N 14 10E
Jonquière 99 48 27N 71 14W
Joplin 101 37 0N 94 25W
Jordan ■ 89 31 0N 36 0E
Jordan, R. 89 31 48N 35 32E
Jorhat 70 26 45N 94 20E
Jos 90 9 53N 8 51E
Joseph Bonaparte G. 78 14 35S 128 50E
Jotunheimen 61 61 35N 8 25E
Juan de Fuca Str. 105 48 15N 124 0W
Juan Fernandez, Is. 108 33 50S 80 0W
Juba 89 4 57N 31 35E
Jubal, Str. of 91 27 30N 34 0E
Júcar, R. 60 40 8N 2 13W
Júcaro 107 21 37N 78 51W
Judea 89 31 35N 34 57E
Julia Cr. 79 20 0S 141 11E
Julianehåb 96 60 43N 46 0W
Jullundur 70 31 20N 75 40E
Jumet 52 50 27N 4 25E
Jumilla 60 38 28N 1 19W
Jumna, R. 70 27 0N 78 30E
Junagadh 70 21 30N 70 30E
Juneau 98 58 26N 134 30W
Junee 79 34 53S 147 35E
Junta, La 100 38 0N 103 30W
Jura 56 46 35N 6 5E
Jura, I. 46 56 0N 5 50W
Jura, Mts. 54 46 40N 6 5E
Jura, Paps of. 46 55 55N 6 0W
Jura, Sd. of 46 55 57N 5 45W
Jutland 61 56 0N 8 0E
Jye-kundo 73 33 0N 96 50E
Jyväskylä 61 62 14N 25 44E

K

K2, Mt. 70 36 0N 77 0E
Kabale 93 1 15S 30 0E
Kabalo 92 6 0S 27 0E
Kabarega Falls 93 2 15N 31 38E
Kabarnet 93 0 31N 35 44E
Kabba 90 7 57N 6 3E
Kabinda 92 6 23S 24 28E
Kabul 71 34 28N 69 18E
Kabwe 92 14 30S 28 29E
Kachin □ 70 26 0N 97 0E
Kadina 79 34 0S 137 43E
Kaduna 90 10 30N 7 21E
Kafanchan 90 9 40N 8 20E
Kafareti 90 10 25N 11 12E
Kafr el Dauwâr 91 31 8N 30 8E
Kafue 92 15 46S 28 9E
Kafue, R. 92 15 30S 26 0E
Kagadi 93 0 58N 30 48E
Kagoshima 76 31 36N 130 40E
Kahe 93 3 30S 37 25E
Kai Is. 72 5 55S 132 45E
Kaiapoi 83 42 24S 172 40E
Kaifeng 73 34 49N 114 30E
Kaikohe 83 35 25S 173 49E
Kaikoura 83 42 25S 173 43E
Kaikoura Ra. 83 41 59S 173 41E
Kaimanawa Mts. 83 39 15S 175 56E
Kainji Res. 90 10 1N 4 40E
Kairuku 79 8 51S 146 35E
Kaiserslautern 52 49 30N 7 43E
Kaitaia 83 35 8S 173 17E
Kaitangata 83 46 17S 169 51E
Kajaani 61 64 17N 27 46E
Kajabbi 79 20 0S 140 1E
Kakamas 94 28 45S 20 33E
Kakamega 93 0 20N 34 46E
Kakanui Mts. 83 45 10S 170 30E
Kakia 94 24 48S 23 22E
Kakinada 70 16 50N 82 11E
Kalahari, Des. 92 24 0S 22 0E
Kalamazoo 104 42 20N 85 35W
Kalemie 93 5 55S 29 9E
Kalgoorlie 78 30 40S 121 22E
Kalimantan Barat □ 72 0 0 110 30E
Kalinin 62 56 55N 35 55E
Kaliningrad 61 54 42N 20 32E
Kalispell 100 48 10N 114 22W
Kalisz 55 51 45N 18 8E
Kaliua 93 5 5S 31 48E
Kalmar 61 56 40N 16 20E
Kama, R. 62 60 0N 53 0E
Kamaishi 76 39 20N 142 0E
Kamba 90 11 50N 3 45E
Kamchatka Pen. 62 57 0N 160 0E
Kamenets-Podolskiy 55 48 45N 26 10E
Kamenjak, C. 58 44 47N 13 55E
Kamenka Bugskaya 55 50 8N 24 16E
Kames 46 55 53N 5 15W
Kamina 92 8 45S 25 0E
Kamloops L. 98 50 45N 120 40W
Kamp-Lintfort 52 51 31N 6 32E
Kampala 93 0 20N 32 30E
Kampen 52 52 33N 5 53E
Kananga 92 5 55S 22 18E
Kanazawa 76 36 30N 136 38E
Kanchenjunga, Mt. 70 27 50N 88 10E
Kanchow 73 25 58N 114 55E
Kandahar 71 31 32N 65 30E
Kandalaksha 61 67 9N 32 30E
Kandi 90 11 7N 2 55E
Kandy 70 7 18N 80 43E
Kangaroo I. 79 35 45S 137 0E
Kangean Is. 72 6 55S 115 23E
Kangnüng 73 37 45N 128 54E
Kaniapiskau, R. 99 57 40N 69 30W
Kanin Pen. 62 68 0N 45 0E
Kankakee 104 41 6N 87 50W
Kankan 88 10 30N 9 15W
Kanker 70 20 10N 81 40E
Kano 90 12 2N 8 30E
Kanowna 78 30 32S 121 31E
Kanoya 76 31 25N 130 50E
Kanpur 70 26 35N 80 20E
Kansas □ 100 38 40N 98 0W
Kansas City 101 39 0N 94 40W
Kansas, R. 101 39 7N 94 36W
Kantché 90 13 31N 8 30E
Kanturk 51 52 10N 8 55W
Kanye 94 25 0S 25 28E
Kaohsiung 73 22 35N 120 16E
Kapela, Mts. 58 44 40N 15 40E
Kapuas Hulu Ra. 72 1 30N 113 30E
Kapuas, R. 72 0 20N 111 40E
Kara Bogaz Gol, Zaliv 71 41 0N 53 30E
Kara Kalpak A.S.S.R. □ 71 43 0N 60 0E
Kara Sea 62 75 0N 70 0E
Karachi 71 24 53N 67 0E
Karaganda 62 49 50N 73 0E
Karakoram 70 35 20N 76 0E
Karakum, Peski 71 39 30N 60 0E
Karamai 73 45 57N 84 30E
Karamea Bight 83 41 22S 171 40E
Karasburg 94 28 0S 18 44E
Karatsu 76 33 30N 130 0E
Karawanken 58 46 30N 14 40E
Karbalā 71 32 47N 44 3E
Karcag 55 47 19N 21 1E
Kareeberge 94 30 50S 22 0E
Karelian A.S.S.R. □ 61 65 30N 32 30E
Kariba Lake 92 16 40S 28 25E
Karikal 70 10 59N 79 50E
Karimata I. 72 1 40S 109 0E
Karkaralinsk 73 49 27N 75 37E
Karkur Tohl 91 22 5N 25 5E
Karl-Marx-Stadt 54 50 50N 12 55E
Karlovac 58 45 31N 15 36E
Karlskrona 61 56 10N 15 35E
Karlsruhe 54 49 3N 8 23E
Karlstad 61 59 23N 13 30E
Karnataka □ 70 13 15N 77 0E
Kars 71 40 40N 43 5E
Karsakpay 62 47 55N 66 40E
Karshi 71 38 53N 65 48E
Karungu 93 0 50S 34 10E
Kasama 93 10 16S 31 9E
Kāshān 71 34 5N 51 30E
Kashgar 73 39 46N 75 52E
Kashima 76 33 7N 130 6E
Kashing 73 30 45N 120 41E
Kashiwazaki 76 37 22N 138 33E
Kasongo 92 4 30S 26 33E
Kassaba 91 22 40N 29 55E
Kassala 89 15 23N 36 26E
Kassel 54 51 19N 9 32E
Kastamonu 71 41 25N 33 43E
Katanning 78 33 40S 117 33E
Katha 70 24 10N 96 30E
Katherîna, Gebel 91 28 30N 33 57E
Katherine 78 14 27S 132 20E
Katihar 70 25 34N 87 36E
Katmandu 70 27 45N 85 12E
Katoomba 79 33 41S 150 19E
Katowice 55 50 17N 19 5E
Katrine, L. 46 56 15N 4 30W
Katsina 90 7 10N 9 20E
Katsuura 76 35 15N 140 20E
Kattakurgan 71 39 55N 66 15E
Kattegat 61 57 0N 11 20E
Katumba 93 7 40S 25 17E
Katwe 93 0 8S 29 52E
Katwijk-aan-Zee 52 52 12N 4 24E
Kaunas 61 54 54N 23 54E
Kawagoe 76 35 55N 139 29E
Kawaguchi 76 35 52N 138 45E
Kawambwa 93 9 48S 29 3E
Kawerau 83 38 7S 176 42E
Kawthoolei □ 70 18 0N 97 30E
Kayah □ 70 19 15N 97 15E
Kayseri 71 38 45N 35 30E
Kazakhstan 62 51 11N 53 0E
Kazan 62 55 48N 49 3E
Kazatin 55 49 45N 28 50E
Kāzerūn 71 29 38N 51 40E
Keady 50 54 15N 6 42W
Kearsley 45 53 33N 2 23W

Kebnekaise, mt Langholm

Kebnekaise, mt. 61 67 54N 18 33E
Kecskemét 55 46 57N 19 35E
Kediri 72 7 51S 112 1E
Keeper Hill 51 52 46N 8 17W
Keetmanshoop 94 26 35S 18 8E
Keewatin □ 98 63 20N 94 40W
Keffi 90 8 55N 7 43E
Keighley 45 53 52N 1 54W
Keimoes 94 28 41S 21 0E
Keith 48 57 33N 2 58W
Kelang 72 3 2N 101 26E
Kellerberrin 78 31 36S 117 38E
Kellogg 100 47 30N 116 5W
Kelowna 105 49 50N 119 25W
Kelso 46 55 36N 2 27W
Kelty 47 56 08N 3 23W
Kem 61 65 0N 34 38E
Kemerovo 62 55 20N 85 50E
Kemi 61 65 44N 24 34E
Kemi, R. 61 67 30N 28 30E
Kemp Coast 112 69 0S 55 0E
Kempsey 79 31 1S 152 50E
Kempten 54 47 42N 10 18E
Kemsing 41 51 18N 0 14E
Kendal 44 54 19N 2 44W
Kendari 72 3 50S 122 30E
Kende 90 11 30N 4 12E
Keng Tung 70 21 0N 99 30E
Kenhardt 94 29 19S 21 12E
Kenilworth 42 52 22N 1 35W
Kenitra 88 34 15N 6 40W
Kenmare 51 51 52N 9 35W
Kenmare, R. 51 51 40N 10 0W
Kennet, R. 40 51 24N 1 7W
Kennewick 100 46 11N 119 2W
Kenosha 104 42 33N 87 48W
Kent □ 40 51 12N 0 40E
Kentucky □ 101 37 20N 85 0W
Kentville 99 45 6N 64 29W
Kenya ■ 93 2 20N 38 0E
Kenya, Mt. 93 0 10S 37 18E
Kerala □ 70 11 0N 76 15E
Kerama-Shotō 76 26 12N 127 22E
Kerang 79 35 40S 143 55E
Kerch 62 45 20N 36 20E
Kerema 79 7 58S 145 50E
Kerguelen I. 23 48 15S 69 10E
Kericho 93 0 22S 35 15E
Kerinci 72 2 5S 101 0E
Kerki 71 37 50N 65 12E
Kermadec Is. 22 31 8S 175 16W
Kermān 71 30 15N 57 1E
*Kermānshāh 71 34 23N 47 0E
Kerrera I. 46 56 24N 5 32W
Kerry □ 51 52 7N 9 35W
Kerulen, R. 73 48 48N 117 0E
Kesh 50 54 31N 7 43W
Keswick 44 54 35N 3 9W
Ketchikan 96 55 25N 131 40W
Kętrzyn 55 54 7N 21 22E
Kettering 45 52 24N 0 44W
Kexbrough 45 53 35N 1 32W
Key West 101 24 40N 82 0W
Keynsham 42 51 25N 2 30W
Khalig el Tina, B. 91 31 20N 32 42E
Kharagpur 70 22 20N 87 25E
Kharga, Oasis de 91 25 0N 30 0E
Kharit, Wadi el 91 24 5N 34 10E
Kharkov 62 49 58N 36 20E
Khartoum 89 15 31N 32 35E
Khashm el Girba 89 14 59N 35 58E
Khasi Hills 70 25 30N 91 30E
Khatanga 62 72 0N 102 20E
Kherson 62 46 35N 32 35E
Khetinsiring 73 32 54N 92 50E
Khilok 73 51 30N 110 45E
Khiva 71 41 30N 60 18E
Kholm 61 57 10N 31 15E
Khong, R. 72 15 0N 106 50E
Khorromshahr 71 30 29N 48 15E
Khotin 55 48 31N 26 27E
Khurasan, prov. 71 34 0N 57 0E
Khust 55 48 10N 23 18E
Khvoy 71 38 35N 45 0E
Khyber Pass 70 34 10N 71 8E
Kiambu 93 1 8S 36 50E
Kiamusze 73 46 45N 130 30E
Kian 73 27 1N 114 58E
Kiangsi □ 73 27 20N 115 40E
Kiangsu □ 73 33 0N 119 50E
Kiaohsien 73 36 20N 120 0E
Kibwezi 93 2 27S 37 57E
Kicking Horse Pass 98 51 28N 116 16W
Kidderminster 41 52 24N 2 13W
Kidsgrove 44 53 6N 2 15W
Kidwelly 42 51 44N 4 20W
Kiel 54 54 16N 10 8E
Kiel B. 54 54 20N 10 20E
Kielce 55 50 58N 20 42E
Kienow 73 27 0N 118 16E
Kienshui 73 23 57N 102 45E
Kiev 55 50 30N 30 28E
Kigali 93 1 5S 30 4E
Kigoma-Ujiji 93 5 30S 30 0E
Kii Chan. 76 33 40N 135 0E
Kikinda 55 45 50N 20 30E
Kikori 79 7 13S 144 15E
Kilbeggan 51 53 22N 7 30W
Kilbirnie 47 55 46N 4 42W
Kilcormac 51 53 11N 7 44W
Kilcreggan 47 55 59N 4 50W
Kilcullen 51 53 8N 6 45W
Kildare 51 53 10N 6 50W
Kildare □ 51 53 10N 6 50W
Kildonan 48 58 10N 3 50W
Kilgetty 42 51 43N 4 43W
Kilifi 93 3 40S 39 48E
Kilimanjaro, Mt. 93 3 7S 37 20E
Kilindini 93 4 4S 39 40E
Kiliya 55 45 28N 29 16E
Kilkee 51 52 41N 9 40W
Kilkeel 50 54 4N 6 0W
Kilkenny 51 52 40N 7 17W
Kilkenny □ 51 52 35N 7 15W
Kill 51 52 11N 7 20W
Killala B. 50 54 20N 9 12W
Killaloe 51 52 48N 8 28W
Killamarsh 45 53 19N 1 19W
Killarney 51 52 2N 9 30W
Killarney, L's. of 51 52 0N 9 30W
Killashandra 50 54 1N 7 32W
Killay 42 51 36N 4 02W
Killenaule 51 52 35N 7 40W
Killiecrankie, Pass of 48 56 44N 3 46W
Killimor 51 53 10N 8 17W
Killiney 51 53 15N 6 8W
Killorglin 51 52 6N 9 48W
Killybegs 50 54 38N 8 26W
Kilmacolm 47 55 54N 4 39W
Kilmacthomas 51 52 13N 7 27W
Kilmallock 51 52 22N 8 35W
Kilmarnock 47 55 36N 4 30W
Kilmaurs 47 55 37N 4 33W
Kilosa 93 6 48S 37 0E
Kilrea 50 54 58N 6 34W
Kilronan 51 53 8N 9 40W
Kilrush 51 52 39N 9 30W
Kilsyth 47 55 58N 4 3W
Kiltamagh 50 53 52N 9 0W
Kilwa Kisiwani 93 8 58S 39 32E
Kilwa Kivinje 93 8 45S 39 25E
Kilwinning 47 55 40N 4 41W
Kimba 79 33 8S 136 23E
Kimberley, Austral. 78 16 20S 127 0E
Kimberley, S. Afr. 94 28 43S 24 46E
Kinabalu, mt. 72 6 0N 116 0E
Kinbrace 48 58 16N 3 56W
Kincardine 47 56 4N 3 43W
Kindu 92 2 55S 25 50E
King George Is. 99 53 40N 80 30W
King I. 79 39 50S 144 0E
King Leopold Ranges 78 17 20S 124 20E
King Sd. 78 16 50S 123 20E
King William's Town 94 32 51S 27 22E
Kinghorn 47 56 4N 3 10W
Kingman 100 35 12N 114 2W
Kings Langley 41 51 42N 0 27W
King's Lynn 40 52 45N 0 25E
King's Worthy 40 51 6N 1 18W
Kingscourt 51 53 55N 6 48W
Kingsclere 40 51 19N 1 15W
Kingston, Can. 104 44 14N 76 30W
Kingston, Jamaica 107 18 0N 76 50W
Kingston, N.Z. 83 45 20S 168 43E
Kingston, U.S.A. 104 41 55N 74 0W
Kingston-upon-Thames 41 51 23N 0 20W
Kingstown 107 13 10N 61 10W
Kingsville 100 27 30N 97 53W
Kingswood, Glos., U.K. 43 51 26N 2 31W
Kingswood, War., U.K. 41 52 20N 1 43W
Kingtehchen (Fowliang) 73 29 8N 117 21E
Kington 42 52 12N 3 2W
Kingussie 48 57 5N 4 2W
Kinleith 83 38 20S 175 56E
Kinloch 83 44 51S 168 20E
Kinlochewe 48 57 37N 5 20W
Kinlochleven 48 56 42N 4 59W
Kinnairds Hd. 48 57 40N 2 0W
Kinnegad 51 53 28N 7 8W
Kinross 47 56 13N 3 25W
Kinsale 51 51 42N 8 31W
Kinsale, Old Hd. of 51 51 37N 8 32W
Kinshasa 92 4 20S 15 15E
Kinston 101 35 18N 77 35W
Kintore 48 57 14N 2 20W
Kintyre 46 55 30N 5 35W
Kintyre, Mull of 46 55 17N 5 4W
Kipini 93 2 30S 40 32E
Kippax 45 53 46N 1 22W
Kirby Muxloe 41 52 37N 1 13W
Kirensk 62 57 50N 107 55E
Kirgiz S.S.R. □ 71 42 0N 75 0E
Kirin 73 43 58N 126 31E
Kirin □ 73 43 50N 125 45E
Kirkburton 45 53 36N 1 42W
Kirkby 45 53 29N 2 54W
Kirkby Lonsdale 44 54 13N 2 36W
Kirkby Moorside 44 54 16N 0 56W
Kirkby Steven 44 54 27N 2 23W
Kirkcaldy 47 56 7N 3 10W
Kirkcolm 46 54 59N 5 4W
Kirkconnel 46 55 23N 4 0W
Kirkcudbright 46 54 50N 4 3W
Kirkham 45 53 47N 2 52W
Kirkheaton 45 53 39N 1 44W
Kirkintilloch 47 55 57N 4 10W
Kirkland Lake 99 48 9N 80 2W
Kirkliston 47 55 55N 3 27W
Kirkūk 71 35 30N 44 21E
Kirkwall 49 58 59N 2 59W
Kirkwood 94 33 22S 25 15E
Kirov 62 58 35N 49 40E
Kirovabad 71 40 45N 46 10E
Kirovsk 61 67 48N 33 50E
Kirtachi 90 12 52N 2 30E
Kiruna 61 67 52N 20 15E
Kiryū 76 36 24N 139 20E
Kisangani 92 0 35N 25 15E
Kishangarh 70 27 50N 70 30E
Kishinev 55 47 0N 28 50E
Kishiwada 76 34 28N 135 22E
Kisi 73 45 21N 131 0E
Kisii 93 0 40S 34 45E
Kiskörös 55 46 37N 19 20E
Kiskunfelégyháza 55 46 42N 19 53E
Kismayu 89 0 20S 42 30E
Kisumu 93 0 3S 34 45E
Kitakyūshū 76 33 50N 130 50E
Kitale 93 1 0N 35 12E
Kitami 76 43 48N 143 54E
Kitchener 99 43 30N 80 30W
Kitgum Matidi 93 3 17N 32 52E
Kitimat 98 54 3N 128 38W
Kitui 93 1 17S 38 0E
Kitwe 92 12 54S 28 7E
Kiukiang 73 29 37N 116 2E
Kizil Avvat 71 39 0N 56 25E
Kizlyar 71 43 51N 46 40E
Kladno 54 50 10N 14 7E
Klagenfurt 54 46 38N 14 20E
Klaipeda 61 55 43N 21 10E
Klamath Falls 105 42 20N 121 50W
Klamath, R. 105 41 40N 124 4W
Klatovy 54 49 23N 13 18E
Klawer 94 31 44S 18 36E
Klerksdorp 94 26 51S 26 38E
Kleve 52 51 46N 6 10E
Klipplaat 94 33 0S 24 22E
Klondike 98 64 0N 139 26W
Klyuchevsk, mt. 62 55 50N 160 30E
Knapdale 46 55 55N 5 30W
Knaresborough 44 54 1N 1 29W
Knighton 42 52 21N 3 2W
Knockmealdown Mts. 51 52 16N 8 0W
Knokke 52 51 20N 3 17E
Knottingley 44 53 42N 1 15W
Knowle 41 52 23N 1 43W
Knowsley 45 53 27N 2 51W
Knoxville 101 35 58N 83 57W
Knoydart, dist. 48 57 3N 5 33W
Knutsford 45 53 18N 2 22W
Knysna 94 34 2S 23 2E
Kobarid 58 46 15N 13 30E
Kobayashi 76 31 56N 130 59E
Kōbe 76 34 45N 135 10E
Koblenz 52 50 21N 7 36E
Kobroor, I. 72 6 10S 134 30E
Kočevje 58 45 39N 14 50E
Kōchi 76 33 30N 133 35E
Kōfu 76 35 40N 138 30E
Kogota 76 38 33N 141 3E
Kohlscheid 52 50 50N 6 6E
Kohtla-Järve 61 59 20N 27 20E
Kokand 71 40 30N 70 57E
Kokchetav 62 53 20N 69 10E
Kokiu 73 23 30N 103 0E
Koko 90 11 28N 4 29E
Koko-Nor 73 37 0N 100 0E
Kokomo 104 40 30N 86 6W
Kokstad 94 30 32S 29 29E
Kola Pen. 61 67 30N 38 0E
Kolar 70 13 12N 78 15E
Kolguyev, I. 62 69 20N 48 30E
Kolhapur 70 16 43N 74 15E
Kolin 54 50 2N 15 9E
Köln = Cologne 52 50 56N 9 58E
Kołobrzeg 54 54 10N 15 35E
Kolomyya 55 48 31N 25 2E
Kolyma, R. 62 64 40N 153 0E
Kolyma Ra. 62 63 0N 157 0E
Kôm Ombo 91 24 25N 32 52E
Komandorskiye, Is. 62 55 0N 167 0E
Komárno 55 47 49N 18 5E
Komatipoort 94 25 25S 31 57E
Kompong Som 72 10 38N 103 30E
Komsberge 94 32 40S 20 45E
Komsomolets I. 62 80 30N 95 0E
Komsomolsk 62 50 30N 137 0E
Kondoa 93 4 55S 35 50E
Kong 88 8 54N 4 36W
Kongmoon 73 22 35N 113 1E
Kongolo 92 5 22S 27 0E
Konjic 58 43 42N 17 58E
Konskie 55 51 15N 20 23E
Konstanz 54 47 39N 9 10E
Kontagora 90 10 23N 5 27E
Konya 71 37 52N 32 35E
Konza 93 1 45S 37 0E
Korčula, I. 58 42 57N 17 0E
Korea Bay 73 39 0N 124 0E
Korea, South ■ 73 36 0N 128 0E
Korea Strait 73 34 0N 129 30E
Korea, North ■ 73 40 0N 127 0E
Koreh Wells 93 0 3N 38 45E
Kōriyama 76 37 24N 140 23E
Korogwe 93 5 5S 38 25E
Körös, R. 55 46 45N 20 20E
Korosten 55 50 57N 28 25E
Kortrijk 52 50 50N 3 17E
Kościan 54 52 5N 16 40E
Kosciusko, Mt. 79 36 27S 148 16E
Kosi, L. 94 27 0S 32 50E
Košice 55 48 42N 21 15E
Koster 94 25 52S 26 54E
Kostrzyn 54 52 24N 17 14E
Koszalin 54 54 12N 16 8E
Kota 70 25 14N 75 49E
Kota Kinabalu 72 6 0N 116 12E
Kotka 61 60 28N 26 58E
Kotlas 62 61 15N 47 0E
Kotovsk 55 47 55N 29 35E
Kouga Mts. 94 33 40S 23 55E
Kounradski 62 47 20N 75 0E
Kovel 55 51 10N 24 20E
Kowloon 73 22 20N 114 15E
Koyuk 98 64 55N 161 20W
Koza 76 26 19N 127 46E
Kra, Isthmus of 72 10 15N 99 30E
Kragujevac 55 44 2N 20 56E
Kraków 55 50 4N 19 57E
Krasnodar 62 45 5N 38 50E
Krasnovodsk 71 40 0N 52 52E
Krasnoyarsk 62 56 8N 93 0E
Kratie 72 12 32N 106 10E
Krefeld 52 51 20N 6 22E
Krishna, R. 70 16 30N 77 0E
Kristiansand 61 58 9N 8 1E
Kristiansund 61 63 7N 7 45E
Krivoy Rog 62 47 51N 33 20E
Krk 58 45 8N 14 40E
Kronshtadt 61 60 5N 29 35E
Kroonstad 94 27 43S 27 19E
Krotoszyn 55 51 42N 17 23E
Krugersdorp 94 26 5S 27 46E
Kruidfontein 94 32 48S 21 59E
Kuala Lumpur 72 3 9N 101 41E
Kucha 73 41 50N 82 30E
Kuching 72 1 33N 110 25E
Kudat 72 6 55N 116 55E
Kufra Oasis 89 24 17N 23 15E
Kufstein 54 47 35N 12 11E
Kuji 76 40 11N 141 46E
Kulunda 73 52 45N 79 15E
Kum Darya 73 41 0N 89 0E
Kumamoto 76 32 45N 130 45E
Kumara 83 42 37S 171 12E
Kumasi 88 6 41N 1 38W
Kumba 90 4 36N 9 24E
Kumbo 90 6 15N 10 36E
Kunlun Shan 73 36 0N 86 30E
Kunming 73 25 11N 102 37E
Kunsan 73 35 59N 126 45E
Kununurra 78 15 40S 128 39E
Kuopio 61 62 53N 27 35E
Kupa, R. 58 45 30N 16 24E
Kupang 72 10 19S 123 39E
Kurashiki 76 34 40N 133 50E
Kure 76 34 14N 132 32E
Kurgan 62 55 26N 65 18E
Kuril Is. 62 45 0N 150 0E
Kurnool 70 15 45N 78 0E
Kurow 83 44 4S 170 29E
Kursk 62 51 42N 36 11E
Kuruman 94 27 28S 23 28E
Kuruman R. 94 27 5S 21 30E
Kurume 76 33 15N 130 30E
Kushiro 76 43 0N 144 25E
Kuskokwim Mts. 98 63 0N 156 0W
Kutaisi 71 42 19N 42 40E
Kutch, G. of 70 22 50N 69 15E
Kutno 55 52 15N 19 23E
Kuwait ■ 71 29 30N 47 30E
Kuwana 76 35 0N 136 43E
Kuybyshev 62 55 27N 78 19E
Kvarner 58 44 50N 14 10E
Kwangchow 73 23 10N 113 10E
Kwangju 73 35 9N 126 55E
Kwangsi-Chuang □ 73 23 30N 108 55E
Kwangtung □ 73 23 45N 114 0E
Kwara □ 90 8 0N 5 0E
Kweichow □ 73 27 20N 107 0E
Kweilin 73 25 16N 110 15E
Kweiyang 73 26 30N 106 35E
Kwidzyn 55 54 45N 18 58E
Kwinana 78 32 15S 115 47E
Kyakhta 73 50 30N 106 25E
Kyle 46 55 32N 4 25W
Kyle of Lochalsh 48 57 17N 5 43W
Kyleakin 48 57 16N 5 44W
Kylestrome 48 58 16N 5 02W
Kyōto 76 35 0N 135 45E
Kyūshū □ 76 33 0N 131 0E
Kyzyl 73 51 50N 94 30E
Kyzyl Kum 71 42 0N 65 0E
Kyzyl Orda 71 44 56N 65 30E

L

Labe, R. 54 50 3N 15 20E
Labrador City 99 52 57N 66 55W
Labrador, Coast of 99 53 20N 61 0W
Labuan, I. 72 5 15N 115 38E
Laccadive Is. = Lakshadweep Is. 70 10 0N 72 30E
Lachine 99 45 30N 73 40W
Lachlan, R. 79 34 22S 143 55E
Ladismith 94 33 28S 21 15E
Ladoga, L. 61 61 15N 30 30E
Lady Grey 94 30 43S 27 13E
Ladybank 47 56 16N 3 8W
Ladybrand 94 29 9S 27 29E
Ladysmith 94 28 32S 29 46E
Lae 79 6 40S 147 2E
Lafayette, Ind., U.S.A. 104 40 25N 86 54W
Lafayette, La., U.S.A. 101 30 18N 92 0W
Lafia 90 8 30N 8 34E
Lafiagi 90 8 52N 5 20E
Lagan, R. 50 54 35N 5 55W
Lagoa dos Patos 108 31 15S 51 0W
Lagos, Nigeria 90 6 25N 3 27E
Lagos, Port. 60 37 5N 8 41W
Lahore 71 31 32N 74 22E
Lahti 61 60 58N 25 40E
Laingsburg 94 33 9S 20 52E
Lairg 48 58 1N 4 24W
Lake Charles 101 30 15N 93 10W
Lake District 44 54 30N 3 10W
Lakeland 101 28 0N 82 0W
Lakewood 104 41 28N 81 50W
Lakshadweep Is. 70 10 0N 72 30E
Lamar 100 38 9N 102 35W
Lambert's Bay 94 32 5S 18 17E
Lambeth 41 51 27N 0 7W
Lambourn 40 51 31N 1 31W
Lamlash 46 55 32N 5 8W
Lammermuir Hills 46 55 50N 2 40W
Lampedusa, I. 58 35 36N 12 40E
Lampeter 42 52 6N 4 6W
Lamu 93 2 10S 40 55E
Lanark 47 55 40N 3 48W
Lancashire □ 44 53 40N 2 30W
Lancaster, U.K. 44 54 3N 2 48W
Lancaster, Calif., U.S.A. 101 34 47N 118 8W
Lancaster, Pa., U.S.A. 104 40 4N 76 19W
Lancaster Sd. 99 74 13N 84 0W
Lanchester 44 54 50N 1 44W
Lanchow, China 73 36 4N 103 44E
Landau 52 49 12N 8 7E
Landeck 54 47 9N 10 34E
Landerneau 56 48 28N 4 17W
Landes □ 56 43 57N 0 48W
Landore 43 51 38N 3 55W
Land's End 43 50 4N 5 43W
Landshut 54 48 31N 12 10E
Langeberge, C. Prov., S. Afr. 94 33 55S 21 20E
Langeberge, C. Prov., S. Afr. 94 28 15S 22 33E
Langholm 46 55 9N 2 59W

* Renamed Bakhtaran

† Renamed Faisalabad
** Now known as Dairen*

Name	Page	Lat.	Long.
Lyonnais	56	45 45N	4 15 E
Lyons	56	45 46N	4 50 E
Lytham St. Anne's	45	53 45N	2 58W
Lyttelton	83	43 35S	172 44 E

M

Name	Page	Lat.	Long.
Maam Cross	51	53 28N	9 32W
Ma'an	71	30 12N	35 44 E
Maanshan	73	31 40N	118 30 E
Maas, R.	52	51 48N	4 55 E
Maastricht	52	50 50N	5 40 E
Mablethorpe	44	53 21N	0 14 E
Mabua	91	30 30N	35 12 E
Macau ■	73	22 16N	113 35 E
Macclesfield	44	53 16N	2 9W
McCook	100	40 15N	100 35W
Macdonnell Ranges	78	23 40S	133 0 E
Macduff	48	57 40N	2 30W
Maceió	108	9 40S	35 41W
Macerata	58	43 19N	13 28 E
Macgillycuddy's Reeks, mts.	51	52 2N	9 45W
Machakos	93	1 30S	37 15 E
Machen	43	51 35N	3 07W
Machrihanish	46	55 25N	5 42W
Machynlleth	42	52 36N	3 51W
* Macias Nguema Biyoga	88	3 30N	8 40 E
Macintyre, R.	79	28 37S	149 40 E
Mackay	79	21 8S	149 11 E
Mackay, L.	78	22 30S	129 0 E
McKeesport	104	40 21N	79 50W
Mackenzie	98	55 20N	123 05W
Mackenzie Bay	98	69 0N	137 30W
Mackenzie Mts.	98	64 0N	130 0W
Mackenzie, R., Austral.	79	23 38S	149 46 E
Mackenzie, R., Can.	98	69 10N	134 20W
McKinley, Mt.	98	63 10N	151 0W
McLaughlin	105	45 50N	100 50W
Maclear	94	31 2S	28 23 E
McLennan	98	55 42N	116 50W
McLeod, L.	78	24 0S	113 50 E
McLure	98	51 2N	120 13W
McMinnville	100	45 16N	123 11W
Macnean, L.	50	54 19N	7 52W
Mâcon	56	46 19N	4 50 E
Macon	101	32 50N	83 37W
McPherson	100	38 25N	97 40W
Macquarie Is.	23	50 0S	160 0 E
Macroom	51	51 54N	8 57W
Madagascar, I.	92	20 0S	47 0 E
Madang	79	5 12S	145 49 E
Madaoua	90	14 5N	6 27 E
Madeira, R.	108	5 30S	61 20W
Madera	106	29 15N	107 55W
Madhya Pradesh □	70	21 50N	81 0 E
Madinat al Shaab	71	12 50N	45 0 E
Madison	101	43 5N	89 25W
Madiun	72	7 38S	111 32 E
Madras	70	13 8N	80 19 E
Madre, Sierra	106	16 0N	93 0W
Madrid	60	40 25N	3 45W
Maebashi	76	36 24N	139 4 E
Maerdy	43	51 40N	3 29W
Maesteg	43	51 36N	3 40W
Maevatanana	92	16 56N	46 49 E
Mafeking	94	25 50S	25 38 E
Mafia I.	93	7 45S	39 50 E
Magadan	62	59 30N	151 0 E
Magadi	93	1 54S	36 19 E
Magdalen Is.	99	47 30N	61 40W
Magdeburg	54	52 8N	11 36 E
Magee, I.	50	54 48N	5 44W
Magelang	72	7 29S	110 13 E
Magellan's Str.	108	52 30S	75 0W
Maggiore, L.	58	46 0N	8 35 E
Maghâgha	91	28 38N	30 50 E
Maghera	50	54 51N	6 40W
Magherafelt	50	54 44N	6 37W
Magherafelt □	50	54 50N	6 40W
Maghull	45	53 31N	2 56W
Magnitogorsk	62	53 27N	59 4 E
Mahābād	71	36 50N	45 45 E
Mahagi	93	2 20N	31 0 E
Mahalapye	94	23 1S	26 51 E
Mahanadi R.	70	20 33N	85 0 E
Maharashtra □	70	19 30N	75 30 E
Mahari Mts.	93	6 20S	30 0 E
Mahé	70	11 42N	75 34 E
Mahenge	93	8 45S	36 35 E
Maheno	83	45 10S	170 50 E
Mahia Pen.	83	39 9S	177 55 E
Mahón	60	39 50N	4 18 E
Mai-Ndombe, L.	92	2 0S	18 0 E
Maiden Newton	43	50 46N	2 35W
Maidenhead	41	51 31N	0 42W
Maidstone	41	51 16N	0 31 E
Maiduguri	90	12 0N	13 20 E
Maimana	71	35 53N	64 38 E
Main Barrier Ra.	79	31 10S	141 20 E
Main, R.	54	50 13N	11 0 E
Maine □	101	45 20N	69 0W
Mainland, I., Orkneys, U.K.	49	59 0N	3 10W
Mainland, I., Shetlands, U.K.	49	60 15N	1 22W
Maintirano	92	18 3S	44 1 E
Mainz	52	50 0N	8 17 E
Maiquetia	107	10 36N	66 57W
Maitland	79	32 44S	151 36 E
Maiyema	90	12 5N	4 25 E
Maizuru	76	35 25N	135 22 E
Majorca, I. = Mallorca, I.	60	39 30N	3 0 E
‡ Majunga	92	15 40S	46 25 E
Makasar, Str. of	72	1 0S	118 20 E
Makeyevka	62	48 0N	38 0 E
Makgadikgadi Salt Pans	92	20 40S	25 45 E
Makhachkala	71	43 0N	47 15 E
Makó	55	46 14N	20 33 E
Makurazaki	76	31 15N	130 20 E
Makurdi	90	7 43N	8 28 E
Mal B.	51	52 50N	9 30W
Malabar Coast	70	11 0N	75 0 E
Malacca, Str. of	72	3 0N	101 0 E
Malad City	100	42 15N	112 20 E
Maladetta, Mt.	60	42 40N	0 30 E
Málaga	60	36 43N	4 23W
Malahide	51	53 26N	6 10W
Malakal	89	9 33N	31 50 E
Malang	72	7 59S	112 35 E
Malanje	92	9 30S	16 17 E
Mälaren	61	59 30N	17 10 E
Malatya	71	38 25N	38 20 E
Malawi ■	93	13 0S	34 0 E
Malaya ■	72	4 0N	102 0 E
Malaya Vishera	61	58 55N	32 25 E
Malbork	55	54 3N	19 10 E
Malcolm	78	28 51S	121 25 E
Maldegem	52	51 14N	3 26 E
Maldive Is. ■	70	2 0N	73 0W
Maldon	41	51 43N	0 41 E
Malhão, Sa. do	60	37 25N	8 0W
Mali ■	88	15 0N	10 0W
Malin Hd.	50	55 18N	7 16W
Malindi	93	3 12S	40 5 E
Mallaig	48	57 0N	5 50W
Mallaranny	50	53 55N	9 46W
Mallawi	91	27 44N	30 44 E
Mallorca, I.	60	39 30N	3 0 E
Mallow	51	52 8N	8 40W
Malmesbury, S. Afr.	94	33 28S	18 41 E
Malmesbury, U.K.	42	51 35N	2 5W
Malmö	61	55 36N	12 59 E
Malta ■	58	35 50N	14 30 E
Malton	44	54 9N	0 48W
Malvern Hills	42	52 0N	2 19W
Mambasa	93	1 22N	29 3 E
Mamfe	90	5 50N	9 15 E
Man, I. of	42	54 15N	4 30W
Manaar, Gulf of	70	8 30N	79 0 E
Managua	106	12 0N	86 20W
Manakara	92	22 8S	48 1 E
Manapouri	83	45 34S	167 39 E
Manapouri, L.	83	45 34S	167 32 E
Manaus	108	3 0S	60 0W
Mancha, La	60	39 10N	2 54W
Manchester, U.K.	45	53 30N	2 15W
Manchester, U.S.A.	104	42 58N	71 29W
Manchouli	73	49 46N	117 24 E
Manda	93	10 30S	34 40 E
Mandal	61	58 2N	7 25 E
Mandalay	70	22 0N	96 10 E
Mandan	100	46 50N	101 0W
Manengouba Mts.	90	5 0N	9 45 E
Manfalût	91	27 20N	30 52 E
Manfredónia	58	41 40N	15 55 E
Mangalore	70	12 55N	74 47 E
Mangaweka	83	39 48S	175 47 E
Mangole I.	72	1 50S	125 55 E
Mangonui	83	35 1S	173 32 E
Mangotsfield	43	51 29N	2 29W
Mangyai	73	38 6N	91 37 E
Mangyshlak Pen.	71	43 40N	52 30 E
Manhattan	101	39 10N	96 40W
Manihiki I.	22	10 24S	161 1W
Manila	72	14 40N	121 3 E
Manipur □	70	24 30N	94 0 E
Manitoba □	98	55 30N	97 0W
Manitoba, L.	98	51 0N	98 45W
Manizales	107	5 5N	75 32W
Manjimup	78	34 15S	116 6 E
Mannheim	54	49 28N	8 29 E
Mannin B.	51	53 27N	10 04W
Manningtree	40	51 56N	1 3 E
Manorhamilton	50	54 19N	8 11W
Mans, Le	56	48 0N	0 10 E
Mansel I.	99	62 0N	79 50W
Mansfield, U.K.	44	53 8N	1 12W
Mansfield, U.S.A.	104	40 45N	82 30W
Mantes-la-Jolie	56	49 0N	1 41 E
Mantiqueira, Serra da	108	22 0S	44 0W
Mantua	58	45 10N	10 47 E
Manukau	83	37 1S	174 55 E
Manyara L.	93	3 40S	35 50 E
Manyoni	93	5 45S	34 55 E
Manzanares	60	39 0N	3 22W
Manzanillo, Cuba	107	20 20N	77 10W
Manzanillo, Mexico	106	19 0N	104 20W
Manzini	94	26 30S	31 25 E
Maputo	94	25 58S	32 32 E
Maputo R.	92	26 35S	32 30 E
Mar del Plata	108	38 0S	57 30W
Mar, Serra do	108	25 30S	49 0W
Mara	93	1 30S	34 32 E
Maracaibo	107	10 40N	71 37W
Maracaibo, L.	107	9 40N	71 30W
Maracay	107	10 15N	67 36W
Maradi	90	13 35N	8 10 E
Marāgheh	71	37 30N	46 12 E
Marajo I.	108	1 0S	49 30W
Maranhão = São Luis	108	2 31S	44 16W
Marañon, R.	108	4 50S	75 35W
Marazion	43	50 8N	5 29W
Marbat	71	17 0N	54 45 E
Marbella	60	36 30N	4 57W
Marble Bar	78	21 9S	119 44 E
March	40	52 33N	0 5 E
Marchena	60	37 18N	5 23W
Marches	58	43 22N	13 10 E
Maree L.	48	57 40N	5 30W
Mareeba	79	16 59S	145 28 E
Margarita I.	107	11 0N	64 0W
Margate, S. Afr.	94	30 50S	30 20 E
Margate, U.K.	40	51 23N	1 24 E
Maria van Diemen, C.	83	34 29S	172 40 E
Mariana Is.	23	17 0N	145 0 E
Marianao	107	23 8N	82 24W
Maribor	58	46 36N	15 40 E
Maricourt	99	61 30N	72 0W
Marietta	101	34 0N	84 30W
Marion	101	40 35N	85 40W
Maritime Alps	56	44 10N	7 10 E
Market Deeping	45	52 40N	0 20W
Market Drayton	45	52 55N	2 30W
Market Harborough	45	52 29N	0 55W
Market Rasen	44	53 24N	0 20W
Market Weighton	44	53 52N	0 40W
Markethill	50	54 18N	6 31W
Markham Mts.	112	83 0S	164 0 E
Markinch	47	56 12N	3 9W
Marl	52	51 39N	7 4 E
Marlborough	42	51 26N	1 44W
Marlborough □	83	41 45S	173 33 E
Marlow	41	51 34N	0 47W
Marmande	56	44 30N	0 10 E
Marmara, Sea of	71	40 45N	28 15 E
Marne, R.	56	48 53N	2 25 E
Maroantsetra	92	15 26S	49 44 E
Maroua	90	10 40N	14 20 E
Marovoay	92	16 6S	46 39 E
Marple	45	53 23N	2 5W
Marquesas Is.	22	9 30S	140 0W
Marquette	101	46 30N	87 21W
Marrakech	88	31 40N	8 0W
Marree	79	29 39S	138 1 E
Marsabit	93	2 18N	38 0 E
Marsala	58	37 48N	12 25 E
Marsden	45	53 36N	1 55W
Marseilles	56	43 18N	5 23 E
Marsh I.	101	29 35N	91 50W
Marshall	101	39 8N	93 15W
Marshall Is.	23	9 0N	171 0 E
Martaban, G. of	70	15 40N	96 30 E
Marte	90	12 23N	13 46 E
Martha's Vineyard	101	41 25N	70 35W
Martigny	54	46 6N	7 3 E
Martigues	56	43 24N	5 4 E
Martinique, I.	107	14 40N	61 0W
Marton	83	40 4S	175 23 E
Martos	60	37 44N	3 58W
Martova = Mantua	58	45 10N	10 47 E
Marugame	76	34 15N	133 55 E
Mary	71	37 40N	61 50 E
Mary Kathleen	79	20 35S	139 48 E
Maryborough, Queens., Austral.	79	25 31S	152 37 E
Maryborough, Vic., Austral.	79	37 0S	143 44 E
Maryland □	101	39 10N	76 40W
Maryport	44	54 43N	3 30W
Masai Steppe	93	4 30S	36 30 E
Masaka	93	0 21S	31 45 E
Masan	73	35 11N	128 32 E
Masasi	93	10 45S	38 52 E
Masbate, I.	72	12 21N	123 36 E
Maseru	94	29 18S	27 30 E
Masham	44	54 15N	1 40W
Mashhad	71	36 20N	59 35 E
Masindi	93	1 40N	31 43 E
Masirah	71	20 25N	58 50 E
Masisi	93	1 23S	28 49 E
Masjed Soleyman	71	31 55N	49 25 E
Mask, L.	51	53 36N	9 24W
Mason City	101	43 9N	93 12W
Massachusetts □	101	42 25N	72 0W
Massif Central	56	45 30N	2 21 E
Masterton	83	40 56S	175 39 E
Masuda	76	34 40N	131 51 E
Masurian Lakes	55	53 30N	21 30 E
Matabeleland North □	92	20 0S	28 0 E
Matadi	92	5 52S	13 31 E
Matagalpa	106	13 10N	85 40W
Matagorda I.	101	28 10N	96 40W
Matamoros, Campeche, Mexico	106	25 53N	97 30W
Matamoros, Coahuila, Mexico	106	25 45N	103 1W
Matanzas	107	23 0N	81 40W
Mataranka	78	14 55S	133 4 E
Matatiele	94	30 20S	28 49 E
Mataura	83	46 11S	168 51 E
Matehuala	106	23 40N	100 50W
Matlock	44	53 8N	1 32W
Mato Grosso, Plat. of	108	15 0S	54 0W
Matopo Hills	92	20 36S	28 20 E
Matrûh	91	31 19N	27 9 E
Matsue	76	35 25N	133 10 E
Matsumoto	76	36 15N	138 0 E
Matsusaka	76	34 34N	136 32 E
Matsuyama	76	33 45N	132 45 E
Matterhorn, mt.	54	45 58N	7 39 E
Maturin	107	9 45N	63 11W
Mau Ranipur	70	25 16N	79 8 E
Maubeuge	56	50 17N	3 57 E
Maumere	72	8 38S	122 13 E
Maumturk Mts.	51	53 32N	9 42W
Mauritania ■	88	20 50N	10 0W
Mauritius ■	23	20 0S	57 0 E
Mawlaik	70	23 40N	94 26 E
Mawson Base	112	67 30N	65 0 E
Mayagüez	107	18 12N	67 9W
Maybole	46	55 21N	4 41W
Mayen	52	50 18N	7 10 E
Mayfield	40	51 1N	0 17 E
Maykop	71	44 35N	40 25 E
Maynooth	51	53 22N	6 38W
Mayo □	50	53 47N	9 7W
Mazar-i-Sharif	71	36 41N	67 0 E
Mazatlán	106	23 10N	106 30W
Mbabane	94	26 18S	31 6 E
Mbala	93	8 46S	31 17 E
Mbale	93	1 8N	34 12 E
Mbandaka	92	0 1S	18 18 E
Mbanga	90	4 30N	9 33 E
Mbarara	93	0 35S	30 25 E
Mberubu	90	6 10N	7 38 E
Mbeya	93	8 54S	33 29 E
Mbulu	93	3 45S	35 30 E
Meath □	51	53 32N	6 40W
Mecca	71	21 30N	39 54 E
Mechelen	52	51 2N	4 29 E
Mechernich	52	50 35N	6 39 E
Medan	72	3 40N	98 38 E
Medellín	107	6 15N	75 35W
Medford	105	42 20N	122 52W
Mediaş	55	46 9N	24 22 E
Medicine Hat	98	50 0N	110 45W
Medina	71	24 35N	39 52 E
Medina-Sidonia	60	36 28N	5 57W
Mediterranean Sea	88	35 0N	15 0 E
Médoc	56	45 10N	0 56W
Medway, R.	40	51 28N	0 45 E
Meekatharra	78	26 32S	118 29 E
Meerut	70	29 1N	77 50 E
Meissen	54	51 10N	13 29 E
Meknès	88	33 57N	5 33W
Mekong, R.	70	18 0N	104 15 E
Melaka	72	2 15N	102 15 E
Melbourne	79	37 50S	145 0 E
Melilla	88	35 21N	2 57W
Melk	54	48 13N	15 20 E
Melksham	42	51 22N	2 9W
Melrose	46	55 35N	2 44W
Meltham	45	53 35N	1 51W
Melton Mowbray	45	52 46N	0 52W
Melun	56	48 32N	2 39 E
Melvich	48	58 33N	3 55W
Melville	98	50 55N	102 50W
Melville I., Austral.	78	11 30S	131 0 E
Melville I., Can.	98	75 30N	111 0W
Melville Pen.	99	68 0N	84 0W
Melvin, L.	50	54 26N	8 10W
Memmingen	54	47 59N	10 12 E
Memphis	101	35 7N	90 0W
Menai Strait	42	53 7N	4 20W
Mendip Hills	42	51 17N	2 40W
Mendocino	105	39 26N	123 50W
Mendoza	108	32 50S	68 52W
Menen	52	50 47N	3 7 E
Mengtsz	73	23 20N	103 20 E
Menindee	79	32 20S	142 25 E
Menorca, I.	60	40 0N	4 0 E
Menston	45	53 53N	1 44W
Mentawai Is.	72	2 0S	99 0 E
Menton	58	43 50N	7 29 E
Menzies	78	29 40S	120 58 E
Meppel	52	52 42N	6 12 E
Meppen	52	52 41N	7 20 E
Merano	58	46 40N	11 10 E
Merced	100	37 18N	120 30W
Mercer	83	37 16S	175 5 E
Mercy C.	99	65 0N	62 30W
Mere	42	51 5N	2 16W
Mergui Arch.	70	12 30N	98 35 E
Mérida, Mexico	106	20 50N	89 40W
Mérida, Spain	60	38 55N	6 25W
Meriden	41	52 27N	1 36W
Merksem	52	51 16N	4 25 E
Merredin	78	31 28S	118 18 E
Merrick	46	55 8N	4 30W
Merse	46	55 40N	2 30W
Mersea I.	40	51 48N	0 55 E
Mersey, R.	44	53 20N	2 56W
Merseyside □	44	53 25N	2 55W
Mersin	71	36 51N	34 36 E
Merthyr Tydfil	43	51 45N	3 23W
Merton	41	51 25N	0 13W
Meru	93	0 3N	37 40 E
Merzig	52	49 26N	6 37 E
Mesa	100	33 20N	111 56W
Mesewa	89	15 35N	39 25 E
Mesopotamia = Al Jazirah	71	33 30N	44 0 E
Messina, Italy	58	38 10N	15 32 E
Messina, S. Afr.	94	22 20S	30 12 E
Messina, Str. of	58	38 5N	15 35 E
Methil	47	56 10N	3 01W
Methven, N.Z.	83	43 38S	171 40 E
Methven, U.K.	47	56 25N	3 35W
Metz	56	49 8N	6 10 E
Meuse, R.	52	50 45N	5 41 E
Mevagissey	43	50 16N	4 48W
Mexborough	45	53 29N	1 18W
Mexicali	106	32 40N	115 30W
México	106	19 20N	99 10W
Mexico ■	106	20 0N	100 0W
Mexico, G. of	106	25 0N	90 0W
Mey	48	58 38N	3 14W
Miami	101	25 52N	80 15W
Miaoli	73	24 33N	120 42 E
Michigan □	101	44 40N	85 40W
Michigan City	104	41 42N	86 56W
Michigan, L.	104	44 0N	87 0W
Michipicoten I.	99	47 40N	85 50W
Mickle Fell	44	54 38N	2 16W
Micklefield	45	53 48N	1 20W
Mid Glamorgan □	42	51 40N	3 25W
Middelburg, Neth.	52	51 30N	3 36 E
Middelburg, C. Prov., S. Afr.	94	31 30S	25 0 E
Middelburg, Trans., S. Afr.	94	25 49S	29 28 E
Middlesboro	101	36 40N	83 40W
Middlesbrough	44	54 35N	1 14W
Middleton	45	53 33N	2 12W
Middleton-in-Teesdale	44	54 38N	2 5W
Middlewich	44	53 12N	2 28W
Midhurst	40	50 59N	0 44W
Midland, Mich., U.S.A.	101	43 37N	84 17W
Midland, Tex., U.S.A.	100	32 0N	102 3W
Midland Junction	78	31 50S	115 58 E
Midleton	51	51 52N	8 12W
Midsomer Norton	42	51 17N	2 29W
Miercurea Ciuc	55	46 21N	25 48 E
Mieres	60	43 18N	5 48W
Mihara	76	34 24N	133 5 E
Mikindani	93	10 15S	40 2 E
Milan	58	45 28N	9 10 E
Milano = Milan	58	45 28N	9 10 E
Milazzo	58	38 13N	15 13 E
Mildenhall	40	52 20N	0 30 E
Mildura	79	34 13S	142 9 E
Miles	79	26 40S	150 23 E
Miles City	100	46 30N	105 50W
Milford Haven	42	51 43N	5 2W
Milford on Sea	40	50 44N	1 36W
Millau	56	44 8N	3 4 E
Millicent	79	37 34S	140 21 E
Millom	44	54 13N	3 16W

* Renamed Bioko
† Now Mafikeng
‡ Renamed Mahajunga

Millport New Alresford

Place	Map	Lat.	Long.
Millport	47	55 45N	4 55W
Millstreet	51	52 4N	9 5W
Milltown Malbay	51	52 51N	9 25W
Milnathort	47	56 14N	3 25W
Milnerton	94	33 54S	18 29E
Milngavie	47	55 57N	4 20W
Milnrow	45	53 36N	2 06W
Milton	83	46 7S	169 59E
Milton Keynes	40	52 3N	0 42W
Milverton	43	51 2N	3 15W
Milwaukee	104	43 9N	87 58W
Minamata	76	32 10N	130 30E
Mindanao, I.	72	8 0N	125 0E
*Mindanao Sea	72	9 0N	124 0E
Minden	54	52 18N	8 54E
Mindoro, I.	72	13 0N	121 0E
Minehead	42	51 12N	3 29W
Mineral Wells	100	32 50N	98 5W
Mingan	99	50 20N	64 0W
Mingulay I.	49	56 50N	7 40W
Minho □	60	41 25N	8 20W
Minho, R.	60	41 58N	8 40W
Minna	90	9 37N	6 30E
Minneapolis	101	44 58N	93 20W
Minnesota □	101	46 40N	94 0W
Minot	100	48 10N	101 15W
Minsk	62	53 52N	27 30E
Mińsk Mazowiecki	55	52 10N	21 33E
Mintlaw	48	57 32N	1 59W
Minûf	91	30 26N	30 52E
Minya Konka, mt.	73	29 36N	101 50E
Miquelon, I.	99	47 2N	56 20W
Mirny	112	66 0S	95 0E
Mirzapur	70	25 10N	82 45E
Mishan	73	45 31N	132 2E
Miskolc	55	48 7N	20 50E
Misoöl, I.	72	2 0S	130 0E
Mississippi □	101	33 0N	90 0W
Mississippi, R.	101	29 0N	89 15W
Missoula	100	47 0N	114 0W
Missouri □	101	38 25N	92 30W
Mistassini L.	99	51 0N	73 40W
Misurata	88	32 10N	15 3E
Mît Ghamr	91	30 42N	31 12E
Mitcheldean	43	51 51N	2 29W
Mitchell, Austral.	79	26 29S	147 58E
Mitchell, U.S.A.	100	43 40N	98 0W
Mitchell, R.	79	37 20S	147 0E
Mitchelstown, Ireland	51	52 16N	8 18W
Mito	76	36 20N	140 30E
Mittelland Kanal	52	52 23N	7 45E
Miyako	76	39 40N	141 75E
Miyako-Jima	76	24 45N	125 20E
Miyakonojō	76	31 32N	131 5E
Miyazaki	76	31 56N	131 30E
Mizen Hd.	51	51 27N	9 50W
Mizoram □	70	23 0N	92 40E
Mjanji	93	0 16N	34 0E
Mjøsa	61	60 40N	11 0E
Mkomanzi R.	94	30 13S	30 57E
Mława	55	53 9N	20 25E
Mme	90	6 18N	10 14E
Moate	51	53 25N	7 43W
Moba	93	7 0S	29 48E
Mobile	101	30 41N	88 3W
Mobutu Sese Seko, L.	93	1 30N	31 0E
Mocha	71	13 36N	43 25E
Mochudi	94	24 27S	26 7E
Modbury	43	50 21N	3 53W
Modderrivier	94	29 2S	24 38E
Módena	58	44 39N	10 55E
Modesto	100	37 43N	121 0W
Moe	79	38 12S	146 19E
Moffat	46	55 20N	3 27W
Mogadiscio	89	2 2N	45 25E
Mogadishu = Mogadiscio	89	2 2N	45 25E
Mogami-gawa, R.	76	38 45N	140 0E
Mogollon Mesa	100	35 0N	111 0W
Mohill	50	53 57N	7 52W
Moidart	48	56 49N	5 41W
Mointy	73	47 40N	73 45E
Moisie	99	50 12N	66 1W
Mokai	83	38 32S	175 56E
Mokpo	73	34 50N	126 30E
Mol	52	51 11N	5 5E
Mold	42	53 10N	3 10W
Moldavian S.S.R. □	55	47 0N	28 0E
Molde	61	62 45N	7 9E
Mole, R.	41	51 13N	0 15W
Molepolole	94	24 28S	25 28E
Molfetta	58	41 12N	16 35E
Molise □	58	41 45N	14 30E
Mollendo	108	17 0S	72 0W
Molopo, R.	94	25 40S	24 30E
Molteno	94	31 22S	26 22E
Molucca Sea	72	4 0S	124 0E
Moluccas Is.	72	1 0S	127 0E
Mombasa	93	4 2S	39 43E
Mombetsu	76	42 27N	142 4E
Mona Passage	107	18 0N	67 40W
Monaco ■	56	43 46N	7 23E
Monadhliath Mts.	48	57 10N	4 4W
Monaghan	50	54 15N	6 58W
Monaghan □	50	54 10N	7 0W
Monarch Mt.	105	51 55N	125 57W
Monasterevan	51	53 10N	7 5W
Moncayo, Sierra del	60	41 48N	1 50W
Mönchengladbach	52	51 12N	6 23E
Monclova	106	26 50N	101 30W
Moncton	99	46 7N	64 51W
Mondego, R.	60	40 28N	8 0W
Mondovì	58	44 23N	7 56E
Moneymore	50	54 42N	6 40W
Monforte	60	39 6N	7 25W
Mongalla	89	5 8N	31 55E
Monghyr	70	25 23N	86 30E
Mongolia ■	73	47 0N	103 0E
Mongu	92	15 16S	23 12E
Moniaive	46	55 11N	3 55W
Monifieth	46	56 30N	2 48W
Monmouth	43	51 48N	2 43W

* Renamed Bohol Sea

Place	Map	Lat.	Long.
Monópoli	58	40 57N	17 18E
Monroe	101	32 32N	92 4W
Monrovia	88	6 18N	10 47W
Mons	52	50 27N	3 58E
Montagu	94	33 45S	20 8E
Montana □	100	47 0N	110 0W
Montargis	56	48 0N	2 43E
Montauban	56	44 0N	1 21E
Montceau-les-Mines	56	46 40N	4 23E
Monte-Carlo	58	43 46N	7 23E
Monte Cristi	107	19 52N	71 39W
Montego B.	107	18 30N	78 0W
Montélimar	56	44 33N	4 45E
Montemorelos	106	25 11N	99 42W
Monterey	105	36 35N	121 57W
Monterrey	106	25 40N	100 30W
Montes Claros	108	16 30S	43 50W
Montevideo	108	34 50S	56 11W
Montgomery, U.K.	42	52 34N	3 9W
Montgomery, U.S.A.	101	32 20N	86 20W
Montluçon	56	46 22N	2 36E
Monto	79	24 52S	151 12E
Montpelier, Idaho, U.S.A.	100	42 15N	111 20W
Montpelier, Vt., U.S.A.	104	44 15N	72 38W
Montpellier	56	43 37N	3 52E
Montréal	104	45 31N	73 34W
Montreux	54	46 26N	6 55E
Montrose, U.K.	48	56 43N	2 28W
Montrose, U.S.A.	100	38 30N	107 52W
Monzón	60	41 52N	0 10E
Moore, L.	78	29 50S	117 35E
Moorfoot Hills	46	55 44N	3 8W
Moorhead	101	47 0N	97 0W
Moorreesburg	94	33 6S	18 38E
Moose Jaw	98	50 24N	105 30W
Moosehead L.	101	45 40N	69 40W
Moosonee	99	51 17N	80 39W
Moradabad	70	28 50N	78 50E
Morar L.	48	56 57N	5 40W
Morava, R.	54	49 50N	16 50E
Moravian Hts.	54	49 30N	15 40E
Moray Firth	48	57 50N	3 30W
Morecambe	44	54 5N	2 52W
Morecambe B.	44	54 7N	3 0W
Moree	79	29 28S	149 54E
Morelia	106	19 40N	101 11W
Morella	60	40 35N	0 5W
Morena, Sierra	60	38 20N	4 0W
Moreton-in-Marsh	42	51 59N	1 42W
Moretonhampstead	43	50 39N	3 45W
Morgan	79	34 0S	139 35E
Morioka	76	39 45N	141 8E
Morlaix	56	48 36N	3 52W
Morley	45	53 45N	1 36W
Moro G.	72	6 30N	123 0E
Morobe	79	7 49S	147 38E
Morocco ■	88	32 0N	5 50W
Morogoro	93	6 50S	37 40E
Morondava	92	20 17S	44 17E
Morotai, I.	72	2 10N	128 30E
Moroto	93	2 28N	34 42E
Morpeth	44	55 11N	1 41W
Morrinsville	83	37 40S	175 32E
Morristown	101	36 18N	83 20W
Morven	48	56 38N	5 44W
Morwell	79	38 10S	146 22E
Mosborough	45	53 19N	1 22W
Moscow	100	46 45N	116 59W
Moscow = Moskva	62	55 45N	37 35E
Mosel, R.	52	50 22N	7 36E
Moselle, R.	56	50 22N	7 36E
Moses Lake	100	47 16N	119 17W
Mosgiel	83	45 53S	170 21E
Moshi	93	3 22S	37 18E
Mosjøen	61	65 51N	13 12E
*Mossâmedes	92	15 7S	12 11E
Mossburn	83	45 41S	168 15E
Mosselbaai	94	34 11S	22 8E
Mossend	47	55 49N	4 00W
Mossley	45	53 31N	2 1W
Mossman	79	16 28S	145 23E
Mostaganem	88	35 54N	0 5E
Mostar	58	43 22N	17 50E
Mostrim	51	53 42N	7 38W
Mosty	55	53 27N	24 38E
Mosul	71	36 20N	43 5E
Motherwell	47	55 48N	4 0W
Motril	60	36 44N	3 37W
Motueka	83	41 7S	173 1E
Moulmein	70	16 30N	97 40E
Mount Barker	78	34 38S	117 40E
Mount Bellew	51	53 28N	8 31W
Mount Gambier	79	37 50S	140 46E
Mount Isa	79	20 42S	139 26E
Mount Lofty Ra.	79	34 35S	139 5E
Mountain Ash	43	51 42N	3 22W
Mountmellick	51	53 7N	7 20W
Mountrath	51	53 0N	7 30W
Mounts Bay	43	50 3N	5 27W
Mourne Mts.	50	54 10N	6 0W
Mourne, R.	50	54 45N	7 39W
Mouscron	52	50 45N	3 12E
Moutohora	83	38 17S	177 32E
Moville	50	55 11N	7 3W
Moy, R.	50	54 5N	8 50W
Moyale	93	3 30N	39 0E
Moyle □	50	55 10N	6 15W
Mozambique	92	15 3S	40 42E
Mozambique ■	92	19 0S	35 0E
Mozambique Chan.	92	20 0S	39 0E
Mpanda	93	6 23S	31 40E
Mporokoso	93	9 25S	30 5E
Mpwapwa	93	6 30S	36 30E
Mtwara	93	10 20S	40 20E
Mubarraz	71	25 29N	49 40E
Mubende	93	0 33N	31 22E
Mubi	90	10 18N	13 16E
Much Wenlock	45	52 36N	2 34W
Muck, I.	48	56 50N	6 15W
Mudgee	79	32 32S	149 31E
Muêda	93	11 36S	39 28E
Muhammad Râs	91	27 50N	34 0E

*Renamed Namibe

Place	Map	Lat.	Long.
Muir of Ord	48	57 30N	4 35W
Muirkirk	46	55 31N	4 6W
Muizenberg	94	34 6S	18 28E
Mukachevo	55	48 27N	22 45E
Mukden = Shenyang	73	41 35N	123 30E
Mulde, R.	54	51 10N	12 48E
Muleba	93	1 50S	31 37E
Mülheim	52	51 26N	6 53E
Mulhouse	56	47 40N	7 20E
Mull I.	48	56 27N	6 0W
Mull, Sound of	48	56 30N	5 50W
Muller Ra.	79	5 30S	143 0E
Mullet Pen.	50	54 10N	10 2W
Mullewa	78	28 29S	115 30E
Mullingar	51	53 31N	7 20W
Multan	71	30 15N	71 30E
Mumbles	43	51 34N	4 0W
Mumbles Hd.	43	51 33N	4 0W
Muna, I.	72	5 0S	122 30E
München = Munich	54	48 8N	11 33E
Muncie	101	40 10N	85 20W
Mundesley	40	52 53N	1 24E
Mungindi	79	28 58S	149 1E
Munich	54	48 8N	11 33E
Münster	52	51 58N	7 37E
Munster □	51	52 20N	8 40W
Mur, R.	54	47 7N	13 55E
Murang'a	93	0 45S	37 9E
Murchison	83	41 49S	172 21E
Murchison, R.	78	26 45S	116 15E
Murchison Ra.	78	20 0S	134 10E
Murcia	60	38 2N	1 10W
Murcia □	60	37 50N	1 30W
Mures R.	55	46 0N	22 0E
Murgon	79	26 15S	151 54E
Müritz, L.	54	53 25N	12 40E
Murmansk	61	68 57N	33 10E
Muroran	76	42 25N	141 0E
Murray Bridge	79	35 6S	139 14E
Murray, R.	79	35 20S	139 22E
Murraysburg	94	31 58S	23 47E
Murrumbidgee, R.	79	34 40S	143 0E
Murshid	91	21 40N	31 10E
Murton	44	54 51N	1 22W
Murupara	83	38 28S	176 42E
Murwara	70	23 46N	80 28E
Murwillumbah	79	28 18S	153 27E
Muş	71	38 45N	41 30E
Musa, G.	91	28 32N	33 59E
Muscat	71	23 40N	58 38E
Musgrave Ras.	78	26 0S	132 0E
Mushin	90	6 32N	3 21E
Muskegon	104	43 15N	86 17W
Muskogee	101	35 50N	95 25W
Musoma	93	1 30S	33 48E
Musselburgh	47	55 57N	3 3W
Musselshell, R.	100	47 21N	107 58W
Muswellbrook	79	32 16S	150 56E
Mût	91	25 28N	28 58E
Mutankiang	73	44 35N	129 30E
Mutsu-Wan	76	41 5N	140 55E
Muzaffarpur	70	26 7N	85 32E
Mwanza	93	2 30S	32 58E
Mwaya	93	9 32S	33 55E
Mweelrea	51	53 37N	9 48W
Mweru, L.	92	9 0S	29 0E
Mwirasandu	93	0 56S	30 22E
Mybster	48	58 27N	3 24W
Myitkyina	70	25 30N	97 26E
Mynydd Prescelly, mt.	42	51 57N	4 48W
Mysore	70	12 17N	76 41E
Mytholmroyd	45	53 43N	1 59W
Mzimvubu, R.	94	31 35S	29 35E

N

Place	Map	Lat.	Long.
Naas	51	53 12N	6 40W
Nababeep	94	29 36S	17 46E
Nabq	91	28 5N	29 23E
Nâbulus	89	32 14N	35 15E
Nacozari	106	30 30N	109 50W
Nafada	90	11 8N	11 20E
Nag Hammâdi	91	26 2N	32 18E
Nagaland □	70	26 0N	94 30E
Nagano	76	36 40N	138 10E
Nagaoka	76	37 27N	138 50E
Nagappattinam	70	10 46N	79 51E
Nagasaki	76	32 47N	129 50E
Nagoya	76	35 10N	136 50E
Nagpur	70	21 8N	79 10E
Nagykanizsa	54	46 28N	17 0E
Naha	76	26 13N	127 42E
Nahariya	89	33 1N	35 5E
Nahiya, Wadi	91	28 55N	31 0E
Nailsea	43	51 25N	2 44W
Nailsworth	43	51 41N	2 12W
Nairn	48	57 35N	3 54W
Nairobi	93	1 17S	36 48E
Naivasha	93	0 40S	36 30E
Najafâbâd	71	32 40N	51 15E
Nakamura	76	33 0N	133 0E
Nakatsu	76	33 40N	131 15E
Nakhichevan A.S.S.R. □	71	39 14N	45 30E
Nakhl	91	29 55N	33 43E
Nakhon Ratchasima (Khorat)	72	14 59N	102 12E
Nakina	99	50 10N	86 40W
Nakuru	93	0 15S	35 5E
Nam Dinh	72	20 25N	106 5E
Nam Tso	73	30 40N	90 30E
Namaland	92	30 0S	18 0E
Namangan	71	41 0N	71 40E
Namasagali	93	1 2N	33 0E
Namatanai	79	3 40S	152 29E
Nambour	79	26 32S	152 58E
Namcha Barwa	73	29 40N	95 10E
Namibia □	92	22 0S	18 9E
Nampa	100	43 40N	116 40W
Nampula	92	15 6S	39 7E
Namur	52	50 27N	4 52E
Namur □	52	50 17N	5 0E
Nan Shan	73	38 30N	99 0E
Nanaimo	105	49 10N	124 0W
Nanango	79	26 40S	152 0E
Nanchang	73	24 26N	117 18E
Nanchung	73	30 47N	105 59E
Nancy	56	48 42N	6 12E
Nander	70	19 10N	77 20E
Nanking	73	32 10N	118 50E
Nannine	78	26 51S	118 18E
Nanning	73	22 48N	108 20E
Nanping	73	26 45N	118 5E
Nantes	56	47 12N	1 33W
Nantucket I.	101	41 16N	70 3W
Nantwich	44	53 5N	2 31W
Nantyglo	43	51 48N	3 10W
Nanyang	73	33 0N	112 32E
Nanyuki	93	0 2N	37 4E
Nao, C. de la	60	38 44N	0 14E
Naoetsu	76	37 12N	138 10E
Napa	100	38 18N	122 17W
Napier	83	39 30S	176 56E
Naples	58	40 50N	14 5E
Napoli = Naples	58	40 50N	14 5E
Naqâda	91	25 53N	32 42E
Nara	76	34 40N	135 49E
Naracoorte	79	36 58S	140 45E
Narberth	42	51 48N	4 45W
Narbonne	56	43 11N	3 0E
Nares Str.	99	81 0N	65 0W
Narmada, R.	70	22 40N	77 30E
Narodnaya, G.	62	65 5N	60 0E
Narok	93	1 20S	33 30E
Narrabri	79	30 19S	149 46E
Narrandera	79	34 42S	146 31E
Narrogin	78	32 58S	117 14E
Narromine	79	32 12S	148 12E
Narva	61	59 10N	28 5E
Narvik	61	68 28N	17 26E
Naseby	83	45 1S	170 10E
Nashua	101	42 50N	71 25W
Nashville	101	36 12N	86 46W
Nasik	70	20 2N	73 50E
Nassau	107	25 0N	77 30W
Nasser City = Kôm Ombo	91	24 25N	32 52E
Nasser, L.	91	23 0N	32 30E
Natal	108	5 47S	35 13W
Natal □	94	28 30S	30 30E
Natashquan	99	50 14N	61 46W
Natchez	101	31 35N	91 25W
Natrûn, W. el.	91	30 25N	30 0E
Natuna Is.	72	4 0N	108 0E
Naturaliste, C.	78	33 32S	115 0E
Nauru I.	23	0 25N	166 0E
Naushahra	71	34 0N	72 0E
Navalcarnero	60	40 17N	4 5W
Navan	51	53 39N	6 40W
Navarra □	60	42 40N	1 40W
Nazareth	89	32 42N	35 17E
Nazas, R.	106	25 20N	104 4W
Naze	76	28 22N	129 27E
Naze, The	40	51 43N	1 19E
Ndala	93	4 45S	33 23E
Ndjamena	88	12 4N	15 8E
Neagh, Lough	50	54 35N	6 25W
Neath	43	51 39N	3 49W
Nebraska □	100	41 30N	100 0W
Nebraska City	101	40 40N	95 52W
Nebrodi Mts.	58	37 55N	14 45E
Neckar, R.	54	48 43N	9 15E
Needles, The	40	50 42N	1 19W
Negoiu, Mt.	55	45 35N	24 31E
Negotin	55	44 16N	22 37E
Negrais C.	70	16 0N	94 30E
Negro, R., Argent.	108	40 0S	64 0W
Negro, R., Boliv.	108	14 11S	63 7W
Negros, I.	72	10 0N	123 0E
Neheim-Hüsten	52	51 27N	7 58E
Neikiang	73	29 35N	105 10E
Neilston	47	55 47N	4 27W
Neisse, R.	54	52 4N	14 46E
Nejd, prov.	71	26 30N	42 0E
Nellore	70	14 27N	79 59E
Nelson, Can.	98	49 30N	117 20W
Nelson, N.Z.	83	41 18S	173 16E
Nelson, U.K.	45	53 50N	2 14W
Nelson □	83	42 11S	172 15E
Nelson Forks	98	59 30N	124 0W
Nelson, R.	98	54 33N	98 2W
Nelspruit	94	25 29S	30 59E
Nemuro	76	43 20N	145 35E
Nemuro-Kaikyō	76	43 30N	145 30E
Nenagh	51	52 52N	8 11W
Nene, R.	40	52 38N	0 7E
Nepal ■	70	28 0N	84 30E
Nepalganj	70	28 0N	81 40E
Nephi	100	39 43N	111 52W
Nephin	50	54 1N	9 21W
Nerchinsk	73	52 0N	116 39E
Neretva, R.	58	43 1N	17 27E
Ness, Loch	48	57 15N	4 30W
Neston	45	53 17N	3 3W
Netherlands ■	52	52 0N	5 30E
Nettilling L.	99	66 30N	71 0W
Neubrandenburg	54	53 33N	13 17E
Neuchâtel	54	47 0N	6 55E
Neuchâtel, L.	54	46 53N	6 50E
Neufchâteau, Belg.	52	49 50N	5 25E
Neufchâteau, France	56	48 21N	5 40E
Neumünster	54	54 4N	9 58E
Neunkirchen	52	49 23N	7 6E
Neusielder, L.	54	47 50N	16 47E
Neuss	52	51 12N	6 39E
Neustadt	52	49 21N	8 10E
Neustrelitz	54	53 22N	13 4E
Neuwied	52	50 26N	7 29E
Nevada □	100	39 20N	117 0W
Nevada, Sierra	60	37 3N	3 15W
Nevel	61	56 0N	29 55E
Nevers	56	47 0N	3 9E
Nevis I.	107	17 0N	62 30W
Nevis, L.	48	57 0N	5 43W
New Alresford	40	51 6N	1 10W

Otaru Pratt

Otaru **76** 43 10N 141 0E
Otavi **92** 19 40S 17 24E
Otford **41** 51 18N 0 11E
Otira Gorge **83** 42 53S 171 33E
Otley **45** 53 54N 1 41W
Otoineppu **76** 44 44N 142 16E
Otorohanga **83** 38 12S 175 14E
Otranto, Str. of **58** 40 15N 18 40E
Ottawa **104** 45 27N 75 42W
Ottawa Is. **99** 59 35N 80 16W
Ottery St. Mary **43** 50 45N 3 16W
Otukpa **90** 7 9N 7 41E
Oturkpo **90** 7 10N 8 15E
Otwock **55** 52 5N 21 20E
Ouachita Mts. **101** 34 50N 94 30W
Ouagadougou **88** 12 25N 1 30W
Oude Rijn, R. **52** 52 12N 4 24E
Oudtshoorn **94** 33 35S 22 14E
Oughter, L. **50** 54 2N 7 30W
Ougrée **52** 50 36N 5 32E
Ouidah **90** 6 25N 2 0E
Oujda □ **88** 33 18N 1 25W
Oulton Broad **40** 52 28N 1 43E
Oulu **61** 65 1N 25 29E
Oulu, L. **61** 64 25N 27 0E
Oundle **40** 52 28N 0 28W
Our, R. **52** 49 55N 6 5E
Ourthe, R. **52** 50 29N 5 35E
Ouse, Great, R. **40** 52 12N 0 7E
Ouse, Little, R. **40** 52 25N 0 20E
Ouse, R., Sussex, U.K. **40** 50 43N 0 3E
Ouse, R., Yorks., U.K. **44** 54 3N 0 7E
Outer Hebrides, Is. **49** 57 30N 7 40W
Ouyen **79** 35 1S 142 22E
Ovar **60** 40 51N 8 40W
Overflakkee **52** 51 44N 4 10E
Overijssel □ **52** 52 25N 6 35E
Oviedo **60** 43 25N 5 50W
Owaka **83** 46 27S 169 40E
Owambo **92** 17 20S 16 30E
Owase **76** 34 7N 136 5E
Owatonna **101** 44 3N 93 17W
Owen Falls **93** 0 30N 33 5E
Owen Stanley Range **79** 8 30S 147 0E
Owens L. **105** 36 20N 118 0W
Owerri **90** 5 29N 7 0E
Owo **90** 7 18N 5 30E
Owosso **101** 43 0N 84 10W
Oxenhope **45** 53 48N 1 57W
Oxford, N.Z. **83** 43 18S 172 11E
Oxford, U.K. **40** 51 45N 1 15W
Oxford □ **40** 51 45N 1 15W
Oxted **41** 51 14N 0 01W
Oykel, R. **48** 57 55N 4 26W
Oyo **90** 7 46N 3 56E
Oyonnax **56** 46 16N 5 40E
Ozamis **72** 8 15N 123 50E
Ozark Plateau **101** 37 20N 91 40W

P

Pa-an **73** 16 45N 97 40E
Paarl **94** 33 45S 18 56E
Pabbay I. **49** 57 46N 7 12W
Pacaraima, Sierra **107** 4 0N 63 0W
Pacific Ocean **22** 10 0N 140 0W
Padang **72** 1 0S 100 20E
Padiham **45** 53 48N 2 20W
Padova = Padua **58** 45 24N 11 52E
Padre I. **101** 27 0N 97 20W
Padstow **43** 50 33N 4 57W
Padua **58** 45 24N 11 52E
Paeroa **83** 37 23S 175 41E
Pag **58** 44 30N 14 50E
Pagalu, I. **88** 1 35S 3 35E
Pahiatua **83** 40 27S 175 50E
Paignton **43** 50 26N 3 33W
Painted Desert **100** 36 40N 111 30W
Paisley **47** 55 51N 4 27W
Paiyin **73** 36 45N 104 4E
Paiyünopo **73** 41 46N 109 58E
Pakanbaru **72** 0 30N 101 15E
Pakhoi **73** 21 30N 109 10E
Pakistan ■ **71** 30 0N 70 0E
Pakse **72** 15 5N 105 52E
Palagruza **58** 42 24N 16 15E
Palapye **92** 22 30S 27 7E
Palawan, I. **72** 10 0N 119 0E
Palembang **72** 3 0S 104 50E
Palencia **60** 42 1N 4 34W
Palermo **58** 38 8N 13 20E
Paletwa **70** 21 30N 92 50E
Palk Strait **70** 10 0N 80 0E
Pallas Green **51** 52 35N 8 22W
Palma **60** 39 33N 2 39E
Palma, Bay of **60** 39 30N 2 39E
Palma, La, Panama **107** 8 15N 78 0W
Palma, La, Spain **60** 37 21N 6 38W
Palmas, C. **88** 4 27N 7 46W
Palmas, G. of **58** 39 0N 8 30E
Palmerston **83** 45 29S 170 43E
Palmerston North **83** 40 21S 175 39E
Palmi **58** 38 21N 15 51E
Palmira **107** 3 32N 76 16W
Palmyra Is. **22** 5 52N 162 5W
Palo Alto **105** 37 25N 122 8W
Palos, C. **60** 37 38N 0 40W
Pamiers **56** 43 7N 1 39E
Pamirs **71** 37 40N 73 0E
Pampa **100** 35 35N 100 58W
Pampas **108** 34 0S 64 0W
Pamplona, Colomb. **107** 7 23N 72 39W
Pamplona, Spain **60** 42 48N 1 38W
Panamá **107** 9 0N 79 25W
Panama ■ **107** 8 48N 79 55W
Panama Canal Zone **107** 9 10N 79 56W
Panama, G. of **96** 8 4N 79 20W
Panay I. **72** 11 10N 122 30E
Pančevo **55** 44 52N 20 41E
Pancorbo Pass **60** 42 32N 3 5W
Pangani **93** 5 25S 38 58E
Pangbourne **40** 51 28N 1 5W
Pangola R. **94** 23 40S 27 43E
Pantar, I. **72** 8 28S 124 10E
Pantelleria, I. **58** 36 52N 12 0E
Panuco **106** 22 0N 98 25W
Panuco, R. **106** 21 30N 98 30W
Panyam **90** 9 27N 9 8E
Paoki **73** 34 25N 107 15E
Paoting **73** 38 50N 115 30E
Paotow **73** 40 45N 110 0E
Papakura **83** 37 4S 174 59E
Papantla **106** 20 45N 97 21W
Papenburg **52** 53 7N 7 25E
Papua, Gulf of **79** 9 0S 144 50E
Papua New Guinea ■ **79** 8 0S 145 0E
Pará □ **108** 3 20S 52 0W
Paragua, R. **107** 6 30N 63 30W
Paraguay ■ **108** 23 0S 57 0W
Paraguay, R. **108** 27 18S 58 38W
Parakou **90** 9 25N 2 40E
Paramaribo **108** 5 50N 55 10W
Paraná **108** 32 0S 60 30W
Paraná, R. **108** 33 43S 59 15W
Pardubice **54** 50 3N 15 45E
Pare Pare **72** 4 0S 119 45E
Parima, Serra **107** 2 30N 64 0W
Paris **56** 48 50N 2 20E
Park Range **100** 40 0N 106 30W
Parkersburg **104** 39 18N 81 31W
Parkes **79** 33 9S 148 11E
Parma **58** 44 50N 10 20E
Parnaiba, R. **108** 3 35S 43 0W
Paroo, R. **79** 30 0S 144 5E
Parry Is. **98** 77 0N 110 0W
Parry Sound **99** 45 20N 80 0W
Partington **45** 53 25N 2 25W
Partry Mts. **51** 53 40N 9 28W
Pasadena, Calif., U.S.A. **105** 34 5N 118 9W
Pasadena, Tex., U.S.A. **101** 29 45N 95 14W
Passage East **51** 52 15N 7 0W
Passage West **51** 51 52N 8 20W
Passau **54** 48 34N 13 27E
Passero, C. **58** 36 42N 15 8E
Patagonia **108** 45 0S 69 0W
Patchway **43** 51 32N 2 34W
Patea **83** 39 45S 174 30E
Pategi **90** 8 50N 5 45E
Pateley Bridge **44** 54 5N 1 45W
Patensie **94** 33 46S 24 49E
Paternò **58** 37 34N 14 53E
Paterson **104** 40 55N 74 10W
Pathfinder Res. **100** 42 30N 107 0W
Patiala **70** 30 23N 76 26E
Patna **70** 25 35N 85 18E
Patrickswell **51** 52 36N 8 42W
Patrington **44** 53 41N 0 1W
Patti **58** 38 8N 14 57E
Pau **56** 43 19N 0 25W
Pauillac **56** 45 11N 0 46W
Paulpietersburg **94** 27 23S 30 50E
Pavia **58** 45 10N 9 10E
Pavlodar **62** 52 33N 77 0E
Pawtucket **104** 41 51N 71 22W
Paz, La, Boliv. **108** 16 20S 68 10W
Paz, La, Mexico **106** 24 10N 110 20W
Peace River **98** 56 15N 117 18W
Peak Hill **78** 32 39S 148 11E
Peak Range **79** 22 50S 148 20E
Peak, The **44** 53 24N 1 53W
Pechenga **61** 69 30N 31 25E
Pechora G. **62** 68 40N 54 0E
Pecos **100** 31 25N 103 35W
Pecos, R. **100** 29 42N 102 30W
Pécs **55** 46 5N 18 15E
Peebles **47** 55 40N 3 12W
Peel **42** 54 14N 4 40W
Peel Fell, mt. **46** 55 17N 2 35W
Pegasus Bay **83** 43 20S 173 10E
Pegu Yoma, mts. **70** 19 0N 96 0E
Pehan **73** 48 17N 120 31E
Pehpei **73** 29 44N 106 29E
Peiping **73** 39 50N 116 20E
Pekalongan **72** 6 53S 109 40E
Peking = Peiping **73** 39 45N 116 25E
Peleng, I. **72** 1 20S 123 30E
Pelly, R. **98** 62 15N 133 30W
Peloro, C. **58** 38 15N 15 40E
Pelvoux, Massif de **56** 44 52N 6 20E
Pematang Siantar **72** 2 57N 99 5E
Pemba, I. **93** 5 0S 39 45E
Pemberton **78** 34 30S 116 0E
Pembroke **42** 51 41N 4 57W
Pen-y-Ghent **44** 54 10N 2 15W
Pen-y-groes **42** 53 3N 4 18W
Penarth **43** 51 26N 3 11W
Pendine **42** 51 44N 4 33W
Pendle Hill **45** 53 53N 2 18W
Pendlebury **45** 53 31N 2 20W
Pendleton **105** 45 35N 118 50W
Pengpu **73** 33 0N 117 25E
Penistone **45** 53 31N 1 38W
Penketh **45** 53 22N 2 37W
Penki **73** 41 20N 132 50E
Pennines **44** 54 50N 2 20W
Pennsylvania □ **101** 40 50N 78 0W
Penong **78** 31 59S 133 5E
Penrhyn Is. **22** 9 0S 150 30W
Penrith, Austral. **79** 33 43S 150 38E
Penrith, U.K. **44** 54 40N 2 45W
Penryn **43** 50 10N 5 7W
Pensacola **101** 30 30N 87 10W
Pensacola Mts. **112** 84 0S 40 0W
Penticton **105** 49 30N 119 30W
Pentland **79** 20 32S 145 25E
Pentland Firth **48** 58 43N 3 10W
Pentland Hills **46** 55 48N 3 25W
Pentre Foelas **42** 53 2N 3 41W
Penygraig **43** 51 36N 3 26W
Penza **62** 53 15N 45 5E
Penzance **43** 50 7N 5 32W
Peoria **101** 40 40N 89 40W
Perche **56** 48 31N 1 1E
Perdido, Mte. **60** 42 40N 0 50E
Pereira **107** 4 49N 75 43W
Peribonca, R. **99** 49 0N 72 25W
Périgueux **56** 45 10N 0 42E
Perm (Molotov) **62** 58 0N 57 10E
Pernambuco = Recife **108** 8 0S 35 0W
Perpignan **56** 42 42N 2 53E
Perranporth **43** 50 21N 5 9W
Perryton **100** 36 28N 100 48W
Pershore **42** 52 7N 2 4W
*Persian Gulf **71** 27 0N 50 0E
Perth, Austral. **78** 31 57S 115 52E
Perth, U.K. **47** 56 24N 3 27W
Peru ■ **108** 8 0S 75 0W
Perúgia **58** 43 6N 12 24E
Pésaro **58** 43 55N 12 53E
Pescara **58** 42 28N 14 13E
Peshawar **71** 34 2N 71 37E
Petange **52** 49 33N 5 55E
Peterborough, Austral. **79** 32 58S 138 51E
Peterborough, U.K. **40** 52 35N 0 14W
Peterculter **48** 57 5N 2 18W
Peterhead **48** 57 30N 1 49W
Peterlee **44** 54 45N 1 18W
Petersburg **101** 37 17N 77 26W
Petersfield **40** 51 0N 0 56W
Peto **106** 20 10N 89 0W
Petone **83** 41 13S 174 53E
Petropavlovsk-Kamchatskiy **62** 53 16N 159 0E
Petrovaradin **55** 45 16N 19 55E
Petrovsk **62** 52 22N 45 19E
Petrozavodsk **61** 61 41N 34 20E
Petton **44** 54 52N 1 36W
Petworth **40** 50 59N 0 37W
Pforzheim **54** 48 53N 8 43E
Phan Rang **72** 11 40N 109 9E
Phan Thiet **72** 11 1N 108 9E
Phanom Dang Raek, mts. **72** 14 45N 104 0E
Phenix City **101** 32 30N 85 0W
Philadelphia **104** 40 0N 75 10W
Philippines ■ **72** 12 0N 123 0E
Philippolis **94** 30 15S 25 16E
Philipstown **94** 30 28S 24 30E
Phnom Penh **72** 11 33N 104 55E
Phoenix **100** 33 30N 112 10W
Phoenix Is. **22** 3 30S 172 0W
Piacenza **58** 45 2N 9 42E
Picardy **56** 50 0N 2 15E
Pickering **44** 54 15N 0 46W
Picton **83** 41 18S 174 3E
Pidurutalagala, mt. **70** 7 10N 80 50E
Piedmont **58** 45 0N 7 30E
Piedmont Plat. **101** 34 0N 81 30W
Piedras Negras **106** 28 35N 100 35W
Pierre **100** 44 23N 100 20W
Piet Retief **94** 27 1S 30 50E
Pietermaritzburg **94** 29 35S 30 25E
Pietersburg **94** 23 54S 29 25E
Pietrosul **55** 47 35N 24 43E
Pikes Peak **100** 38 50N 105 10W
Piketberg **94** 32 55S 18 40E
Pilanesberg **94** 25 14S 27 4E
Pilbara Cr. **78** 21 15S 118 22E
Pilcomayo, R. **108** 25 21S 57 42W
Pilica, R. **55** 51 25N 20 45E
Pilling **45** 53 55N 2 54W
Pilsen = Plzen **54** 49 45N 13 22E
Pimba **79** 31 18S 136 46E
Pine Bluff **101** 34 10N 92 0W
Pine Creek **78** 13 50S 131 49E
Pinetown **94** 29 48S 30 54E
Pingliang **73** 35 20N 106 40E
Pingsiang **73** 22 2N 106 55E
Pingtung **73** 22 36N 120 30E
Pinjarra **78** 32 37S 115 52E
Piombino **58** 42 54N 10 30E
Piotrków Trybunalski **55** 51 23N 19 43E
Pirmasens **52** 49 12N 7 30E
Pisa **58** 43 43N 10 23E
Pisciotta **58** 40 7N 15 12E
Pisek **54** 49 19N 14 10E
Pistóia **58** 43 57N 10 53E
Pitcairn I. **22** 25 5S 130 5W
Piteå **61** 65 20N 21 25E
Pitești **55** 44 52N 24 54E
Pitlochry **48** 56 43N 3 43W
Pitt I. **105** 53 30N 129 50W
Pittsburgh **104** 40 25N 79 55W
Pizzo **58** 38 44N 16 10E
Placentia **99** 47 20N 54 0W
Plainview **100** 34 10N 101 40W
Plasencia **60** 40 3N 6 8W
Plata, La **108** 35 0S 57 55W
Plata, La, Río de **108** 35 0S 56 40W
Plattsburgh **101** 44 41N 73 30W
Plauen **54** 50 29N 12 9E
Plenty, Bay of **83** 37 45S 177 0E
Plockton **48** 57 20N 5 40W
Ploiești **55** 44 57N 26 5E
Plumbridge **50** 54 46N 7 15W
Plymouth **43** 50 23N 4 9W
Plympton **43** 50 24N 4 2W
Plynlimon **42** 52 29N 3 47W
Plzen **54** 49 45N 13 22E
Po, R. **58** 44 57N 12 4E
Pocatello **100** 42 50N 112 25W
Pocklington **44** 53 56N 0 48W
Pofadder **94** 29 10S 19 22E
Pointe-à-Pitre **107** 16 10N 61 30W
Pointe-Noire **92** 4 48S 12 0E
Poitiers **56** 46 35N 0 20E
Poland ■ **55** 52 0N 20 0E
Polegate **40** 50 49N 0 15E
Polesworth **41** 52 37N 1 37W
Polillo I. **72** 14 56N 122 0E
Polperro **43** 50 19N 4 31W
Pombal **60** 39 55N 8 40W
Ponca City **101** 36 40N 97 5W
Ponce **107** 18 1N 66 37W
Pondicherry **70** 11 59N 79 50E
Pondoland **94** 31 10S 29 30E
Ponferrada **60** 42 32N 6 35W
Pongola, R. **94** 27 15S 32 13E
Pont-à-Mousson **56** 45 54N 6 1E
Pontardawe **43** 51 43N 3 51W
Pontardulais **43** 51 42N 4 3W
Pontedera **58** 43 40N 10 37E
Pontefract **45** 53 42N 1 19W
Ponteland **44** 55 03N 1 45W
Pontevedra **60** 42 26N 8 40W
Pontianak **72** 0 3S 109 15E
Pontnewydd **43** 51 40N 3 01W
Pontycymmer **43** 51 36N 3 35W
Pontypool **43** 51 42N 3 1W
Pontypridd **43** 51 36N 3 21W
Poole **43** 50 42N 2 2W
Poolewe **48** 57 45N 5 38W
Pooley Bridge **44** 54 37N 2 49W
Popayán **107** 2 27N 76 36W
Poperinge **52** 50 51N 2 42E
Popocatepetl, vol. **106** 19 10N 98 40W
Porbandar **70** 21 44N 69 43E
Porcher I. **105** 53 50N 130 30W
Pori **61** 61 29N 21 48E
Porkkala **61** 59 59N 24 26E
Porlock **42** 51 13N 3 36W
Port Alberni **105** 49 40N 124 50W
Port Alfred **94** 33 36S 26 55E
Port Angeles **105** 48 7N 123 30W
Port Arthur **101** 30 0N 94 0W
Port Askaig **46** 55 51N 6 8W
Port-au-Prince **107** 18 40N 72 20W
Port Augusta **79** 32 30S 137 50E
Port-Cartier **99** 50 2N 66 50W
Port Chalmers **83** 45 49S 170 30E
Port Coquitlam **105** 49 15N 122 45W
Port Darwin **78** 12 24S 130 45E
Port Davey **79** 43 16S 145 55E
Port de Paix **107** 19 50N 72 50W
Port Elizabeth **94** 33 58S 25 40E
Port Erin **42** 54 5N 4 45W
Port Étienne = Nouadhibou **88** 21 0N 17 0W
Port Fairy **79** 38 22S 142 12E
Port Glasgow **46** 55 57N 4 40W
Port Harcourt **90** 4 40N 7 10E
Port Hedland **78** 20 25S 118 35E
Port Huron **104** 43 0N 82 28W
Port Kelang **72** 3 0N 101 23E
Port Laoise **51** 53 2N 7 20W
Port Lincoln **79** 34 42S 135 52E
Port Macquarie **79** 31 25S 152 54E
Port Moresby **79** 9 24S 147 8E
Port Musgrave **79** 11 55S 141 50E
Port Nelson **98** 57 3N 92 36W
Port Nolloth **94** 29 17S 16 52E
Port Nouveau-Quebec **99** 58 30N 65 50W
Port of Ness **48** 58 29N 6 13W
Port of Spain **107** 10 40N 61 20W
Port Phillip B. **79** 38 10S 144 50E
Port Pirie **79** 33 10S 137 58E
Port Said **91** 31 16N 32 18E
Port St. Mary **42** 54 5N 4 45W
Port Shepstone **94** 30 44S 30 28E
Port Sudan **89** 19 32N 37 9E
Port Sunlight **45** 53 22N 3 0W
Port Talbot **43** 51 35N 3 48W
Port William **46** 54 46N 4 35W
Portadown **50** 54 27N 6 26W
Portaferry **50** 54 23N 5 32W
Portage La Prairie **98** 49 58N 98 18W
Portarlington **51** 53 10N 7 10W
Portballintrae **50** 55 13N 6 32W
Portglenone **50** 54 53N 6 30W
Porthmadog **42** 52 55N 4 13W
Portishead **43** 51 29N 2 46W
Portknockie **48** 57 40N 2 52W
Portland, Austral. **79** 38 20S 141 35E
Portland, Me., U.S.A. **104** 43 40N 70 15W
Portland, Oreg., U.S.A. **105** 45 35N 122 40W
Portland, Bill of **43** 50 31N 2 27W
Portmahomack **48** 57 50N 3 50W
Portmarnock **51** 53 25N 6 10W
Portnahaven **46** 55 40N 6 30W
Porto = Oporto **60** 41 8N 8 40W
Pôrto Alegre **108** 30 5S 51 3W
Porto Empèdocle **58** 37 18N 13 30E
Porto Novo **90** 6 23N 2 42E
Porto Tórres **58** 40 50N 8 23E
Portoferráio **58** 42 50N 10 20E
Portoscuso **58** 39 12N 8 22E
Portpatrick **46** 54 50N 5 7W
Portree **48** 57 25N 6 11W
Portrush **50** 55 13N 6 40W
Portsmouth, U.K. **40** 50 48N 1 6W
Portsmouth, Ohio, U.S.A. **101** 38 45N 83 0W
Portsmouth, Va., U.S.A. **101** 36 50N 76 20W
Portstewart **50** 55 12N 6 43W
Porttipahta **61** 68 5N 26 30E
Portugal ■ **60** 40 0N 7 0W
Portuguesa R. **107** 9 0N 68 20W
Portumna **51** 53 5N 8 12W
Postmasburg **94** 28 18S 23 5E
Potchefstroom **94** 26 41S 27 7E
Potenza **58** 40 40N 15 50E
Potgietersrus **94** 24 10S 29 3E
Potomac, R. **101** 38 0N 76 23W
Potow **73** 38 8N 116 31E
Potsdam **54** 52 23N 13 4E
Potters Bar **41** 51 42N 0 11W
Poulaphouca Res. **51** 53 8N 6 30W
Poulton le Fylde **45** 53 51N 2 59W
Póvoa de Varzim **60** 41 25N 8 46W
Powell **100** 44 45N 108 45W
Powell Creek **78** 18 6S 133 46E
Powys □ **42** 52 20N 3 20W
Poyang Hu **73** 29 10N 116 10E
Poynton **45** 53 21N 2 07W
Požarevac **55** 44 35N 21 18E
Poznan **54** 52 25N 16 55E
Prague **54** 50 5N 14 22E
Praha = Prague **54** 50 5N 14 22E
Prato **58** 43 53N 11 5E
Pratt **100** 37 40N 98 45W

* Now known as The Gulf

Preesall **Ryukyu Is**

Place	Page	Lat.	Long.
Preesall	**45**	53 55N	2 58W
Premier Downs	**78**	30 30S	126 30E
Prenzlau	**54**	53 19N	13 51E
Přerov	**55**	49 28N	17 27E
Prescot	**45**	53 27N	2 49W
Prescott	**100**	34 35N	112 30W
Presidio	**100**	29 30N	104 20W
Prestatyn	**42**	53 20N	3 24W
Presteigne	**42**	52 17N	3 0W
Preston, U.K.	**45**	53 46N	2 42W
Preston, U.S.A.	**100**	42 10N	111 55W
Preston, C.	**78**	20 51S	116 12E
Prestonpans	**47**	55 58N	3 0W
Prestwich	**45**	53 32N	2 18W
Prestwick	**46**	55 30N	4 38W
Pretoria	**94**	25 44S	28 12E
Price	**100**	39 40N	110 48W
Prieska	**94**	29 40S	22 42E
Prince Albert	**98**	53 15N	105 50W
Prince Albert Pen.	**98**	72 30N	116 0W
Prince Charles I.	**99**	67 47N	76 12W
Prince Edward I. □.	**99**	46 30N	63 30W
Prince Edward Is.	**23**	45 15S	39 0E
Prince George	**98**	53 55N	122 50W
Prince of Wales I.	**98**	73 0N	99 0W
Prince Patrick I.	**98**	77 0N	120 0W
Prince Rupert	**98**	54 20N	130 20W
Princes Risborough	**41**	51 43N	0 50W
Princess Charlotte B.	**79**	14 25S	144 0E
Princetown	**43**	50 33N	4 0W
Prins Albert	**94**	33 12S	22 2E
Pripyat, R.	**55**	51 30N	30 0E
Progreso	**106**	21 20N	89 40W
Prokopyevsk	**62**	54 0N	87 3E
Provence	**56**	43 40N	5 46E
Providence	**104**	41 41N	71 15W
Provo	**100**	40 16N	111 37W
Prudhoe	**44**	54 57N	1 52W
Prudhoe Bay	**98**	70 20N	148 20W
Pruszków	**55**	52 9N	20 49E
Prut, R.	**55**	46 3N	28 10E
Przemyśl	**55**	49 50N	22 45E
Przeworsk	**55**	50 6N	22 32E
Przhevalsk	**71**	42 30N	78 20E
Pskov	**61**	57 50N	28 25E
Puddletown	**43**	50 45N	2 21W
Pudsey	**45**	53 47N	1 40W
Puebla □	**106**	18 30N	98 0W
Pueblo	**100**	38 20N	104 40W
Pueblonuevo	**60**	38 16N	5 16W
Puerto de Santa María	**60**	36 36N	6 13W
Puerto Montt	**108**	41 22S	72 40W
Puerto Rico ■	**107**	18 15N	66 45W
Puertollano	**60**	38 43N	4 7W
Puget Sd.	**105**	47 15N	122 30W
Puigcerdá	**60**	42 24N	1 50E
Pukekohe	**83**	37 12S	174 55E
Pula	**58**	44 54N	13 57E
Pulantien	**73**	39 25N	122 0E
Pulborough	**40**	50 58N	0 30W
Pullman	**100**	46 49N	117 10W
Punakha	**70**	27 42N	89 52E
Pune	**70**	18 29N	73 57E
Punjab □	**70**	31 0N	76 0E
Punta Arenas	**108**	53 0S	71 0W
Purfleet	**41**	51 29N	0 15E
Puri	**70**	19 50N	85 58E
Purisima, La	**106**	26 10N	112 4W
Purley	**41**	51 29N	1 4W
Purus, R.	**108**	5 25S	64 0W
Pusan	**73**	35 5N	129 0E
Putaruru	**83**	38 2S	175 50E
Putumayo, R.	**108**	1 30S	70 0W
Puy-de-Dôme	**56**	45 46N	2 57E
Puy-de-Sancy	**56**	45 32N	2 41E
Puy, Le	**56**	45 2N	3 53E
Puyallup	**100**	47 10N	122 22W
Pwllheli	**42**	52 54N	4 26W
Pyatigorsk	**71**	44 2N	43 0E
Pyŏngyang	**73**	39 0N	125 30E
Pyramid L.	**105**	40 0N	119 30W
Pyrenees	**56**	42 45N	0 18E

Q

Place	Page	Lat.	Long.
Qal'at el Mudauwara	**71**	29 28N	36 3E
Qalyûb	**91**	30 12N	31 11E
Qâra	**91**	29 38N	26 30E
Qasr Farâfra	**91**	27 0N	28 1E
Qatar ■	**71**	25 30N	51 15E
Qazvin	**71**	36 15N	50 0E
Qena	**91**	26 10N	32 43E
Qena, Wadi	**91**	26 12N	32 44E
Qom	**71**	34 40N	51 0E
Quackenbrück	**52**	52 40N	7 59E
Quang Tri	**72**	16 45N	107 13E
Quantock Hills, The	**42**	51 8N	3 10W
Queanbeyan	**79**	35 17S	149 14E
Québec	**99**	46 52N	71 13W
Québec □	**99**	50 0N	70 0W
Queen Charlotte	**105**	53 15N	132 2W
Queen Charlotte Is.	**98**	53 20N	132 10W
Queen Elizabeth Is.	**98**	76 0N	95 0W
Queen Maud G.	**98**	68 15N	102 30W
Queenborough	**41**	51 24N	0 46E
Queensbury	**45**	53 46N	1 50W
Queensferry	**47**	56 0N	3 25W
Queensland □	**79**	15 0S	142 0E
Queenstown, Austral.	**79**	42 4S	145 35E
Queenstown, N.Z.	**83**	45 1S	168 40E
Queenstown, S. Afr.	**94**	31 52S	26 52E
Quelimane	**92**	17 53S	36 58E
Querétaro	**106**	20 40N	100 23W
Quesnel	**98**	53 0N	122 30W
Quetta	**71**	30 15N	66 55E
Quezon City	**72**	14 38N	121 0E
Qui Nhon	**72**	13 40N	109 13E
Quibdó	**107**	5 42N	76 40W
Quilon	**70**	8 50N	76 38E
Quilpie	**79**	26 35S	144 11E
Quimper	**56**	48 0N	4 9W
Quimperlé	**56**	47 53N	3 33W
Quincy	**104**	42 14N	71 0W
Quintanar de la Sierra	**60**	41 57N	2 55W
Quito	**108**	0 15S	78 35W
Quneitra	**89**	33 7N	35 48E
Quoich, L.	**48**	57 4N	5 20W
Quorn	**79**	32 25S	138 0E
Quruq Tagh, mts.	**73**	41 30N	90 0E
Qûs	**91**	25 55N	32 50E
Quseir	**91**	26 7N	34 16E

R

Place	Page	Lat.	Long.
Raahe	**61**	64 40N	24 28E
Raasay I.	**48**	57 25N	6 4W
Raba	**72**	8 36S	118 55E
Rabat	**88**	34 2N	6 48W
Rabaul	**79**	4 24S	152 18E
Rabigh	**71**	22 50N	39 5E
Raciborz	**55**	50 7N	18 18E
Racine	**104**	42 41N	87 51W
Radcliffe	**45**	53 35N	2 19W
Radekhov	**55**	50 25N	24 32E
Radium Hill	**79**	32 30S	140 42E
Radlett	**41**	51 41N	0 19W
Radnor Forest	**42**	52 17N	3 10W
Radom	**55**	51 23N	21 12E
Radomsko	**55**	51 5N	19 28E
Radstock	**42**	51 17N	2 25W
Radyr	**43**	51 32N	3 16W
Raetihi	**83**	39 25S	175 17E
Rafah	**91**	31 18N	34 14E
Raglan	**83**	37 55S	174 55E
Ragusa	**58**	36 56N	14 42E
Raichur	**70**	16 10N	77 20E
Raigarh	**70**	21 56N	83 25E
Rainford	**45**	53 31N	2 48W
Rainham	**41**	51 22N	0 36E
Rainhill	**45**	53 24N	2 46W
Rainier, Mt.	**100**	46 50N	121 50W
Raipur	**70**	21 17N	81 45E
Rajahmundry	**70**	17 1N	81 48E
Rajasthan □	**70**	26 45N	73 30E
Rajkot	**70**	22 15N	70 56E
Raleigh	**101**	35 46N	78 38W
Ramallah	**89**	31 55N	35 10E
Rambouillet	**56**	48 40N	1 48E
Ramla	**89**	31 55N	34 52E
Ramnad	**70**	9 25N	78 55E
Ramoutsa	**94**	24 50S	25 52E
Rampart	**98**	65 0N	150 15W
Ramsbottom	**45**	53 36N	2 20W
Ramsey, Cambs., U.K.	**40**	52 27N	0 6W
Ramsey, I. of M., U.K.	**42**	54 20N	4 21W
Ramsey I.	**42**	51 52N	5 21W
Ramsgate	**40**	51 20N	1 25E
Ranchi	**70**	23 19N	85 27E
Randers	**61**	56 29N	10 1E
Randfontein	**94**	26 8S	27 45E
Rangitaiki, R.	**83**	37 54S	176 49E
Rangoon	**70**	16 45N	96 20E
Rangwe	**93**	0 38S	34 35E
Rannoch	**48**	56 41N	4 20W
Rannoch, L.	**48**	56 41N	4 20W
Rantemario	**72**	3 15S	119 57E
Rapid City	**100**	44 0N	103 0W
Ras al Hadd	**71**	22 30N	59 50E
Ras Bânâs	**91**	23 57N	35 59E
Ras en Naqb	**91**	30 0N	35 37E
Rashid = Rosetta	**91**	31 21N	30 22E
Rasht	**71**	37 20N	49 40E
Rath Luirc (Charleville)	**51**	52 21N	8 40W
Rathcoole	**51**	53 17N	6 29W
Rathdowney	**51**	52 52N	7 36W
Rathdrum	**51**	52 57N	6 13W
Rathkeale	**51**	52 32N	8 57W
Rathlin I.	**50**	55 18N	6 14W
Rathmelton	**50**	55 3N	7 35W
Rathmore, Cork, Ireland	**51**	51 30N	9 21W
Rathmore, Kerry, Ireland	**51**	52 5N	9 12W
Rathnew	**51**	53 00N	6 05W
Ratlam	**70**	23 20N	75 0E
Rattray Hd.	**48**	57 38N	1 50W
Raukumara Ra.	**83**	38 5S	177 55E
Ravenglass	**45**	54 21N	3 25W
Ravenna	**58**	44 28N	12 15E
Ravensburg	**54**	47 48N	9 38E
Ravenshoe	**79**	17 37S	145 29E
Ravensthorpe	**78**	33 35S	120 2E
Rawalpindi	**71**	33 38N	73 8E
Rawdon	**45**	53 52N	1 40W
Rawene	**83**	35 25S	173 32E
Rawlinna	**78**	30 58S	125 28E
Rawlins	**100**	41 50N	107 20W
Rawlinson Range	**78**	24 40S	128 30E
Rawmarsh	**45**	53 27N	1 20W
Rawtenstall	**45**	53 42N	2 18W
Ray, C.	**99**	47 33N	59 15W
Rayleigh	**41**	51 36N	0 38E
Raz, Pte. du	**56**	48 2N	4 47W
Ré, Île de	**56**	46 12N	1 30W
Reading, U.K.	**41**	51 27N	0 57W
Reading, U.S.A.	**104**	40 20N	75 53W
Rebun-Tō	**76**	45 23N	141 2E
Recife	**108**	8 0S	35 0W
Recklinghausen	**52**	51 36N	7 10E
Red Deer	**98**	52 20N	113 50W
Red, R.	**101**	31 0N	91 40W
Red Sea	**71**	25 0N	36 0E
Redbourne	**41**	51 47N	0 23W
Redbridge	**41**	51 35N	0 7E
Redcar	**44**	54 37N	1 4W
Redcliff Bay	**43**	51 27N	2 50W
Redding	**105**	40 30N	122 25W
Redditch	**42**	52 18N	1 57W
Redhill	**41**	51 14N	0 10W
Redland	**49**	59 6N	3 4W
Redlands	**100**	34 0N	117 11W
Redondela	**60**	42 15N	8 38W
Redruth	**43**	50 14N	5 14W
Ree, L.	**51**	53 35N	8 0W
Reefton	**83**	42 6S	171 51E
Reepham	**40**	52 46N	1 6E
Regensburg	**54**	49 1N	12 7E
Regina	**98**	50 27N	104 35W
Registan □	**71**	30 15N	65 0E
Reichenbach	**54**	50 36N	12 19E
Reigate	**41**	51 14N	0 11W
Reims	**56**	49 15N	4 0E
Reinga, C.	**83**	34 25S	172 43E
Reitz	**94**	27 48S	28 29E
Remscheid	**52**	51 11N	7 12E
Renfrew	**47**	55 52N	4 24W
Reni	**55**	45 28N	28 15E
Renishaw	**45**	53 18N	1 21W
Renkum	**52**	51 58N	5 43E
Renmark	**79**	34 11S	140 43E
Rennes	**56**	48 7N	1 41W
Reno	**105**	39 30N	119 50W
Reno, R.	**58**	44 37N	12 17E
Renton	**47**	55 58N	4 35W
Requena	**60**	39 30N	1 4W
Resolution I., Can.	**99**	61 30N	65 0W
Resolution I., N.Z.	**83**	45 40S	166 40E
Réunion, I.	**23**	22 0S	56 0E
Reus	**60**	41 10N	1 5E
Revelstoke	**98**	51 0N	118 10W
Revilla Gigedo, Is. de	**22**	18 40N	112 0W
Rewari	**70**	28 15N	76 40E
Rexburg	**100**	43 55N	111 50W
Reykjavík	**96**	64 10N	21 57E
Reynolds Ra.	**78**	22 30S	133 0E
Rheine	**52**	52 17N	7 25E
Rheinland-Pfalz □	**54**	50 50N	7 0E
Rheydt	**52**	51 10N	6 24E
Rhine, R.	**54**	51 52N	6 20E
Rhode Island □	**101**	41 38N	71 37W
* Rhodesia ■	**92**	20 0S	30 0E
Rhondda	**42**	51 39N	3 30W
Rhône, R.	**56**	43 28N	4 42E
Rhossilli	**42**	51 34N	4 18W
Rhyl	**42**	53 19N	3 29W
Rhymney	**43**	51 45N	3 17W
Rhynie	**48**	57 20N	2 50W
Riau Arch.	**72**	0 30N	104 20E
Ribatejo □	**60**	39 15N	8 30W
Ribble, R.	**44**	54 13N	2 20W
Ribeirão Prêto	**108**	21 10S	47 50W
Riccarton	**83**	43 32S	172 37E
Richards B.	**94**	28 48S	32 6E
Richfield	**100**	38 50N	112 0W
Richland	**105**	46 15N	119 15W
Richmond, Austral.	**79**	20 43S	143 8E
Richmond, N.Z.	**83**	41 4S	173 12E
Richmond, S. Afr.	**94**	29 51S	30 18E
Richmond, N. Yorks., U.K.	**44**	54 24N	1 43W
Richmond, Surrey, U.K.	**41**	51 28N	0 18W
Richmond, Calif., U.S.A.	**105**	38 0N	122 21W
Richmond, Va., U.S.A.	**101**	37 33N	77 27W
Rickmansworth	**41**	51 38N	0 28W
Ried	**54**	48 14N	13 30E
Rietfontein	**94**	26 44S	20 1E
Rieti	**58**	42 23N	12 50E
Riga	**61**	56 53N	24 8E
Riga, G. of	**61**	57 40N	23 45E
Rijau	**90**	11 8N	5 17E
Rijeka	**58**	45 20N	14 21E
Rijssen	**52**	52 19N	6 30E
Rijswijk	**52**	52 4N	4 22E
Rimini	**58**	44 3N	12 33E
Rimouski	**99**	48 27N	68 30W
Ringwood	**40**	50 50N	1 48W
Rio de Janeiro	**108**	23 0S	43 12W
Rio Grande del Norte, R.	**106**	26 0N	97 0W
Rio Grande do Sul □	**108**	30 0S	53 0W
Rio Muni □	**88**	1 30N	10 0E
Riom	**56**	45 54N	3 7E
Ripley	**44**	54 3N	1 34W
Ripon	**44**	54 8N	1 31W
Risca	**43**	51 36N	3 6W
Rishton	**45**	53 46N	2 26W
Riva del Garda	**58**	45 53N	10 50E
Riversdal	**94**	34 7S	21 15E
Riverside	**100**	34 0N	117 22W
Riverton, Can.	**98**	51 1N	97 0W
Riverton, N.Z.	**83**	46 21S	168 0E
Riviera	**54**	44 0N	8 30E
Riviera di Levante	**58**	44 23N	9 15E
Rivière-du-Loup	**99**	47 50N	69 30W
Riyadh	**71**	24 41N	46 42E
Rizzuto, C.	**58**	38 54N	17 5E
Roanne	**56**	46 3N	4 4E
Roanoke	**101**	37 19N	79 55W
Robertson	**94**	33 46S	19 50E
Robertson Ra.	**78**	23 15S	121 0E
Roberttown	**45**	53 41N	1 42W
Robinson Ranges	**78**	25 40S	118 0E
Robla, La	**60**	42 50N	5 41W
Robson, Mt.	**98**	53 10N	119 10W
Roca, C.	**60**	38 40N	9 31W
Rochdale	**45**	53 36N	2 10W
Roche-sur-Yon, La	**56**	46 40N	1 26W
Rochefort	**56**	45 56N	0 57W
Rochelle, La	**56**	46 10N	1 9W
Rochester, Kent, U.K.	**41**	51 22N	0 30E
Rochester, Northum., U.K.	**46**	55 16N	2 16W
Rochester, Minn., U.S.A.	**101**	44 1N	92 28W
Rochester, N.Y., U.S.A.	**104**	43 10N	77 40W
Rochford	**41**	51 36N	0 42E
Rochfortbridge	**51**	53 25N	7 19W
Rock Hill	**101**	34 55N	81 2W
Rock Island	**101**	41 30N	90 35W
Rock Sprs.	**100**	41 40N	109 10W
Rockford	**101**	42 20N	89 0W
Rockhampton	**79**	23 22S	150 32E
Rockingham Forest	**44**	52 28N	0 42W
Rockland	**101**	44 0N	69 0W
Rocky Ford	**100**	38 7N	103 45W
Rocky Mount	**100**	35 55N	77 48W
Rocky Mts.	**98**	55 0N	121 0W
Rodel	**48**	57 45N	6 57W
Ródhos	**71**	36 15N	28 10E
Roding R.	**41**	51 31N	0 7E
Roebourne	**78**	20 44S	117 9E
Roermond	**52**	51 12N	6 0E
Roeselare	**52**	50 57N	3 7E
Roggeveldberge	**94**	32 10S	20 10E
Rojo, C.	**106**	21 33N	97 20W
Roma	**79**	26 32S	148 49E
Roma = Rome	**58**	41 54N	12 30E
Roman	**55**	46 57N	26 55E
Romania ■	**55**	46 0N	25 0E
Romans	**56**	45 3N	5 3E
Rome, Italy	**58**	41 54N	12 30E
Rome, Ga., U.S.A.	**101**	34 20N	85 0W
Rome, N.Y., U.S.A.	**104**	43 14N	75 29W
Romiley	**45**	53 24N	2 08W
Romilly	**56**	48 31N	3 44E
Romney Marsh	**40**	51 0N	1 0E
Romsey	**40**	51 0N	1 29W
Romsley	**41**	52 25N	2 03W
Rona I.	**48**	57 33N	6 0W
Roncesvalles, Pass	**60**	43 1N	1 19W
Ronda	**60**	36 46N	5 12W
Ronse	**52**	50 45N	3 35E
Roodepoort-Maraisburg	**94**	26 8S	27 52E
Roosendaal	**52**	51 32N	4 29E
Roosevelt I.	**112**	79 0S	161 0W
Roosevelt, R.	**108**	7 35S	60 20W
Roraima, Mt.	**107**	5 10N	60 40W
Rosa, Monte	**54**	45 57N	7 53E
Rosario, Argent.	**108**	33 0S	60 50W
Rosario, Venez.	**106**	10 19N	72 19W
Roscommon	**51**	53 38N	8 11W
Roscommon □	**50**	53 40N	8 15W
Roscrea	**51**	52 58N	7 50W
Roseburg	**105**	43 10N	123 10W
Rosehearty	**48**	57 42N	2 8W
Rosenheim	**54**	47 51N	12 9E
Rosetown	**98**	51 35N	107 59W
Rosetta	**91**	31 21N	30 22E
Rosetta Mouth	**91**	31 30N	30 20E
Rosneath	**47**	56 1N	4 49W
Ross	**83**	42 53S	170 49E
Ross on Wye	**42**	51 55N	2 34W
Ross Sea	**112**	74 0S	178 0E
Rossan Pt.	**50**	54 42N	8 47W
Rosslare	**51**	52 17N	6 23W
Rostock	**54**	54 4N	12 9E
Rostov	**62**	57 14N	39 25E
Roswell	**100**	33 26N	104 32W
Rosyth	**47**	56 2N	3 26W
Rothbury	**46**	55 19N	1 55W
Rother, R.	**40**	50 59N	0 40E
Rotherham	**45**	53 26N	1 21W
Rothes	**48**	57 31N	3 12W
Rothesay	**46**	55 50N	5 3W
Rothwell	**45**	53 46N	1 29W
Roto	**79**	33 0S	145 30E
Rotoroa Lake	**83**	41 55S	172 39E
Rotorua	**83**	38 9S	176 16E
Rotorua, L.	**83**	38 5S	176 18E
Rotterdam	**52**	51 55N	4 30E
Rotuma, I.	**23**	12 25S	177 5E
Roubaix	**56**	50 40N	3 10E
Rouen	**56**	49 27N	1 4E
Rousay, I.	**49**	59 10N	3 2W
Roussillon	**56**	45 24N	4 49E
Rouxville	**94**	30 11S	26 50E
Rouyn	**99**	48 20N	79 0W
Rovaniemi	**61**	66 29N	25 41E
Rovereto	**58**	45 53N	11 3E
Rovigo	**58**	45 4N	11 48E
Rovno	**55**	50 40N	26 10E
Rowlands Gill	**44**	54 55N	1 44W
Roxburgh, N.Z.	**83**	45 33S	169 19E
Roxburgh, U.K.	**46**	55 34N	2 30W
Royston, U.K.	**40**	52 3N	0 1W
Royston, U.K.	**45**	53 36N	1 27W
Royton	**45**	53 34N	2 7W
Ruabon	**42**	53 0N	3 3W
Ruahine Ra.	**83**	39 55S	176 2E
Ruapehu	**83**	39 17S	175 35E
Ruapuke I.	**83**	46 46S	168 31E
Rubery	**41**	52 24N	1 59W
Rubha Hunish, C.	**48**	57 42N	6 20W
Rudok	**73**	33 30N	79 40E
Rufiji, R.	**93**	7 50S	38 15E
Rugby	**41**	52 23N	1 16W
Rugeley	**45**	52 47N	1 56W
Rügen	**54**	54 22N	13 25E
Ruhr, R.	**54**	51 25N	6 44E
Rum Jungle	**78**	13 0S	130 59E
Rumania ■	**55**	46 0N	25 0E
Rumford	**101**	44 30N	70 30W
Rumoi	**76**	43 56N	141 39W
Rumuruti	**93**	0 17N	36 32E
Runanga	**83**	42 25S	171 15E
Runcorn	**45**	53 20N	2 44W
Runka	**90**	12 28N	7 20E
Rupat, I.	**72**	1 45N	101 40E
Rur, R.	**52**	50 54N	6 24E
Ruschuk = Ruse	**55**	43 48N	25 59E
Ruse	**55**	43 48N	25 59E
Rush	**51**	53 31N	6 7W
Rushden	**40**	52 17N	0 37W
Russellville	**101**	35 15N	93 0W
Rustenburg	**94**	25 41S	27 14E
Rutherglen	**47**	55 50N	4 11W
Ruthin	**42**	53 7N	3 20W
Rutshuru	**93**	1 13S	29 25E
Ruzomberok	**55**	49 3N	19 17E
Rwanda ■	**93**	2 0S	30 0E
Ryan, L.	**46**	55 0N	5 2W
* Rybinsk	**62**	58 5N	38 50E
† Rybinsk Res.	**62**	58 30N	38 0E
Ryde	**40**	50 44N	1 9W
Rye	**40**	50 57N	0 46E
Rye, R.	**44**	54 12N	0 53W
Ryukyu Is.	**76**	26 0N	128 0E

** Renamed Zimbabwe*

** Renamed Andropov*
† Renamed Andropov Res.

Rzeszów Shawinigan

Rzeszów 55 50 5N 21 58 E
Rzhev 61 56 20N 34 20 E

S

's Gravenhage = Hague, The 52 52 7N 4 17 E
Saale, R. 54 51 57N 11 56 E
Saar (Sarre), □ 52 49 20N 6 45 E
Saarbrücken 52 49 15N 6 58 E
Saarburg 52 49 36N 6 32 E
Saaremaa 61 58 30N 22 30 E
Saarlouis 52 49 19N 6 45 E
Sabadell 60 41 28N 2 7 E
Sabah □ 72 6 0N 117 0 E
Sabi, R. 92 18 50S 31 40 E
Sabie 94 25 4S 30 48 E
Sabinas 106 27 50N 101 10W
Sabinas Hidalgo 106 26 40N 100 10W
Sabine Mts. 58 42 15N 12 50 E
Sable, C., Can. 99 43 29N 65 38W
Sable, C., U.S.A. 101 25 5N 81 0W
Sable I. 99 44 0N 60 0W
Sabzevār 71 36 15N 57 40 E
Sacramento 105 38 39N 121 30 E
Sacramento Mts. 100 32 30N 105 30W
Sacramento, R. 105 38 3N 121 56W
Sacriston 44 54 49N 1 38W
Sadaba 60 42 19N 1 12W
Sadiya 70 27 50N 95 40 E
Sado 76 38 0N 138 25 E
Sadon 70 25 28N 98 0 E
Safad 89 32 58N 35 28 E
Safâga 91 26 42N 34 0 E
Saffron Walden 40 52 2N 0 15 E
Safi 88 31 2N 35 28 E
Saga 76 33 15N 130 16 E
Sagaing 70 23 30N 95 30 E
Saginaw 104 43 26N 83 55W
Saguenay, R. 99 48 22N 71 0W
Sagunto 60 39 42N 0 18W
Sahagún 60 42 18N 5 2W
Sahara 88 23 0N 5 0 E
Saharan Atlas 88 34 9N 3 29 E
Saharanpur 70 29 58N 77 33 E
Sahiwal 71 30 45N 73 8 E
Saigon = Ho Chi Minh City 72 10 58N 106 40 E
Saiki 76 32 58N 131 57 E
Saimaa. 61 61 15N 28 15 E
St. Abb's Head 46 55 55N 2 10W
St. Agnes 43 50 18N 5 13W
St. Albans, U.K. 41 51 44N 0 19W
St. Albans, U.S.A. 101 44 49N 73 7W
St. Alban's Head 43 50 34N 2 3W
St. Andrews, N.Z. 83 44 33S 171 10 E
St. Andrews, U.K. 46 56 20N 2 48W
St. Anne's 45 53 45N 3 2W
St. Asaph 42 53 15N 3 27W
St. Austell 43 50 20N 4 48W
St. Bee's Hd. 44 54 30N 3 38 E
St. Boniface 98 49 53N 97 5W
St. Bride's B. 42 51 48N 5 15W
St-Brieuc 56 48 30N 2 46W
St. Catharines 104 43 10N 79 15W
St. Catherine's Pt. 40 50 34N 1 18W
St. Charles 101 38 46N 90 30W
St-Claude 56 46 22N 5 52 E
St. Clears 42 51 48N 4 30W
St. Cloud 101 45 30N 94 11W
St. Columb Major 43 50 26N 4 56W
St. Combs 48 57 40N 1 55W
St. Croix, I. 107 17 45N 64 45W
St. David's 42 51 54N 5 16W
St. David's Head 42 51 54N 5 16W
St-Dizier 56 48 40N 5 0 E
St. Elias, Mt. 98 60 20N 141 59W
St-Étienne 56 45 27N 4 22 E
St. Francis C. 94 34 14S 24 49 E
St. Gallen 54 47 25N 9 20 E
St-Gaudens 56 43 6N 0 44 E
St. George, Austral. 79 28 1S 148 41 E
St. George, U.S.A. 100 37 10N 113 35W
St-Georges 52 50 37N 4 20 E
St. George's 107 12 5N 61 43W
St. George's Channel 32 52 0N 6 0W
St. Gotthard P. 54 46 33N 8 33 E
St. Govan's Hd. 42 51 35N 4 56W
St. Helena B. 94 32 40S 18 10 E
St. Helena, I. 22 15 55S 5 44W
St. Helens, U.K. 45 53 28N 2 44W
St. Helens, U.S.A. 100 45 55N 122 50W
St. Helier 43 49 11N 2 6W
St-Hyacinthe 104 45 40N 72 58W
St. Ives, Cambs., U.K. 40 52 20N 0 5W
St. Ives, Cornwall, U.K. 43 50 13N 5 29W
St. John 99 45 20N 66 8W
St. John's 99 47 35N 52 40W
St. Joseph 101 39 40N 94 50W
St. Just 43 50 7N 5 41W
St. Kilda 83 45 53S 170 31 E
St. Kitts-Nevis ■ 107 17 20N 62 40W
St. Lawrence, Gulf of 99 48 25N 62 0W
St. Lawrence, I. 96 63 0N 170 0W
St. Lawrence, R. 99 49 30N 66 0W
St. Louis, Senegal 88 16 8N 16 27W
St. Louis, U.S.A. 101 38 40N 90 12W
St. Lucia, C. 94 28 32S 32 29 E
St. Lucia ■ 107 14 0N 60 50W
St. Lucia, Lake 94 28 5S 32 30 E
St. Magnus B. 49 60 25N 1 35W
St-Malo 56 48 39N 2 1W
St.-Martin, I. 107 18 0N 63 0W
St. Marys 79 41 32S 148 11 E
St. Mary's I. 43 49 55N 6 17W
St. Mawes 43 50 10N 5 1W
St. Monance 46 56 13N 2 46W
St. Moritz 54 46 30N 9 51 E
St-Nazaire 56 47 17N 2 12W
St. Neots 40 52 14N 0 16W
St-Niklaas 52 51 10N 4 8 E
St. Paul 101 44 54N 93 5W
St. Peter Port 43 49 27N 2 31W
St. Petersburg 101 27 45N 82 40W
St.-Pierre, I. 99 46 48N 56 12W
St-Quentin 56 49 50N 3 16 E
St-Raphaël 56 43 25N 6 46 E
St.-Servan-sur-Mer 56 48 38N 2 0W
St. Thomas, I. 107 18 21N 64 55W
St.-Trond 52 50 48N 5 12 E
St-Tropez 56 43 17N 6 38 E
St.-Vallier 56 45 11N 4 50 E
St. Vincent ■ 107 13 10N 61 10W
Saintes 56 45 45N 0 37W
Saintfield 50 54 28N 5 50W
Saintonge 56 45 40N 0 50W
Säkahka 71 30 0N 40 8 E
Sakai 76 34 30N 135 30 E
Sakata 76 38 38N 138 19 E
Sakhalin, I. 62 51 0N 143 0 E
Sala 61 59 58N 16 35 E
Salālah 71 16 56N 53 59 E
Salamanca 60 40 58N 5 39W
Saldaña 60 42 32N 4 48W
Saldanha 94 33 0S 17 58 E
Saldanha Bay 94 33 6S 18 0 E
Sale, Austral. 79 38 6S 147 6 E
Sale, U.K. 45 53 26N 2 19W
Salekhard 62 66 30N 66 25 E
Salem, India 70 11 40N 78 11 E
Salem, U.S.A. 105 45 0N 123 0W
Salen 48 56 42N 5 48W
Salerno 58 40 40N 14 44 E
Salford 45 53 30N 2 17W
Salida 100 38 35N 106 0W
Salima 92 13 47S 34 28 E
Salina 100 38 50N 97 40W
Salina Cruz 106 16 10N 95 10W
Salina, I. 58 38 35N 14 50 E
Salinas, R. 105 36 45N 121 48W
*Salisbury, Zimb. 92 17 50S 31 2 E
Salisbury, U.K. 42 51 4N 1 48W
Salisbury, Md., U.S.A. 101 38 20N 75 38W
Salisbury, N.C., U.S.A. 101 35 42N 80 29W
Salisbury Plain 42 51 13N 1 50W
Salmon, R. 100 45 51N 116 46W
Salonta 55 46 49N 21 42 E
Salop □ 45 52 36N 2 45W
Salt Lake City 100 40 45N 111 58W
Salta 108 24 47S 65 25W
Saltash 43 50 25N 4 13W
Saltburn by the Sea 44 54 35N 0 58W
Saltcoats 47 55 38N 4 47W
Saltfleet 44 53 25N 0 11 E
Saltillo 106 25 30N 100 57W
Salton Sea 100 33 20N 115 50W
Salûm 91 31 31N 25 7 E
Salûm, G. of 91 31 30N 25 9 E
Saluzzo 58 44 39N 7 29 E
Salvador 108 13 0S 38 30W
Salvador ■ 106 13 50N 89 0W
Salween, R. 70 16 31N 97 37 E
Salzburg 54 47 48N 13 2 E
Salzburg □ 54 47 15N 13 0 E
Salzgitter 54 52 2N 10 22 E
Samâlût 91 28 20N 30 42 E
Samar, I. 72 12 0N 125 0 E
Samarkand 71 39 40N 66 55 E
Sambiase 58 38 58N 16 16 E
Sambor 55 49 30N 23 10 E
Sambre, R. 52 50 27N 4 52 E
Same 93 4 2S 37 38 E
Samshui 73 23 7N 112 58 E
Samsun 71 41 15N 36 15 E
San Andres Mts. 100 33 0N 106 45W
San Angelo 100 31 30N 100 30W
San Antonio 100 29 30N 98 30W
San Antonio, C. 106 21 50N 84 57W
San Antônio Falls 108 9 30S 65 0W
San Benedetto 58 45 2N 10 57 E
San Bernardino 105 34 7N 117 18W
San Carlos 106 29 0N 101 10W
San Cristóbal 107 7 46N 72 14W
San Diego 105 32 43N 117 10W
San Felipe 107 10 20N 68 44W
San Fernando, Mexico 106 30 0N 115 10W
San Fernando, Trin. 107 10 20N 61 30W
San Francisco 105 37 47N 122 30W
San Francisco de Macorís 107 19 19N 70 15W
San Joaquin R. 105 37 4N 121 51W
San Jorge, G. of, Spain 60 40 50N 0 55W
San Jorge, G. of, Argent. 108 46 0S 66 0W
San José, C. Rica 107 10 0N 84 2W
San José, Guat. 106 14 0N 90 50W
San Jose 105 37 20N 121 53W
San Juan 107 18 28N 66 37W
San Juan Mts. 100 38 30N 108 30W
San Lucas C. 106 22 50N 110 0W
San Luis Obispo 100 35 21N 120 38W
San Luis Potosi 106 22 9N 100 59W
San Marcos 100 29 53N 98 0W
San Marino ■ 58 43 56N 12 25 E
San Mateo 105 37 32N 122 19W
San Matias, G. of 108 41 30S 64 0W
San Miguel 106 9 40N 65 11W
San Pedro Sula 106 15 30N 88 0W
San, R. 55 50 25N 22 20 E
San Remo 58 43 48N 7 47 E
San Salvador 106 13 40N 89 20W
San Salvador (Watlings) I. 107 24 0N 74 40W
San Sebastián 60 43 17N 1 58W
San Severo 58 41 41N 15 23 E
San Vicente de la Barquera 60 43 30N 4 29W
San'a 71 15 27N 44 12 E
Sanandaj 71 35 25N 47 7 E
Sancti-Spiritus 107 21 52N 79 33W
Sandakan 72 5 53N 118 10 E
Sanday, I. 49 59 15N 2 30W
Sandbach 44 53 9N 2 23W
Sandbank 47 55 58N 4 57W
Sandhurst 41 51 21N 0 48W
Sandoa 92 9 48S 23 0 E
Sandomierz 55 50 40N 21 43 E
Sandoway 70 18 20N 94 30 E
Sandown 40 50 39N 1 9W
Sandpoint 100 48 20N 116 40W
Sandstone 78 27 59S 119 16 E
Sandusky 104 41 25N 82 40W
Sandwich 40 51 16N 1 21 E
Sandy 40 52 07N 0 17W
Sanford Mt. 98 62 30N 143 0W
Sangli 70 16 55N 74 33 E
Sangre de Cristo Mts. 100 37 0N 105 0W
Sangsang 73 29 30N 86 0 E
Sankuru, R. 92 4 17S 20 25 E
Sanlucár la Mayor 60 37 26N 6 18W
Sanok 55 49 35N 22 10 E
Sanquhar 46 55 21N 3 56W
Santa Ana 105 33 48N 117 55W
Santa Barbara 105 34 25N 119 40W
Santa Catalina, I. 100 33 20N 118 30W
Santa Clara, Cuba 107 22 20N 80 0W
Santa Clara, U.S.A. 100 37 21N 122 0W
Santa Cruz 105 36 55N 122 1W
Santa Cruz I. 105 34 0N 119 45W
Santa Cruz, Is. 23 10 30S 166 0 E
Santa Fe, Argent. 108 31 35S 60 41W
Santa Fe, U.S.A. 100 35 40N 106 0W
Santa Lucia Range 105 36 0N 121 20W
Santa Maria 100 34 58N 120 29W
Santa Marta 107 11 15N 74 13W
Santa Monica 100 34 0N 118 30W
Santa Rosa 105 38 26N 122 43W
Santa Rosa I. 100 34 0N 120 6W
Santander 60 43 27N 3 51W
Santarém, Brazil 108 2 25S 54 42W
Santarém, Port. 60 39 12N 8 42W
Santiago, Chile 108 33 24S 70 50W
Santiago, Dom. Rep. 107 19 30N 70 40W
Santiago, Spain 60 42 52N 8 37W
Santiago de Cuba 107 20 0N 75 49W
Santo Domingo 107 18 30N 70 0W
Santos 108 24 0S 46 20W
Santry 51 53 24N 6 15W
São Francisco, R. 108 10 30S 36 24W
São Luis 108 2 39S 44 15W
São Paulo 108 23 40S 46 50W
Sao Roque, C. 108 5 30S 35 10W
São Tomé, I. 88 0 10N 7 0 E
Saône, R. 56 46 25N 4 50 E
Sapporo 76 43 0N 141 15 E
Sarajevo 58 43 52N 18 26 E
Sarasota 101 27 10N 82 30W
Saratoga Springs 104 43 5N 73 47W
Saratov 62 51 30N 46 2 E
Sarawak □ 72 2 0N 113 0 E
Sardinia, I. 58 40 0N 9 0 E
Sarina 79 21 22S 149 13 E
Sark, I. 43 49 25N 2 20W
Sarny 55 51 17N 26 40 E
Sarrebourg 56 48 43N 7 3 E
Sarreguemines 56 49 1N 7 4 E
Sarthe, R. 56 47 33N 0 31W
Sasebo 76 33 10N 129 43 E
Saskatchewan □ 98 54 40N 106 0W
Saskatchewan, R. 98 53 12N 99 16W
Saskatoon 98 52 10N 106 38W
Sasolburg 94 26 46S 27 49 E
Sássari 58 40 44N 8 33 E
Sassnitz 54 54 29N 13 39 E
Sátoraljaújhely 55 48 25N 21 41 E
Satpura Ra. 70 21 40N 75 0 E
Satsuna-Shotō 76 30 0N 130 0 E
Satu Mare 55 47 46N 22 55 E
Saudi Arabia ■ 71 26 0N 44 0 E
Sault Ste. Marie 99 46 30N 84 20W
Saumur 56 47 15N 0 5W
Saundersfoot 42 51 43N 4 42W
Sauri 90 11 50N 6 44 E
Sava, R. 55 44 50N 20 26 E
Savalou 90 7 57N 2 4 E
Savannah 101 32 4N 81 4W
Savé 90 8 2N 2 17 E
Savona 58 44 19N 8 29 E
Savoy □ 56 45 26N 6 35 E
Sawahlunto 72 0 52S 100 52 E
Sawara 76 35 55N 140 30 E
Sawatch Mts. 100 38 30N 106 30W
Sawbridgeworth 40 51 49N 0 10 E
Sawel, Mt. 50 54 48N 7 5W
Sawu, I. 72 10 35S 121 50 E
Sawu Sea 72 9 30S 121 50 E
Saxmundham 40 52 13N 1 29 E
Sayda 89 33 35N 35 25 E
Saynshand 73 44 55N 110 11 E
Sazin 70 35 35N 73 30 E
Sca Fell 44 54 27N 3 14W
Scalby 44 54 18N 0 26W
Scalloway 49 60 9N 1 16W
Scalpay, I. 48 57 18N 6 0W
Scammon Bay 98 62 0N 165 49W
Scandinavia 61 64 0N 12 0 E
Scapa Flow 49 58 52N 3 6W
Scarba, I. 46 56 10N 5 42W
Scarborough 44 54 17N 0 24W
Scariff 51 52 55N 8 32W
Scarinish 48 56 30N 6 48W
Scarp, I. 48 58 1N 7 8W
Schaffhausen 54 47 42N 8 39 E
Schefferville 99 54 48N 66 50W
Schelde, R. 52 51 10N 4 20 E
Schenectady 104 42 50N 73 58W
Schiedam 52 51 55N 4 25 E
Schiermonnikoog, I. 52 53 30N 6 15 E
Schio 58 45 42N 11 21 E
Schleswig 54 54 32N 9 34 E
Schleswig-Holstein □ 54 54 10N 9 40 E
Schouten Is. 72 1 0S 136 0 E
Schouwen, I. 52 51 43N 3 45 E
Schwangyashan 73 46 35N 131 15 E
Schwarzrand 94 26 0S 17 0 E
Schweinfurt 54 50 3N 10 12 E
Schweizer Reneke 94 27 11S 25 18 E
Schwerin 54 53 37N 11 22 E
Schwyz 54 47 2N 8 39 E
Sciacca 58 37 30N 13 3 E
Scilla 58 38 18N 15 44 E
Scilly, Isles of 43 49 55N 6 15W
Scone 46 56 25N 3 26W
Scotia Sea 22 56 5S 56 0W
Scotland □ 46 57 0N 4 0W
Scott, I. 112 67 0S 179 0 E
Scottsbluff 100 41 55N 103 35W
Scottsdale 79 41 9S 147 31 E
Scourie 48 58 20N 5 10W
Scrabster 48 58 36N 3 31W
Scranton 104 41 22N 75 41W
Scunthorpe 44 53 35N 0 38W
Seaford 40 50 46N 0 8 E
Seaforth, L. 48 57 52N 6 36W
Seaham 44 54 51N 1 20W
Seahouses 46 55 35N 1 39W
Seaton 43 50 42N 3 3W
Seaton Burn 44 55 3N 1 37W
Seaton Delaval 44 55 5N 1 33W
Seattle 105 47 41N 122 15W
Secretary I. 83 45 15S 166 56 E
Secunderabad 70 17 28N 78 30 E
Sedalia 101 38 40N 93 18W
Sedbergh 44 54 20N 2 31W
Seddon 83 41 40S 174 7 E
Seddonville 83 41 33S 172 1 E
Sedgefield 44 54 40N 1 27W
Seeheim 94 26 32S 17 52 E
Segovia 60 40 57N 4 10W
Segura, R. 60 38 9N 0 40W
Seil, I. 46 56 17N 5 37W
Seine, R. 56 49 28N 0 15 E
Sekenke 93 4 18S 34 11 E
Sekondi-Takoradi 88 5 0N 1 48W
Sekuma 94 24 36S 23 57 E
Selaru, I. 72 8 18S 131 0 E
Selby 44 53 47N 1 5W
Selkirk 46 55 33N 2 50W
Selkirk Mts. 98 51 15N 117 40W
Selma 101 32 30N 87 0W
Selsey 40 50 44N 0 47W
Selsey Bill 40 50 44N 0 47W
Selvas 108 6 30S 67 0W
Selwyn 79 21 30S 140 29 E
Selwyn Ra. 79 21 10S 140 0 E
Semarang 72 7 0S 110 26 E
Semipalatinsk 62 50 30N 80 10 E
Semois, R. 52 49 53N 4 44 E
Sendai, Kagoshima, Japan 76 31 50N 130 20 E
Sendai, Miyagi, Japan 76 38 15N 141 0 E
Senegal ■ 88 14 30N 14 30W
Senekal 94 28 18S 27 36 E
Senga Hill 93 9 19S 31 11 E
Senigállia 58 43 42N 13 12 E
Senja I. 61 69 15N 17 30 E
Sennâr 89 13 30N 33 35 E
Senta 55 45 55N 20 3 E
Sept-Îles 99 50 13N 66 22W
Sequoia Nat. Park 100 36 30N 118 30W
Seraing 52 50 35N 5 32 E
Seram, I. 72 3 10S 129 0 E
Seremban 72 2 43N 101 53 E
Serengeti Plain 93 2 40S 35 0 E
Serov 62 59 36N 60 35 E
Serowe 92 22 25S 26 43 E
Sesheke 92 17 29S 24 13 E
Sestao 60 43 18N 3 0W
Setana 76 42 26N 139 51 E
Sète 56 43 25N 3 42 E
Sétif 88 36 9N 5 26 E
Seto Naikai 76 34 20N 133 30 E
Settle 44 54 5N 2 18W
Setúbal 60 38 30N 8 58W
Setúbal, B. of 60 38 40N 8 56W
Sevastopol 62 44 35N 33 30 E
Sevenoaks 41 51 16N 0 11 E
Severn, R. 42 51 35N 2 38W
Severnaya Zemlya 62 79 0N 100 0 E
Seville 60 37 23N 6 0W
Seward 98 60 0N 149 40W
Seward Pen. 98 65 0N 164 0W
Seychelles, Is. 23 5 0S 56 0 E
Seyne 56 44 21N 6 22 E
Sfax 88 34 49N 10 48 E
Sfîntu Gheorghe 55 45 52N 25 48 E
Sgurr Mor 48 57 42N 5 0W
Shaba 92 8 0S 25 0 E
Shackleton 112 78 30S 36 1W
Shaftesbury 43 51 0N 2 12W
Shagamu 90 6 51N 3 39 E
Shahjahanpur 70 27 54N 79 57 E
Shāhrūd 71 36 30N 55 0 E
Shaki 90 8 41N 3 21 E
Shan □ 70 21 30N 98 30 E
Shanga 90 9 1N 5 2 E
Shanghai 73 31 10N 121 25 E
Shangjao 73 28 25N 117 57 E
Shangkiu 73 34 28N 115 42 E
Shangshui 73 33 42N 114 34 E
Shanklin 40 50 39N 1 9W
Shannon 83 40 33S 175 25 E
Shannon, R. 51 53 10N 8 10W
Shansi □ 73 37 30N 112 15 E
Shantar Is. 62 55 9N 137 40 E
Shantow (Swatow) 73 23 25N 116 40 E
Shantung □ 73 36 0N 117 30 E
Shaohing 73 30 0N 120 32 E
Shaoyang 73 27 10N 111 30 E
Shap 44 54 32N 2 40W
Shapinsay, I. 49 59 2N 2 50W
Sharjah 71 25 23N 55 26 E
Shark B. 78 11 20S 130 35 E
Sharlston 45 53 42N 1 23W
Sharma 91 27 52N 35 27 E
Sharpness 43 51 43N 2 28W
Shasi 73 30 16N 112 20 E
Shasta, Mt. 105 41 30N 122 0W
Shasta Res. 105 40 50N 122 15W
Shaw 45 53 34N 2 05W
Shawinigan 104 46 35N 72 50W

* Renamed Harare

Shawnee — Sugarloaf Pt

Name	Map	Lat.	Long.
Shawnee	101	35 15N	97 0W
Sheboygan	104	43 46N	87 45W
Sheelin, Lough	50	53 48N	7 20W
Sheerness	41	51 26N	0 47 E
Sheffield	45	53 23N	1 28W
Shefford	40	52 2N	0 20W
Shelburne	99	43 47N	65 20W
Shelby	100	48 30N	111 59W
Shelekhov G.	62	59 30N	157 0 E
Shellharbour	79	34 31 S	150 51 E
Shenandoah	101	40 50N	95 25W
Shendam	90	9 10N	9 30 E
Shensi □	73	34 50N	109 25 E
Shenyang (Mukden)	73	41 35N	123 30 E
Shepetovka	55	50 10N	27 10 E
Shephelah	89	31 30N	34 43 E
Shepparton	79	36 23 S	145 26 E
Shepton Mallet	42	51 11N	2 31W
Sherborne	43	50 56N	2 31W
Sherbro I.	88	7 30N	12 40W
Sherbrooke	104	45 28N	71 57W
Sherburn	45	53 47N	1 15W
Sheridan	100	44 50N	107 0W
Sheringham	40	52 56N	1 11 E
Sherman	101	33 40N	96 35W
Sheslay	98	58 17N	131 45W
Shetland Is.	49	60 30N	1 30W
Shevchenko	71	44 25N	51 20 E
Shibata	76	37 57N	139 20 E
Shibetsu	76	44 10N	142 23 E
Shibîn El Kôm	91	30 31N	30 55 E
Shiel, L.	48	56 48N	5 32W
Shieldaig	48	57 31N	5 39W
Shigatse	73	29 10N	89 0 E
Shihkiachwang	73	38 0N	114 32 E
Shikarpur	71	27 57N	68 39 E
Shikoku □	76	33 30N	133 30 E
Shildon	44	54 37N	1 39W
Shillelagh	51	52 46N	6 32W
Shimabara	76	32 48N	130 20 E
Shimada	76	34 49N	138 19 E
Shimanovsk	73	52 15N	127 30 E
Shimizu	76	35 0N	138 30 E
Shimo-Jima	76	32 15N	130 7 E
Shimoga	70	13 57N	75 32 E
Shimonoseki	76	33 58N	131 0 E
Shin, L.	48	58 7N	4 30W
Shin, R.	48	57 58N	4 26W
Shiney Row	44	54 53N	1 29W
Shingu	76	33 40N	135 55 E
Shinjō	76	38 46N	140 18 E
Shinyanga	93	3 45 S	33 27 E
Shiogama	76	38 19N	141 1 E
Shipki La	70	31 45N	78 40 E
Shipley	45	53 50N	1 47W
Shipston-on-Stour	42	52 4N	1 38W
Shir Kūh	71	31 45N	53 30 E
Shirāz	71	29 42N	52 30 E
Shirehampton	43	51 29N	2 41W
Shiremoor	44	55 01N	1 28W
Shirwa, L.	92	15 55 S	35 40 E
Shiukwan	73	24 58N	113 3 E
Shiyata	91	29 25N	25 7 E
Shizuoka	76	35 0N	138 30 E
Shoeburyness	40	51 31N	0 49 E
Sholapur	70	17 43N	75 56 E
Shoreham-by-Sea	40	50 50N	0 17W
Shoshone	100	43 0N	114 27W
Shoshone Mts.	105	39 30N	117 30W
Shoshong	92	22 56 S	26 31 E
Shotton Colliery	44	54 44N	1 18W
Shotts	47	55 49N	3 47W
Shreveport	101	32 30N	93 50W
Shrewsbury	45	52 42N	2 45W
Shwebo	70	22 30N	95 45 E
Si Kiang, R.	73	22 20N	113 20 E
Siakwan	73	25 45N	100 10 E
Sialkot	71	32 32N	74 30 E
* Siam, G. of	72	11 30N	101 0 E
Sian	73	34 2N	109 0 E
Siang K.	73	27 10N	112 45 E
Siangfan	73	32 15N	112 2 E
Siangtan	73	28 0N	112 55 E
Siao Hingan Ling	73	49 0N	127 0 E
Sibâi, Gebel el	91	25 45N	34 10 E
Sibaya, L.	94	27 20 S	32 45 E
Sibenik	58	43 48N	15 54 E
Siberut, I.	72	1 30 S	99 0 E
Sibiu	55	45 45N	24 9 E
Sibolga	72	1 50N	98 45 E
Sibuyan, I.	72	12 25N	122 40 E
Sichang	73	28 0N	102 10 E
Sîdi Abd el Rahmân	91	30 55N	28 41 E
Sîdi Barrâni	91	31 32N	25 58 E
Sidi-Bel-Abbès	88	35 13N	0 10W
Sidi Haneish	91	31 10N	27 35 E
Sidi Omar	91	31 24N	24 57 E
Sidlaw Hills	48	56 32N	3 10W
Sidmouth	43	50 40N	3 13W
Sidney	100	47 51N	104 7W
Sidra, G. of	88	31 40N	18 30 E
Siedlce	55	52 10N	22 20 E
Siegburg	52	50 48N	7 12 E
Siegen	52	50 52N	8 2 E
Siena	58	43 20N	11 20 E
Sienyang	73	34 20N	108 48 E
Sierra Leone ■	88	9 0N	12 0W
Sikhote Alin Ra.	62	46 0N	136 0 E
Sikkim □	70	27 50N	88 50 E
Sil, R.	60	42 23N	7 30W
Silchar	70	24 49N	92 48 E
Silesia	54	51 0N	16 30 E
Silloth	44	54 53N	3 25W
Silsden	45	53 55N	1 55W
Silver City	100	32 50N	108 18W
Silvermine Mts.	51	52 47N	8 15W
Simeulue, I.	72	2 45N	95 45 E
Simikot	70	30 0N	81 50 E
Simla	70	31 2N	77 15 E
Simonstown	94	34 14 S	18 26 E
Simplon Pass	54	46 15N	8 0 E
Simpson Des.	79	25 0 S	137 0 E
Sinai	91	29 0N	34 0 E
Sinai, Mt. = Musa, G.	91	28 32N	33 59 E
Sinaloa	106	25 50N	108 20W
Sines	60	37 56N	8 51W
Singapore ■	72	1 17N	103 51 E
Singida	93	4 49 S	34 48 E
Singtai	73	37 2N	114 30 E
Sining	73	36 35N	101 50 E
Sinkiang-Uighur □	73	42 0N	86 0 E
Sinnûris	91	29 26N	30 31 E
Sinop	71	42 1N	35 11 E
Sinsiang	73	35 15N	113 55 E
Sintai	73	30 59N	105 0 E
Sintra	60	38 47N	9 25W
Sinŭiju	73	40 5N	124 24 E
Sinyang	73	32 6N	114 2 E
Sioux City	101	42 32N	96 25W
Sioux Falls	101	43 35N	96 40W
Sipora, I.	72	2 15 S	99 44 E
Siracusa	58	37 4N	15 17 E
Siret, R.	55	47 58N	26 5 E
Sirsa	70	29 33N	75 4 E
Sisak	58	45 30N	16 21 E
Sitka	96	57 9N	134 58W
Sittard	52	51 0N	5 52 E
Sittingbourne	41	51 20N	0 43 E
Siwa	91	29 11N	25 31 E
Siwa Oasis	91	29 10N	25 30 E
Sizewell	40	52 13N	1 38 E
Sjælland	61	55 30N	11 30 E
Skagen	61	68 37N	14 27 E
Skagerrak	61	57 30N	9 0 E
Skagway	98	59 30N	135 20W
Skeena, R.	98	54 9N	130 5W
Skegness	44	53 9N	0 20 E
Skellefte, R.	61	65 30N	18 30 E
Skellefteå	61	64 45N	20 58 E
Skelmanthorpe	45	53 35N	1 39W
Skelmersdale	45	53 34N	2 49W
Skelmorlie	47	55 52N	4 53W
Skerries	51	53 35N	6 7W
Skibbereen	51	51 33N	9 16W
Skiddaw	44	54 39N	3 9W
Skien	61	59 12N	9 35 E
Skierniewice	55	51 58N	20 19 E
Skikda	88	36 50N	6 58 E
Skipton	44	53 57N	2 1W
Skokholm, I.	42	51 42N	5 16W
Skomer, I.	42	51 44N	5 19W
Skye, I.	48	57 15N	6 10W
Slaithwaite	45	53 37N	1 53W
Slaney, R.	51	52 52N	6 45W
Slavkov = Austerlitz	54	49 10N	16 52 E
Slea Hd.	51	52 7N	10 30W
Sleaford	44	53 0N	0 22W
Sleat, Sd. of	48	57 5N	5 47W
Sliedrecht	52	51 50N	4 45 E
Slieve Aughty	51	53 4N	8 30W
Slieve Bloom	51	53 4N	7 40W
Slieve Donard	50	54 10N	5 57W
Slieve Gamph	50	54 6N	9 0W
Slieve Gullion	50	54 8N	6 26W
Slieve League	50	54 40N	8 42W
Slieve Mish	51	52 12N	9 50W
Slievenamon	51	52 25N	7 37W
Sligo	50	54 17N	8 28W
Sligo □	50	54 10N	8 35W
Sligo B.	50	54 20N	8 40W
Slough	41	51 30N	0 35W
Slovakian Ore Mts.	55	50 25N	13 0 E
Slovensko	55	48 30N	19 0 E
Slyne Hd.	51	53 25N	10 10W
Smederevo	55	44 40N	20 57 E
Smethwick	41	52 29N	1 58W
Smithers	105	54 45N	127 10W
Smithfield	94	30 13 S	26 32 E
Smolensk	62	54 45N	32 0 E
Snaefell	44	54 18N	4 26W
Snake, R.	105	46 12N	119 2W
Sneek	52	53 2N	5 40 E
Sneem	51	51 50N	9 55W
Sneeuberge	94	32 0 S	24 55 E
Snizort, L.	48	57 33N	6 28W
Snodland	41	51 19N	0 26 E
Snohetta	61	62 19N	9 16 E
Snowdon	42	53 4N	4 8W
Snowy, R.	79	37 46 S	148 30 E
Sobat, R.	89	8 32N	32 40 E
Soche (Yarkand)	73	38 24N	77 20 E
Socorro	100	34 30N	106 58W
Socotra, I.	89	12 30N	54 0 E
Söderhamn	61	61 18N	17 10 E
Soekmekaar	94	23 30 S	29 55 E
Soest	52	52 9N	5 19 E
Sohâg	91	26 27N	31 43 E
Soham	40	52 20N	0 20 E
Soignies	52	50 35N	4 5 E
Soissons	56	49 25N	3 19 E
Sokółka	55	53 25N	23 30 E
Sokoto	90	13 2N	5 16 E
Solai	93	0 2N	36 12 E
Soledad	107	10 55N	74 46W
Solent, The	40	50 45N	1 25W
Solihull	41	52 26N	1 47W
Solingen	52	51 10N	7 4 E
Solomon Is.	79	6 0 S	155 0 E
Solomon Sea	79	7 0 S	150 0 E
Solway Firth	32	54 45N	3 38W
Sombor	55	45 46N	19 17 E
Sombrerete	106	23 40N	103 40W
Somerset	101	37 5N	84 40W
Somerset □	42	51 9N	3 0W
Somerset East	94	32 42 S	25 35 E
Somerset, I.	98	73 30N	93 0W
Somerset West	94	34 8 S	18 50 E
Somerton	43	51 3N	2 45W
Someş, R.	55	47 15N	23 45 E
Songea	93	10 40 S	35 40 E
Songkhla	72	7 13N	100 37 E
Sonning Common	41	51 28N	0 54W
Sonora, R.	106	28 30N	111 33W
Soochow	73	31 18N	120 41 E
Sopron	54	47 41N	16 37 E
Sorgono	58	40 0N	9 0 E
Soria	60	41 43N	2 32W
Soroti	93	1 43N	33 35 E
Soröya	61	70 35N	22 45 E
Sorrento	58	40 38N	14 23 E
Sorsogon	72	13 0N	124 0 E
Sosnowiec	55	50 20N	19 10 E
Sŏul	73	37 31N	127 6 E
Sound, The	61	56 7N	12 30 E
South Africa, Rep. of, ■	94	30 0 S	25 0 E
South Auckland & Bay of Plenty □	83	38 30 S	177 0 E
South Australia □	78	32 0 S	139 0 E
South Bend	104	41 38N	86 20W
South Benfleet	41	51 33N	0 34 E
South Carolina □	101	33 45N	81 0W
South China Sea	72	7 0N	107 0 E
South Dakota □	100	45 0N	100 0W
South Dorset Downs	43	50 40N	2 26W
South Downs	40	50 53N	0 10W
South Elmsall	45	53 36N	1 17W
South Esk, R.	48	56 44N	3 3W
South Foreland	40	51 7N	1 23 E
South Georgia	112	54 30 S	37 0W
South Glamorgan □	42	51 30N	3 20W
South Hayling	40	50 47N	0 56W
South Hetton	44	54 48N	1 24W
South Hiendley	45	53 36N	1 21W
South Invercargill	83	46 26 S	168 23 E
South Island	83	44 0 S	170 0 E
South Kirkby	45	53 35N	1 25W
South Korea ■	73	36 0N	128 0 E
South Ockendon	41	51 30N	0 18 E
South Orkney Is.	112	63 0 S	45 0W
South Oxhey	41	51 37N	0 23W
South Pagai, I.	72	3 28 S	100 20 E
South Pole	112	90 0 S	0 0 E
South Ronaldsay, I.	49	58 46N	2 58W
S. Sandwich Is.	112	57 0 S	27 0W
South Shetland Is.	112	62 0 S	59 0W
South Shields	44	54 59N	1 26W
South Taranaki Bight	83	39 40 S	174 5 E
South Tyne, R.	44	54 46N	2 25W
South West Africa ■ = Namibia	92	22 0 S	18 9 E
South West Cape	83	47 16 S	167 31 E
South Yemen ■	71	15 0N	48 0 E
South Yorkshire □	44	53 30N	1 20W
Southampton	40	50 54N	1 23W
Southampton I.	99	64 30N	84 0W
Southborough	40	51 10N	0 15 E
Southbridge	83	43 48 S	172 16 E
Southend-on-Sea	41	51 32N	0 42 E
Southern Alps	83	43 41 S	170 11 E
Southern Cross	78	31 12 S	119 15 E
Southern Ocean	23	62 0 S	160 0W
Southern Uplands	46	55 30N	3 3W
Southland □	83	45 51 S	168 13 E
Southport, Austral.	79	27 58 S	153 25 E
Southport, U.K.	45	53 38N	3 1W
Southwark	41	51 29N	0 5W
Southwell	44	53 4N	0 57W
Southwold	40	52 19N	1 41 E
Sovetsk	61	55 6N	21 50 E
Sowerby Bridge	45	53 42N	1 55W
Spa	52	50 29N	5 53 E
Spain ■	60	40 0N	5 0W
Spalding	45	52 47N	0 9W
Spandau	54	52 35N	13 7 E
Spanish Fork	100	40 10N	111 37W
Sparks	100	39 30N	119 45W
Spartivento, C.	58	37 56N	16 4 E
Spean Bridge	48	56 53N	4 55W
Spean, R.	48	56 55N	4 59W
Speke	45	53 21N	2 51W
Spencer G.	79	34 0 S	137 20 E
Spennymoor	44	54 43N	1 35W
Spenser Mts.	83	42 15 S	172 45 E
Sperrin Mts.	50	54 50N	7 0W
Spey, R.	48	57 26N	3 25W
Speyer, R.	52	49 18N	7 52 E
Spézia, La	58	44 8N	9 50 E
Spiddal	51	53 14N	9 19W
Spilsby	44	53 10N	0 6 E
Spinazzola	58	40 58N	16 5 E
Spithead	40	50 43N	0 56W
Split	58	43 31N	16 26 E
Spokane	105	47 45N	117 25W
Spree, R.	54	52 32N	13 13 E
Springburn	83	43 40 S	171 32 E
Springfield, N.Z.	83	43 19 S	171 56 E
Springfield, Ill., U.S.A.	101	39 48N	89 40W
Springfield, Mass., U.S.A.	104	42 8N	72 37W
Springfield, Mo., U.S.A.	101	37 15N	93 20W
Springfield, Ohio, U.S.A.	104	39 58N	83 48W
Springfield, Oreg., U.S.A.	105	44 7N	123 0W
Springfontein	94	30 15 S	25 40 E
Springs	94	26 13 S	28 25 E
Springsure	79	24 8 S	148 6 E
Springville	100	40 14N	111 35W
Spurn Hd.	44	53 34N	0 8 E
Sredinnyy Ra.	62	57 0N	160 0 E
Srepok, R.	72	13 33N	106 16 E
Sretensk	73	52 10N	117 40 E
Sri Lanka ■	70	7 30N	80 50 E
Srinagar	70	34 12N	74 50 E
Stadskanaal	52	53 4N	6 48 E
Staffa, I.	48	56 26N	6 21W
Stafford	45	52 49N	2 9W
Stafford □	45	52 53N	2 10W
Staincross	45	53 35N	1 30W
Staines	41	51 26N	0 30W
Stainland	45	53 40N	1 53W
Stalbridge	43	50 57N	2 22W
Stalybridge	45	53 29N	1 56W
Stamford	45	52 39N	0 29W
Standerton	94	26 55 S	29 13 E
Standish	45	53 35N	2 39W
Stanger	94	29 18 S	31 21 E
Stanley, Falk. Is.	108	51 40 S	58 0W
Stanley, U.K.	44	54 53N	1 42W
Stannington	45	53 23N	1 33W
Stanovoy Ra.	62	55 0N	130 0 E
Stanthorpe	79	28 36 S	151 59 E
Stapleford	45	52 56N	1 16W
Staplehurst	40	51 9N	0 35 E
Staraya Russa	61	57 58N	31 10 E
Stargard	54	53 29N	13 19 E
Start Pt.	43	50 13N	3 38W
Staten, I.	108	54 40 S	64 0W
Staunton	101	38 7N	79 4W
Stavelot	52	50 23N	5 55 E
Stawell	79	37 5 S	142 47 E
Steenwijk	52	52 47N	6 7 E
Steinkjer	61	63 59N	11 31 E
Stella Land	94	26 45 S	24 50 E
Stellenbosch	94	33 58 S	18 50 E
Stelvio P.	58	46 32N	10 27 E
Stendal	54	52 36N	11 50 E
Stenhousemuir	47	56 2N	3 46W
Stepps	47	55 53N	4 09W
Sterkstroom	94	31 32 S	26 32 E
Sterling	100	40 40N	103 15W
Stettin = Szczecin	54	53 27N	14 27 E
Stettler	98	52 19N	112 40W
Stevenage	40	51 54N	0 11W
Stevenston	47	55 38N	4 46W
Stewart I.	83	46 58 S	167 54 E
Stewarton	47	55 40N	4 30W
Steyning	40	50 54N	0 19W
Steynsburg	94	31 15 S	25 49 E
Steytlerville	94	33 17 S	24 19 E
Stikine, R.	98	58 0N	131 12W
Stillwater	101	36 5N	97 3W
Stirling	47	56 7N	3 57W
Stirling Ra.	78	34 0 S	118 0 E
Stockbridge	40	51 7N	1 30W
Stockerau	54	48 24N	16 12 E
Stockholm	61	59 20N	18 3 E
Stockport	45	53 25N	2 11W
Stocksbridge	45	53 30N	1 36W
Stockton Heath	45	53 22N	2 35W
Stockton-on-Tees	44	54 34N	1 20W
Stoke-on-Trent	44	53 1N	2 11W
Stoke Poges	41	51 31N	0 35W
Stokenchurch	41	51 39N	0 54W
Stolberg	52	50 48N	6 13 E
Stone	45	52 55N	2 10W
Stonehaven	48	56 58N	2 11W
Stonehenge	40	51 9N	1 45W
Stonehouse, Glos., U.K.	43	51 45N	2 18W
Stonehouse, Strathclyde, U.K.	47	55 42N	4 0W
Stonewall	98	50 10N	97 19W
Storavan	61	65 45N	18 10 E
Stormberg	94	31 16 S	26 17 E
Stornoway	48	58 12N	6 23W
Storrington	40	50 54N	0 27W
Storsjön	61	60 35N	16 45 E
Storuman, L.	61	65 5N	17 10 E
Stour (Gt. Stour), R.	40	51 15N	0 57 E
Stour, R., Dorset, U.K.	43	50 48N	2 7W
Stour, R., Heref. & Worcs., U.K.	42	52 25N	2 13W
Stour, R., Suffolk, U.K.	40	51 55N	1 5 E
Stourbridge	41	52 28N	2 8W
Stourport	42	52 21N	2 18W
Stow-on-the-Wold	42	51 55N	1 42W
Stowmarket	40	52 11N	1 0 E
Strabane	50	54 50N	7 28W
Strabane □	50	54 45N	7 25W
Strachur	46	56 10N	5 5W
Stradbally, Laois, Ireland	51	53 2N	7 10W
Stradbally, Waterford, Ireland	51	52 7N	7 28W
Stralsund	54	54 17N	13 5 E
Strand	94	34 9 S	18 48 E
Strangford	50	54 23N	5 34W
Strangford, L.	50	54 30N	5 37W
Stranorlar	50	54 52N	7 47W
Stranraer	46	54 54N	5 0W
Strasbourg	56	48 35N	7 42 E
Stratford	83	39 20 S	174 19 E
Stratford-on-Avon	42	52 12N	1 42W
Strath Earn	47	56 20N	3 50W
Strath Spey	48	57 15N	3 40W
Strathaven	47	55 40N	4 4W
Strathbogie, Dist.	48	57 25N	2 45W
Strathclyde □	46	56 0N	4 50W
Strathmore	48	58 20N	4 40W
Stratton	43	50 49N	4 31W
Street	42	51 7N	2 43W
Stretford	45	53 27N	2 19W
Strokestown	50	53 47N	8 6W
Strómboli, I.	58	38 48N	15 12 E
Stromeferry	48	57 20N	5 33W
Stronsay, I.	49	59 8N	2 38W
Stroud	43	51 44N	2 12W
Stuart L.	105	54 30N	124 30W
Sturminster Newton	43	50 56N	2 18W
Stutterheim	94	32 33 S	27 28 E
Stuttgart, Ger.	54	48 46N	9 10 E
Stuttgart, U.S.A.	101	34 30N	91 33W
Styr, R.	55	51 4N	25 20 E
Styria	54	47 26N	15 0 E
Suakin	89	19 0N	37 20 E
Suanhwa	73	40 35N	115 0 E
Subotica	55	46 6N	19 29 E
Suchou	73	31 15N	120 40 E
Suck, R.	51	53 17N	8 10W
Sucre	108	19 0 S	65 15W
Sudan ■	89	15 0N	30 0 E
Sudan, The	88	11 0N	9 0 E
Sudbury, Can.	99	46 30N	81 0W
Sudbury, U.K.	40	52 2N	0 44 E
Sudeten Highlands	54	50 20N	16 45 E
Sudr	91	29 40N	32 42 E
Sueca	60	39 12N	0 21W
Suez	91	29 58N	32 31 E
Suez Canal	91	31 0N	32 20 E
Suez, G. of	91	28 40N	33 0 E
Suffolk □	40	52 16N	1 0 E
Sugarloaf Pt.	79	32 22 S	152 30 E

* *Renamed Thailand, G. of*

Sühbaatar □ Torremolinos

Sühbaatar □ 73 46 54N 113 25E
Suihwa 73 46 40N 126 57E
Suir, R. 51 52 31N 7 59W
Sukhumi 71 43 0N 41 0E
Sukkur 71 27 50N 68 46E
Sulaimaniya 71 35 35N 45 29E
Sulawesi, I. 72 2 0S 120 0E
Sulina 55 45 10N 29 40E
Sulitjelma 61 61 7N 16 8E
Sulu Arch. 72 6 0N 121 0E
Sulu Sea 72 8 0N 120 0E
Suluq 89 31 44N 20 14E
Sumatera, I. 72 0 40N 100 20E
Sumba, I. 72 9 45S 119 35E
Sumbawa, I. 72 8 34S 117 17E
Sumbawanga □ 93 8 0S 31 30E
Sumter 101 33 55N 80 10W
Sunart, L. 48 56 42N 5 43W
Sunbury on Thames 41 51 24N 0 24W
Sunda Str. 72 6 20S 105 30E
Sundarbans, The 70 22 0N 89 0E
Sundays, R. 94 32 10S 24 40E
Sunderland 44 54 54N 1 22W
Sundsvall 61 62 23N 17 17E
Sungari, R. 73 44 30N 126 20E
Sunninghill 41 51 25N 0 40W
Suō-Nada 76 33 50N 131 30E
Superior, Nebr., U.S.A. 100 40 3N 98 2W
Superior, Wis., U.S.A. 101 46 45N 92 5W
Superior, L. 101 47 40N 87 0W
Sūr 89 33 19N 35 16E
Surabaja 72 7 17S 112 45E
Surakarta 72 7 35S 110 48E
Surat 70 21 12N 72 55E
Surinam ■ 108 4 0N 56 15W
Surrey □ 40 51 16N 0 30W
Susa 58 45 8N 7 3E
Susanville 105 40 28N 120 40W
Sutherland 94 32 33S 20 40E
Sutherland Falls 83 44 48S 167 46E
Sutterton 45 52 54N 0 8W
Sutton 41 51 22N 0 13W
Sutton Bridge 45 52 46N 0 12E
Sutton Coldfield 41 52 33N 1 50W
Sutton-in-Ashfield 44 52 8N 1 16W
Suva 23 17 40S 178 8E
Suwa 76 36 2N 138 8E
Svalbard 112 78 0N 17 0E
Svealand □ 61 59 55N 15 0E
Sverdlovsk 62 56 50N 60 30E
Sverdrup Is. 98 79 0N 97 0W
Svobodnyy 73 51 20N 128 0E
Swadlincote 45 52 47N 1 34W
Swaffham 40 52 38N 0 42E
Swakopmund 92 22 37S 14 30E
Swale, R. 44 54 18N 1 20W
Swan Hill 79 35 20S 143 33E
Swan, R. 78 32 3S 115 35E
Swanage 43 50 36N 1 59W
Swanley 41 51 23N 0 10E
Swanlinbar 50 54 11N 7 42W
Swanscombe 41 51 27N 0 13E
Swansea 43 51 37N 3 57W
Swansea Bay 42 51 34N 3 55W
Swartberge 94 33 20S 22 0E
Swaziland ■ 94 26 30S 31 30E
Sweden ■ 61 67 0N 15 0E
Sweetwater 100 32 30N 100 28W
Swellendam 94 34 1S 20 26E
Swift Current 98 50 20N 107 45W
Swilly, L. 50 55 12N 7 35W
Swindon 42 51 33N 1 47W
Swinford 50 53 57N 8 57W
Świnoujście 54 53 54N 14 16E
Swinton 45 53 31N 2 21W
Switzerland ■ 54 46 30N 8 0E
Swords 51 53 27N 6 15W
Sydney, Austral. 79 33 53S 151 10E
Sydney, Can. 99 46 7N 60 7W
Sylhet 70 24 54N 91 52E
Syracuse 104 43 4N 76 11W
Syrdarya, R. 62 45 0N 65 0E
Syre 48 58 22N 4 14W
Syria ■ 89 35 0N 38 0E
Syston 45 52 42N 1 5W
Szczecin 54 53 27N 14 27E
Szczecinek 54 53 43N 16 41E
Szechwan □ 73 30 15N 103 15E
Székesfehérvár 55 47 15N 18 25E
Szolnok 55 47 10N 20 15E
Szombathely 54 47 14N 16 38E

T

Tablas, I. 72 12 25N 122 2E
Table Mt. 94 34 0S 18 22E
Tábor 54 49 25N 14 39E
Tabora 93 5 2S 32 57E
Tabrīz 71 38 7N 46 20E
Tacloban 72 11 15N 124 58E
Tacna 108 18 0S 70 20W
Tacoma 105 47 15N 122 30W
Tadcaster 45 53 53N 1 16W
Tademaït, Plateau du 88 28 30N 2 30E
Tadmor 83 41 27S 172 45E
Tadzhik S.S.R. □ 71 35 30N 70 0E
Taegu 73 35 50N 128 37E
Taejōn 73 36 20N 127 28E
Taganrog 62 47 12N 38 50E
Taghmon 51 52 19N 6 40W
Tagus = Tajo, R. 60 39 44N 5 50W
Tahakopa 83 46 30S 169 23E
Tahcheng 73 46 50N 83 1E
Tahiti, I. 22 17 37S 149 27W
Tahoe, L. 105 39 0N 120 9W
Tahta 91 26 44N 31 32E
T'ai Hu 73 31 10N 120 0E
Taichung 73 24 10N 120 35E
T'aihang Shan 73 35 40N 113 20E
Taihape 83 39 41S 175 48E
Taimyr Pen. 62 75 0N 100 0E
Tain 48 57 49N 4 4W
Tainan 73 23 0N 120 15E
Taipei 73 25 2N 121 30E
Taitung 73 22 43N 121 4E
Taiwan ■ 73 23 30N 121 0E
Taiyüan 73 38 0N 112 30E
Tajima 76 35 19N 135 8E
Tajo, R. 60 40 35N 1 52W
Tak 72 16 52N 99 8E
Takaka 83 40 51S 172 50E
Takamatsu 76 34 20N 134 5E
Takaoka 76 36 40N 137 0E
Takapuna 83 36 47S 174 47E
Takasaki 76 36 20N 139 0E
Takayama 76 36 18N 137 11E
Takefu 76 35 50N 136 10E
Talai 73 45 30N 124 20E
Talata Mafara 90 12 38N 6 4E
Talaud Is. 72 4 30N 127 10E
Talavera de la Reina 60 39 55N 4 46W
Talca 108 35 20S 71 46W
Taldy Kurgan 73 45 10N 78 45E
Talgarth 42 51 59N 3 15W
Tali 73 25 45N 100 5E
Taliabu, I. 72 1 45S 125 0E
Talkeetna 98 62 20N 150 0W
Tallaght 51 53 17N 6 22W
Tallahassee 101 30 25N 84 15W
Tallinn 61 59 29N 24 58E
Tallow 51 52 6N 8 0W
Talybont 42 52 29N 3 59W
Tama Abu Ra. 72 3 10N 115 0E
Tamale 88 9 22N 0 50W
Tamanrasset 88 22 56N 5 30E
Tamar, R. 43 50 33N 4 15W
* Tamatave 92 18 10S 49 25E
Tambov 62 52 45N 41 20E
Tame, R. 41 52 43N 1 45W
Tamgak, Mts. 88 19 12N 8 35E
Tamil Nadu □ 70 11 0N 77 0E
Tampa 101 27 57N 82 30W
Tampere 61 61 30N 23 50E
Tampico 106 22 20N 97 50W
Tamsagbulag 73 47 14N 117 21E
Tamu 70 24 13N 94 12E
Tamworth, Austral. 79 31 0S 150 58E
Tamworth, U.K. 41 52 38N 1 41W
Tanacross 98 63 40N 143 30W
Tanami Des. 78 18 50S 132 0E
Tananarive □ 92 19 0S 47 0E
Tando Adam 71 25 45N 68 40E
Tandragee 50 54 22N 6 23W
Tane-ga-Shima 76 30 30N 131 0E
Taneatua 83 38 4S 177 1E
Tanezrouft 88 23 9N 0 11E
Tanga 93 5 5S 39 2E
Tanganyika, L. 93 6 40S 30 0E
Tangier 88 35 50N 5 49W
Tanglha Shan 73 33 0N 90 0E
Tangshan 73 39 40N 118 10E
Tanimbar Is. 72 7 30S 131 30E
Tanjungbalai 72 2 55N 99 44E
Tanta 91 30 45N 30 57E
Tanzania ■ 93 6 40S 34 0E
Taonan 73 45 30N 122 20E
Tapa Shan 73 31 45N 109 30E
Tapajós, R. 108 4 30S 56 10W
Tapanui 83 45 56S 169 18E
Tapuaenuku, Mt. 83 41 55S 173 50E
Tarabulus = Tripoli 88 34 31N 33 52E
Taranaki □ 83 39 5S 174 51E
Taransay, I. 48 57 54N 7 0W
Táranto 58 40 30N 17 11E
Taranto, G. of 58 40 0N 17 15E
Tararua Range 83 40 45S 175 25E
Tarawera 83 39 2S 176 36E
Tarbagatai Ra. 62 48 30N 83 0E
Tarbat Ness 48 57 52N 3 48W
Tarbert, Ireland 51 52 34N 9 22W
Tarbert, U.K. 48 57 54N 6 49W
Tarbes 56 43 15N 0 3E
Tarbet 46 56 13N 4 44W
Tarcoola 78 30 44S 134 36E
Taree 79 31 50S 152 30E
Tarf Shaqq al Abd 91 26 50N 36 6E
Tarfa, Wadi el 91 28 16N 31 15E
Tarifa 60 36 1N 5 36W
Tarkastad 94 32 0S 26 16E
Tarleton 45 53 41N 2 50W
Tarn, R. 56 44 5N 1 2E
Tarnobrzeg □ 55 50 40N 22 0E
Tarnów 55 50 3N 21 0E
Tarporley 44 53 10N 2 42W
Tarragona 60 41 5N 1 17E
Tarrasa 60 41 26N 2 1E
Tartu 61 58 25N 26 58E
Tashauz 71 42 0N 59 20E
Tashigong 73 33 0N 79 30E
Tashkent 71 41 20N 69 10E
Tashkurghan 71 36 45N 67 40E
Tasman Bay 83 40 59S 173 25E
Tasman Mts. 83 41 3S 172 25E
Tasman Sea 83 36 0S 160 0E
Tasmania, I., □ 79 49 0S 146 30E
Tatabánya 55 47 32N 18 25E
Tatarsk 62 55 20N 75 50E
Tateyama 76 35 0N 139 50E
Tatsaitan 73 37 55N 95 0E
Tat'ung 73 40 9N 113 19E
Taumarunui 83 38 53S 175 15E
Taung 94 27 33S 24 47E
Taunggyi 70 20 50N 97 0E
Taunton 43 51 1N 3 7W
Taupo 83 38 41S 176 7E
Taupo, L. 83 38 46S 175 55E
Tauranga 83 37 35S 176 11E
Taurianova 58 38 22N 16 1E
Taurus Mts. 71 37 0N 35 0E
Tavani 98 62 10N 93 30W
Tavira 60 37 8N 7 40W
Tavistock 43 50 33N 4 9W
Tavoy 70 14 7N 98 18E
Taw, R. 43 50 58N 3 58W
Tawau 72 4 20N 117 55E
Tay, Firth of 47 56 25N 3 8W
Tay, L. 48 56 30N 4 10W
Tay, R. 46 56 37N 3 38W
Taylor Mt. 100 35 16N 107 50W
Tayma 71 27 35N 38 45E
Taynuilt 46 56 25N 5 15W
Tayport 47 56 27N 2 52W
Tayside □ 48 56 25N 3 30W
Tbilisi 71 41 50N 44 50E
Te Anau L. 83 45 15S 167 45E
Te Aroha 83 37 32S 175 44E
Te Awamutu 83 38 1S 175 20E
Te Kuiti 83 38 20S 175 11E
Te Puke 83 37 46S 176 22E
Tebay 44 54 25N 2 35W
Tebing Tinggi 72 3 38S 102 1E
Tecuci 55 45 51N 27 27E
Tees, R. 44 54 36N 1 25W
Teesdale 44 54 37N 2 10W
Tegal 72 6 52S 109 8E
Tegina 90 10 5N 6 11E
Tegucigalpa 106 14 10N 87 0W
Tehrān 71 35 44N 51 30E
Tehuantepec 106 16 10N 95 19W
Tehuantepec, Gulf of 106 15 50N 95 0W
Tehuantepec, Isthmus of 106 17 0N 94 30W
Teifi, R. 42 52 4N 4 14W
Teign, R. 43 50 41N 3 42W
Teignmouth 43 50 33N 3 30W
Tekapo, L. 83 43 53S 170 33E
Tel Aviv-Jaffa 89 32 4N 34 48E
Tela 106 15 40N 87 28W
Telford 45 52 42N 2 31W
Telukbetung 72 5 29S 105 17E
Tembuland □ 94 31 35S 28 0E
Teme, R. 42 52 23N 2 15W
Temple 101 31 5N 97 28W
Temple B. 79 12 15S 143 3E
Templemore 51 52 48N 7 50W
Temuco 108 38 50S 72 50W
Temuka 83 44 14S 171 17E
Tenasserim 70 12 6N 99 3E
Tenbury Wells 42 52 18N 2 35W
Tenby 42 51 40N 4 42W
Tenerife, I. 88 28 20N 16 40W
T'enghsien 73 35 8N 117 9E
Tennant Creek 78 19 30S 134 0E
Tennessee □ 101 36 0N 86 30W
Tennessee, R. 101 34 30N 86 20W
Tenryū 76 34 52N 137 55E
Tenterden 40 51 4N 0 42E
Tenterfield 79 29 0S 152 0E
Tepic 106 21 30N 104 54W
Téramo 58 42 40N 13 40E
Teresina 108 5 2S 42 45W
Termez 71 37 0N 67 15E
Términi Imerese 58 37 59N 13 51E
Térmoli 58 42 0N 15 0E
Terne, R. 61 65 50N 24 12E
Terni 58 42 34N 12 38E
Ternopol 55 49 30N 25 40E
Terracina 58 41 17N 13 12E
Terralba 58 39 42N 8 38E
Terschelling, I. 52 53 25N 5 20E
Teruel 60 40 22N 1 8W
Teshio 76 44 53N 141 44E
Teshio-Gawa, R. 76 44 53N 141 45E
Tessaoua 90 13 47N 7 56E
Test, R. 40 51 7N 1 30W
Tetbury 42 51 37N 2 9W
Tete 92 16 13S 33 33E
Teton, R. 100 47 58N 111 0W
Tettenhall 41 52 35N 2 7W
Tetuan 88 35 30N 5 25W
Teviot, R. 46 55 21N 2 51W
Teviotdale 46 55 25N 2 50W
Tewkesbury 42 51 59N 2 8W
Texarkana 101 33 25N 94 3W
Texas □ 100 31 40N 98 30W
Texel, I. 52 53 5N 4 50E
Tezpur 70 26 40N 92 45E
Thaba Nchu 94 29 10S 26 52E
Thaba Putsoa, mt. 94 29 45S 28 0E
Thabana Ntlenyana, Mt. 94 29 30S 29 9E
Thabazimbi 94 24 40S 26 4E
Thailand (Siam) ■ 72 16 0N 102 0E
Thame, R. 40 51 35N 1 8W
Thames 83 37 7S 175 34E
Thanet, I. of 40 51 21N 1 20E
Thanh Hoa 72 19 48N 105 46E
Thar (Great Indian) Desert 70 28 25N 72 0E
Thargomindah 79 27 58S 143 46E
Thatcham 40 51 24N 1 17W
Thaxted 40 51 57N 0 20E
The Dalles 105 45 40N 121 11W
The Pas 98 53 45N 101 15W
Theodore 79 24 55S 150 3E
Thermopolis 100 43 35N 108 10W
Thessaloniki 89 40 38N 23 0E
Thetford 40 52 25N 0 44E
Thetford Mines 104 46 8N 71 18W
Theydon Bois 41 51 40N 0 06E
Thika 93 1 1S 37 5E
Thionville 56 49 20N 6 10E
Thirsk 44 54 15N 1 20W
Thomastown 51 52 32N 7 10W
Thompson, R., Can. 98 50 15N 121 24W
Thompson, R., U.S.A. 105 39 46N 93 37W
Thornaby on Tees 44 54 36N 1 19W
Thornbury 43 51 36N 2 31W
Thorne 44 53 36N 0 56W
Thorney 40 52 37N 0 8W
Thornhill 46 55 15N 3 46W
Thornley 44 54 45N 1 26W
Thornliebank 47 55 48N 4 18W
Thornton 45 53 52N 3 1W
Thrapston 40 52 24N 0 32W
Throckley 44 54 58N 1 49W
Thrybergh 45 53 27N 1 18W
Thuin 52 50 20N 4 17E
Thule 99 77 30N 69 0W
Thun 54 46 45N 7 38E
Thunder B. 98 48 20N 89 0W
Thurcroft 45 53 24N 1 13W
Thurles 51 52 40N 7 53W
Thurlstone 45 53 31N 1 37W
Thurmaston 41 52 40N 1 8W
Thurso 48 58 34N 3 31W
Thurso, R. 48 58 36N 3 30W
Tiber, R. 58 41 44N 12 14E
Tiberias 89 32 47N 35 32E
Tibesti 88 21 0N 17 30E
Tibet 73 32 30N 86 0E
Tibooburra 79 29 26S 142 1E
Tiburón, I. 106 29 0N 112 30W
Ticino, R. 58 45 9N 9 14E
Tideswell 45 53 17N 1 46W
Tiel 52 51 53N 5 26E
Tielt 52 51 0N 3 20E
Tien Shan 73 42 0N 80 0E
Tienen 52 50 48N 4 57E
Tienshui 73 34 30N 105 34E
Tientsin 73 39 10N 117 0E
Tierra del Fuego, I. 108 54 0S 69 0W
Tighnabruaich 46 55 55N 5 13W
Tigris, R. 71 37 0N 42 30E
Tijuana 106 32 30N 117 3W
Tikhvin 61 59 35N 33 30E
Tiko 90 4 4N 9 20E
Tiksi 62 71 50N 129 0E
Tilburg 52 51 31N 5 6E
Tilbury 41 51 27N 0 24E
Tillicoultry 47 56 9N 3 44W
Timaru 83 44 23S 171 14E
Timau 93 0 4N 37 15E
Timbuktu = Tombouctou 88 16 50N 3 0W
Timişoara 55 45 43N 21 15E
Timmins 99 48 28N 81 25W
Timor, I. 72 9 0S 125 0E
Timor Sea 72 10 0S 127 0E
Tindouf 88 27 50N 8 4W
Tingley 45 53 44N 1 35W
Tintagel 43 50 40N 4 45W
Tintern 42 51 42N 2 41W
Tipperary 51 52 28N 8 10W
Tipperary □ 51 52 37N 7 55W
Tipton 41 52 32N 2 4W
Tiptree 40 51 48N 0 46E
Tirân 91 27 56N 34 35E
Tiraspol 55 46 55N 29 35E
Tiree, I. 48 56 31N 6 55W
Tîrgu Mureş 55 46 31N 24 38E
Tiruchchirappalli 70 10 45N 78 45E
Tisa, R. 55 45 15N 20 17E
Tisza, R. 55 47 38N 20 44E
Titicaca, L. 108 15 30S 69 30W
Titilagarh 70 20 15N 83 5E
Titiwa 90 12 14N 12 53E
Titovo Užice 55 43 55N 19 50E
Tiverton 43 50 54N 3 30W
Tlacotalpán 106 18 37N 95 40W
Tlaxcala 106 19 20N 98 14W
Tlaxiaco 106 17 10N 97 40W
Toba 76 34 30N 136 45E
Toba, L. 72 2 40N 98 50E
Tobago, I. 107 11 10N 60 30W
Tobercurry 50 54 3N 8 43W
Tobruk 89 32 7N 23 55E
Tocantins, R. 108 14 30S 49 0W
Tochigi 76 36 25N 139 45E
Todmorden 45 53 43N 2 7W
Togo ■ 88 6 15N 1 35E
Tokaj 55 48 8N 21 27E
Tokanui 83 46 34S 168 56E
Tokara Is. 76 29 0N 129 0E
Tokarahi 83 44 56S 170 39E
Tokelau Is. 22 9 0S 172 0W
Tokunoshima 76 27 56N 128 55E
Tokushima 76 34 4N 134 34E
Tokuyama 76 34 0N 131 50E
Tōkyō 76 35 45N 139 45E
Tolaga Bay 83 38 21S 178 20E
Toledo, Spain 60 39 50N 4 2W
Toledo, U.S.A. 104 41 37N 83 33W
Toledo, Mts. 60 39 33N 4 20W
Tolima □ 107 3 45N 75 15W
Tolosa 60 43 8N 2 5W
Toluca 106 19 20N 99 50W
Tomakomai 76 42 38N 141 36E
Tomatin 48 57 20N 4 0W
Tombouctou 88 16 50N 3 0W
Tomini 72 0 30N 120 30E
Tomintoul 48 57 15N 3 22W
Tomsk 62 56 30N 85 12E
Tonbridge 40 51 12N 0 18E
Tone R. 43 50 59N 3 15W
Tonga Is. ■ 22 20 0S 173 0W
Tongaat 94 29 33S 31 9E
Tongaland 94 27 0S 32 0E
Tongeren 52 50 47N 5 28E
Tongking, G. of 72 20 0N 108 0E
Tonlé Sap 72 13 0N 104 0E
Tonopah 100 38 4N 117 12W
Tonyrefail 43 51 35N 3 26W
Tooele 100 40 30N 112 20W
Toome 50 54 45N 6 28W
Toowoomba 79 27 32S 151 56E
Topeka 101 39 3N 95 40W
Topolobampo 106 25 40N 109 10W
Topsham 43 50 40N 3 27W
Torbay 43 50 26N 3 31W
Tordesillas 60 41 30N 5 0W
Torhout 52 51 5N 3 7E
Tormes, R. 60 41 7N 6 0W
Torne, R. 61 65 50N 24 12E
Tornio 61 65 50N 24 12E
Toronto 104 43 39N 79 20W
Tororo 93 0 45N 34 12E
Torpoint 43 50 23N 4 12W
Torquay 43 50 27N 3 31W
Tôrre de Moncorvo 60 41 12N 7 8W
Torrelavega 60 43 20N 4 5W
Torremolinos 60 36 38N 4 30W

* Renamed Toamasina

* Renamed Mutare
† Renamed Burkina Faso

141

Vltava, R **Wyndham**

Vltava, R. **54** 49 35N 14 10E
Vogelkop **72** 1 25S 133 0E
Vogels Berg, mt. **54** 50 37N 9 30E
Voi **93** 3 25S 38 32E
Volga, R. **62** 52 20N 48 0E
Volgograd **62** 48 40N 44 25E
Völklingen **52** 49 15N 6 50E
Volksrust **94** 27 24S 29 53E
Vologda **62** 59 25N 40 0E
Volsk **62** 52 5N 47 28E
Volta, L. **88** 7 30N 0 15E
Volta, R. **88** 8 0N 0 10W
Voorburg **52** 52 5N 4 24E
Vorarlberg □ **54** 47 20N 10 0E
Vorkuta **62** 67 48N 64 20E
Voronezh **62** 51 40N 39 10E
Vosges □ **56** 48 12N 6 20E
Vostok I. **22** 10 5S 152 23W
Vrbas, R. **58** 45 8N 17 29E
Vrede **94** 27 24S 29 6E
Vredendal **94** 31 41S 18 35E
Vršac **55** 45 8N 21 18E
Vryburg **94** 26 55S 24 45E
Vryheid **94** 27 54S 30 47E
Vulcano, I. **58** 38 25N 14 58E
Vyborg **61** 60 43N 28 47E
Vyrnwy, L. **42** 52 48N 3 30W
Vyrnwy, R. **42** 52 43N 3 15W
Vyshniy Volochek **61** 57 30N 34 30E

W

Waal, R. **52** 51 59N 4 8E
Waalwijk **52** 51 42N 5 4E
Wabash **101** 40 48N 85 46W
Wabrzeźno **55** 53 16N 18 57E
Waco **101** 31 33N 97 5W
Wâd Medanî **89** 14 28N 33 30E
Waddenzee **52** 53 6N 5 10E
Waddington, Mt. **105** 51 23N 125 15W
Wadebridge **43** 50 31N 4 51W
Wadi Gemâl **91** 24 35N 35 10E
Wadi Halfa **91** 21 53N 31 19E
Wageningen **52** 51 58N 5 40E
Wagga Wagga **79** 35 7S 147 24E
Wagin, Austral. **78** 33 17S 117 25E
Wagin, Nigeria **90** 12 42N 7 10E
Waigeo, I. **72** 0 20S 130 40E
Waihi **83** 37 23S 175 52E
Waikaremoana L. **83** 38 49S 177 9E
Waikari **83** 42 58S 172 41E
Waikato, R. **83** 37 23S 174 43E
Waikokopu **83** 39 3S 177 52E
Waikouaiti **83** 45 36S 170 41E
Waimakariri, R. **83** 42 23S 172 42E
Waimarino **83** 40 40S 175 20E
Waimate **83** 44 53S 171 3E
Waiouru **83** 39 28S 175 41E
Waipara **83** 43 3S 172 46E
Waipawa **83** 39 56S 176 38E
Waipiro **83** 38 2S 176 22E
Waipu **83** 35 59S 174 29E
Waipukurau **83** 40 1S 176 33E
Wairakei **83** 38 37S 176 6E
Wairarapa I. **83** 41 14S 175 15E
Wairau, R. **83** 41 32S 174 7E
Wairoa **83** 39 3S 177 25E
Waitaki, R. **83** 44 23S 169 55E
Waitara **83** 38 59S 174 15E
Waiuku **83** 37 15S 174 45E
Wajima **76** 37 30N 137 0E
Wajir **93** 1 42N 40 20E
Wakasa B. **76** 35 45N 135 30E
Wakatipu, L. **83** 45 5S 168 33E
Wakayama **76** 34 15N 135 15E
Wake I. **23** 19 18N 166 36E
Wakefield, N.Z. **83** 41 24S 173 5E
Wakefield, U.K. **45** 53 41N 1 31W
Wakkanai **76** 45 28N 141 35E
Wakkerstroom **94** 27 24S 30 10E
Wałbrzych **54** 50 45N 16 18E
Walbury Hill **40** 51 22N 1 28W
Walcheren, I. **52** 51 30N 3 35E
Waldbröl **52** 50 52N 7 36E
Wales □ **42** 52 30N 3 30W
Walgett **79** 30 0S 148 5E
Walkden **45** 53 37N 2 24W
Walker **44** 54 58N 1 32W
Walla Walla **100** 46 3N 118 25W
Wallace **100** 47 30N 116 0W
Wallachia **55** 44 35N 25 0E
Wallaroo **79** 33 56S 137 39E
Wallasey **45** 53 26N 3 2W
Wallingford **40** 51 36N 1 08W
Walls **49** 60 14N 1 32W
Wallsend **44** 54 59N 1 30W
Walmer, S. Afr. **94** 33 57S 25 35E
Walmer, U.K. **40** 51 12N 1 23E
Walmley **41** 52 31N 1 47W
Walsall **41** 52 36N 1 59W
Walsenburg **100** 37 42N 104 45W
Walsingham **40** 52 53N 0 53E
Waltham Abbey **41** 51 40N 0 1E
Waltham Forest **41** 51 37N 0 2E
Walton-le-Dale **45** 53 45N 2 41W
Walton-on-Thames **41** 51 21N 0 22W
Walton-on-the-Naze **40** 51 52N 1 17E
Walvis Bay **92** 23 0S 14 28E
Wanaka **83** 44 42S 169 7E
Wanaka L. **83** 44 33S 169 7E
Wandoan **79** 26 5S 149 55E
Wandsworth **41** 51 28N 0 15W
Wanganui **83** 39 35S 175 3E
Wangaratta **79** 36 21S 146 19E
Wanhsien **73** 36 45N 107 24E
* Wankie **92** 18 18S 26 30E
Wantage **40** 51 35N 1 25W
Ward, Ireland **51** 53 25N 6 19W
Ward, N.Z. **83** 41 49S 174 11E
Warden **94** 27 50S 29 0E
Wardha **70** 20 45N 78 39E
Wareham **43** 50 41N 2 8W
Warendorf **52** 51 57N 8 0E
Warkworth **83** 36 24S 174 41E
Warley **41** 52 30N 2 0W
Warlingham **41** 51 18N 0 03W
Warmbad **94** 24 51S 28 19E
Warminster **42** 51 12N 2 11W
Warnemünde **54** 54 9N 12 5E
Warner Range, Mts. **105** 41 30S 120 20W
Warrego, R. **79** 30 24S 145 21E
Warren **101** 41 52N 79 10W
Warrenpoint **50** 54 7N 6 15W
Warrenton **94** 28 9S 24 47E
Warri **90** 5 30N 5 41E
Warrina **79** 28 12S 135 50E
Warrington **45** 53 25N 2 38W
Warrnambool **79** 38 25S 142 30E
Warsaw **55** 52 13N 21 0E
Warsop **44** 53 13N 1 9W
Warszawa = Warsaw **55** 52 13N 21 0E
Warta, R. **54** 52 35N 14 39E
Warwick, Austral. **79** 28 10S 152 1E
Warwick, U.K. **42** 52 17N 1 36W
Warwick □ **42** 52 20N 1 30W
Wasatch, Mt., Ra. **100** 40 30N 111 15W
Wash, The **32** 52 58N 0 20W
Washington, U.K. **44** 54 55N 1 30W
Washington, U.S.A. **104** 38 52N 77 0W
Washington □ **100** 47 45N 120 30W
Watchet **42** 51 10N 3 20W
Waterbeach **40** 52 16N 0 11E
Waterberg **94** 24 14S 28 0E
Waterbury **104** 41 32N 73 0W
Waterford **51** 52 16N 7 8W
Waterford □ **51** 52 10N 7 40W
Waterford Harb. **51** 52 10N 6 58W
Waterloo **101** 42 27N 92 20W
Waterlooville **40** 50 52N 1 01W
Watertown **104** 43 58N 75 57W
Waterval-Boven **94** 25 40S 30 18E
Waterville **51** 51 49N 10 10W
Watford **41** 51 38N 0 23W
Wath upon Dearne **45** 53 30N 1 21W
Watlings I. **107** 24 0N 74 35W
Watlington **41** 51 38N 1 0W
Watsa **93** 3 4N 29 30E
Watson Lake **98** 60 6N 128 49W
Watton **40** 52 35N 0 50E
Watubella Is. **72** 4 28S 131 54E
Wau **89** 7 45N 28 1E
Waukegan **101** 42 22N 87 54W
Wausau **101** 44 57N 89 40W
Wave Hill **78** 17 32S 131 0E
Waveney, R. **40** 52 24N 1 20E
Waverley **83** 39 46S 174 37E
Waxahachie **101** 32 22N 96 53W
Waycross **101** 31 12N 82 25W
Weald, The **40** 51 7N 0 9E
Wear, R. **44** 54 55N 1 22W
Weardale **44** 54 44N 2 5W
Weatherford **100** 32 45N 97 48W
Weaver, R. **45** 53 17N 2 35W
Weddell Sea **112** 72 30S 40 0W
Wednesbury **41** 52 33N 2 1W
Wednesfield **41** 52 36N 2 3W
Weenen **94** 28 48S 30 7E
Weert **52** 51 15N 5 43E
Weifang **73** 36 47N 119 10E
Weihai **73** 37 30N 122 10E
Weimar **54** 51 0N 11 20E
Weipa **79** 12 24S 141 50E
Weiser **100** 44 10N 117 0W
Welkom **94** 28 0S 26 50E
Welland, R. **45** 52 43N 0 10W
Wellesley Is. **79** 17 20S 139 30E
Wellingborough **40** 52 18N 0 41W
Wellington, Austral. **79** 32 35S 148 59E
Wellington, N.Z. **83** 41 19S 174 46E
Wellington, S. Afr. **94** 33 38S 18 57E
Wellington, U.K. **44** 52 42N 2 31W
Wellington, U.S.A. **100** 37 15N 97 25W
Wellington □ **83** 40 8S 175 36E
Wells, Norfolk, U.K. **40** 52 57N 0 51E
Wells, Somerset, U.K. **42** 51 12N 2 39W
Wels **54** 48 9N 14 1E
Welshpool **42** 52 40N 3 9W
Welwyn Garden City **41** 51 49N 0 11W
Wem **45** 52 52N 2 45W
Wemyss Bay **47** 55 52N 4 54W
Wenatchee **105** 47 30N 120 17W
Wenchow **73** 28 0N 120 35E
Wendover **41** 51 46N 0 45W
Wensleydale **44** 54 18N 2 0W
Wensum, R. **40** 52 37N 1 22E
Wentworth, Austral. **79** 34 2S 141 54E
Wentworth, U.K. **45** 53 28N 1 25W
Wepener **94** 29 42S 27 3E
Werne **52** 51 38N 7 38E
Werra, R. **54** 51 0N 10 0E
Wesel **52** 51 39N 6 34E
Weser, R. **54** 53 33N 8 30E
West Bengal □ **70** 25 0N 90 0E
West Beskids, mts. **55** 49 30N 19 20E
West Bromwich **41** 52 32N 2 1W
West Calder **47** 55 51N 3 34W
West Glamorgan □ **42** 51 40N 3 55W
West Indies **96** 15 0N 70 0W
West Kilbride **47** 55 41N 4 50W
West Kirby **45** 53 22N 3 11W
West Linton **47** 55 45N 3 24W
West Mersea **40** 51 46N 0 55E
West Midlands □ **45** 52 30N 1 55W
West Nicholson **92** 21 2S 29 20E
West Palm Beach **101** 26 44N 80 3W
West Schelde R. **52** 51 23N 3 50E
West Siberian Plain **62** 62 0N 75 0E
West Sussex □ **40** 50 55N 0 30W
West Virginia □ **101** 39 0N 81 0W
West Yorkshire □ **44** 53 45N 1 40W
Westerdal R. **61** 60 30N 14 0E
Westerham **41** 51 16N 0 5E
Western Australia □ **78** 25 0S 118 0E
Western Desert **91** 27 40N 26 30E
Western Germany ■ **54** 50 0N 8 0E
Western Ghats **70** 15 30N 74 30E
Western Isles □ **49** 57 30N 7 10W
† Western Malaysia □ **72** 4 0N 102 0E
Western Samoa ■ **22** 14 0S 172 0W
Westerwald **54** 50 39N 8 0E
Westhoughton **45** 53 34N 2 30W
Westland □ **83** 43 33S 169 59E
Westland Bight **83** 42 55S 170 5E
Westmeath □ **51** 53 30N 7 30W
Westminster **41** 51 30N 0 07W
Weston **45** 53 55N 2 44W
Weston-super-Mare **42** 51 20N 2 59W
Weston upon Trent **45** 52 50N 2 02W
Westport, Ireland **50** 53 44N 9 31W
Westport, N.Z. **83** 41 46S 171 37E
Westray Firth **49** 59 15N 3 0W
Westray, I. **49** 59 18N 3 0W
Westward Ho! **43** 51 2N 4 16W
Wetar, I. **72** 7 30S 126 30E
Wetherby **45** 53 56N 1 23W
Wetteren **52** 51 0N 3 53E
Wetzlar **54** 50 33N 8 30E
Wewak **79** 3 38S 143 41E
Wexford **51** 52 20N 6 28W
Wexford □ **51** 52 20N 6 25W
Wexford Harb. **51** 52 20N 6 25W
Wey, R. **41** 51 19N 0 29W
Weybridge **41** 51 22N 0 28W
Weyburn **98** 49 40N 103 50W
Weymouth **43** 50 36N 2 28W
Whakatane **83** 37 57S 177 1E
Whaley Bridge **45** 53 20N 2 0W
Whalley **45** 53 49N 2 25W
Whalsay, I. **49** 60 22N 1 0W
Whangamomona **83** 39 8S 174 44E
Whangarei **83** 35 43S 174 21E
Wharfe, R. **44** 53 55N 1 30W
Wheatley Hill **44** 54 45N 1 23W
Wheeling **104** 40 2N 80 41W
Whernside, Mt. **44** 54 14N 2 24W
Whickham **44** 54 56N 1 41W
Whiddy, I. **51** 51 41N 9 30W
Whiston **45** 53 25N 2 45W
Whitburn **47** 55 52N 3 41W
Whitby **44** 54 29N 0 37W
Whitchurch, Hants., U.K. **40** 51 14N 1 20W
Whitchurch, Salop, U.K. **44** 52 58N 2 42W
White Cliffs **83** 43 26S 171 55E
White Mts., Czech. **55** 49 0N 17 50E
White Mts., U.S.A. **100** 37 30N 118 15W
White, Mts. **101** 44 15N 71 15W
White Nile, R. **89** 9 30N 31 40E
White Russia **62** 53 30N 27 0E
White Sea **61** 66 30N 38 0E
Whitefield **45** 53 33N 2 18W
Whitehaven **44** 54 33N 3 35W
Whitehead **50** 54 45N 5 42W
Whitehorse **98** 60 43N 135 3W
Whithorn **46** 54 55N 4 25W
Whitianga **83** 36 47S 175 41E
Whitley Bay **44** 55 4N 1 28W
Whitney, Mt. **105** 36 35N 118 14W
Whitstable **40** 51 21N 1 2E
Whittington, Derby, U.K. **45** 53 17N 1 26W
Whittington, Salop, U.K. **45** 52 53N 3 0W
Whittle-le-Woods **45** 53 40N 2 37W
Whittlesey **40** 52 34N 0 8W
Whitworth **45** 53 40N 2 11W
Whyalla **79** 33 2S 137 30E
Wichita **101** 37 40N 97 29W
Wichita Falls **100** 33 57N 98 30W
Wick **48** 58 26N 3 5W
Wickersley **45** 53 25N 1 17W
Wickford **41** 51 37N 0 31E
Wickham Market **40** 52 9N 1 21E
Wicklow **51** 53 0N 6 2W
Wicklow □ **51** 52 59N 6 25W
Wicklow Hd. **51** 52 59N 6 3W
Wicklow Mts. **51** 53 0N 6 30W
Wide Open **44** 55 02N 1 36W
Widnes **45** 53 22N 2 44W
Wieliczka **55** 50 0N 20 5E
Wien = Vienna **54** 48 12N 16 22E
Wiener Neustadt **54** 47 49N 16 16E
Wiesbaden **52** 50 7N 8 17E
Wigan **45** 53 33N 2 38W
Wigmore **41** 52 19N 2 51W
Wigston **41** 52 35N 1 6W
Wigton **44** 54 50N 3 9W
Wigtown **46** 54 52N 4 27W
Wigtown B. **46** 54 46N 4 15W
Wilcannia **79** 31 30S 143 26E
Wilge R. **94** 25 40S 29 10E
Wilhelm II Coast **112** 67 0S 90 0E
Wilhelmshaven **54** 53 30N 8 9E
Wilkes Barre **104** 41 15N 75 52W
Wilkes Land **112** 69 0S 120 0E
Wilkie **98** 52 27N 108 42W
Willebroek **52** 51 4N 4 22E
Willemstad **107** 12 5N 69 0W
Willenhall **41** 52 36N 2 3W
Williams Lake **105** 52 10N 122 10W
Williamsport **104** 41 18N 77 1W
Willington **44** 54 43N 1 42W
Williston, S. Afr. **94** 31 20S 20 53E
Williston, U.S.A. **100** 48 10N 103 35W
Williton **42** 51 9N 3 20W
Willmar **101** 45 5N 95 0W
Willowmore **94** 33 15S 23 30E
Wilmington, Del., U.S.A. **104** 39 45N 75 32W
Wilmington, N.C., U.S.A. **101** 34 14N 77 54W
Wilmslow **45** 53 19N 2 14W
Wilnecote **41** 52 36N 1 40W
Wilson's Promontory **79** 38 55S 146 25E
Wilton **42** 51 5N 1 52W
Wiltshire □ **42** 51 20N 2 0W
Wiluna **78** 26 36S 120 14E
Wimborne Minster **43** 50 48N 2 0W
Wimmera **79** 36 30S 142 0E
Winburg **94** 28 30S 27 2E
Wincanton **43** 51 3N 2 24W
Winchester, U.K. **40** 51 4N 1 19W
Winchester, U.S.A. **101** 39 14N 78 8W
Wind River Range, Mts. **100** 43 0N 109 30W
Windermere **44** 54 24N 2 56W
Windermere, L. **44** 54 20N 2 57W
Windhoek **92** 22 35S 17 4E
Windorah **79** 25 24S 142 36E
Windsor, Can. **104** 42 18N 83 0W
Windsor, U.K. **41** 51 28N 0 36W
Windward Is. **107** 13 0N 63 0W
Windward Passage **107** 20 0N 74 0W
Windygates **47** 56 12N 3 1W
Wingate **44** 54 44N 1 23W
Winkleigh **43** 50 49N 3 57W
Winnemucca **100** 41 0N 117 45W
Winnipeg **98** 49 54N 97 9W
Winnipeg, L. **98** 52 0N 97 0W
Winnipegosis L. **98** 52 30N 100 0W
Winona **101** 44 2N 91 45W
Winschoten **52** 53 9N 7 3E
Winsford **44** 53 12N 2 31W
Winslow, U.K. **40** 51 57N 0 52W
Winslow, U.S.A. **100** 35 2N 110 41W
Winston-Salem **101** 36 7N 80 15W
Winterswijk **52** 51 58N 6 43E
Winterthur **54** 47 30N 8 44E
Winton, Austral. **79** 22 24S 143 3E
Winton, N.Z. **83** 46 8S 168 20E
Wirksworth **44** 53 5N 1 34W
Wirral **44** 53 25N 3 0W
Wisbech **40** 52 39N 0 10E
Wisconsin □ **101** 44 30N 90 0W
Wishaw **47** 55 46N 3 55W
Wismar **107** 5 59N 58 18W
Witbank **94** 25 51S 29 14E
Witham **41** 51 48N 0 39E
Witheridge **43** 50 55N 3 43W
Withernsea **44** 53 43N 0 2W
Withnell **45** 53 42N 2 33W
Witney **40** 51 47N 1 29W
Witten **52** 51 26N 7 19E
Wittenberge **54** 53 0N 11 44E
Wittenburg **54** 53 30N 11 4E
Wittenoom **78** 22 15S 118 20E
Witwatersrand **94** 26 0S 27 0E
Wiveliscombe **43** 51 2N 3 20W
Wivenhoe **40** 51 51N 0 59E
Wokam, I. **72** 5 45S 134 28E
Woking **41** 51 18N 0 33W
Wokingham **41** 51 25N 0 50W
Wolin **54** 53 50N 14 37E
Wollaston Pen. **98** 69 30N 115 0W
Wollongong **79** 34 25S 150 54E
Wolmaransstad **94** 27 12S 26 13E
Wolverhampton **41** 52 35N 2 6W
Wolverton **40** 52 3N 0 48W
Wolviston **44** 54 39N 1 25W
Wombwell **45** 53 31N 1 23W
Wondai **79** 26 20S 151 49E
Wonthaggi **79** 38 37S 145 37E
Woodbridge **40** 52 6N 1 19E
Woodhall Spa **44** 53 10N 0 12W
Woodroffe, Mt. **78** 26 20S 131 45E
Woodstock **40** 51 51N 1 20W
Woodville **83** 40 20S 175 53E
Woolacombe **42** 51 10N 4 12W
Wooler **46** 55 33N 2 0W
Woomera **79** 31 11S 136 47E
Woonsockett **104** 44 5N 98 15W
Wootton Bassett **42** 51 32N 1 55W
Worcester, S. Afr. **94** 33 39S 19 27E
Worcester, U.K. **42** 52 12N 2 12W
Worcester, U.S.A. **104** 42 14N 71 49W
Workington **44** 54 39N 3 34W
Worksop **44** 53 19N 1 9W
Wormerveer **52** 52 30N 4 46E
Wormit **47** 56 26N 2 59W
Worms **54** 49 37N 8 21E
Worms Head **42** 51 33N 4 19W
Worsbrough **45** 53 31N 1 29W
Worsley **45** 53 30N 2 23W
Worthing **40** 50 49N 0 21W
Wotton-under-Edge **43** 51 37N 2 20W
Wragby **44** 53 17N 0 18W
Wrangel I. **62** 71 0N 180 0E
Wrangell **98** 56 30N 132 25W
Wrangell Mts. **98** 61 40N 143 30W
Wrath, C. **48** 58 38N 5 0W
Wrekin, The **42** 52 41N 2 35W
Wrexham **42** 53 5N 3 0W
Writtle **41** 51 44N 0 27E
Wrocław **54** 51 5N 17 5E
Wrotham **41** 51 18N 0 20E
Wroughton **42** 51 31N 1 47W
Wroxham **40** 52 42N 1 23E
Wuchow **73** 23 26N 111 19E
Wuchung **73** 38 4N 106 12E
Wuhan **73** 30 35N 114 15E
Wuhu **73** 31 18N 118 20E
Wukari **90** 7 57N 9 42E
Wulumuchi **73** 43 40N 87 50E
Wum **90** 6 40N 10 2E
Wuppertal **52** 51 15N 7 8E
Würzburg **54** 49 46N 9 55E
Wusih **73** 31 30N 120 30E
Wusu **73** 44 27N 84 37E
Wutunghliao **73** 29 25N 104 0E
Wuwei **73** 37 55N 102 48E
Wuyi Shan **73** 26 40N 116 30E
Wuyun **73** 46 16N 129 37E
Wyandra **79** 27 12S 145 56E
Wye **40** 51 11N 0 56E
Wye, R. **42** 51 36N 2 40W
Wymondham **40** 52 34N 1 7E
Wynberg **94** 34 2S 18 28E
Wyndham, Austral. **78** 15 33S 128 3E
Wyndham, N.Z. **83** 46 20S 168 51E

** Renamed Hwange*

† Now part of Malaysia

Wyoming □ **100** 42 48N 109 0W
Wyre, R. **45** 53 52N 2 57W

X

Xingu, R. **108** 2 25S 52 35W

Y

Yaan **73** 30 0N 102 59E
Yabassi **90** 4 30N 9 57E
Yablonovyy Ra. **62** 53 0N 114 0E
Yakima **105** 46 42N 120 30W
Yakima, R. **105** 47 0N 120 30W
Yaku-Jima **76** 30 20N 130 30E
Yakutsk **62** 62 5N 129 40E
Yalgoo **78** 28 16S 116 39E
Yallourn **79** 38 10S 146 18E
Yalung K. **73** 32 0N 100 0E
Yamagata **76** 38 15N 140 15E
Yamaguchi **76** 34 10N 131 32E
Yamdena **72** 7 45S 131 20E
Yampi Sd. **78** 16 8S 123 38E
Yan **90** 10 5N 12 11E
Yanam **70** 16 47N 82 15E
Yangchuan **73** 38 0N 113 29E
Yangtze Chiang, R. **73** 27 30N 99 30E
Yaoundé **88** 3 50N 11 35E
Yaqui, R. **106** 28 28N 109 30W
Yaraka **79** 24 53S 144 3E
Yare, R. **40** 52 36N 1 28E
Yarmouth **40** 50 42N 1 29W
Yaroslavl **62** 57 35N 39 55E
Yate **43** 51 32N 2 26W
Yatsushiro **76** 32 30N 130 40E
Yawatahama **76** 33 27N 132 24E
Yayama-rettō **76** 24 30N 123 40E
Yazd **71** 31 55N 54 27E
Yeadon **45** 53 52N 1 40W
Yedintsy **55** 48 5N 27 20E
Yehsien **73** 37 12N 119 58E
Yell, I. **49** 60 35N 1 5W
Yellow Sea **73** 35 0N 123 0E
Yellowknife **98** 62 27N 114 21W
Yellowstone National Park **100** 44 35N 110 0W
Yellowstone, R. **100** 47 58N 103 59W
Yemen ■ **71** 15 0N 44 0E
Yenangyaung **70** 20 30N 95 0E
Yenbo' **71** 24 0N 38 5E
Yenisey, R. **62** 68 0N 86 30E
Yeniseysk **62** 58 39N 92 4E
Yent'ai **73** 37 35N 121 25E
Yeovil **43** 50 57N 2 38W
Yeppoon **79** 23 5S 150 47E
Yerevan **71** 40 10N 44 20E
Yes Tor **43** 50 41N 3 59W
Yeu, I. d' **56** 46 42N 2 20W
Yilan **73** 24 47N 121 44E
Yinchwan **73** 38 30N 106 20E
Yingkow **73** 40 43N 122 9E
Yiyang **73** 28 45N 112 16E
Yogyakarta **72** 7 49S 110 22E
Yokkaichi **76** 35 0N 136 30E
Yokohama **76** 35 27N 139 39E
Yokosuka **76** 35 20N 139 40E
Yokote **76** 39 20N 140 30E
Yola **90** 9 10N 12 29E
Yonago **76** 35 25N 133 19E
Yonezawa **76** 37 57N 140 4E
Yonkers **104** 40 57N 73 51W
York, Austral. **78** 31 52S 116 47E
York, U.K. **44** 53 58N 1 7W
York, U.S.A. **104** 39 57N 76 43W
York, C. **79** 10 42S 142 31E
Yorke Pen. **79** 34 50S 137 40E
Yorkshire Wolds **44** 54 0N 0 30W
Yorkton **98** 51 11N 102 28W
Yosemite National Park **100** 38 0N 119 30W
Youghal **51** 51 58N 7 51W
Youghal Har. **51** 51 55N 7 50W
Young **79** 34 19S 148 18E
Youngstown **104** 41 7N 80 41W
Ypres **52** 50 50N 2 52E
Ysabel, I. **79** 8 0S 158 40E
Ystalyfera **43** 51 46N 3 48W
Ystradgynlais **43** 51 47N 3 45W
Ythan, R. **48** 57 26N 2 12W
Yuba City **100** 39 12N 121 37W
Yūbari **76** 43 4N 141 59E
Yucatán □ **106** 21 30N 86 30W
Yucatán Channel **106** 22 0N 86 30W
Yugoslavia ■ **58** 44 0N 20 0E
Yukon, R. **98** 65 30N 150 0W
Yukon Territory □ **98** 63 0N 135 0W
Yuma **100** 32 45N 114 37W
Yumen **73** 41 13N 96 55E
Yun Ho **73** 35 0N 117 0E
Yungtsi **73** 34 50N 110 25E
Yunnan □ **73** 25 0N 102 30E
Yutze **73** 37 45N 112 45E

Z

Zaandam **52** 52 26N 4 49E
Zabrze **55** 50 18N 18 46E
Zacapa **106** 14 59N 89 31W
Zacatecas **106** 22 49N 102 34W
Zacoalco **106** 20 10N 103 40W
Zadar **58** 44 8N 15 8E
Zafra **60** 38 26N 6 30W
Zagań **54** 51 39N 15 22E
Zagazig **91** 30 40N 31 12E
Zagnanado **90** 7 18N 2 28E
Zagreb **58** 45 50N 16 0E
Zagros Mts. **71** 33 45N 47 0E
Zāhedān **71** 29 30N 60 50E
Zahlah **71** 33 52N 35 50E
Zaïre, R. **92** 1 30N 28 0E
Zaïre, Rep. of ■ **92** 3 0S 23 0E
Zambèze, R. **92** 18 46S 36 16E
Zambia ■ **92** 15 0S 28 0E
Zamboanga **72** 6 59N 122 3E
Zamora, Mexico **106** 20 0N 102 21W
Zamora, Spain **60** 41 30N 5 45W
Zanesville **104** 39 56N 82 2W
Zanjan **71** 36 40N 48 35E
Zanthus **78** 31 2S 123 34E
Zanzibar **93** 6 12S 39 12E
Zaporozhye **62** 47 50N 35 10E
Zaragoza **60** 41 39N 0 53W
Zaria **90** 11 0N 7 40E
Żary **54** 51 37N 15 10E
Zastron **94** 30 18S 27 7E
Zawyet Shammâs **91** 31 30N 26 37E
Zâwyet Um el Rakham **91** 31 18N 27 1E
Zâwyet Ungeîla **91** 31 23N 26 42E
Zduńska Wola **55** 51 37N 18 59E
Zeehan **79** 41 52S 145 25E
Zeeland **52** 51 41N 5 40E
Zeerust **94** 25 31S 26 4E
Zeist **52** 52 5N 5 15E
Zemun **55** 44 51N 20 25E
Zerbst **54** 51 59N 12 8E
Zhabinka **55** 52 13N 24 2E
Zhdanov **62** 47 5N 37 31E
Zhigansk **62** 66 35N 124 10E
Zhitomir **55** 50 20N 28 40E
Zhmerinka **55** 49 2N 28 10E
Ziel, Mt. **78** 23 20S 132 30E
Zielona Góra **54** 51 57N 15 31E
Žilina **55** 49 12N 18 42E
Zilling Tso **73** 31 40N 89 0E
Zimbabwe **92** 20 16S 31 0E
Zimbabwe ■ (formerly Rhodesia) **92** 19 0S 29 0E
Zinder **90** 13 48N 9 0E
Zion Nat. Park **100** 37 25N 112 50W
Zolochev **55** 49 45N 24 58E
Zrenjanin **55** 45 22N 20 23E
Zug **54** 47 10N 8 31E
Zuid-Holland □ **52** 52 0N 4 35E
Zululand **94** 43 19N 2 15E
Zumbo **92** 15 35S 30 26E
Zürich **54** 47 22N 8 32E
Zuru **90** 11 27N 5 4E
Zutphen **52** 52 9N 6 12E
Zvolen **55** 48 33N 19 10E
Zweibrücken **52** 49 15N 7 20E
Zwickau **54** 50 43N 12 30E
Zwolle **52** 52 31N 6 6E
Żyrardów **55** 52 3N 20 35E
Zyryanovsk **73** 49 50N 84 57E

Acknowledgement is made to the following for providing the photographs used in this atlas.

Aerofilms Ltd.; Air France; Air India; G. Atkinson; Australian Information Service; Brazilian Embassy, London; Brazilian Tourist Office, London; G. P. Chapman; Chile Tourist Office, London; Danish Embassy, London; Fiat (England) Ltd.; Finnish Tourist Bureau; Harland & Wolff Ltd.; R. J. Harrison Church; Japan Information Centre, London; Meteorological Office, London; Moroccan Tourist Office, London; N.A.S.A.; Netherlands National Tourist Office, London; Novosti Press Agency; M. Rentsch; S.A.S.; W. B. Smith; Society For Anglo-Chinese Understanding, London; Transafrica Pix.; Wiggins Teape Ltd.; B. M. Willett; Z.E.F.A. (U.K.) Ltd.

The Earth from Space

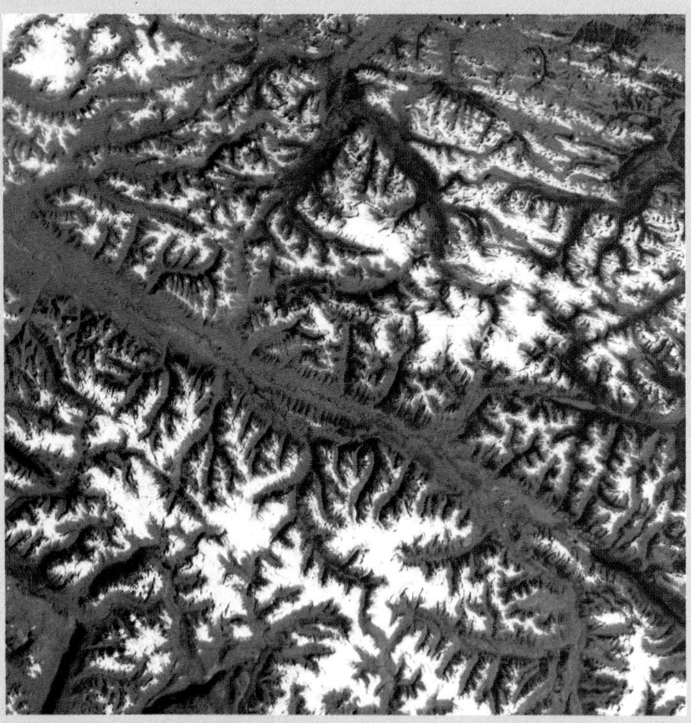

1. ROCKY MOUNTAINS (CANADA)

This is an area of high precipitation, much of which falls as snow which covers the high mountains in this image, emphasizing the drainage pattern. The conspicuous straight valley bisecting the area is part of the Rocky Mountain Trench, a great fault zone that extends from Alaska to Montana.

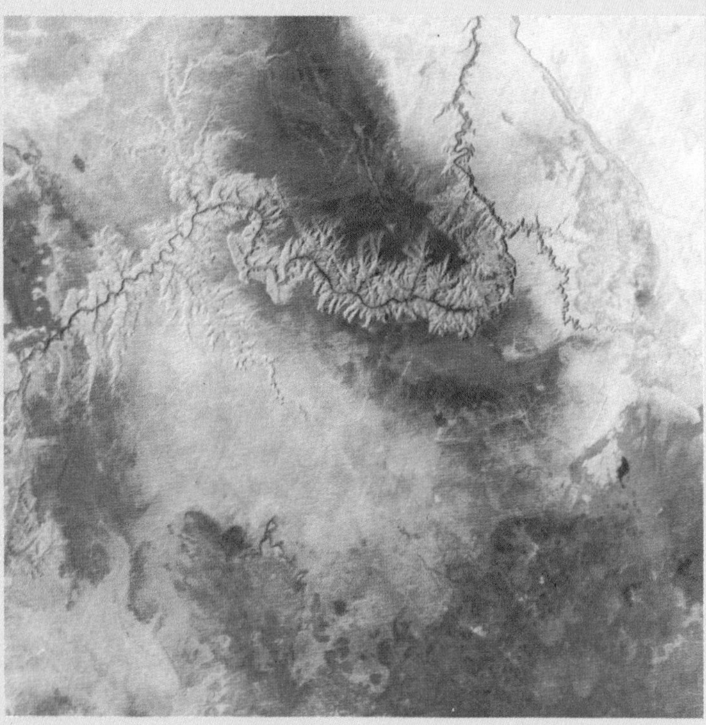

2. GRAND CANYON (U.S.A.)

The Colorado river is shown here flowing through the Grand Canyon in Arizona, which at this point is 1.5km deep and 20km wide. The reddish area distinguishes the Kaibab Plateau, an area over 2700m high, on which higher precipitation permits vegetation growth.

A Landsat satellite launched and controlled by NASA in the USA travels around the earth at a height of 917km, and "photographs" every point of the world once every 18 days. The view from the satellite is broken into four component bands of the spectrum, bands 4, 5, 6, and 7, converted into electrical signals and transmitted

3. TAKLA MAKAN DESERT (CHINA)

Snow cover spreads over the marshland around the Tarim River in the north-east and onto the desert sand-dunes. It is not easy to appreciate the size of these dunes from this image; individual dune ridges are 1½-3km wide and extend 8-32km.

4. MISSISSIPPI DELTA (U.S.A.)

The blue colouring of the farmland in this area is caused by extensive flooding of the Mississippi River. The large quantity of sediment transported by the river flows into the Gulf of Mexico building up the delta and appearing as a light blue mass in the sea.

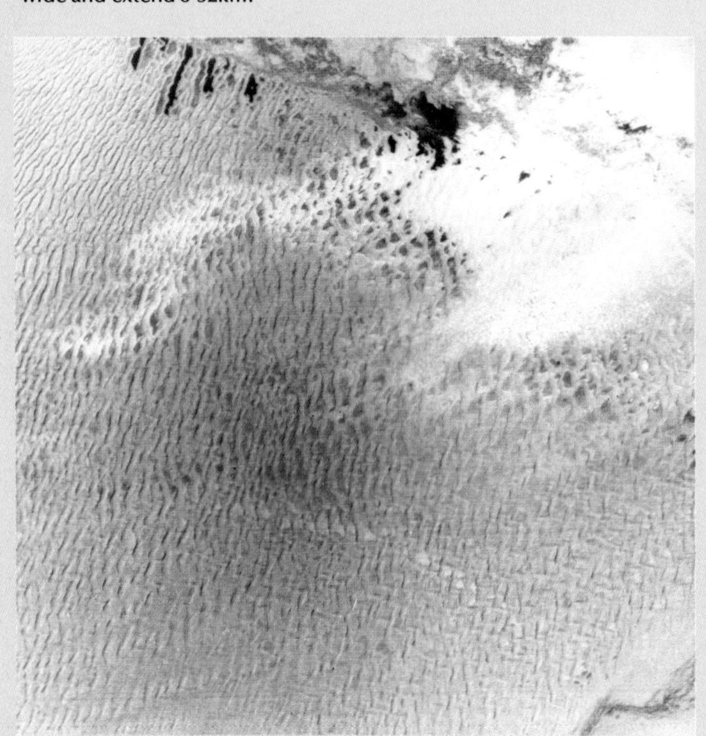

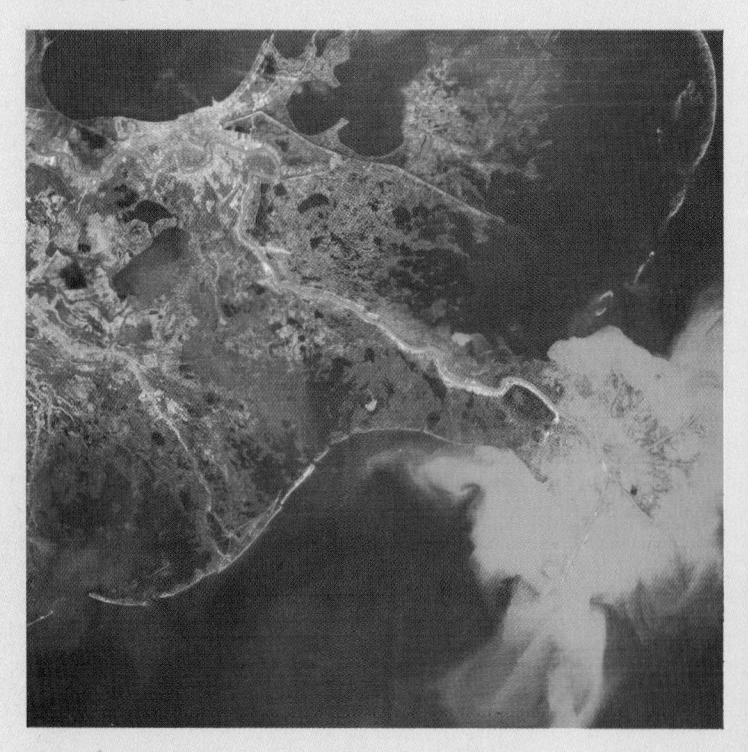

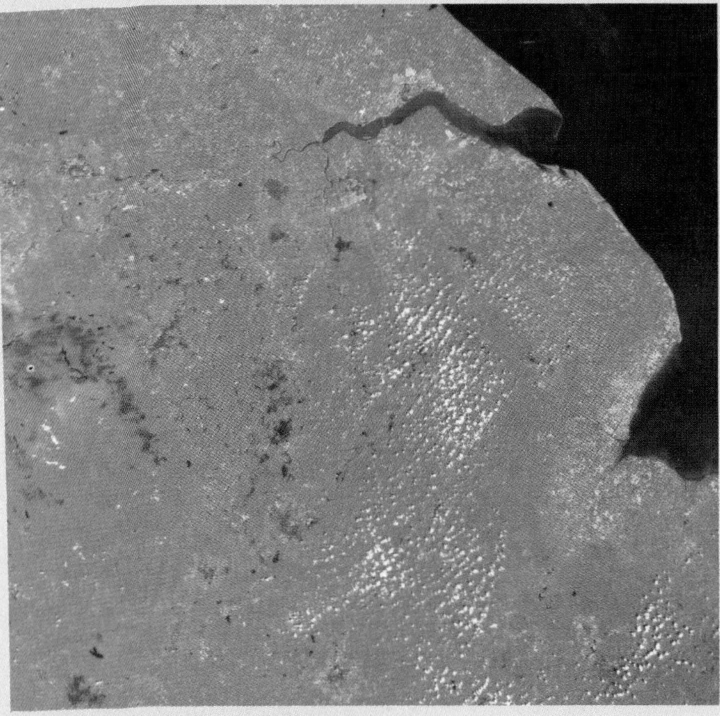

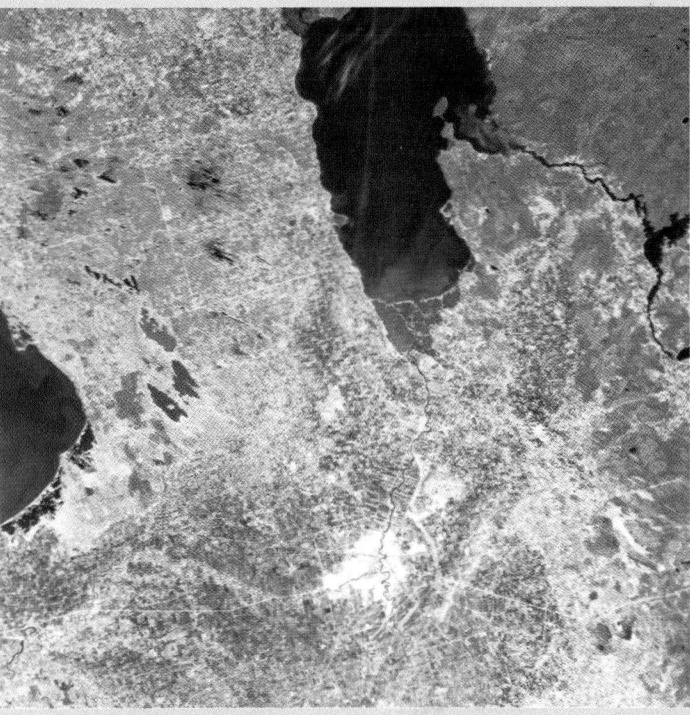

5. EAST MIDLANDS (ENGLAND)

This image covers an area from the Pennines to the North Sea, with the Humber estuary and The Wash. The urban centres are blue-grey, the bare highlands and mine-workings black. The agricultural areas are bright red except round the Wash where the waterlogged Fenlands have a bluish tinge.

back to earth. Here it is reassembled into false-colour photographic images where vegetation appears bright red, soil as yellow-brown, water as black and concrete or bare-rock as blue-grey. The scale of these satellite photographs is approximately 1:1 600 000, each showing an area about 170km by 170km.

6. WINNIPEG (CANADA)

The town of Winnipeg is surrounded by a cultivated plain which is divided up by farm roads into 1.6km wide sections resulting in a chequerboard pattern. The contrasts between the flat plain and neighbouring mountains are emphasized by the differing landuse.

7. GEZIRA (SUDAN)

The White Nile here flows through flat, clay plains in a region of only 20cm of rain per year. The Gezira scheme developed irrigation in this area permitting cultivation of the land. The strip fields show up as red where cotton is being grown and yellow where they are lying fallow.

8. NEW YORK (U.S.A.)

New York city is clearly visible as a purple area in the north-east of this picture. The parks show up as bright red patches within the city. The white patches are where infertile sandy soils of the coastal plain, the sandy offshore bar, and spit inhibit vegetation growth.

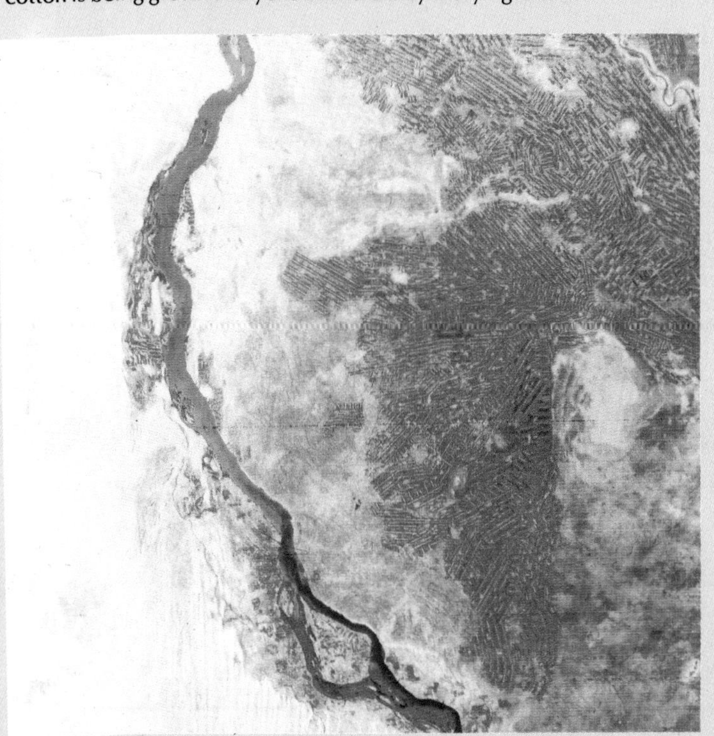

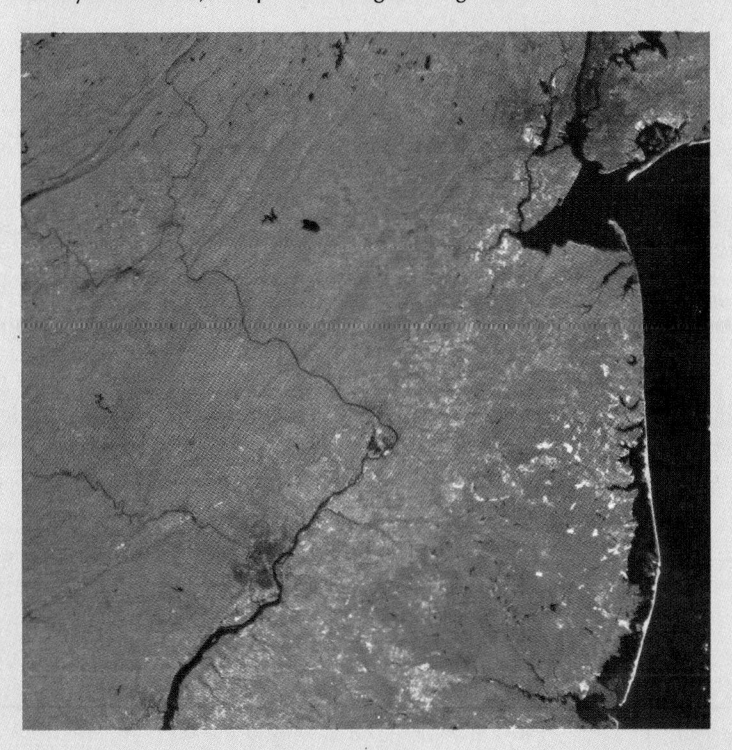